Student Solutions Manual

to accompany
D. Franklin Wright's
Introductory Algebra
Sixth Edition

Solutions by A. Elizabeth Allen and Thom Clark

Hawkes Learning Systems

ISBN 13: 978-1-932628-65-4
ISBN 10: 1-932628-65-7

Table of Contents

Chapter R: Review of Basic Topics 1

Chapter 1:   Integers and Real Numbers 16

Chapter 2:   Fractions with Variables and Algebraic Expressions........................ 29

Chapter 3:   Solving Equations and Inequalities.................................... 52

Chapter 4:   Graphing Linear Equations and Inequalities in Two Variables 92

Chapter 5:   Systems of Linear Equations 137

Chapter 6:   Exponents and Polynomials.. 181

Chapter 7:   Factoring Polynomials and Solving Quadratic Equations.............................. 211

Chapter 8:   Rational Expressions.. 242

Chapter 9:   Real Numbers and Radicals.. 283

Chapter 10:   Quadratic Equations... 310

Appendix   358

[PAGE INTENTIONALLY LEFT BLANK.]

Section R.1
Solutions to Odd Exercises

1. **a.** Base: 2
    b. Exponent: 3
    c. $2^3 = 2 \cdot 2 \cdot 2 = 8$

3. **a.** Base: 4
    b. Exponent: 2
    c. $4^2 = 4 \cdot 4 = 16$

5. **a.** Base: 9
    b. Exponent: 2
    c. $9^2 = 9 \cdot 9 = 81$

7. **a.** Base: 11
    b. Exponent: 2
    c. $11^2 = 11 \cdot 11 = 121$

9. $2^4 = 16$ or $4^2 = 16$

11. $5^2 = 25$

13. $7^2 = 49$

15. $6^2 = 36$

17. $10^3 = 1000$

19. $2^3 = 8$

21. $6^5 = 6 \cdot 6 \cdot 6 \cdot 6 \cdot 6$

23. $11^3 = 11 \cdot 11 \cdot 11$

25. $2^3 \cdot 3^2 = 2 \cdot 2 \cdot 2 \cdot 3 \cdot 3$

27. $2 \cdot 3^2 \cdot 11^2 = 2 \cdot 3 \cdot 3 \cdot 11 \cdot 11$

29. $3^3 \cdot 7^3 = 3 \cdot 3 \cdot 3 \cdot 7 \cdot 7 \cdot 7$

31. $8^2 = 64$

33. $11^2 = 121$

35. $20^2 = 400$

37. $13^2 = 169$

39. $30^2 = 900$

41. $52^2 = 2704$

43. $25^4 = 390,625$

45. $125^3 = 1,953,125$

47. $5^7 = 78,125$

49. $6 + 5 \cdot 3 = 6 + 15 = 21$

51. $20 - 4 \div 4 = 20 - 1 = 19$

53. $32 - 14 + 10 = 18 + 10 = 28$

55. $18 \div 2 - 1 - 3 \cdot 2 = 9 - 1 - 6 = 8 - 6 = 2$

57. $2 + 3 \cdot 7 - 10 \div 2 = 2 + 21 - 5 = 23 - 5 = 18$

59. $(2 + 3 \cdot 4) \div 7 - 2 = (2 + 12) \div 7 - 2$
$$= 14 \div 7 - 2$$
$$= 2 - 2 = 0$$

61. $14(2 + 3) - 65 - 5 = 14(5) - 65 - 5$
$$= 70 - 65 - 5$$
$$= 5 - 5 = 0$$

63. $2 \cdot 5^2 - 8 \div 2 = 2 \cdot 25 - 8 \div 2 = 2 \cdot 25 - 4$
$$= 50 - 4 = 46$$

65. $(2^3 + 2) \div 5 + 7^2 \div 7 = (8 + 2) \div 5 + 49 \div 7$
$$= (10) \div 5 + 49 \div 7$$
$$= 2 + 7 = 9$$

67. $(4 + 3)^2 + (2 + 3)^2 = 7^2 + 5^2 = 49 + 25 = 74$

69. $8 \div 2 \cdot 4 - 16 \div 4 \cdot 2 + 3 \cdot 2^2 = 4 \cdot 4 - 4 \cdot 2 + 3 \cdot 4$
$$= 16 - 8 + 12$$
$$= 8 + 12 = 20$$

71. $(10 + 1)\left[(5 - 2)^2 + 3(4 - 3)\right] = 11\left[3^2 + 3 \cdot 1\right]$
$$= 11[9 + 3]$$
$$= 11 \cdot 12 = 132$$

73. $100 + 2\left[3\left(4^2 - 6\right) + 2^3\right] = 100 + 2\left[3(16 - 6) + 8\right]$
$$= 100 + 2\left[3 \cdot 10 + 8\right]$$
$$= 100 + 2\left[30 + 8\right]$$
$$= 100 + 2 \cdot 38$$
$$= 100 + 76 = 176$$

75. $16 + 3\left(17 + 2^3 \div 2^2 - 4\right) = 61$

77. $12^3 - 1\left[\left(3^2 + 5\right)^2\right] = 1532$

79. $30 \div 2 - 11 + 2\left(5 - 1\right)^3 = 132$

Writing and Thinking About Mathematics

81. The result is an 'Error'. Answers will vary. For example:

An error results because 0^0 is undefined.

[End of Section R.1]

Section R.2
Solutions to Odd Exercises

1. A prime number is a counting number greater than 1 that has exactly two different factors: 1 and itself.

3. A composite number is a counting number with more than two different factors, or divisors.

5. Answers may vary. Some prime numbers greater than 50 are: 53, 59, 61, 67, 71, 79, 83.

7. 13 and 17 are the prime numbers in the set.

9. 2 is the only prime number in the set.

For Exercises 11–20, there may be more than one correct answer. Multiple possibilities are shown here, when applicable.

11. 8, 9 or 6, 12 or 4, 18 or 3, 24 or 2, 36

13. 2, 34 or 4, 17

15. 2, 251

17. 2, 85 or 5, 34 or 10, 17

19. 2, 222 or 3, 148 or 4, 111 or 12, 37

21. $52 = 2^2 \cdot 13$

23. $616 = 2^3 \cdot 7 \cdot 11$

25. $308 = 2^2 \cdot 7 \cdot 11$

27. 79 is prime.

29. $289 = 17^2$

31. $125 = 5^3$

33. $400 = 2^4 \cdot 5^2$

35. $120 = 2^3 \cdot 3 \cdot 5$

37. $231 = 3 \cdot 7 \cdot 11$

39. $1692 = 2^2 \cdot 3^2 \cdot 47$

41. a. 5, 10, 15, 20, 25, 30, 35, 40, 45, 50, 55, 60
    b. 6, 12, 18, 24, 30, 36, 42, 48, 54, 60, 66, 72
    c. 10, 20, 30, 40, 50, 60, 70, 80, 90, 100, 110, 120
    d. 15, 30, 45, 60, 75, 90, 105, 120, 135, 150, 165, 180

In Exercises 43–76, use the prime factorization of each number to determine the LCM.

43. 3, 5, and 7 are prime numbers, so the LCM is:
    $3 \cdot 5 \cdot 7 = 105$

45. $6 = 2 \cdot 3$ and $10 = 2 \cdot 5$, so the LCM is:
    $2 \cdot 3 \cdot 5 = 30$

47. 2, 3, and 11 are prime numbers, so the LCM is:
    $2 \cdot 3 \cdot 11 = 66$

49. $4 = 2^2$, $14 = 2 \cdot 7$, and $35 = 5 \cdot 7$, so the LCM is:
    $2^2 \cdot 5 \cdot 7 = 140$

51. $50 = 2 \cdot 5^2$ and $75 = 3 \cdot 5^2$, so the LCM is:
    $2 \cdot 3 \cdot 5^2 = 150$

53. $20 = 2^2 \cdot 5$ and $90 = 2 \cdot 3^2 \cdot 5$, so the LCM is:
    $2^2 \cdot 3^2 \cdot 5 = 180$

55. $28 = 2^2 \cdot 7$ and $98 = 2 \cdot 7^2$, so the LCM is:
    $2^2 \cdot 7^2 = 196$

57. $10 = 2 \cdot 5$, $15 = 3 \cdot 5$, and $35 = 5 \cdot 7$, so the LCM is:
    $2 \cdot 3 \cdot 5 \cdot 7 = 210$

59. 15 and 45 are each factors of 90, so the LCM is 90.

61. 20 and 50 are each factors of 100, so the LCM is 100.

63. $10 = 2 \cdot 5$, $15 = 3 \cdot 5$, and $25 = 5^2$, so the LCM is:
    $2 \cdot 3 \cdot 5^2 = 150$

65. $26 = 2 \cdot 13$, $28 = 2^2 \cdot 7$ and $91 = 7 \cdot 13$, so the LCM is:
    $2^2 \cdot 7 \cdot 13 = 364$

67. $35 = 5 \cdot 7$, $40 = 2^3 \cdot 5$ and $72 = 2^3 \cdot 3^2$, so the LCM is:
    $2^3 \cdot 3^2 \cdot 5 \cdot 7 = 2520$

69. $12 = 2^2 \cdot 3$, $21 = 3 \cdot 7$ and $44 = 2^2 \cdot 11$, so the LCM is:
$$2^2 \cdot 3 \cdot 7 \cdot 11 = 924$$

71. $99 = 3^2 \cdot 11$, $121 = 11^2$ and $231 = 3 \cdot 7 \cdot 11$, so the LCM is:
$$3^2 \cdot 7 \cdot 11^2 = 7623$$

73. $48 = 2^4 \cdot 3$, $120 = 2^3 \cdot 3 \cdot 5$, $144 = 2^4 \cdot 3^2$, and $192 = 2^6 \cdot 3$, so the LCM is:
$$2^6 \cdot 3^2 \cdot 5 = 2880$$

75. $40 = 2^3 \cdot 5$, $56 = 2^3 \cdot 7$, $160 = 2^5 \cdot 5$, and $196 = 2^2 \cdot 7^2$, so the LCM is:
$$2^5 \cdot 5 \cdot 7^2 = 7840$$

[End of Section R.2]

Section R.3
Solutions to Odd Exercises

1. **a.** A fraction is proper if the numerator is less than the denominator.
    b. A fraction is improper if the numerator is greater than the denominator.

3. Answers may vary. One possible answer is: Fractions may be used to indicate equal parts of a whole, or to indicate division.

5. To reduce a fraction to lowest terms, use the fact that 1 is equal to any counting number divided by itself to divide out all common factors in both the numerator and denominator.

7. **a.** There are 25 squares total, 13 of which are shaded. So $\dfrac{13}{25}$ of the figure is shaded.
    b. Of the 25 squares, 12 are left blank. So $\dfrac{12}{25}$ of the figure is not shaded.

9. **a.** There are 5 sections total, 2 of which are shaded. So $\dfrac{2}{5}$ of the figure is shaded.
    b. Of the 5 sections, 3 are left blank. So $\dfrac{3}{5}$ of the figure is not shaded.

11. Since $4 \cdot 3 = 12$, multiply the numerator and the denominator by 3:
$$\frac{3}{4} = \frac{3}{4} \cdot \frac{3}{3} = \frac{9}{12}$$

13. Since $7 \cdot 2 = 14$, multiply the numerator and the denominator by 2:
$$\frac{6}{7} = \frac{6}{7} \cdot \frac{2}{2} = \frac{12}{14}$$

15. Since $16 \cdot 5 = 80$, multiply the numerator and the denominator by 5:
$$\frac{3}{16} = \frac{3}{16} \cdot \frac{5}{5} = \frac{15}{80}$$

17. Since $26 \cdot 2 = 52$, multiply the numerator and the denominator by 2:
$$\frac{7}{26} = \frac{7}{26} \cdot \frac{2}{2} = \frac{14}{52}$$

19. Since $1 \cdot 3 = 3$, multiply the numerator and the denominator by 3:
$$\frac{18}{1} = \frac{18}{1} \cdot \frac{3}{3} = \frac{54}{3}$$

21. $\dfrac{7}{8} \cdot \dfrac{4}{21} = \dfrac{7 \cdot 4}{8 \cdot 21} = \dfrac{\cancel{7} \cdot \cancel{4}}{2 \cdot \cancel{4} \cdot \cancel{7} \cdot 3} = \dfrac{1}{2 \cdot 3} = \dfrac{1}{6}$

23. $9 \cdot \dfrac{7}{24} = \dfrac{9 \cdot 7}{24} = \dfrac{\cancel{3} \cdot 3 \cdot 7}{\cancel{3} \cdot 8} = \dfrac{3 \cdot 7}{8} = \dfrac{21}{8}$

25. $\dfrac{9}{10} \cdot \dfrac{35}{40} \cdot \dfrac{65}{15} = \dfrac{9 \cdot 35 \cdot 65}{10 \cdot 40 \cdot 15} = \dfrac{\cancel{3} \cdot 3 \cdot \cancel{5} \cdot 7 \cdot \cancel{5} \cdot 13}{2 \cdot \cancel{5} \cdot 40 \cdot \cancel{3} \cdot \cancel{5}}$
$= \dfrac{3 \cdot 7 \cdot 13}{2 \cdot 40} = \dfrac{273}{80}$

27. $\dfrac{75}{8} \cdot \dfrac{16}{36} \cdot 9 \cdot \dfrac{7}{25} = \dfrac{75 \cdot 16 \cdot 9 \cdot 7}{8 \cdot 36 \cdot 25} = \dfrac{3 \cdot \cancel{25} \cdot \cancel{2} \cdot \cancel{8} \cdot \cancel{9} \cdot 7}{\cancel{8} \cdot \cancel{2} \cdot 2 \cdot \cancel{9} \cdot \cancel{25}}$
$= \dfrac{3 \cdot 7}{2} = \dfrac{21}{2}$

29. $\dfrac{3}{4} \cdot \dfrac{7}{2} \cdot \dfrac{22}{54} \cdot 18 = \dfrac{3 \cdot 7 \cdot 22 \cdot 18}{4 \cdot 2 \cdot 54} = \dfrac{\cancel{3} \cdot 7 \cdot \cancel{2} \cdot 11 \cdot \cancel{9} \cdot \cancel{2}}{4 \cdot \cancel{2} \cdot \cancel{9} \cdot \cancel{3} \cdot \cancel{2}}$
$= \dfrac{7 \cdot 11}{4} = \dfrac{77}{4}$

31. $\dfrac{5}{8} \div \dfrac{3}{5} = \dfrac{5}{8} \cdot \dfrac{5}{3} = \dfrac{5 \cdot 5}{8 \cdot 3} = \dfrac{25}{24}$

33. $\dfrac{2}{3} \div \dfrac{1}{5} = \dfrac{2}{3} \cdot \dfrac{5}{1} = \dfrac{2 \cdot 5}{3 \cdot 1} = \dfrac{10}{3}$

35. $\dfrac{3}{14} \div \dfrac{3}{14} = \dfrac{3}{14} \cdot \dfrac{14}{3} = \dfrac{\cancel{3} \cdot \cancel{14}}{\cancel{14} \cdot \cancel{3}} = 1$

37. $\dfrac{3}{4} \div \dfrac{4}{3} = \dfrac{3}{4} \cdot \dfrac{3}{4} = \dfrac{3 \cdot 3}{4 \cdot 4} = \dfrac{9}{16}$

39. $\dfrac{15}{20} \div 3 = \dfrac{15}{20} \cdot \dfrac{1}{3} = \dfrac{15}{20 \cdot 3} = \dfrac{\cancel{5} \cdot \cancel{3}}{4 \cdot \cancel{5} \cdot \cancel{3}} = \dfrac{1}{4}$

41. $\dfrac{25}{40} \div 10 = \dfrac{25}{40} \cdot \dfrac{1}{10} = \dfrac{25 \cdot 1}{40 \cdot 10} = \dfrac{\cancel{5} \cdot \cancel{5}}{8 \cdot \cancel{5} \cdot 2 \cdot \cancel{5}} = \dfrac{1}{8 \cdot 2} = \dfrac{1}{16}$

43. $\dfrac{7}{8} \div 0$ is undefined.

45. $0 \div \dfrac{5}{6} = 0 \cdot \dfrac{6}{5} = \dfrac{0 \cdot 6}{5} = \dfrac{0}{5} = 0$

47. $\dfrac{16}{35} \div \dfrac{2}{7} = \dfrac{16}{35} \cdot \dfrac{7}{2} = \dfrac{16 \cdot 7}{35 \cdot 2} = \dfrac{\cancel{2} \cdot 8 \cdot \cancel{7}}{5 \cdot \cancel{7} \cdot \cancel{2}} = \dfrac{8}{5}$

49. $\dfrac{15}{24} \div \dfrac{25}{18} = \dfrac{15}{24} \cdot \dfrac{18}{25} = \dfrac{15 \cdot 18}{24 \cdot 25} = \dfrac{\cancel{3} \cdot \cancel{5} \cdot \cancel{2} \cdot 9}{\cancel{3} \cdot \cancel{2} \cdot 4 \cdot \cancel{5} \cdot 5}$

$= \dfrac{9}{4 \cdot 5} = \dfrac{9}{20}$

51. You have spent $9 out of your $20, so you have spent $\dfrac{9}{20}$ of your money. You did not spend $\$20 - \$9 = \$11$, so you still have $\dfrac{11}{20}$ of your money.

53. The number of miles downhill is given by:

$\dfrac{1}{5} \cdot 60 = \dfrac{60}{5} = \dfrac{\cancel{5} \cdot 12}{\cancel{5}} = 12$ miles. Then $60 - 12 = 48$

miles will not be downhill. So, $\dfrac{48}{60} = \dfrac{\cancel{12} \cdot 4}{\cancel{12} \cdot 5} = \dfrac{4}{5}$ of the trip will not be downhill.

55. **a.** The product is $\dfrac{2}{5}$.

b. The other number is given by:

$\dfrac{2}{5} \div \dfrac{5}{6} = \dfrac{2}{5} \cdot \dfrac{6}{5} = \dfrac{2 \cdot 6}{5 \cdot 5} = \dfrac{12}{25}$

57. **a.** The capacity of the airplane is greater than 90, because 90 is only $\dfrac{9}{10}$ the capacity of the airplane, and $\dfrac{9}{10}$ is less than 1.

b. The product would be less than 90.

c. Dividing 90 by $\dfrac{9}{10}$ yields the capacity of the airplane:

$90 \div \dfrac{9}{10} = 90 \cdot \dfrac{10}{9} = \dfrac{90 \cdot 10}{9} = \dfrac{\cancel{9} \cdot 100}{\cancel{9}} = 100$

59. **a.** More than half will vote for the new court, because $\dfrac{3}{5}$ of $\dfrac{7}{10}$ of all members live near the club and are in favor. This is

$\dfrac{3}{5} \cdot \dfrac{7}{10} = \dfrac{3 \cdot 7}{5 \cdot 10} = \dfrac{21}{50}$

of the members. Also, $\dfrac{1}{3}$ of $\dfrac{3}{10}$ of all members do not live near the club and are in favor. This is

$\dfrac{1}{3} \cdot \dfrac{3}{10} = \dfrac{1 \cdot \cancel{3}}{\cancel{3} \cdot 10} = \dfrac{1}{10} = \dfrac{1}{10} \cdot \dfrac{5}{5} = \dfrac{5}{50}$

of the members. So a total of

$\dfrac{21}{50} + \dfrac{5}{50} = \dfrac{26}{50}$

members are in favor, and this is more than one half.

b. By part **a.**, the vote will pass with $\dfrac{26}{50}$ of all members voting in favor. That is,

$\dfrac{26}{50} \cdot 250 = \dfrac{26 \cdot 250}{50} = \dfrac{26 \cdot 5 \cdot 50}{50} = 130$

members will vote in favor. Since $250 \div 2 + 1 = 126$ votes (a majority) are required for the question to pass, the question will pass by 5 votes.

[End of Section R.3]

Section R.4
Solutions to Odd Exercises

1. **a.** $\dfrac{1}{3} \cdot \dfrac{1}{3} = \dfrac{1 \cdot 1}{3 \cdot 3} = \dfrac{1}{9}$

   b. $\dfrac{1}{3} + \dfrac{1}{3} = \dfrac{1+1}{3} = \dfrac{2}{3}$

3. **a.** $\dfrac{2}{7} \cdot \dfrac{2}{7} = \dfrac{2 \cdot 2}{7 \cdot 7} = \dfrac{4}{49}$

   b. $\dfrac{2}{7} + \dfrac{2}{7} = \dfrac{2+2}{7} = \dfrac{4}{7}$

5. **a.** $\dfrac{5}{4} \cdot \dfrac{5}{4} = \dfrac{5 \cdot 5}{4 \cdot 4} = \dfrac{25}{16}$

   b. $\dfrac{5}{4} + \dfrac{5}{4} = \dfrac{5+5}{4} = \dfrac{10}{4} = \dfrac{5 \cdot 2}{2 \cdot 2} = \dfrac{5}{2}$

7. **a.** $\dfrac{8}{9} \cdot \dfrac{8}{9} = \dfrac{8 \cdot 8}{9 \cdot 9} = \dfrac{64}{81}$

   b. $\dfrac{8}{9} + \dfrac{8}{9} = \dfrac{8+8}{9} = \dfrac{16}{9}$

9. **a.** $\dfrac{9}{12} \cdot \dfrac{9}{12} = \dfrac{9 \cdot 9}{12 \cdot 12} = \dfrac{9 \cdot \cancel{3} \cdot \cancel{3}}{4 \cdot \cancel{3} \cdot 4 \cdot \cancel{3}} = \dfrac{9}{4 \cdot 4} = \dfrac{9}{16}$

   b. $\dfrac{9}{12} + \dfrac{9}{12} = \dfrac{9+9}{12} = \dfrac{18}{12} = \dfrac{\cancel{6} \cdot 3}{\cancel{6} \cdot 2} = \dfrac{3}{2}$

11. $\dfrac{3}{14} + \dfrac{3}{14} = \dfrac{3+3}{14} = \dfrac{6}{14} = \dfrac{\cancel{2} \cdot 3}{\cancel{2} \cdot 7} = \dfrac{3}{7}$

13. $\dfrac{1}{10} + \dfrac{3}{10} = \dfrac{1+3}{10} = \dfrac{4}{10} = \dfrac{\cancel{2} \cdot 2}{\cancel{2} \cdot 5} = \dfrac{2}{5}$

15. $\dfrac{14}{25} - \dfrac{6}{25} = \dfrac{14-6}{25} = \dfrac{8}{25}$

17. $\dfrac{4}{15} - \dfrac{1}{15} = \dfrac{4-1}{15} = \dfrac{3}{15} = \dfrac{\cancel{3}}{\cancel{3} \cdot 5} = \dfrac{1}{5}$

19. $\dfrac{3}{8} - \dfrac{5}{16} = \dfrac{2}{2} \cdot \dfrac{3}{8} - \dfrac{5}{16} = \dfrac{6}{16} - \dfrac{5}{16} = \dfrac{6-5}{16} = \dfrac{1}{16}$

21. $\dfrac{2}{7} + \dfrac{4}{21} + \dfrac{1}{3} = \dfrac{3}{3} \cdot \dfrac{2}{7} + \dfrac{4}{21} + \dfrac{1}{3} \cdot \dfrac{7}{7} = \dfrac{6}{21} + \dfrac{4}{21} + \dfrac{7}{21}$

   $= \dfrac{6+4+7}{21} = \dfrac{17}{21}$

23. $\dfrac{5}{6} - \dfrac{1}{2} = \dfrac{5}{6} - \dfrac{3}{3} \cdot \dfrac{1}{2} = \dfrac{5}{6} - \dfrac{3}{6} = \dfrac{5-3}{6} = \dfrac{2}{6} = \dfrac{\cancel{2} \cdot 1}{\cancel{2} \cdot 3} = \dfrac{1}{3}$

25. $\dfrac{5}{7} - \dfrac{5}{14} = \dfrac{2}{2} \cdot \dfrac{5}{7} - \dfrac{5}{14} = \dfrac{10}{14} - \dfrac{5}{14} = \dfrac{10-5}{14} = \dfrac{5}{14}$

27. $\dfrac{20}{35} - \dfrac{24}{42} = \dfrac{\cancel{5} \cdot 4}{\cancel{5} \cdot 7} - \dfrac{\cancel{6} \cdot 4}{\cancel{6} \cdot 7} = \dfrac{4}{7} - \dfrac{4}{7} = 0$

29. $\dfrac{78}{100} - \dfrac{7}{10} = \dfrac{78}{100} - \dfrac{7}{10} \cdot \dfrac{10}{10} = \dfrac{78}{100} - \dfrac{70}{100} = \dfrac{78-70}{100}$

   $= \dfrac{8}{100} = \dfrac{2 \cdot \cancel{4}}{25 \cdot \cancel{4}} = \dfrac{2}{25}$

31. $\dfrac{2}{3} + \dfrac{3}{4} - \dfrac{5}{6} = \dfrac{4}{4} \cdot \dfrac{2}{3} + \dfrac{3}{3} \cdot \dfrac{3}{4} - \dfrac{2}{2} \cdot \dfrac{5}{6} = \dfrac{8}{12} + \dfrac{9}{12} - \dfrac{10}{12}$

   $= \dfrac{8+9-10}{12} = \dfrac{7}{12}$

33. $\dfrac{2}{27} - \dfrac{1}{54} + \dfrac{1}{45} = \dfrac{10}{10} \cdot \dfrac{2}{27} - \dfrac{5}{5} \cdot \dfrac{1}{54} + \dfrac{6}{6} \cdot \dfrac{1}{45}$

   $= \dfrac{20}{270} - \dfrac{5}{270} + \dfrac{6}{270} = \dfrac{20-5+6}{270}$

   $= \dfrac{21}{270} = \dfrac{\cancel{3} \cdot 7}{\cancel{3} \cdot 90} = \dfrac{7}{90}$

35. $\dfrac{5}{6} - \dfrac{50}{60} + \dfrac{1}{3} = \dfrac{5}{6} - \dfrac{5 \cdot \cancel{10}}{6 \cdot \cancel{10}} + \dfrac{1}{3} = \dfrac{5}{6} - \dfrac{5}{6} + \dfrac{1}{3} = \dfrac{1}{3}$

37. $\dfrac{1}{2} + \dfrac{3}{4} + \dfrac{1}{100} = \dfrac{50}{50} \cdot \dfrac{1}{2} + \dfrac{25}{25} \cdot \dfrac{3}{4} + \dfrac{1}{100}$

   $= \dfrac{50}{100} + \dfrac{75}{100} + \dfrac{1}{100} = \dfrac{50+75+1}{100}$

   $= \dfrac{126}{100} = \dfrac{\cancel{2} \cdot 63}{\cancel{2} \cdot 50} = \dfrac{63}{50}$

39. $\dfrac{1}{5} + \dfrac{3}{10} - \dfrac{4}{15} = \dfrac{6}{6} \cdot \dfrac{1}{5} + \dfrac{3}{3} \cdot \dfrac{3}{10} - \dfrac{2}{2} \cdot \dfrac{4}{15} = \dfrac{6}{30} + \dfrac{9}{30} - \dfrac{8}{30}$

   $= \dfrac{6+9-8}{30} = \dfrac{7}{30}$

41. $1 - \dfrac{13}{16} = \dfrac{16}{16} - \dfrac{13}{16} = \dfrac{16-13}{16} = \dfrac{3}{16}$

43. $1 - \dfrac{5}{8} = \dfrac{8}{8} - \dfrac{5}{8} = \dfrac{8-5}{8} = \dfrac{3}{8}$

45. $\dfrac{3}{5}+\dfrac{7}{15}+\dfrac{5}{6}=\dfrac{6}{6}\cdot\dfrac{3}{5}+\dfrac{2}{2}\cdot\dfrac{7}{15}+\dfrac{5}{5}\cdot\dfrac{5}{6}=\dfrac{18}{30}+\dfrac{14}{30}+\dfrac{25}{30}$

$$=\dfrac{18+14+25}{30}=\dfrac{57}{30}=\dfrac{\cancel{3}\cdot19}{\cancel{3}\cdot10}=\dfrac{19}{10}$$

47. $\dfrac{7}{24}+\dfrac{7}{16}+\dfrac{7}{12}=\dfrac{2}{2}\cdot\dfrac{7}{24}+\dfrac{3}{3}\cdot\dfrac{7}{16}+\dfrac{4}{4}\cdot\dfrac{7}{12}$

$$=\dfrac{14}{48}+\dfrac{21}{48}+\dfrac{28}{48}$$

$$=\dfrac{14+21+28}{48}$$

$$=\dfrac{63}{48}=\dfrac{\cancel{3}\cdot21}{\cancel{3}\cdot16}=\dfrac{21}{16}$$

49. **a.** He plans to spend

$$\dfrac{1}{3}+\dfrac{1}{10}=\dfrac{10}{10}\cdot\dfrac{1}{3}+\dfrac{3}{3}\cdot\dfrac{1}{10}=\dfrac{10}{30}+\dfrac{3}{30}=\dfrac{13}{30}$$

of his income on rent and food.

    b. The amount of money he plans to spend is

$$\dfrac{13}{30}\cdot3300=\dfrac{13\cdot3300}{30}=\dfrac{13\cdot\cancel{30}\cdot110}{\cancel{30}}=1430$$

dollars each month.

51. The width of the face of the watch is

$$\dfrac{3}{4}+\dfrac{1}{10}+\dfrac{1}{10}=\dfrac{5}{5}\cdot\dfrac{3}{4}+\dfrac{2}{2}\cdot\dfrac{1}{10}+\dfrac{2}{2}\cdot\dfrac{1}{10}=\dfrac{15}{20}+\dfrac{2}{20}+\dfrac{2}{20}$$

$$=\dfrac{15+2+2}{20}=\dfrac{19}{20}\text{ in.}$$

The length of the face of the watch is

$$\dfrac{1}{2}+\dfrac{1}{10}+\dfrac{1}{10}=\dfrac{5}{5}\cdot\dfrac{1}{2}+\dfrac{1}{10}+\dfrac{1}{10}=\dfrac{5}{10}+\dfrac{1}{10}+\dfrac{1}{10}=\dfrac{7}{10}\text{ in.}$$

So the watch face is $\dfrac{19}{20}\times\dfrac{7}{10}$ in.

[End of Section R.4]

Section R.5
Solutions to Odd Exercises

1. Using place-value concept: $\dfrac{5}{10} = 0.5$, so

$$7\dfrac{5}{10} = 7 + \dfrac{5}{10} = 7 + 0.5 = 7.5$$

3. $\dfrac{27}{1000} = 0.027$, so

$$14\dfrac{27}{1000} = 14 + \dfrac{27}{1000} = 14 + 0.027 = 14.027$$

5. $0.02 = \dfrac{2}{100}$, so

$$23.02 = 23 + 0.02 = 23 + \dfrac{2}{100} = 23\dfrac{2}{100}$$

7. $0.001 = \dfrac{1}{1000}$, so

$$52.001 = 52 + 0.001 = 52 + \dfrac{1}{1000} = 52\dfrac{1}{1000}$$

9. Sixth tenths is 0.6.

11. Two and twenty-four hundredths is 2.24.

For Exercises 13 – 16, use the place-value concept and recall that the name for the fractional part comes from the last position on the right in the decimal name.

13. 0.2 is two tenths.

15. 260.003 is two hundred sixty and three thousandths.

17. 72.06 to the nearest tenth is 72.1.

19. 0.295 to the nearest hundredth is 0.30.

For Exercises 21 – 32, adding and subtracting decimal numbers requires that you be careful lining up the corresponding place-values.

21. Add:
$$\begin{array}{r} 0.6 \\ 0.4 \\ + \; 0.4 \\ \hline 1.4 \end{array}$$

23. Add:
$$\begin{array}{r} 0.79 \\ 4.92 \\ + \; 0.05 \\ \hline 5.76 \end{array}$$

25. Subtract:
$$\begin{array}{r} 5.40 \\ - \; 3.76 \\ \hline 1.64 \end{array}$$

27. Subtract:
$$\begin{array}{r} 39.60 \\ - \; 13.71 \\ \hline 25.89 \end{array}$$

29. Add:
$$\begin{array}{r} 57.300 \\ 52.080 \\ + \; 38.005 \\ \hline 147.385 \end{array}$$

31. Subtract:
$$\begin{array}{r} 21.007 \\ - \; 1.543 \\ \hline 19.464 \end{array}$$

33. Multiply: $(0.2)(0.2) = 0.04$

35. Multiply:
$$\begin{array}{r} 0.137 \\ \times \; 0.08 \\ \hline 0.01096 \end{array}$$

37. Divide: $28 \div 5.6 = 5$

$$5.6\overline{)28}$$

$$56\overline{)280} \qquad \text{Move the decimal one place.}$$

$$\begin{array}{r} 5 \\ 56\overline{)280} \\ \underline{280} \\ 0 \end{array} \qquad \text{Divide}$$

39. Divide: $2.7\overline{)5.483} \approx 2.03$, to the nearest hundredth.

$$2.7\overline{)5.483}$$

$27\overline{)54.83}$ Move the decimals one place.

$$\begin{array}{r} 2.0307 \\ 27\overline{)54.8300} \end{array}$$ Divide.

$$\begin{array}{r} 54 \\ \hline 083 \\ 81 \\ \hline 200 \\ 189 \end{array}$$ Stop here and round

41. $28000 - 21000 + 450 + 2240$

$$\begin{aligned} &= 7000 + 450 + 2240 \\ &= 7450 + 2240 \\ &= 9690 \end{aligned}$$

Marshall needs $9690 in cash to buy the car.

43. $\begin{array}{r} 0.254 \\ -0.017 \\ \hline 0.237 \end{array}$

Pluto's eccentricity is 0.237 more.

45. $1200 + 0.08 \cdot 1200 = 1200 + 96 = 1296$
$1296 is the total amount paid for the television.

47. $11.3\overline{)150.6} \approx 13.3$, to the nearest tenth.

$11.3\overline{)150.6}$ Move the decimals and divide.

$$\begin{array}{r} 13.32 \\ 113\overline{)1506.00} \\ \hline 113 \\ \hline 376 \\ 339 \\ \hline 370 \\ 339 \\ \hline 310 \\ 226 \end{array}$$ Stop here and round.

The bicyclist's average speed was 13.3 mph.

49. **a.** The perimeter is the sum of the lengths of each side: $25.2 + 25.2 + 13.4 + 13.4 = 77.2$
The perimeter is 77.2 inches.

    b. The area is the product of the length and the width: $(25.2)(13.4) = 337.68$
The area is 337.68 sq in.

For Exercises 51–70, remember the following:
    When converting a decimal to a percent, move the decimal two places to the right.
    When converting a percent to a decimal, move the decimal two places to the left.
    When converting a fraction to a percent, convert to decimal form, then move the decimal.

51. $0.03 = 3\%$

53. $3.0 = 300\%$

55. $1.08 = 108\%$

57. $6\% = 0.06$

59. $3.2\% = 0.032$

61. $120\% = 1.2$

63. $\dfrac{3}{20} = \dfrac{5}{5} \cdot \dfrac{3}{20} = \dfrac{15}{100} = 0.15 = 15\%$

65. $\dfrac{24}{25} = \dfrac{4}{4} \cdot \dfrac{24}{25} = \dfrac{96}{100} = 0.96 = 96\%$

67. $1\dfrac{5}{8} = 1 + \dfrac{5}{8} = 1 + .625 = 1.625 = 162.5\%$

69. $0.009 = 0.9\%$

In Exercises 71 – 80, use your graphing calculator to find the value, accurate to three decimal places.

71. $8.35 + 9.76 + 13.002 = 31.112$

73. $897.54 - 789.5 = 108.040$

75. $3.52(0.12)(2.5) = 1.056$

77. $5.7 \div 0.214 \approx 26.636$

79. $\dfrac{67.15}{3.45} \approx 19.464$

[End of Section R.5]

Chapter R Review
Solutions to All Exercises

1. $3 \cdot 3 \cdot 3 \cdot 5 \cdot 5 = 3^3 \cdot 5^2$

2. $13 \cdot 13 \cdot 13 \cdot 17 = 13^3 \cdot 17$

3. $2 \cdot 2 \cdot 11 \cdot 11 = 2^2 \cdot 11^2$

4. $7 \cdot 11 \cdot 11 \cdot 19 \cdot 19 \cdot 19 = 7 \cdot 11^2 \cdot 19^3$

5. $12^2 = 144$

6. $8^2 = 64$

7. $6^7 = 279,936$

8. $2^{15} = 32,768$

9. $19 + 2 \cdot 7 = 19 + 14 = 33$

10. $25 + 3 \cdot 4 = 25 + 12 = 37$

11. $40 - 4 \div 4 = 40 - 1 = 39$

12. $20 - 5 \div 5 = 20 - 1 = 19$

13. $4(2+3) \div 5 \cdot 2 = 4 \cdot 5 \div 5 \cdot 2 = 20 \div 5 \cdot 2 = 4 \cdot 2 = 8$

14. $(34 + 10) \div 11 \cdot 2 = 44 \div 11 \cdot 2 = 4 \cdot 2 = 8$

15. $32 \div 2^4 + 27 \div 3^2 = 32 \div 16 + 27 \div 9 = 2 + 3 = 5$

16. $2 \cdot 5^3 - 8 \div 2^2 = 2 \cdot 125 - 8 \div 4 = 250 - 2 = 248$

17. $(4+5)^2 + (4+3)^2 = 9^2 + 7^2 = 81 + 49 = 130$

18. $40 - 9(3^2 - 2^3) = 40 - 9(9 - 8) = 40 - 9 \cdot 1 = 31$

19. $50 + 2\left[3(4^2 - 6) - 5^2\right] = 50 + 2[3(16 - 6) - 25]$
$= 50 + 2(3 \cdot 10 - 25)$
$= 50 + 2(30 - 25)$
$= 50 + 2 \cdot 5 = 50 + 10$
$= 60$

20. $(12 - 2)\left[5(7 - 3) + (5 - 3)^2\right] = 10\left[5 \cdot 4 + 2^2\right]$
$= 10(20 + 4)$
$= 10 \cdot 24$
$= 240$

21.
$$
\begin{array}{r}
75 \\
5\overline{)375} \\
\underline{35} \\
25 \\
\underline{25} \\
0
\end{array}
$$

22.
$$
\begin{array}{r}
304 \\
3\overline{)912} \\
\underline{9} \\
012 \\
\underline{12} \\
0
\end{array}
$$

23. 2, 3, 5, 7, 11, 13, 17, 19, 23, 29

24. A composite number is a counting number with more than two different factors, or divisors.

25. 89 is prime.

26. 189 is composite. $3^3 \cdot 7 = 189$

27. 299 is composite. $13 \cdot 23 = 299$

28. 301 is composite. $7 \cdot 43 = 301$

29. $175 = 5^2 \cdot 7$

30. $154 = 2 \cdot 7 \cdot 11$

31. $195 = 3 \cdot 5 \cdot 13$

32. $2475 = 3^2 \cdot 5^2 \cdot 11$

33. $27 = 3^3$, $30 = 2 \cdot 3 \cdot 5$, and $35 = 5 \cdot 7$, so:
LCM $= 2 \cdot 3^3 \cdot 5 \cdot 7 = 1890$

34. $10 = 2 \cdot 5$, $27 = 3^3$, and $30 = 2 \cdot 3 \cdot 5$, so:
LCM $= 2 \cdot 3^3 \cdot 5 = 270$

35. $15 = 3 \cdot 5$, $24 = 2^3 \cdot 3$, and $40 = 2^3 \cdot 5$, so:
LCM $= 2^3 \cdot 3 \cdot 5 = 120$

36. $18 = 2 \cdot 3^2$, $36 = 2^2 \cdot 3^2$, and $78 = 2 \cdot 3 \cdot 13$, so:
$$\text{LCM} = 2^2 \cdot 3^2 \cdot 13 = 468$$

37. $8 = 2^3$, $10 = 2 \cdot 5$, $20 = 2^2 \cdot 5$, and $60 = 2^2 \cdot 3 \cdot 5$, so:
$$\text{LCM} = 2^3 \cdot 3 \cdot 5 = 120$$

38. $11 = 11$, $22 = 2 \cdot 11$, $66 = 2 \cdot 3 \cdot 11$, and $77 = 7 \cdot 11$, so:
$$\text{LCM} = 2 \cdot 3 \cdot 7 \cdot 11 = 462$$

39. $121 = 11^2$, $169 = 13^2$, and $225 = 3^2 \cdot 5^2$, so:
$$\text{LCM} = 3^2 \cdot 5^2 \cdot 11^2 \cdot 13^2 = 4,601,025$$

40. $25 = 5^2$, $49 = 7^2$, and $169 = 13^2$, so:
$$\text{LCM} = 5^2 \cdot 7^2 \cdot 13^2 = 207,025$$

41. $\dfrac{1}{3}$ of $36 = \dfrac{1}{3} \cdot 36 = \dfrac{36}{3} = 12$

42. $\dfrac{2}{3}$ of $36 = \dfrac{2}{3} \cdot 36 = \dfrac{2 \cdot 36}{3} = 2 \cdot 12 = 24$

43. $\dfrac{1}{10}$ of $\dfrac{3}{4} = \dfrac{1}{10} \cdot \dfrac{3}{4} = \dfrac{3}{10 \cdot 4} = \dfrac{3}{40}$

44. Since $8 \cdot 3 = 24$, multiply the numerator and the denominator by 3:
$$\frac{7}{8} = \frac{7}{8} \cdot \frac{3}{3} = \frac{21}{24}$$

45. Since $17 \cdot 3 = 51$, multiply the numerator and the denominator by 3:
$$\frac{15}{17} = \frac{15}{17} \cdot \frac{3}{3} = \frac{45}{51}$$

46. Since $16 \cdot 4 = 64$, multiply the numerator and the denominator by 4:
$$\frac{1}{16} = \frac{1}{16} \cdot \frac{4}{4} = \frac{4}{64}$$

47. $\dfrac{15}{51} = \dfrac{\cancel{3} \cdot 5}{\cancel{3} \cdot 17} = \dfrac{5}{17}$

48. $\dfrac{72}{84} = \dfrac{\cancel{12} \cdot 6}{\cancel{12} \cdot 7} = \dfrac{6}{7}$

49. $\dfrac{20}{46} = \dfrac{\cancel{2} \cdot 10}{\cancel{2} \cdot 23} = \dfrac{10}{23}$

50. $\dfrac{6}{57} = \dfrac{\cancel{3} \cdot 2}{\cancel{3} \cdot 19} = \dfrac{2}{19}$

51. $\dfrac{55}{44} \cdot \dfrac{8}{26} = \dfrac{5 \cdot \cancel{11} \cdot \cancel{2} \cdot \cancel{4}}{\cancel{4} \cdot \cancel{11} \cdot \cancel{2} \cdot 13} = \dfrac{5}{13}$

52. $\dfrac{52}{90} \cdot \dfrac{65}{26} = \dfrac{\cancel{2} \cdot \cancel{2} \cdot \cancel{13} \cdot \cancel{5} \cdot 13}{\cancel{2} \cdot \cancel{5} \cdot 9 \cdot \cancel{2} \cdot \cancel{13}} = \dfrac{13}{9}$

53. $\dfrac{16}{25} \div \dfrac{4}{5} = \dfrac{16}{25} \cdot \dfrac{5}{4} = \dfrac{\cancel{4} \cdot 4 \cdot \cancel{5}}{\cancel{5} \cdot 5 \cdot \cancel{4}} = \dfrac{4}{5}$

54. $\dfrac{19}{27} \div \dfrac{38}{54} = \dfrac{19}{27} \cdot \dfrac{54}{38} = \dfrac{\cancel{19} \cdot \cancel{2} \cdot \cancel{27}}{\cancel{27} \cdot \cancel{2} \cdot \cancel{19}} = 1$

55. $\dfrac{18}{35} \cdot \dfrac{1}{3} \cdot \dfrac{21}{12} = \dfrac{\cancel{2} \cdot \cancel{3} \cdot \cancel{3} \cdot 3 \cdot \cancel{7}}{5 \cdot \cancel{7} \cdot \cancel{3} \cdot \cancel{3} \cdot \cancel{2} \cdot 2} = \dfrac{3}{5 \cdot 2} = \dfrac{3}{10}$

56. $\dfrac{4}{9} \cdot \dfrac{2}{5} \cdot \dfrac{2}{3} = \dfrac{4 \cdot 2 \cdot 2}{9 \cdot 5 \cdot 3} = \dfrac{16}{135}$

57. $\dfrac{35}{33} \div \dfrac{7}{3} = \dfrac{35}{33} \cdot \dfrac{3}{7} = \dfrac{5 \cdot \cancel{7} \cdot \cancel{3}}{\cancel{3} \cdot 11 \cdot \cancel{7}} = \dfrac{5}{11}$

58. $\dfrac{3}{8} \div 4 = \dfrac{3}{8} \cdot \dfrac{1}{4} = \dfrac{3}{8 \cdot 4} = \dfrac{3}{32}$

59. $\dfrac{2}{5} \cdot \dfrac{3}{4} \cdot \dfrac{11}{12} \cdot \dfrac{20}{26} = \dfrac{\cancel{2} \cdot \cancel{3} \cdot 11 \cdot \cancel{4} \cdot \cancel{5}}{\cancel{5} \cdot \cancel{4} \cdot \cancel{3} \cdot 4 \cdot \cancel{2} \cdot 13} = \dfrac{11}{4 \cdot 13} = \dfrac{11}{52}$

60. $\dfrac{7}{16} \cdot \dfrac{9}{14} \cdot \dfrac{30}{45} \cdot \dfrac{4}{13} = \dfrac{\cancel{7} \cdot \cancel{9} \cdot \cancel{2} \cdot 3 \cdot \cancel{5} \cdot \cancel{4}}{\cancel{4} \cdot 4 \cdot \cancel{2} \cdot \cancel{7} \cdot \cancel{9} \cdot 5 \cdot 13} = \dfrac{3}{4 \cdot 13} = \dfrac{3}{52}$

61. $\dfrac{2}{3} + \dfrac{2}{3} = \dfrac{2+2}{3} = \dfrac{4}{3}$

62. $\dfrac{3}{5} + \dfrac{3}{5} = \dfrac{3+3}{5} = \dfrac{6}{5}$

63. $\dfrac{9}{16} - \dfrac{5}{16} = \dfrac{9-5}{16} = \dfrac{4}{16} = \dfrac{1 \cdot \cancel{4}}{4 \cdot \cancel{4}} = \dfrac{1}{4}$

64. $\dfrac{7}{10} - \dfrac{3}{10} = \dfrac{7-3}{10} = \dfrac{4}{10} = \dfrac{\cancel{2} \cdot 2}{\cancel{2} \cdot 5} = \dfrac{2}{5}$

65. $\dfrac{9}{100} + \dfrac{3}{10} = \dfrac{9}{100} + \dfrac{30}{100} = \dfrac{9+30}{100} = \dfrac{39}{100}$

66. $\dfrac{7}{50} + \dfrac{2}{25} = \dfrac{7}{50} + \dfrac{4}{50} = \dfrac{7+4}{50} = \dfrac{11}{50}$

67. $\dfrac{1}{2} + \dfrac{3}{4} + \dfrac{3}{100} = \dfrac{50}{100} + \dfrac{75}{100} + \dfrac{3}{100} = \dfrac{50+75+3}{100}$

$= \dfrac{128}{100} = \dfrac{\cancel{4} \cdot 32}{\cancel{4} \cdot 25} = \dfrac{32}{25}$

68. $\dfrac{2}{3} + \dfrac{1}{9} + \dfrac{5}{18} = \dfrac{12}{18} + \dfrac{2}{18} + \dfrac{5}{18} = \dfrac{12+2+5}{18} = \dfrac{19}{18}$

69. $\dfrac{9}{10} - \dfrac{1}{20} = \dfrac{18}{20} - \dfrac{1}{20} = \dfrac{18-1}{20} = \dfrac{17}{20}$

70. $\dfrac{11}{12} - \dfrac{1}{2} = \dfrac{11}{12} - \dfrac{6}{12} = \dfrac{11-6}{12} = \dfrac{5}{12}$

71. $\dfrac{5}{8} - \dfrac{5}{16} = \dfrac{10}{16} - \dfrac{5}{16} = \dfrac{10-5}{16} = \dfrac{5}{16}$

72. $\dfrac{9}{16} - \dfrac{9}{32} = \dfrac{18}{32} - \dfrac{9}{32} = \dfrac{18-9}{32} = \dfrac{9}{32}$

73. $\dfrac{9}{45} + \dfrac{1}{36} = \dfrac{36}{180} + \dfrac{5}{180} = \dfrac{36+5}{180} = \dfrac{41}{180}$

74. $\dfrac{1}{8} + \dfrac{5}{12} = \dfrac{3}{24} + \dfrac{10}{24} = \dfrac{3+10}{24} = \dfrac{13}{24}$

75. $1 - \dfrac{1}{9} = \dfrac{9}{9} - \dfrac{1}{9} = \dfrac{9-1}{9} = \dfrac{8}{9}$

76. $1 - \dfrac{3}{10} = \dfrac{10}{10} - \dfrac{3}{10} = \dfrac{10-3}{10} = \dfrac{7}{10}$

77. $\dfrac{3}{24} + \dfrac{1}{16} + \dfrac{5}{12} = \dfrac{6}{48} + \dfrac{3}{48} + \dfrac{20}{48} = \dfrac{6+3+20}{48} = \dfrac{29}{48}$

78. $\dfrac{4}{5} + \dfrac{7}{15} + \dfrac{1}{2} = \dfrac{24}{30} + \dfrac{14}{30} + \dfrac{15}{30} = \dfrac{24+14+15}{30} = \dfrac{53}{30}$

79. $\dfrac{4}{9} + \dfrac{2}{3} + \dfrac{7}{15} = \dfrac{20}{45} + \dfrac{30}{45} + \dfrac{21}{45} = \dfrac{20+30+21}{45} = \dfrac{71}{45}$

80. $\dfrac{2}{5} + \dfrac{7}{15} + \dfrac{5}{6} = \dfrac{12}{30} + \dfrac{14}{30} + \dfrac{25}{30} = \dfrac{12+14+25}{30}$

$= \dfrac{51}{30} = \dfrac{\cancel{3} \cdot 17}{\cancel{3} \cdot 10} = \dfrac{17}{10}$

81. Fourteen hundredths is 0.14.

82. Seven and six tenths is 7.6.

83. Three hundred and three thousandths is 300.003.

84. Three hundred three thousandths is 0.303.

85. 75.09 is seventy-five and nine hundredths.

86. 6.0023 is six and twenty-three ten-thousandths.

87. $86.34 + 70.58 + 16.7 = 173.62$

88. $100.54 - 37.6 = 62.94$

89. $18.3(7.55) = 138.165$

90. $122.8 \div 3.1 \approx 39.61$

91. $\$45,000 - \$15,250 + \$2700 + \$560 = \$33,010$

92. $P = 2(3.45) + 2(4.3) = 6.9 + 8.6 = 15.5$ inches.

93. $0.035 = 3.5\%$

94. $1.23 = 123\%$

95. $83\% = 0.83$

96. $0.5\% = 0.005$

97. $\dfrac{1}{20} = 5\%$

98. $\dfrac{3}{8} = 37.5\%$

99. $\dfrac{45}{100} = 45\%$

100. $\dfrac{7}{10} = 70\%$

[End of Chapter R Review]

Chapter R Test
Solutions to All Exercises

1. $3 \cdot 3 \cdot 5 \cdot 5 \cdot 5 \cdot 5 = 3^2 \cdot 5^4$

2. $13^2 = 169$ is the square of 13.

3. $17 + 25 \div 5 \cdot 2 - 3^2 = 17 + 5 \cdot 2 - 9$
    $$= 17 + 10 - 9 = 18$$

4. $8^2 - 2(16 \div 4 \cdot 2 + 4) = 64 - 2(4 \cdot 2 + 4)$
    $$= 64 - 2(8 + 4)$$
    $$= 64 - 2 \cdot 12$$
    $$= 64 - 24 = 40$$

5. $26 + 3(7^2 - 3 \cdot 5) - 11^2 = 26 + 3(49 - 15) - 121$
    $$= 26 + 3 \cdot 34 - 121$$
    $$= 26 + 102 - 121 = 7$$

6. $24 \div 3 \cdot 2 - 18 \div 3^2 + 2 \cdot 5 = 8 \cdot 2 - 18 \div 9 + 10$
    $$= 16 - 2 + 10 = 24$$

7. 2, 3, 5, 7, 11, 13, 17, 19, 23, 29, 31, 37

8. $940 = 2^2 \cdot 5 \cdot 47$

9. $\text{LCM} = 2^3 \cdot 3^2 \cdot 5 = 360$
    $$9 = 3^2$$
    $$15 = 3 \cdot 5$$
    $$24 = 2^3 \cdot 3$$

10. $\text{LCM} = 2^3 \cdot 3 \cdot 5^2 = 600$
    $$20 = 2^2 \cdot 5$$
    $$30 = 2 \cdot 3 \cdot 5$$
    $$40 = 2^3 \cdot 5$$
    $$50 = 2 \cdot 5^2$$

11. $\dfrac{1}{10} + \dfrac{3}{10} = \dfrac{1+3}{10} = \dfrac{4}{10} = \dfrac{2}{5}$

12. $\dfrac{5}{8} - \dfrac{3}{8} = \dfrac{5-3}{8} = \dfrac{2}{8} = \dfrac{1}{4}$

13. $\dfrac{25}{49} \cdot \dfrac{14}{30} = \dfrac{\cancel{5} \cdot 5 \cdot \cancel{2} \cdot \cancel{7}}{\cancel{7} \cdot 7 \cdot \cancel{2} \cdot 3 \cdot \cancel{5}} = \dfrac{5}{7 \cdot 3} = \dfrac{5}{21}$

14. $\dfrac{7}{18} \cdot \dfrac{3}{5} \cdot \dfrac{4}{9} \cdot \dfrac{1}{14} = \dfrac{\cancel{7} \cdot \cancel{3} \cdot \cancel{2} \cdot \cancel{2} \cdot 1}{\cancel{2} \cdot \cancel{3} \cdot 3 \cdot 5 \cdot 9 \cdot \cancel{2} \cdot \cancel{7}} = \dfrac{1}{135}$

15. $\dfrac{24}{75} \div \dfrac{14}{15} = \dfrac{24}{75} \cdot \dfrac{15}{14} = \dfrac{\cancel{2} \cdot \cancel{3} \cdot 4 \cdot 3 \cdot \cancel{5}}{\cancel{5} \cdot \cancel{5} \cdot 5 \cdot \cancel{2} \cdot 7} = \dfrac{12}{35}$

16. $\dfrac{16}{19} \div \dfrac{8}{38} = \dfrac{16}{19} \cdot \dfrac{38}{8} = \dfrac{\cancel{2} \cdot \cancel{2} \cdot \cancel{2} \cdot 2 \cdot 2 \cdot \cancel{19}}{\cancel{19} \cdot \cancel{2} \cdot \cancel{2} \cdot \cancel{2}} = 4$

17. $130.7 + 62.75 + 83.02 = 276.47$

18. $89.05 - 39.76 = 49.29$

19. $6.1(2.03) = 12.383$

20. $\begin{array}{r} 107.5 \\ \times\ 3.02 \\ \hline 324.65 \end{array}$

21. $\begin{array}{r} 3.1 \\ 16\overline{)49.6} \\ \underline{48} \\ 16 \\ \underline{16} \\ 0 \end{array}$

22. $5.2\overline{)31.876} = \begin{array}{r} 6.13 \\ 52\overline{)318.76} \\ \underline{312} \\ 67 \\ \underline{52} \\ 156 \\ \underline{156} \\ 0 \end{array}$

23. Seventeen and thirteen hundredths is 17.13.

24. Six and one hundred seven thousandths is 6.107.

25. 5.107 is five and one hundred seven thousandths.

26. 100.001 is one hundred and one thousandth.

27. $4.32\% = 0.0432$

28. $0.057 = 5.7\%$

29. $0.03 = 3\%$

30. $1.02 = 102\%$

31. $\dfrac{1}{4} = 0.25 = 25\%$

32. $\dfrac{1}{50} = 0.02 = 2\%$

33. If 40 is $\dfrac{5}{8}$ of the capacity, then the capacity is

$$40 \div \dfrac{5}{8} = 40 \cdot \dfrac{8}{5} = \dfrac{40 \cdot 8}{5} = 8 \cdot 8 = 64 \text{ passengers.}$$

34. $\left(\dfrac{3}{8} \cdot 64\right) \div \left(\dfrac{2}{5} \cdot \dfrac{15}{16}\right) = (3 \cdot 8) \div \left(\dfrac{3}{8}\right)$

$$= 24 \div \dfrac{3}{8}$$
$$= 24 \cdot \dfrac{8}{3}$$
$$= 8 \cdot 8 = 64$$

35. Area $= 17.2 \cdot 23.1 = 397.32$ sq. feet.

36. Average miles per gallon $= \dfrac{250}{12} = 20.833 \approx 20.8$

37. $15^4 = 50,625$

38. $893.6 + 175.82 + 645.33 = 1714.75$

39. $8476 + 7542 - 8319.7 = 7698.3$

40. $393.635 \div 21.05 = 18.7$

[End of Chapter R Test]

Section 1.1
Solutions to Odd Exercises

For Exercises 1– 24, first, draw a number line. Mark at equal lengths the integers that would include the numbers shown in each set.

1. $\{1, 2, 5, 6\}$

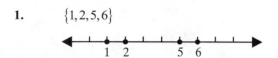

3. $\{2, -3, 0, -1\}$

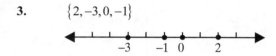

5. $\left\{0, -1, \dfrac{7}{4}, 3, 1\right\}$

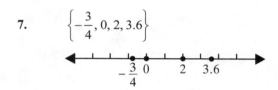

7. $\left\{-\dfrac{3}{4}, 0, 2, 3.6\right\}$

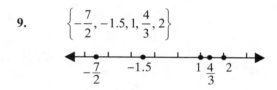

9. $\left\{-\dfrac{7}{2}, -1.5, 1, \dfrac{4}{3}, 2\right\}$

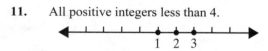

11. All positive integers less than 4.

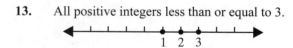

13. All positive integers less than or equal to 3.

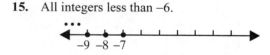

15. All integers less than -6.

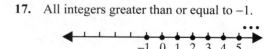

17. All integers greater than or equal to -1.

18.

19. All negative integers greater than or equal to 2.
    No Solution

21. All integers more than 6 units from 0

23. All integers less than 3 units from 0.

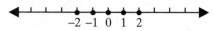

25. $|-10| = 10$

27. $-|-4| = -4$

29. $-|11.3| = -11.3$

31. $4 < 6$

33. $-2 > -4$

35. $5 = -(-5)$

37. $5.6 < -(-8.7)$

39. $-\dfrac{3}{4} > -1$

41. $\dfrac{1}{3} < \dfrac{1}{2}$

43. $-\dfrac{1}{2} < -\dfrac{1}{3}$

45. $-\dfrac{2}{8} = -\dfrac{1}{4}$

For Exercises 47 – 72, test the validity of the statement $x < y$ for two numbers x and y by determining whether there is some positive number z for which $x + z = y$. If such a number z can be found, then $x < y$.

To compare two fractions, convert them so that they have a common denominator. Then, compare the numerators using the steps described above.

If the expression contains an absolute value symbol, write the equivalent of the expression without the absolute value symbol, then compare the two resulting values. Remember that any negative number is less than any positive number.

47. $0 = -0$ is <u>true</u>, as the opposite of zero is itself.

49. $-22 < -16$ is <u>true</u> because there is a positive number, namely 6, such that $-22 + 6 = -16$.

51. <u>False</u>; $-9 < -8.5$.

53. $-17 \le 17$ is <u>true</u> since a negative number is always less than a positive number.

55. $4.7 \ge 3.5$ is <u>true</u> because there is a positive number, namely 1.2, such that $3.5 + 1.2 = 4.7$.

57. $|-5| = 5$ is <u>true</u> because -5 and 5 are both five units away from 0 on the number line.

59. $-|-7| \ge -|7|$ is <u>true</u>, because $-|-7| = -7$ and $-|7| = -7$. Since $-7 = -7$, $-7 \ge -7$ also.

61. $|-8| \ge 4$ is <u>true</u> as $|-8| = 8$ and $8 \ge 4$.

63. <u>False</u>; $-|-3| > -|4|$.

65. <u>False</u>; $\left|-\dfrac{5}{2}\right| > 2$.

67. $\dfrac{2}{3} < |-1|$ is <u>true</u>, because $|-1| = 1$ and $\dfrac{2}{3} < 1$.

69. $|-1.6| < |-2.1|$ is <u>true</u>, because $|-1.6| = 1.6$ and $|-2.1| = 2.1$, and $1.6 < 2.1$.

71. $|2.5| = \left|-\dfrac{5}{2}\right|$ is <u>true</u>, as $|2.5| = 2.5$ and $\left|-\dfrac{5}{2}\right| = \dfrac{5}{2} = 2.5$.

73. $\{-4, 4\}$ because $|-4| = 4$ and $|4| = 4$.

75. $\{-9, 9\}$ because $|-9| = 9$ and $|9| = 9$.

77. $\{0\}$ because $|0| = 0$.

79. There are no numbers that satisfy the equation $-2 = |x|$, because the absolute value of any real number is either positive or zero.

81. $\{-3.5, 3.5\}$ because $|-3.5| = 3.5$ and $|3.5| = 3.5$.

83. $|x| \le 4$

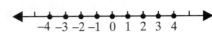

85. $|x| > 5$

87. $|x| < 7$

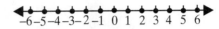

89. $|x| \le x$

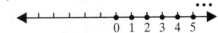

91. $|x| = x$

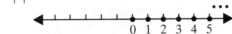

Calculator Problems

93. $|34 - 80| = 46$

95. $-|10 - 16| = -|-6| = -6$

97. $\dfrac{|-6| - |-3|}{|-6 - 3|} = \dfrac{6 - 3}{|-9|} = \dfrac{3}{9} = \dfrac{1}{3}$

Writing and Thinking About Mathematics

99. Answers will vary. One possible explanation is:

$-y$ means opposite, or negative, of y, so if y represents a negative number, then its opposite, or $-y$, would be positive. For example, if $y = -2$, then $-y = -(-2) = 2$.

[End of Section 1.1]

Section 1.2
Solutions to Odd Exercises

1. $4 + 9 = 13$

3. $(-9) + 5 = -4$

5. $(-9) + 9 = 0$

7. $11 + (-6) = 5$

9. $-18 + 5 = -13$

11. $-5 + (-3) = -(5 + 3) = -8$

13. $(-2) + (-8) = -(2 + 8) = -10$

15. $17 + (-17) = 0$

17. $21 + (-4) = 17$

19. $(-12) + (-17) = -(12 + 17) = -29$

21. $-4 + (-5) = -(4 + 5) = -9$

23. $9 + (-12) = -3$

25. $38 + (-16) = 22$

27. $(-33) + (-21) = -(33 + 21) = -54$

29. $-3 + 4 + (-8) = 1 + (-8) = -7$

31. $(-9) + (-2) + (-5) = -(9 + 2 + 5) = -(11 + 5) = -16$

33. $-13 + (-1) + (-12) = -(13 + 1 + 12) = -(14 + 12)$
$= -26$

35. $27 + (-14) + (-13) = 13 + (-13) = 0$

37. $-43 + (-16) + 27 = -59 + 27 = -32$

39. $-38 + 49 + (-6) = 11 + (-6) = 5$

41. $\begin{array}{r} -21 \\ \underline{-62} \\ -83 \end{array}$

43. $\begin{array}{r} -15 \\ 8 \\ \underline{19} \\ 12 \end{array}$

45. $\begin{array}{r} -163 \\ 204 \\ \underline{-73} \\ -32 \end{array}$ Tip: Combine the two negative terms as -236 and then combine with 204.

47. $\begin{array}{r} -16 \\ 34 \\ 2 \\ \underline{-31} \\ -11 \end{array}$

49. $\begin{array}{r} 52 \\ -33 \\ 29 \\ \underline{-9} \\ 39 \end{array}$

51. $x = -2$ is a solution, as it makes the equation true:
$$(-2) + 4 \overset{?}{=} 2$$
$$2 = 2$$

53. $x = -4$ is a solution, as it makes the equation true:
$$-10 + (-4) \overset{?}{=} -14$$
$$-14 = -14$$

55. $y = -6$ is a solution, as it makes the equation true:
$$17 + (-6) \overset{?}{=} 11$$
$$11 = 11$$

57. $z = 18$ is a solution, as it makes the equation true:
$$18 + (-12) \overset{?}{=} 6$$
$$6 = 6$$

59. $y = -2$ is NOT a solution.

$$|-2| + (-10) \overset{?}{=} -12$$
$$2 + (-10) \overset{?}{=} -12$$
$$-8 \neq -12$$

61. $x = -10$ is a solution, as it makes the equation true:

$$-10 + (-5) \overset{?}{=} -15$$
$$-15 = -15$$

63. $z = -72$ is NOT a solution.

$$42 + |-72| \overset{?}{=} -30$$
$$42 + 72 \overset{?}{=} -30$$
$$114 \neq -30$$

Calculator Problems

Reminder: The TI – 84+, and other calculators, distinguish between the negative sign "–" and the subtraction sign by placing them on different keys. Be certain to use the two keys appropriately.

65. $47 + (-29) + 66 = 84$

67. $(2932) + 4751 + (-3876) = 3807$

69. $(-16,945) + (-27,302) + (-53,467) = -97,714$

Writing and Thinking About Mathematics

71. The only way the sum of two absolute values can be zero is in the case of: $|0| + |0| = 0$.

[End of Section 1.2]

Section 1.3
Solutions to Odd Exercises

For exercises 1–10, recall that the additive inverse of a number is the opposite of that number.

1. -11 since $11+(-11)=0$

3. $-(-6)=6$ since $-6+6=0$

5. -47 since $47+(-47)=0$

7. $-0=0$ since the opposite of 0 is 0

9. $-(-52)=52$ since $-52+52=0$

11. $8-3=5$

13. $-4-6=-10$

15. $3-(-4)=3+4=7$

17. $-8-(-11)=-8+11=3$

19. $-14-2=-16$

21. $16-(-8)=16+8=24$

23. $27-(+42)=27-42=-15$

25. $-23-(-7)=-23+7=-16$

27. $-21-(+36)=-21-36=-57$

29. $-27-(+27)=-27-27=-54$

31. $-5-(-6)=-5+6=1$

33. $-10-(-3)=-10+3=-7$

35. $-13-13=-26$

37. $(-4+(-17))-(-12+6)=-21-(-6)$
$$=-21+6=-15$$

39. $-6+(-4)-5=-6-4-5=-10-5=-15$

41. $6+(-3)+(-4)=6-3-4=3-4=-1$

43. $-5-2-(-4)=-5-2+4=-7+4=-3$

45. $-7-(-2)+6=-7+2+6=-5+6=1$

47. $97-16-(81)=97-16-81=81-81=0$

49. $-6+(-2) \; \underset{?}{\rule{1em}{0.4pt}} \; 3+(-8)$
$$-6-2 \; \underset{?}{\rule{1em}{0.4pt}} \; 3-8$$
$$-8 \; < \; -5$$

51. $5-8 \; \underset{?}{\rule{1em}{0.4pt}} \; 8-5$
$$-3 \; < \; 3$$

53. $11+(-3) \; \underset{?}{\rule{1em}{0.4pt}} \; 11-3$
$$11-3 \; \underset{?}{\rule{1em}{0.4pt}} \; 8$$
$$8 = 8$$

55. $-8-(-8) \; \underset{?}{\rule{1em}{0.4pt}} \; -14-13$
$$-8+8 \; \underset{?}{\rule{1em}{0.4pt}} \; -27$$
$$0 \; > \; -27$$

57. $-151-86 \; \underset{?}{\rule{1em}{0.4pt}} \; -(107+141)$
$$-237 \; > \; -248$$

59. $58° - 76° = -18°$

61. $\$39 - \$47 = -\$8$

63. $-282 \text{ ft} - 14{,}495 \text{ ft} = -14{,}777 \text{ ft}$

65. $1727 - 1642 = 85$ years old

67. $x=-8$ is a solution:
$$(-8)+5 \overset{?}{=} -3$$
$$-8+5 \overset{?}{=} -3$$
$$-3 = -3$$
The equation is true when $x=-8$.

69. $y=-2$ is a solution:
$$15-(-2) \overset{?}{=} 17$$
$$15+2 \overset{?}{=} 17$$
$$17 = 17$$
The equation is true when $y=-2$.

71. $x = 4$ is NOT a solution:

$$4 - 3 \overset{?}{=} -7$$
$$1 \neq -7$$

The equation is false when $x = 4$.

73. $x = 5$ is a solution:

$$-9 - 5 \overset{?}{=} -14$$
$$-14 = -14$$

The equation is true when $x = 5$.

75. $x = 1$ is NOT a solution:

$$1 - 2 \overset{?}{=} -3$$
$$-1 \neq -3$$

The equation is false when $x = 1$.

77. $x = -4$ is a solution:

$$|-4| - (-10) \overset{?}{=} 14$$
$$4 + 10 \overset{?}{=} 14$$
$$14 = 14$$

The equation is true when $x = -4$.

79. $z = -5$ is a solution:

$$|-5| - 5 \overset{?}{=} 0$$
$$5 - 5 \overset{?}{=} 0$$
$$0 = 0$$

The equation is true when $z = -5$.

81. $x = -28$ is a solution:

$$|-28| - |-3| \overset{?}{=} 25$$
$$28 - 3 \overset{?}{=} 25$$
$$25 = 25$$

The equation is true when $x = -28$.

Calculator Problems

83. $>$ is the correct symbol because

$$648 - (-396) = 1044$$

is greater than $124 - 163 = -39$.

85. $>$ is the correct symbol because

$$-43,931 - (-28,677) = -15,254 \text{ is greater than}$$

$$-(13,665 + 21,426) = -35,090.$$

87. Harry's weight change can be represented by:

$$-5 + 3 + (-2) + (-4) + 1 = -5 + 3 - 2 - 4 + 1 = -7$$

So, he lost 7 lb. Since his starting weight was 210, his weight after 5 weeks was $210 - 7 = 203$ lb.

89. The net yardage can be represented by:

$$5 + (-3) + 15 + 7 + 12 + (-4) + (-2) + 20 + (-5) + 6$$
$$= 5 - 3 + 15 + 7 + 12 - 4 - 2 + 20 - 5 + 6$$
$$= 51$$

So, he gained 51 yards.

[End of Section 1.3]

Section 1.4
Solutions to Odd Exercises

1. −12. The product of a positive number and a negative number is negative.

3. 48. The product of two positive numbers is positive.

5. 56. The product of two negative numbers is positive.

7. −21. The product of a positive number and a negative number is negative.

9. 56. The product of two negative numbers is positive.

11. 26. The product of two negative numbers is positive.

13. −70. The product of a negative and a positive number is negative.

15. −24. The product of three negative numbers is always negative. In general, an odd number of negative numbers being multiplied is always negative. The product of an even number of negative numbers is always positive.

17. −288. The product of one negative number and two positive numbers is the same as the product of a negative and a positive number, which is negative.

19. 0. Any number multiplied by 0 is 0.

21. −9. The product of a positive number and a negative number is negative.

23. −7.31. The product of a positive number and a negative number is negative.

25. 4. The quotient of two negative numbers is positive.

27. −6. The quotient of two negative numbers is positive.

29. 2. The quotient of two negative numbers is positive.

31. 0. Zero divided by any non-zero number is 0.

33. Undefined. Division by zero is always undefined.

35. −3. The quotient of a positive number divided by a negative number is negative.

37. −17. The quotient of a negative number divided by a positive number is negative.

39. 5. The quotient of two negative numbers is positive.

41. −0.6. The quotient of a negative number divided by a positive number is negative.

43. 20. The quotient of two negative numbers is positive.

45. 0.375. The quotient of two negative numbers is positive.

47. True. $(-4)(6) = -24$, $3 \cdot 8 = 24$, and $-24 < 24$.

49. True. $(-12)(6) = -72$, and $9(-8) = -72$.

51. True. $6(-3) = -18$, $(-14) + (-4) = -18$, and $-18 \geq -18$.

53. True. $(-7) + 0 = -7$, $(-7) \cdot 0 = 0$, and $-7 \leq 0$.

55. True. $(-4)(9) = -36$, $-24 + (-12) = -36$, and $-36 = -36$.

57. The mean of the following set is:
$$\{-10, 15, 16, -17, -34, -42\}$$
$$\left(-10 + 15 + 16 + (-17) + (-34) + (-42)\right) \div 6$$
$$= -72 \div 6$$
$$= -12$$

59. The average of the temperatures is found by adding all twenty values, then dividing by 20. The sum of the twenty values is 100, so the average is $100 \div 20 = 5$ degrees.

61. The mean score on the exam is given by:
$$\left(1 \cdot 67 + 4 \cdot 73 + 3 \cdot 77 + 2 \cdot 80 + 3 \cdot 88 + 2 \cdot 93\right) \div 15$$
$$= 1200 \div 15$$
$$= 80$$

63. The average blood calcium level is found by adding all twenty values, then dividing by 20. The sum of the twenty values is 185.4, so the average is $185.4 \div 20 = 9.3$ mg/dL (to the nearest tenth).

65. The mean height is given by:
$$\left(2 \cdot 68 + 3 \cdot 69 + 8 \cdot 72 + 5 \cdot 73 + 2 \cdot 74 + 75 + 78\right) \div 22$$
$$= 1585 \div 22$$
$$\approx 72 \text{ inches}$$

67. The average time people over the age of 18 spend watching TV is given by:

$$(26 + 30 + 44 + 23 + 28 + 38) \div 6 = 189 \div 6$$
$$= 31.5 \text{ hours}$$

69. The approximate mean size of the Great Lakes is given by:

$$(18,000 + 22,000 + 32,000 + 17,000 + 20,000) \div 5$$
$$= 109,000 \div 5$$
$$= 21,800$$

The mean size is approximately 22,000 sq mi.

Calculator Problems

71. $(-4,613)(-45)(-166) = (207,585)(-166)$
$$= -34,459,110$$

73. $(-77,459) \div 29 = -2671$

75. $(-35 - 45 - 56) \div 3 = (-136) \div 3 \approx -45.33$

77. $72 \div (15 - 22) = 72 \div (-7) \approx -10.29$

79. $-15.3 \div (-5.4) \approx 2.83$

81. $(-2.5)(-3.41)(-10.6) = -90.365$

Writing and Thinking About Mathematics

83. See page 109 of the textbook.

[End of Section 1.4]

Section 1.5
Solutions to Odd Exercises

1. **a.** The integers are $\{-1, 0, 4\}$.

b. The rational numbers are
$$\left\{-3.56, -\frac{5}{8}, -1, 0, 0.7, 4, \frac{13}{2}\right\}.$$

c. The irrationals are $\left\{-\sqrt{8}, \pi, \sqrt[3]{25}\right\}$.

d. All numbers in the list are real numbers.

3. $7 + 3 = 3 + 7$

5. $19 \cdot 4 = 4 \cdot 19$

7. $6(5 + 8) = 6(5) + 6(8)$ or $6(5 + 8) = 30 + 48$

9. $2 \cdot (3x) = (2 \cdot 3)x$

11. $3 + (x + 7) = (3 + x) + 7$

13. $6 \cdot 0 = 0 \cdot 6 = 0$

15. $0 + (x + 7) = x + 7$

17. $2(x - 12) = 2x - 24$

19. $6.3 + (-6.3) = 0$

21. The commutative property of addition.

23. The multiplicative identity.

25. The associative property of addition.

27. The commutative property of multiplication.

29. The commutative property of multiplication.

31. The multiplicative inverse.

33. The additive inverse.

35. The multiplicative identity.

37. The zero factor law.

39. The associative property of addition.

41. $6 \cdot x = x \cdot 6$ illustrates the commutative property of multiplication. For $x = 4$, the statement is true:
$$6 \cdot 4 = 24 = 4 \cdot 6$$

43. $8 + (5 + y) = (8 + 5) + y$ illustrates the associative property of addition. For $y = -2$, the statement is true:
$$8 + (5 + (-2)) = 11 = (8 + 5) + (-2)$$

45. $5(x + 18) = 5x + 90$ illustrates the distributive property of multiplication over addition. For $x = 4$, the statement is true:
$$5(4 + 18) = 110 = 5 \cdot 4 + 90$$

47. $(6 \cdot y) \cdot 9 = 6 \cdot (y \cdot 9)$ illustrates the associative property of multiplication. For $y = -2$, the statement is true:
$$(6 \cdot (-2)) \cdot 9 = -108 = 6 \cdot ((-2) \cdot 9)$$

49. $z + (-34) = -34 + z$ illustrates the commutative property of addition. For $z = 3$, the statement is true:
$$3 + (-34) = -31 = -34 + 3$$

51. $2(3 + x) = 2(x + 3)$ illustrates the commutative property of addition. For $x = 4$, the statement is true:
$$2(3 + 4) = 14 = 2(4 + 3)$$

53. $5 + (x - 15) = (x - 15) + 5$ illustrates the commutative property of addition. For $x = 4$, the statement is true:
$$5 + (4 - 15) = -6 = (4 - 15) + 5$$

55. $(3x) \cdot 5 = 3 \cdot (x \cdot 5)$ illustrates the associative property of multiplication. For $x = 4$, the statement is true:
$$(3 \cdot 4) \cdot 5 = 60 = 3 \cdot (4 \cdot 5)$$

[End of Section 1.5]

Chapter 1 Review
Solutions to All Exercises

1. $\{-3, 0, 3, 4\}$:

2. $\left\{-4, -1, -\dfrac{1}{4}, 2.5\right\}$:

3. All positive integers less than or equal to 5:

4. All integers greater than -2:

5. $|-6| = 6$

6. $-|7.3| = -7.3$

7. $-|-4.1| = -4.1$

8. $|-10| = 10$

9. True

10. True

11. False; $|-2.3| > 0$

12. True

13. $|x| = 5.$ $|5| = 5$ and $|-5| = 5$

14. $|x| = -5.$ The absolute value of any number is always positive, so there is no solution.

15. $12 = |x|.$ $|12| = 12$ and $|-12| = 12$

16. $3.5 = |y|.$ $|3.5| = 3.5$ and $|-3.5| = 3.5$

17. $|x| \le 3$

18. $|x| > 3$

19. $|y| = y$

20. $|y| < 2$

21. $6 + 9 = 15$

22. $15 + 7 = 22$

23. $8 + (-3) = 5$

24. $(-9) + 6 = -3$

25. $18 + (-18) = 0$

26. $75 + (-75) = 0$

27. $-23 + (-10) = -33$

28. $-7 + (-14) = -21$

29. $-12 + (-1) + (-5) = -13 + (-5) = -18$

30. $-20 + (-3) + (-2) = -23 + (-2) = -25$

31. $-42 + 20 + 3 = -22 + 3 = -19$

32. $-65 + (-4) + 40 = -69 + 40 = -29$

33. $-8 + 23 + (-7) = 15 + (-7) = 8$

34. $-25 + 13 + (-18) = -12 + (-18) = -30$

35. $-22 + (-10) + (-13) = -32 + (-13) = -45$

36. $-6 + 20 + 32 = 14 + 32 = 46$

37. Substitute $x = -3$ into $x + 5 = 2$:

$$(-3) + 5 \overset{?}{=} 2$$
$$2 = 2$$

So $x = -3$ is a solution.

38. Substitute $x = 2$ into $x + (-8) = -10$:

$$2 + (-8) \overset{?}{=} -10$$
$$-6 \neq -10$$

So $x = 2$ is not a solution.

39. Substitute $y = -20$ into $-10 + |y| = 10$:

$$-10 + |-20| \overset{?}{=} 10$$
$$-10 + 20 \overset{?}{=} 10$$
$$10 = 10$$

So $y = -20$ is a solution.

40. Substitute $y = 3$ into $|y| + (-4) = -1$:

$$|3| + (-4) \overset{?}{=} -1$$
$$3 - 4 \overset{?}{=} -1$$
$$-1 = -1$$

So $y = 3$ is a solution.

41. $-4 - 3 = -7$

42. $6 - 8 = -2$

43. $-15 - 4 = -19$

44. $10 - 18 = -8$

45. $-7 + (-2) - (-5) = -9 + 5 = -4$

46. $-10 + (-7) - 14 = -17 - 14 = -31$

47. $25 - (17) - (-13) = 8 + 13 = 21$

48. $98 - 16 - (32) = 82 - 32 = 50$

49. $-100 + 126 - 20 = 26 - 20 = 6$

50. $(-5) - (-5) + 22 = -5 + 5 + 22 = 0 + 22 = 22$

51. $32 - (40) = -8$

52. $-24 - (-7) = -24 + 7 = -17$

53. $-16 - (-25) = -16 + 25 = 9$

54. $-29 - (-36) = -29 + 36 = 7$

55. $(-13 + (-10)) - (16 + (-30)) = -23 - (-14)$
$$= -23 + 14 = -9$$

56. $17 - (-20) = 17 + 20 = 37$

57. $-3° - 30° = -33°$

58. $24,000 - 30,000 = -6,000$ feet

59. Substitute $x = -1$ into $x - 3 = -4$:

$$-1 - 3 \overset{?}{=} -4$$
$$-4 = -4$$

So $x = -1$ is a solution.

60. Substitute $x = -17$ into $12 - x = -5$:

$$12 - (-17) \overset{?}{=} -5$$
$$12 + 17 \overset{?}{=} -5$$
$$29 \neq -5$$

So $x = -17$ is not a solution.

61. $(-5) \cdot 7 = -35$

62. $8(-3.2) = -25.6$

63. $(-9) \cdot (-10) = 90$

64. $(-7) \cdot (-13) = 91$

65. $-3 \cdot 17 \cdot 5 = -225$

66. $-4 \cdot 3 \cdot (-2) = 24$

67. $8(-9) \cdot 0 = 0$

68. $-6(-7)(-13) = -546$

69. $\dfrac{-10}{-2} = 5$

70. $\dfrac{-51}{-3} = 17$

71. $\dfrac{-16}{0}$ is undefined.

72. $\dfrac{0}{-16} = 0$

73. $\dfrac{3.2}{-8} = -0.4$

74. $\dfrac{-4.5}{9} = -0.5$

75. $\dfrac{-70}{3.5} = -20$

76. $\dfrac{100}{-2.5} = -40$

77. $\dfrac{30 + 25 + 70 + 85 + 100}{5} = \dfrac{310}{5} = 62$

78. $\dfrac{(2 \cdot 82) + (3 \cdot 88)}{5} = \dfrac{428}{5} = 85.6$

79. $\dfrac{2 \cdot 60 + 1 \cdot 61 + 6 \cdot 63 + 4 \cdot 66 + 4 \cdot 70 + 3 \cdot 71}{20}$

$= \dfrac{1316}{20} = 65.8$

80. $\dfrac{9 + 4 + 7 + 9 + 15 + 5}{6} = \dfrac{49}{6} \approx 8.167 \approx 8$

81. Commutative Property of Addition.

82. Associative Property of Addition.

83. Additive Identity.

84. Additive Inverse.

85. Associative Property of Multiplication.

86. Commutative Property of Multiplication.

87. Multiplicative Identity.

88. Multiplicative Inverse.

89. Distributive Property of Multiplication over Addition.

90. Distributive Property of Multiplication over Addition.

91. $100.9 - (-32.8) + 104.73 = 238.43$

92. $-76.35 - (-14.7) + (-13) = -74.65$

93. $7.2(-8.3) = -59.76$

94. $93.75(17.2)(-25.1) = -40,473.75$

95. $83.54 \div 16.3 \approx 5.13$

96. $-17.15 \div 0.15 \approx -114.33$

97. $\dfrac{98.5 - (-10)}{108.5} = 1.00$

98. $\dfrac{-17.3 - (-20)}{-2.7} = -1.00$

99. $\dfrac{|16.3| + |-16.3|}{-2} = -16.30$

100. $\dfrac{-|-23.5| + |-20|}{-7} = 0.50$

[End of Chapter 1 Review]

Chapter 1 Test
Solutions to All Exercises

1. From the given set of numbers:
    a. The integers are $\{-5, -1, 0\}$.
    b. The rational numbers are
    $$\left\{-5, -1, -\frac{1}{3}, 0, \frac{1}{2}, 3\frac{1}{4}, 7.121212...\right\}.$$
    c. The irrational numbers are $\{-\pi\}$.
    d. All of them are real numbers.

2. **a.** True. All integers are rational numbers because any integer n can be written as $\frac{n}{1}$.
    b. False. There are many rational numbers that are not integers. For example: $\frac{2}{3}, \frac{5}{4}, -\frac{23}{9}$.

3. Recall that $<$ is "less than" and $>$ is "greater than."
    a. $-4 < -2$
    b. $-(-2) > 0$, since $-(-2) = 2$.
    c. $|-9| = |9|$, since $|-9| = 9$.

4. $\left\{-2, -0.4, |-1|, 2\frac{1}{3}, 4.1\right\}$:

5. $|x| = x$ if x is positive or zero and $|x| = -x$ if x is negative. $|x|$ is the distance the x lies from zero on the real number line.

6. -7 and 7, because $|-7| = 7$ and $|7| = 7$.

7. **a.** $|x| < 3$: $\{-2, -1, 0, 1, 2\}$
    b. $|x| \geq 9$: $\{..., -11, -10, -9, 9, 10, 11...\}$

8. $15 - 45 + 17 = -30 + 17 = -13$

9. $13 - (-16) + (-6) = 13 + 16 - 6 = 29 - 6 = 23$

10. $-29 - (-47) = -29 + 47 = 18$

11. $(-7)(-16) = 7 \cdot 16 = 112$

12. $3(-42) = -126$

13. $41(-5)(-3)(0) = 0$

14. $\frac{-57}{19} = -3$

15. $\frac{-140}{-20} = 7$

16. $\frac{4.5}{-15} = -0.3$

17. Additive Identity.

18. Distributive Property of Multiplication over Addition.

19. Commutative Property of Multiplication.

20. Associative Property of Addition.

21. Multiplicative Inverse.

22. Zero Factor Law.

23. $(-14 + (-16)) - ((-6) \cdot (-5)) = (-14 - 16) - 30$
    $$= (-30) - 30 = -60$$

24. $(-10 + 3)(-10 - 3) = -7(-13) = 91$

25. Net change in the stock market can be represented as $32 + 140 - 30 - 53 + 63 = 152$. The market gained 152 points.

26. The difference in temperature is $5 - (-10) = 5 + 10 = 15$. The temperature rose $15°F$.

27. The average of $\{-12, -15, -32, -31\}$ is:
    $$\frac{-12 + (-15) + (-32) + (-31)}{4} = \frac{-90}{4} = -22.5$$

28.

mph	Frequency	Product
19	4	76
20	5	100
22	7	154
25	3	75
28	1	28
Total	20	433

Mean $= \frac{433}{20} \approx 21.7$ to nearest tenth

[End of Chapter One Test]

Section 2.1
Solutions to Odd Exercises

1. Like terms are grouped:
    $\left\{-5, \dfrac{1}{6}, 8\right\}, \{7x, 9x\}, \{3y\}$

3. Like terms are grouped:
    $\{5xy, -6xy\}, \{-x^2, 2x^2\}, \{3x^2y, 5x^2y\}$

5. Like terms are grouped:
    $\{24, 8.3, -6\}, \{1.5xyz, -1.4xyz, xyz\},$
    $\{5xy^2z\}, \{2xyz^2\}$

7. $(-8)^2 = 64$

9. $-11^2 = -121$

11. $8x + 7x = (8+7)x = 15x$

13. $5x - 2x = (5-2)x = 3x$

15. $-n - n = -(n+n)$
    $= -(1+1)n$
    $= -2n$

17. $6y^2 - y^2 = (6-1)y^2 = 5y^2$

19. $23x^2 - 11x^2 = (23-11)x^2 = 12x^2$

21. $4x + 2 + 3x = 4x + 3x + 2$
    $= (4x + 3x) + 2$
    $= (4+3)x + 2$
    $= 7x + 2$

23. $2x - 3y - x = 2x - x - 3y$
    $= (2x - x) - 3y$
    $= (2-1)x - 3y$
    $= x - 3y$

25. $2x^2 + 5y + 6x^2 - 2y = 2x^2 + 6x^2 + 5y - 2y$
    $= (2x^2 + 6x^2) + (5y - 2y)$
    $= 8x^2 + 3y$

27. $3(n+1) - n = 3n + 3 - n$
    $= 3n - n + 3$
    $= (3-1)n + 3$
    $= 2n + 3$

29. $5(a-b) + 2a - 3b = 5a - 5b + 2a - 3b$
    $= 5a + 2a - 5b - 3b$
    $= (5+2)a - (5+3)b$
    $= 7a - 8b$

31. $3(2x+y) + 2(x-y) = 6x + 3y + 2x - 2y$
    $= 6x + 2x + 3y - 2y$
    $= (6+2)x + (3-2)y$
    $= 8x + y$

33. $2x + 3x^2 - 3x - x^2 = 2x - 3x + 3x^2 - x^2$
    $= (2-3)x + (3-1)x^2$
    $= -x + 2x^2$
    $= 2x^2 - x$

35. $2(n^2 - 3n) + 4(-n^2 + 2n) = 2n^2 - 6n - 4n^2 + 8n$
    $= 2n^2 - 4n^2 - 6n + 8n$
    $= (2-4)n^2 - (6-8)n$
    $= -2n^2 + 2n$

37. $3x^2 + 4xy - 5xy + y^2 = 3x^2 + (4-5)xy + y^2$
    $= 3x^2 - xy + y^2$

39. $\dfrac{x+5x}{6} + x = \dfrac{6x}{6} + x = x + x = 2x$

41. $y - \dfrac{2y+4y}{3} = y - \dfrac{6y}{3}$
    $= y - 2y$
    $= -y$

43. **a.** $5x + 4 - 2x = (5x - 2x) + 4 = 3x + 4$
    b. For $x = 4$, $3(4) + 4 = 16$

45. **a.** $x - 10 - 3x + 2 = x - 3x - 10 + 2 = -2x - 8$
    b. For $x = 4$, $-2(4) - 8 = -16$

47. **a.** $3(y-1) + 2(y+2) = 3y - 3 + 2y + 4 = 5y + 1$
    b. For $y = 3$, $5(3) + 1 = 15 + 1 = 16$

49. **a.** $-5(x+y)+2(x-y) = -5x-5y+2x-2y$
$$= -3x-7y$$
   b. For $x=4$ and $y=3$,
$$-3(4)-7(3) = -12-21 = -33$$

51. **a.** $8.3x^2 - 5.7x^2 + x^2 = (8.3-5.7+1)x^2$
$$= 3.6x^2$$
   b. For $x=4$, $3.6(4)^2 = 57.6$

53. **a.** $5ab+b^2-2ab+b^3 = 5ab-2ab+b^2+b^3$
$$= (5-2)ab+b^2+b^3$$
$$= 3ab+b^2+b^3$$
   b. For $a=-2$ and $b=-1$,
$$3(-2)(-1)+(-1)^2+(-1)^3 = 6$$

55. **a.** $2.4(x+1)+1.3(x-1) = 2.4x+2.4+1.3x-1.3$
$$= 3.7x+1.1$$
   b. For $x=4$, $3.7(4)+1.1 = 14.8+1.1 = 15.9$

57. **a.** $\dfrac{3a+5a}{-2}+12a = \dfrac{8a}{-2}+12a$
$$= -4a+12a$$
$$= 8a$$
   b. For $a=-2$, $8(-2) = -16$

59. **a.** $\dfrac{-4b-2b}{-3}+\dfrac{2b+5b}{7} = \dfrac{-6b}{-3}+\dfrac{7b}{7} = 2b+b = 3b$
   b. For $b=-1$, $3(-1) = -3$

61. **a.** $2x+3\big[x-2(9+x)\big] = 2x+3\big[x-18-2x\big]$
$$= 2x+3\big[-x-18\big]$$
$$= 2x-3x-54$$
$$= -x-54$$
   b. For $x=4$, $-(4)-54 = -58$

[End of Section 2.1]

Section 2.2
Solutions to Odd Exercises

1. Since $6 \cdot 8 = 48$, multiply the numerator and the denominator by 8:

$$\frac{5}{6} = \frac{5}{6} \cdot \frac{8}{8} = \frac{40}{48}$$

3. Since $9 \cdot 7b = 63b$, multiply the numerator and the denominator by $7b$:

$$\frac{0}{9} = \frac{0}{9} \cdot \frac{7b}{7b} = \frac{0}{63b}$$

5. Since $5 \cdot 7y = 35y$, multiply the numerator and the denominator by $7y$:

$$\frac{-3}{5} = \frac{-3}{5} \cdot \frac{7y}{7y} = -\frac{21y}{35y}$$

7. $\dfrac{18}{45} = \dfrac{\not{9} \cdot 2}{\not{9} \cdot 5} = \dfrac{2}{5}$

9. $\dfrac{150xy}{350y} = \dfrac{\not{50} \cdot 3 \cdot x \cdot \not{y}}{\not{50} \cdot 7 \cdot \not{y}} = \dfrac{3x}{7}$

11. $\dfrac{-12a}{-100ab} = \dfrac{\not{4} \cdot 3 \cdot \not{a}}{\not{4} \cdot 25 \cdot \not{a} \cdot b} = \dfrac{3}{25b}$

13. $\dfrac{-30y}{45y} = \dfrac{-1 \cdot \not{15} \cdot 2 \cdot \not{y}}{\not{15} \cdot 3 \cdot \not{y}} = -\dfrac{2}{3}$

15. $\dfrac{-28}{56x} = \dfrac{-1 \cdot \not{28} \cdot 1}{\not{28} \cdot 2x} = -\dfrac{1}{2x}$

17. $\dfrac{12xyz}{-35xy} = \dfrac{12 \cdot \not{x} \cdot \not{y} \cdot z}{-1 \cdot 35 \cdot \not{x} \cdot \not{y}} = -\dfrac{12z}{35}$

19. $\dfrac{1}{2}$ of $\dfrac{3}{4}$ is equivalent to $\dfrac{1}{2} \cdot \dfrac{3}{4}$. This simplifies to:

$$\frac{1}{2} \cdot \frac{3}{4} = \frac{1 \cdot 3}{2 \cdot 4} = \frac{3}{8}$$

21. $\dfrac{7}{8}$ of 40 is equivalent to $\dfrac{7}{8} \cdot \dfrac{40}{1}$. This simplifies to:

$$\frac{7}{8} \cdot \frac{40}{1} = \frac{7 \cdot \not{8} \cdot 5}{\not{8} \cdot 1} = \frac{35}{1} = 35$$

23. $\dfrac{-3}{8} \cdot \dfrac{4}{9} = \dfrac{-1 \cdot 3 \cdot 4}{8 \cdot 9} = \dfrac{-1 \cdot \not{3} \cdot \not{2} \cdot \not{2}}{2 \cdot \not{2} \cdot \not{2} \cdot \not{3} \cdot 3} = -\dfrac{1}{6}$

25. $\dfrac{9}{10x} \div \dfrac{10}{9x} = \dfrac{9}{10x} \cdot \dfrac{9x}{10} = \dfrac{9 \cdot 9 \cdot \not{x}}{10 \cdot \not{x} \cdot 10} = \dfrac{81}{100}$

27. $\dfrac{-16y}{7} \div \dfrac{16}{y} = \dfrac{-16y}{7} \cdot \dfrac{y}{16} = \dfrac{-1 \cdot \not{16} \cdot y \cdot y}{7 \cdot \not{16}} = -\dfrac{y^2}{7}$

29. $\dfrac{-15x}{2} \cdot \dfrac{6x}{25} = \dfrac{-1 \cdot 3 \cdot \not{5} \cdot x \cdot \not{2} \cdot 3 \cdot x}{\not{2} \cdot \not{5} \cdot 5} = -\dfrac{9x^2}{5}$

31. $\dfrac{26b}{51a} \cdot \dfrac{17b}{13a} = \dfrac{2 \cdot \not{13} \cdot b \cdot \not{17} \cdot b}{3 \cdot \not{17} \cdot a \cdot \not{13} \cdot a} = \dfrac{2b^2}{3a^2}$

33. $\dfrac{-15}{4} \cdot \dfrac{5}{6} \cdot \dfrac{16}{9} = \dfrac{-1 \cdot \not{3} \cdot 5 \cdot 5 \cdot \not{4} \cdot \not{2} \cdot 2}{\not{4} \cdot \not{2} \cdot \not{3} \cdot 3 \cdot 3} = -\dfrac{50}{9}$

35. $\dfrac{9a}{15} \cdot \dfrac{10}{3b} \cdot \dfrac{1}{5a} = \dfrac{\not{3} \cdot \not{3} \cdot \not{a} \cdot 2 \cdot \not{5} \cdot 1}{\not{3} \cdot \not{5} \cdot \not{3} \cdot b \cdot 5 \cdot \not{a}} = \dfrac{2}{5b}$

37. $\dfrac{-35y}{12} \cdot \dfrac{-6}{14y} \cdot \dfrac{4y}{3} = \dfrac{5 \cdot \not{7} \cdot \not{y} \cdot \not{6} \cdot \not{2} \cdot \not{2} \cdot y}{\not{2} \cdot \not{6} \cdot \not{2} \cdot \not{7} \cdot \not{y} \cdot 3} = \dfrac{5y}{3}$

39. $0 \div \dfrac{37y}{21x} = 0 \cdot \dfrac{21x}{37y} = 0$

41. $\dfrac{6xy}{15x} \div 0$ is undefined.

43. $\dfrac{92}{7a} \div \dfrac{46a}{77} = \dfrac{92}{7a} \cdot \dfrac{77}{46a} = \dfrac{\not{46} \cdot 2 \cdot 11 \cdot \not{7}}{\not{7} \cdot a \cdot \not{46} \cdot a} = \dfrac{22}{a^2}$

45. $\dfrac{21}{30} \div (-7) = \dfrac{21}{30} \cdot \dfrac{-1}{7} = \dfrac{-1 \cdot \not{3} \cdot \not{7}}{\not{3} \cdot 10 \cdot \not{7}} = -\dfrac{1}{10}$

47. $\dfrac{45a}{30b} \cdot \dfrac{12ab}{18b} \div \left(\dfrac{10ac}{9c}\right) = \dfrac{45a \cdot 12ab}{30b \cdot 18b} \cdot \dfrac{9c}{10ac}$

$$= \frac{\not{5} \cdot \not{9} \cdot \not{a} \cdot \not{2} \cdot \not{6} \cdot a \cdot \not{b} \cdot 9 \cdot \not{c}}{\not{5} \cdot \not{6} \cdot \not{b} \cdot \not{2} \cdot \not{9} \cdot b \cdot 10 \cdot \not{a} \cdot \not{c}}$$

$$= \frac{9a}{10b}$$

49. $\dfrac{-35m}{21n}\cdot\dfrac{27n}{40mn}\cdot\dfrac{-9m}{40mn}=\dfrac{\cancel{5}\cdot\cancel{7}\cdot\cancel{m}\cdot\cancel{3}\cdot 9\cdot\cancel{n}\cdot 9\cdot\cancel{m}}{\cancel{3}\cdot\cancel{7}\cdot\cancel{n}\cdot\cancel{5}\cdot 8\cdot\cancel{m}\cdot n\cdot 40\cdot\cancel{m}\cdot n}$

$$=\dfrac{81}{320n^2}$$

51. $\dfrac{-9x}{40y}\cdot\dfrac{35x}{15y}\cdot\dfrac{5}{10}\cdot\dfrac{8xy}{30xyz}$

$$=\dfrac{-1\cdot\cancel{3}\cdot\cancel{3}\cdot\cancel{x}\cdot\cancel{5}\cdot 7\cdot x\cdot\cancel{5}\cdot\cancel{8}\cdot x\cdot\cancel{y}}{\cancel{5}\cdot\cancel{8}\cdot\cancel{y}\cdot\cancel{3}\cdot\cancel{5}\cdot y\cdot 10\cdot\cancel{3}\cdot 10\cdot\cancel{x}\cdot y\cdot z}$$

$$=-\dfrac{7x^2}{100y^2z}$$

53. $\left(\dfrac{19}{2}\div\dfrac{3}{4}\right)\cdot\dfrac{13}{2}=\dfrac{19}{2}\cdot\dfrac{4}{3}\cdot\dfrac{13}{2}=\dfrac{19\cdot\cancel{2}\cdot\cancel{2}\cdot 13}{\cancel{2}\cdot 3\cdot\cancel{2}}=\dfrac{247}{3}$

55. $\dfrac{11}{4}\cdot\left(\dfrac{8}{5}\div\dfrac{11}{6}\right)=\dfrac{11}{4}\cdot\dfrac{8}{5}\cdot\dfrac{6}{11}=\dfrac{\cancel{11}\cdot 2\cdot\cancel{4}\cdot 6}{\cancel{4}\cdot 5\cdot\cancel{11}}=\dfrac{12}{5}$

57. If $\dfrac{3}{5}$ times a number is $\dfrac{-5}{8}$, then $\dfrac{-5}{8}\div\dfrac{3}{5}$ is that

number. So $\dfrac{-5}{8}\div\dfrac{3}{5}=\dfrac{-5}{8}\cdot\dfrac{5}{3}=\dfrac{-25}{24}$ is the number.

59. $\dfrac{1}{3}$ of 6 is $\dfrac{1}{3}\cdot 6=\dfrac{1}{3}\cdot\dfrac{6}{1}=\dfrac{2\cdot\cancel{3}}{\cancel{3}}=2$. So the height of the

milk is 2 inches.

61. $\dfrac{7}{10}$ of 140 is $\dfrac{7}{10}\cdot 140=\dfrac{7}{10}\cdot\dfrac{140}{1}=\dfrac{7\cdot 14\cdot\cancel{10}}{\cancel{10}\cdot 1}=98$. So

the person is 98 pounds water.

63. If $\dfrac{1}{5}$ of the land area of the world is 11,707,000, then

$11,707,000\div\dfrac{1}{5}=11,707,000\cdot 5=58,535,000$ sq mi

is approximately the total area of the world.

65. If 5 students received a grade of A, then $40-5=35$
students did not receive a grade of A. Then the
fraction of the class that did not receive a grade of A

is $\dfrac{35}{40}=\dfrac{\cancel{5}\cdot 7}{\cancel{5}\cdot 8}=\dfrac{7}{8}$.

67. a. The capacity of the airplane is greater than 180,

because 180 is only $\dfrac{9}{10}$ the capacity of the

airplane, and $\dfrac{9}{10}$ is less than 1.

b. Dividing 180 by $\dfrac{9}{10}$ yields the capacity of the

airplane: $180\div\dfrac{9}{10}=180\cdot\dfrac{10}{9}=\dfrac{\cancel{9}\cdot 20\cdot 10}{\cancel{9}}=200$.

Writing and Thinking About Mathematics

69. Various responses are possible.
Example:
> Zero does not have a reciprocal because the
> product of a number and its reciprocal is 1,
> while the product of zero and any number is 0.

Another example:
> Since $\dfrac{1}{0}$ is not defined, zero has no reciprocal.

[End of Section 2.2]

Section 2.3
Solutions to Odd Exercises

1. **a.** The factors of 6 are 1, 2, 3, 6.
   b. The first six multiples of 6 are 6, 12, 18, 24, 30, 36.

3. **a.** The factors of 15 are 1, 3, 5, 15.
   b. The first six multiples of 15 are 15, 30, 45, 60, 75, 90.

5. The LCM (least common multiple) of 24, 15 and 10 is the smallest number that each will divide. By factoring each of the numbers, we can construct the LCM:
$$24 = 2^3 \cdot 3$$
$$15 = 3 \cdot 5$$
$$10 = 2 \cdot 5$$
So, the LCM is $2^3 \cdot 3 \cdot 5 = 120$.

7. The LCM of $8x$, $10y$ and $20xy$ is found by factoring each of the expressions and using the factors that occur most in each of the three terms:
$$8x = 2^3 \cdot x$$
$$10y = 2 \cdot 5 \cdot y$$
$$20xy = 2^2 \cdot 5 \cdot x \cdot y$$
So, the LCM is $2^3 \cdot 5 \cdot x \cdot y = 40xy$.

9. The LCM of $14x^2$, $21xy$ and $35xy^2$ is found by factoring each of the expressions and using the factors that occur most in each of the three terms:
$$14x^2 = 2 \cdot 7 \cdot x^2$$
$$21xy = 3 \cdot 7 \cdot x \cdot y$$
$$35xy^2 = 5 \cdot 7 \cdot x \cdot y^2$$
So, the LCM is $2 \cdot 3 \cdot 5 \cdot 7 \cdot x^2 \cdot y^2 = 210x^2y^2$.

11. These two fractions have a common denominator, so just add the numerators and use the common denominator in the sum:
$$\frac{2}{9} + \frac{5}{9} = \frac{2+5}{9} = \frac{7}{9}$$

13. $\dfrac{24}{23} - \dfrac{1}{23} = \dfrac{24-1}{23} = \dfrac{23}{23} = 1$

15. $\dfrac{11}{12} - \dfrac{13}{12} = \dfrac{11-13}{12} = \dfrac{-2}{12} = -\dfrac{1}{6}$

17. First, obtain the LCM of the denominators of the fractions. The LCM of 6 and 10 can be found by factoring 6 and 10 and using prime factors that occur most between the two numbers:
$$6 = 2 \cdot 3$$
$$10 = 2 \cdot 5$$
So, the LCM of 6 and 10 is $2 \cdot 3 \cdot 5 = 30$.
$$\frac{5}{6} + \frac{7}{10} = \frac{5 \cdot 5}{30} + \frac{7 \cdot 3}{30} = \frac{25}{30} + \frac{21}{30} = \frac{46}{30} = \frac{23}{15}$$

19. $\dfrac{11}{15a} + \dfrac{5}{15a} = \dfrac{11+5}{15a} = \dfrac{16}{15a}$

21. $\dfrac{-19}{25x} + \dfrac{9}{25x} = \dfrac{-19+9}{25x} = \dfrac{-10}{25x} = -\dfrac{2}{5x}$

23. The LCM of $5y$ and $10y$ is $10y$.
$$\frac{7}{10y} - \frac{1}{5y} = \frac{7}{10y} - \frac{2}{10y} = \frac{7-2}{10y} = \frac{5}{10y} = \frac{1}{2y}$$

25. The LCM of 9 and 12 is 36.
$$\frac{8}{9} + \left(-\frac{5}{12}\right) = \frac{8 \cdot 4}{36} - \frac{5 \cdot 3}{36} = \frac{32}{36} - \frac{15}{36}$$
$$= \frac{32-15}{36} = \frac{17}{36}$$

27. The LCM of $24x$ and $36x$ is $72x$.
$$\frac{11}{24x} + \frac{5}{36x} = \frac{11 \cdot 3}{72x} + \frac{5 \cdot 2}{72x} = \frac{33+10}{72x} = \frac{43}{72x}$$

29. The LCM of y and 3 is $3y$.
$$\frac{5}{y} - \frac{1}{3} = \frac{5 \cdot 3}{3y} - \frac{1 \cdot y}{3y} = \frac{15-y}{3y}$$

31. The LCM of x and 4 is $4x$.
$$\frac{4}{x} + \frac{1}{4} = \frac{4 \cdot 4}{4x} + \frac{1 \cdot x}{4x} = \frac{16+x}{4x}$$

33. The LCM of 5 and 4 is 20.
$$\frac{a}{5} - \frac{3}{4} = \frac{a \cdot 4}{20} - \frac{3 \cdot 5}{20} = \frac{4a-15}{20}$$

35. The LCM of 4 and 8 is 8.
$$\frac{3}{4} + \left(-\frac{x}{8}\right) = \frac{3 \cdot 2}{8} - \frac{x}{8} = \frac{6-x}{8}$$

37. The LCM of $3a$ and $6a$ is $6a$.
$$\frac{2}{3a} + \frac{1}{6a} = \frac{2 \cdot 2}{6a} + \frac{1}{6a} = \frac{4+1}{6a} = \frac{5}{6a}$$

39. The LCM of $3x$ and $4x$ is $12x$.

$$\frac{1}{3x} + \left(-\frac{5}{4x}\right) = \frac{1\cdot 4}{12x} - \frac{5\cdot 3}{12x} = \frac{4-15}{12x} = -\frac{11}{12x}$$

41. The LCM of $4a$ and $8a$ is $8a$.

$$\frac{3}{4a} - \frac{6}{8a} = \frac{3\cdot 2}{8a} - \frac{6}{8a} = \frac{6-6}{8a} = \frac{0}{8a} = 0$$

43. The LCM of 7 and 14 is 14.

$$\frac{5x}{7} - \frac{10x}{14} = \frac{5x\cdot 2}{14} - \frac{10x}{14} = \frac{10x-10x}{14} = \frac{0}{14} = 0$$

45. **a.** $3a+3a = 6a$

  b. $3a\cdot 3a = 9a^2$

  c. $\dfrac{1}{3a} + \dfrac{1}{3a} = \dfrac{1+1}{3a} = \dfrac{2}{3a}$

  d. $\dfrac{1}{3a} \cdot \dfrac{1}{3a} = \dfrac{1}{3a\cdot 3a} = \dfrac{1}{9a^2}$

  e. $\dfrac{1}{3a} \div \dfrac{1}{3a} = \dfrac{1}{3a} \cdot \dfrac{3a}{1} = \dfrac{3a}{3a} = 1$

47. **a.** $-7ab - 7ab = -14ab$

  b. $(-7ab)\cdot(-7ab) = 49a^2b^2$

  c. $\dfrac{1}{7ab} + \dfrac{1}{7ab} = \dfrac{1+1}{7ab} = \dfrac{2}{7ab}$

  d. $\dfrac{1}{7ab} \cdot \dfrac{1}{7ab} = \dfrac{1}{7ab\cdot 7ab} = \dfrac{1}{49a^2b^2}$

  e. $\dfrac{1}{7ab} \div \dfrac{1}{7ab} = \dfrac{1}{7ab} \cdot \dfrac{7ab}{1} = \dfrac{7ab}{7ab} = 1$

49. **a.** $\dfrac{4x}{3} + \dfrac{4x}{3} = \dfrac{8x}{3}$

  b. $\dfrac{4x}{3} \cdot \dfrac{4x}{3} = \dfrac{4x\cdot 4x}{3\cdot 3} = \dfrac{16x^2}{9}$

  c. $\dfrac{4x}{3} - \dfrac{4x}{3} = \dfrac{4x-4x}{3} = \dfrac{0}{3} = 0$

  d. $\dfrac{4x}{3} \cdot \dfrac{3}{4x} = \dfrac{4x\cdot 3}{3\cdot 4x} = 1$

  e. $\dfrac{4x}{3} \div \dfrac{3}{4x} = \dfrac{4x}{3} \cdot \dfrac{4x}{3} = \dfrac{4x\cdot 4x}{3\cdot 3} = \dfrac{16x^2}{9}$

51. $\dfrac{6}{51}\cdot\dfrac{5}{50}\cdot\dfrac{4}{49}\cdot\dfrac{3}{48}\cdot\dfrac{2}{47}\cdot\dfrac{1}{46} = \dfrac{2}{17}\cdot\dfrac{1}{10}\cdot\dfrac{4}{49}\cdot\dfrac{1}{16}\cdot\dfrac{2}{47}\cdot\dfrac{1}{46}$

$$= \frac{2\cdot 4\cdot 2}{17\cdot 2\cdot 5\cdot 49\cdot 4\cdot 2\cdot 2\cdot 47\cdot 46}$$

$$= \frac{1}{18,009,460}$$

53. Over all, this problem is a division ("quotient") of two numbers. The first is a sum and the second is a difference (subtraction): Simplify within the parentheses first.

$$\left(\frac{4}{5} + \frac{11}{15}\right) \div \left(\frac{7}{12} - \frac{4}{15}\right) = \left(\frac{4\cdot 3}{15} + \frac{11}{15}\right) \div \left(\frac{7\cdot 5}{12\cdot 5} - \frac{4\cdot 4}{15\cdot 4}\right)$$

$$= \frac{12+11}{15} \div \frac{35-16}{60} = \frac{23}{15} \div \frac{19}{60}$$

$$= \frac{23}{15} \cdot \frac{60}{19} = \frac{92}{19}$$

55. **a.** The fraction of her income to be used for rent and food is the sum of the two fractions:

$$\frac{1}{3} + \frac{1}{10} = \frac{10}{30} + \frac{3}{30} = \frac{13}{30}$$

  b. The dollar amount of her $2700 income that she plans to spend on rent and food is

$$\frac{13}{30} \cdot \frac{2700}{1} = \frac{13\cdot 90}{1\cdot 1} = \$1170$$

[End of Section 2.3]

Section 2.4
Solutions to Odd Exercises

1. $(60.4)(1.8) = 108.72$

3. $(1.23)(-6.2) = -7.626$

5. $(-5.12)(-30.7) = 157.184$

7. $(6.7)(3.2)(7.9) = 169.376$

9. $5.1\overline{)77.01}$

$51\,\overline{)770.1}$ Move the decimals one place.

$51\overline{)770.1}$ Place the decimal point.

$\begin{array}{r} 15.1 \\ 51\overline{)770.1} \end{array}$ Divide.

$\quad \underline{51}$
$\quad 260$
$\quad \underline{255}$
$\quad\quad 51$
$\quad\quad \underline{51}$
$\quad\quad\quad 0$

11. $4.6\overline{)16.192}$

$46\,\overline{)161.92}$ Move the decimals one place.

$46\overline{)161.92}$ Place the decimal point.

$\begin{array}{r} 3.52 \\ 46\overline{)161.92} \end{array}$ Divide.

$\quad \underline{138}$
$\quad 239$
$\quad \underline{230}$
$\quad\quad 92$
$\quad\quad \underline{92}$
$\quad\quad\quad 0$

13. $0.256\overline{)0.09728}$

$256\overline{)97.28}$ Move the decimals 3 places.

$256\overline{)97.28}$ Place the decimal point.

$\begin{array}{r} 0.38 \\ 256\overline{)97.28} \end{array}$ Divide.

$\quad \underline{768}$
$\quad 2048$
$\quad \underline{2048}$
$\quad\quad\quad 0$

15. $15.3\overline{)0.3825}$

$153\overline{)3.825}$ Move the decimals one place.

$153\overline{)3.825}$ Place the decimal point.

$\begin{array}{r} 0.025 \\ 153\overline{)3.825} \end{array}$ Divide.

$\quad \underline{306}$
$\quad 765$
$\quad \underline{765}$
$\quad\quad 0$

17. $\begin{array}{r} .375 \\ 8\overline{)3.000} \end{array}$

$\quad \underline{24}$
$\quad 60$
$\quad \underline{56}$
$\quad 40$
$\quad \underline{40}$
$\quad\quad 0$

19. $\begin{array}{r} 0.05 \\ 20\overline{)1.00} \end{array}$

$\quad \underline{100}$
$\quad\quad 0$

21. $\begin{array}{r} 0.\overline{66} \\ 3\overline{)2.00} \end{array}$

$\quad \underline{18}$
$\quad 20$
$\quad \underline{18}$
$\quad 20$
$\quad \vdots$

23.
$$30\overline{)7.00}^{0.2\overline{3}}$$
$$\underline{60}$$
$$100$$
$$\underline{90}$$
$$100$$
$$\vdots$$

25. $\dfrac{7}{16} = 0.4375$

27. $-\dfrac{6}{11} = -0.\overline{54}$

29. $1\dfrac{5}{27} = \dfrac{32}{27} = 1.\overline{185}$

31. $6\dfrac{7}{12} = \dfrac{79}{12} = 6.58\overline{3}$

33. The average of the following set is:
$$\{86.46, 95.35, 99.07, 100.34\}$$
$$\dfrac{86.46 + 95.35 + 99.07 + 100.34}{4}$$
$$= \dfrac{381.22}{4} = 95.305 \approx 95.31$$

35. The average amount per item that she spent is given by:
$$\dfrac{1456.70 + 815 + 975}{3} = \dfrac{3246.70}{3} \approx 1082.23$$

37. $0.57 + 0.73 = 1.3$
$$1.3 = 1\dfrac{3}{10} = \dfrac{13}{10}$$

39. $-2.78 - 1.93 + 10.002 = 5.292$
$$5.292 = 5\dfrac{292}{1000} = 5\dfrac{73}{250} = \dfrac{1323}{250}$$

41. $11.67 - 25.9 - 50 = -64.23$
$$-64.23 = -64\dfrac{23}{100} = -\dfrac{6423}{100}$$

43. $\dfrac{3}{4} + \dfrac{1}{10} = \dfrac{17}{20}$

45. $\dfrac{3}{7} + \dfrac{1}{3} + \dfrac{2}{21} = \dfrac{18}{21} = \dfrac{6}{7}$

47. $\dfrac{3}{10} - \dfrac{27}{100} - \dfrac{133}{1000} = -\dfrac{103}{1000}$

[End of Section 2.4]

Section 2.5
Solutions to Odd Exercises

1.   **a.** $24 \div 4 \cdot 6 = 6 \cdot 6$
                $= 36$
   b. $24 \cdot 4 \div 6 = 96 \div 6$
                $= 16$

3. $15 \div (-3) \cdot 3 - 10 = -5 \cdot 3 - 10$
                        $= -15 - 10$
                        $= -25$

5. $3^3 \div (-9) \cdot (4 - 2^2) + 5(-2)$
              $= 27 \div (-9) \cdot (4 - 4) + 5(-2)$
              $= -3 \cdot (4 - 4) + 5(-2)$
              $= -3 \cdot 0 + 5(-2)$
              $= 0 + (-10)$
              $= -10$

7. $14 \cdot 3 \div (-2) - 6(4) = 42 \div (-2) - 24$
                          $= -21 - 24$
                          $= -45$

9. $-10 + 15 \div (-5) \cdot 3^2 - 10^2$
          $= -10 + (-3) \cdot 9 - 100$
          $= -10 + (-27) - 100$
          $= -137$

11. $2 - 5\big[(-20) \div (-4) \cdot 2 - 40\big]$
            $= 2 - 5[5 \cdot 2 - 40]$
            $= 2 - 5[10 - 40]$
            $= 2 - 5[-30]$
            $= 2 + 150$
            $= 152$

13. $(7 - 10)\big[49 \div (-7) + 20 \cdot 3 - (-10)\big]$
           $= (-3)\big[-7 + 60 - (-10)\big]$
           $= (-3)[63]$
           $= -189$

15. $8 - 9\Big[(-39) \div (-13) + 7(-2) - (-2)^2\Big]$
         $= 8 - 9\big[3 + (-14) - 4\big]$
         $= 8 - 9[3 - 14 - 4] = 8 - 9[-15]$
         $= 8 + 135 = 143$

17. $|16 - 20| \cdot \Big[32 \div |3 - 5| - 5^2\Big]$
          $= |-4| \cdot \big[32 \div |-2| - 25\big]$
          $= 4 \cdot \big[32 \div 2 - 25\big]$
          $= 4 \cdot \big[16 - 25\big] = 4[-9]$
          $= -36$

19. $\dfrac{(-10) + (-2)}{6^2 - 30} \div |2 - 4| = \dfrac{-12}{36 - 30} \div |-2|$
                             $= \dfrac{-12}{6} \div 2$
                             $= -2 \div 2 = -1$

21. $\dfrac{3}{8} \cdot \dfrac{4}{5} + \dfrac{1}{15} = \dfrac{3 \cdot 4}{8 \cdot 5} + \dfrac{1}{15}$
               $= \dfrac{12}{40} + \dfrac{1}{15} = \dfrac{3}{10} + \dfrac{1}{15}$
               $= \dfrac{3}{10} \cdot \dfrac{3}{3} + \dfrac{1}{15} \cdot \dfrac{2}{2}$
               $= \dfrac{9}{30} + \dfrac{2}{30} = \dfrac{11}{30}$

23. $\dfrac{1}{3} \div \dfrac{1}{2} - \dfrac{5}{6} \cdot \dfrac{3}{4} = \dfrac{1}{3} \cdot \dfrac{2}{1} - \dfrac{5 \cdot 3}{6 \cdot 4}$
               $= \dfrac{1 \cdot 2}{3 \cdot 1} - \dfrac{15}{24} = \dfrac{2}{3} - \dfrac{15}{24}$
               $= \dfrac{2}{3} \cdot \dfrac{8}{8} - \dfrac{15}{24} = \dfrac{16}{24} - \dfrac{15}{24}$
               $= \dfrac{1}{24}$

25. $\left(\dfrac{5}{6}\right)^2 \div \dfrac{5}{12} - \dfrac{3}{8} = \dfrac{25}{36} \cdot \dfrac{12}{5} - \dfrac{3}{8}$
               $= \dfrac{5}{3} - \dfrac{3}{8} = \dfrac{5}{3} \cdot \dfrac{8}{8} - \dfrac{3}{8} \cdot \dfrac{3}{3}$
               $= \dfrac{40}{24} - \dfrac{9}{24} = \dfrac{31}{24}$

27. $\dfrac{7}{6} \cdot 2^2 - \dfrac{2}{3} \div 3\dfrac{1}{5} = \dfrac{7}{6} \cdot 4 - \dfrac{2}{3} \div \dfrac{16}{5} = \dfrac{28}{6} - \dfrac{2}{3} \cdot \dfrac{5}{16}$

$$= \dfrac{28}{6} - \dfrac{5}{24} = \dfrac{28}{6} \cdot \dfrac{4}{4} - \dfrac{5}{24}$$

$$= \dfrac{112}{24} - \dfrac{5}{24} = \dfrac{107}{24}$$

29. $\left(-\dfrac{3}{4}\right) \div \left(-\dfrac{3}{5}\right) \cdot \dfrac{7}{8} + \dfrac{3}{16} = \left(-\dfrac{3}{4}\right) \cdot \left(-\dfrac{5}{3}\right) \cdot \dfrac{7}{8} + \dfrac{3}{16}$

$$= \dfrac{15}{12} \cdot \dfrac{7}{8} + \dfrac{3}{16}$$

$$= \dfrac{105}{96} + \dfrac{3}{16} = \dfrac{105}{96} + \dfrac{3}{16} \cdot \dfrac{6}{6}$$

$$= \dfrac{105}{96} + \dfrac{18}{96} = \dfrac{123}{96} = \dfrac{41}{32}$$

31. $\left(-\dfrac{9}{10}\right) + \dfrac{5}{8} \cdot \dfrac{4}{5} \div \dfrac{6}{10} + \dfrac{2}{3} = \left(-\dfrac{9}{10}\right) + \dfrac{5}{8} \cdot \dfrac{4}{5} \cdot \dfrac{10}{6} + \dfrac{2}{3}$

$$= \left(-\dfrac{9}{10}\right) + \dfrac{5 \cdot 4 \cdot 10}{8 \cdot 5 \cdot 6} + \dfrac{2}{3}$$

$$= \left(-\dfrac{9}{10}\right) + \dfrac{5}{6} + \dfrac{2}{3}$$

$$= \left(-\dfrac{9}{10}\right) \cdot \dfrac{3}{3} + \dfrac{5}{6} \cdot \dfrac{5}{5} + \dfrac{2}{3} \cdot \dfrac{10}{10}$$

$$= \left(-\dfrac{27}{30}\right) + \dfrac{25}{30} + \dfrac{20}{30}$$

$$= \dfrac{-27 + 25 + 20}{30}$$

$$= \dfrac{18}{30} = \dfrac{3}{5}$$

33. $-2\dfrac{1}{2} \cdot 3\dfrac{1}{5} \div \dfrac{3}{4} - \dfrac{7}{10} = -\dfrac{5}{2} \cdot \dfrac{16}{5} \cdot \dfrac{4}{3} - \dfrac{7}{10}$

$$= \dfrac{-5 \cdot 16 \cdot 4}{2 \cdot 5 \cdot 3} - \dfrac{7}{10} \cdot \dfrac{3}{3}$$

$$= \dfrac{-320}{30} - \dfrac{21}{30} = -\dfrac{341}{30}$$

35. $\left(\dfrac{1}{2} - 1\dfrac{3}{4}\right) \div \left(\dfrac{2}{3} + \dfrac{3}{4}\right) = \left(\dfrac{1}{2} - \dfrac{7}{4}\right) \div \left(\dfrac{2}{3} \cdot \dfrac{4}{4} + \dfrac{3}{4} \cdot \dfrac{3}{3}\right)$

$$= \left(\dfrac{1}{2} \cdot \dfrac{2}{2} - \dfrac{7}{4}\right) \div \left(\dfrac{8}{12} + \dfrac{9}{12}\right)$$

$$= \left(\dfrac{2}{4} - \dfrac{7}{4}\right) \div \dfrac{17}{12} = -\dfrac{5}{4} \cdot \dfrac{12}{17}$$

$$= -\dfrac{5 \cdot 12}{4 \cdot 17} = -\dfrac{15}{17}$$

37. $-15 \div \left(\dfrac{1}{4} - \dfrac{7}{8}\right) = -15 \div \left(\dfrac{1}{4} \cdot \dfrac{2}{2} - \dfrac{7}{8}\right)$

$$= -15 \div \left(\dfrac{2}{8} - \dfrac{7}{8}\right) = -15 \div \left(-\dfrac{5}{8}\right)$$

$$= -15 \cdot \left(-\dfrac{8}{5}\right) = 3 \cdot 8 = 24$$

39. $\left(1\dfrac{1}{10} - 3\dfrac{1}{5}\right) \div 3\dfrac{1}{2} + 1\dfrac{3}{10} = \left(\dfrac{11}{10} - \dfrac{16}{5}\right) \div \dfrac{7}{2} + \dfrac{13}{10}$

$$= \left(\dfrac{11}{10} - \dfrac{16}{5} \cdot \dfrac{2}{2}\right) \div \dfrac{7}{2} + \dfrac{13}{10}$$

$$= \left(\dfrac{11 - 32}{10}\right) \div \dfrac{7}{2} + \dfrac{13}{10}$$

$$= -\dfrac{21}{10} \div \dfrac{7}{2} + \dfrac{13}{10}$$

$$= -\dfrac{21}{10} \cdot \dfrac{2}{7} + \dfrac{13}{10}$$

$$= -\dfrac{21 \cdot 2}{10 \cdot 7} + \dfrac{13}{10}$$

$$= -\dfrac{6}{10} + \dfrac{13}{10} = \dfrac{7}{10}$$

41. $\left(3\dfrac{2}{5} + 5\dfrac{3}{4} + 6\dfrac{1}{10}\right) \div 3 = \left(\dfrac{17}{5} + \dfrac{23}{4} + \dfrac{61}{10}\right) \div 3$

$$= \left(\dfrac{17}{5} \cdot \dfrac{4}{4} + \dfrac{23}{4} \cdot \dfrac{5}{5} + \dfrac{61}{10} \cdot \dfrac{2}{2}\right) \div 3$$

$$= \left(\dfrac{68}{20} + \dfrac{115}{20} + \dfrac{122}{20}\right) \div 3$$

$$= \dfrac{68 + 115 + 122}{20} \div 3$$

$$= \dfrac{305}{20} \cdot \dfrac{1}{3} = \dfrac{305}{60}$$

$$= \dfrac{61}{12} = 5\dfrac{1}{12}$$

43. $\left(\dfrac{3}{10}\right)^2 - \left(\dfrac{7}{8}\right)^2 = \dfrac{9}{100} - \dfrac{49}{64}$

$\qquad = \dfrac{9}{100} \cdot \dfrac{16}{16} - \dfrac{49}{64} \cdot \dfrac{25}{25}$

$\qquad = \dfrac{144}{1600} - \dfrac{1225}{1600} = -\dfrac{1081}{1600}$

45. $3.4 \div 4 + 5 \cdot 8.32 = 42.45$

47. $0.75 \div 1.5 + 7 \cdot 3.1^2 = 67.77$

49. $6.32 \cdot 8.4 \div 16.8 + 3.5^2 = 15.41$

51. Follow rules for order of operations, reducing to simplest form as you proceed. Remember that the subtraction is performed last.

$\dfrac{3}{4a} \div \dfrac{3}{16a} - \dfrac{2b}{3} \cdot \dfrac{3}{4b} = \dfrac{3}{4a} \cdot \dfrac{16a}{3} - \dfrac{2b}{3} \cdot \dfrac{3}{4b}$

$\qquad = \dfrac{4}{1} - \dfrac{1}{2}$

$\qquad = \dfrac{8}{2} - \dfrac{1}{2}$

$\qquad = \dfrac{7}{2}$

53. Carryout the multiplication first, according to the rules for order of operations.

$\dfrac{-5}{7y} - \dfrac{1}{2y} \cdot \dfrac{2}{3} - \dfrac{1}{21y} = \dfrac{-5}{7y} - \dfrac{1}{3y} - \dfrac{1}{21y}$

$\qquad = \dfrac{-15}{21y} - \dfrac{7}{21y} - \dfrac{1}{21y}$

$\qquad = \dfrac{-23}{21y}$

55. First simplify the expressions in the parentheses. Then, divide.

$\left(\dfrac{7}{10x} + \dfrac{3}{5x}\right) \div \left(\dfrac{1}{2x} + \dfrac{3}{7x}\right)$

$\qquad = \left(\dfrac{7}{10x} + \dfrac{3 \cdot 2}{10x}\right) \div \left(\dfrac{1 \cdot 7}{14x} + \dfrac{3 \cdot 2}{14x}\right)$

$\qquad = \left(\dfrac{7+6}{10x}\right) \div \left(\dfrac{7+6}{14x}\right)$

$\qquad = \left(\dfrac{13}{10x}\right) \cdot \left(\dfrac{14x}{13}\right)$

$\qquad = \left(\dfrac{1}{5}\right) \cdot \left(\dfrac{7}{1}\right) = \dfrac{7}{5}$

57. First simplify the expressions in the parentheses. Then, divide.

$\left(\dfrac{3x}{2} - 5x\right) \div \left(\dfrac{x}{3} + \dfrac{x}{15}\right) = \left(\dfrac{3x}{2} - \dfrac{5x \cdot 2}{2}\right) \div \left(\dfrac{x \cdot 5}{15} + \dfrac{x}{15}\right)$

$\qquad = \left(\dfrac{3x - 10x}{2}\right) \div \left(\dfrac{5x + x}{15}\right)$

$\qquad = \left(\dfrac{-7x}{2}\right) \cdot \left(\dfrac{15}{6x}\right)$

$\qquad = \dfrac{-7x \cdot 15}{2 \cdot 6x} = \dfrac{-35}{4}$

59. Apply the exponent first, then carryout the multiplications, according to the rules for order of operations.

$\dfrac{a}{8} \cdot \left(\dfrac{1}{3}\right)^2 + \dfrac{2a}{16} \cdot \dfrac{1}{9} = \dfrac{a}{8} \cdot \dfrac{1}{9} + \dfrac{2a}{16 \cdot 9}$

$\qquad = \dfrac{a}{8 \cdot 9} + \dfrac{a}{8 \cdot 9}$

$\qquad = \dfrac{2a}{72} = \dfrac{a}{36}$

Writing and Thinking about Mathematics

61. In the expression, $\left(24 - 2^4\right) + 6\left(3 - 5\right) \div \left(3^2 - 9\right)$, the divisor $\left(3^2 - 9\right)$ is equal to 0. Since division by 0 is undefined, the expression cannot be evaluated.

63. Any number between -1 and 0 is negative, but when you square a negative number, the result is positive. Since a positive number is always greater than a negative number, the result is always larger than the original number.

[End of Section 2.5]

Section 2.6
Solutions to Odd Exercises

1. Four times the variable x, or four times a number.

3. Two times the variable x plus one, or one more than twice a number.

5. Seven times a number less 5.3, or the difference between seven times a number and 5.3.

7. Negative two times the difference between a number and eight.

9. Five times the sum of twice a number and three.

11. Six times the difference between a number and 1.

13. $3x + 7$ is the sum of three times a number and seven; $3(x+7)$ is three times the sum of a number and seven.

15. $7x - 3$ is the difference between seven times a number and three; $7(x-3)$ is seven times the difference between a number and three.

17. $x + 6$

19. $x - 4$

21. $3x - 5$

23. $\dfrac{x-3}{7}$, or $(x-3) \div 7$

25. $3(x-8)$

27. $3x - 5$

29. $8 - 2x$

31. $8(x-6) + 4$

33. $3x - 5x$

35. $2(x-7) - 6$

37. $2(17 + x) + 9$

39. **a.** $x - 6$
    b. $6 - x$

41. $24d$

43. $60m$

45. $365y$

47. $0.11x$

49. $60h + 20$

51. If m is the number of miles driven in one day, the daily cost of renting the car is:
$$20 + 0.15m$$

53. If the weekly sales is x dollars, then 9% of the weekly sales is $0.09x$. The weekly salary is:
$$250 + 0.09x$$

55. The perimeter of a rectangle is given by $2w + 2l$. If the length is 3 cm less than twice the width, then the perimeter can be given by:
$$2w + 2(2w - 3) = 6w - 6$$

[End of Section 2.6]

Chapter 2 Review
Solutions to All Exercises

1. $9x + 3x + x = (9 + 3 + 1)x = 13x$

2. $3y + 4y - y = (3 + 4 - 1)y = 6y$

3. $18x^2 + 7x^2 = (18 + 7)x^2 = 25x^2$

4. $-a^2 - 5a^2 + 3a^2 = (-1 - 5 + 3)a^2 = -3a^2$

5. $4x + 2 - 7x = (4 - 7)x + 2 = -3x + 2$

6. $-5y - 6 + 10y = (-5 + 10)y - 6 = 5y - 6$

7. $4(n + 1) - n = 4n + 4 - n = (4 - 1)n + 4 = 3n + 4$

8. $5(x - 6) - 4 = 5x - 30 - 4 = 5x - 34$

9. $2y^2 + 5y - y^2 + 2y = (2 - 1)y^2 + (5 + 2)y = y^2 + 7y$

10. $12a + 3a^2 - 8a + 4a^2 = (3 + 4)a^2 + (12 - 8)a$
    $$= 7a^2 + 4a$$

11. **a.** $3(x + 5) + 2(x - 5) = 3x + 15 + 2x - 10$
    $$= (3 + 2)x + 5 = 5x + 5$$
    b. For $x = -1$:
    $$5(-1) + 5 = -5 + 5 = 0$$

12. **a.** $-4(y + 6) + 2(y - 1) = -4y - 24 + 2y - 2$
    $$= (-4 + 2)y - 26$$
    $$= -2y - 26$$
    b. For $y = 4$:
    $$-2(4) - 26 = -8 - 26 = -34$$

13. **a.** $5a^2 + 3a - 6a^2 + a + 7 = (5 - 6)a^2 + (3 + 1)a + 7$
    $$= -a^2 + 4a + 7$$
    b. For $a = -2$:
    $$-(-2)^2 + 4(-2) + 7 = -4 - 8 + 7 = -5$$

14. **a.** $x^2 + 6x + 3x - 10 = x^2 + (6 + 3)x - 10$
    $$= x^2 + 9x - 10$$
    b. For $x = -1$:
    $$(-1)^2 + 9(-1) - 10 = 1 - 9 - 10 = -18$$

15. **a.** $3a^2 - 4a + 2$ is already simplified.
    b. For $a = -2$:
    $$3(-2)^2 - 4(-2) + 2 = 3 \cdot 4 + 8 + 2$$
    $$= 12 + 8 + 2 = 22$$

16. **a.** $5y^2 - 8y + 26$ is already simplified.
    b. For $y = 4$:
    $$5(4)^2 - 8(4) + 26 = 5 \cdot 16 - 32 + 26$$
    $$= 80 - 32 + 26 = 74$$

17. **a.** $5x + x^2 - 2x + x^2 - 13 = (1 + 1)x^2 + (5 - 2)x - 13$
    $$= 2x^2 + 3x - 13$$
    b. For $x = -1$:
    $$2(-1)^2 + 3(-1) - 13 = 2 - 3 - 13 = -14$$

18. **a.** $20 - 17 + 13x - 20 + 4x = (13 + 4)x - 17$
    $$= 17x - 17$$
    b. For $x = -1$:
    $$17(-1) - 17 = -17 - 17 = -34$$

19. **a.** $12y + 6y^2 + y^2 - 136 - 3y$
    $$= (6 + 1)y^2 + (12 - 3)y - 136$$
    $$= 7y^2 + 9y - 136$$
    b. For $y = 4$:
    $$7(4)^2 + 9(4) - 136 = 7 \cdot 16 + 36 - 136$$
    $$= 112 + 36 - 136 = 12$$

20. **a.** $100 - 10x^2 + 90x^2 + 20x$
    $$= (-10 + 90)x^2 + 20x + 100$$
    $$= 80x^2 + 20x + 100$$
    b. For $x = -1$:
    $$80(-1)^2 + 20(-1) + 100 = 80 - 20 + 100 = 160$$

21. $\dfrac{-7}{8} \cdot \dfrac{3}{14} = \dfrac{-1 \cdot \cancel{7} \cdot 3}{8 \cdot 2 \cdot \cancel{7}} = \dfrac{-1 \cdot 3}{8 \cdot 2} = -\dfrac{3}{16}$

22. $\dfrac{-5}{16} \cdot \dfrac{4}{15} = \dfrac{-1 \cdot \cancel{5} \cdot \cancel{4}}{\cancel{4} \cdot 4 \cdot 3 \cdot \cancel{5}} = \dfrac{-1}{3 \cdot 4} = -\dfrac{1}{12}$

23. $\dfrac{6}{31} \div \dfrac{9}{62} = \dfrac{6}{31} \cdot \dfrac{62}{9} = \dfrac{2 \cdot \cancel{3} \cdot 2 \cdot \cancel{31}}{\cancel{31} \cdot \cancel{3} \cdot 3} = \dfrac{2 \cdot 2}{3} = \dfrac{4}{3}$

24. $\dfrac{15}{27} \div \dfrac{5}{9} = \dfrac{15}{27} \cdot \dfrac{9}{5} = \dfrac{\cancel{3} \cdot \cancel{5} \cdot \cancel{9}}{\cancel{3} \cdot \cancel{9} \cdot \cancel{5}} = 1$

25. $\dfrac{9a}{15} \div \dfrac{35a}{12} = \dfrac{9a}{15} \cdot \dfrac{12}{35a} = \dfrac{9 \cdot \cancel{a} \cdot \cancel{3} \cdot 4}{\cancel{3} \cdot 5 \cdot 35 \cdot \cancel{a}} = \dfrac{9 \cdot 4}{5 \cdot 35} = \dfrac{36}{175}$

26. $\dfrac{6x}{32} \div \dfrac{5x}{8} = \dfrac{6x}{32} \cdot \dfrac{8}{5x} = \dfrac{\cancel{2} \cdot 3 \cdot \cancel{x} \cdot \cancel{8}}{\cancel{2} \cdot 2 \cdot \cancel{8} \cdot 5 \cdot \cancel{x}} = \dfrac{3}{2 \cdot 5} = \dfrac{3}{10}$

27. $\dfrac{1}{2}$ of $\dfrac{8}{11} = \dfrac{1}{2} \cdot \dfrac{8}{11} = \dfrac{\cancel{2} \cdot 4}{\cancel{2} \cdot 11} = \dfrac{4}{11}$

28. $\dfrac{2}{3}$ of $\dfrac{21}{24} = \dfrac{2}{3} \cdot \dfrac{21}{24} = \dfrac{\cancel{2} \cdot \cancel{3} \cdot 7}{\cancel{3} \cdot \cancel{2} \cdot 3 \cdot 4} = \dfrac{7}{3 \cdot 4} = \dfrac{7}{12}$

29. $35 - 6 = 29$ did not receive a grade of A. So $\dfrac{29}{35}$ of the class did not receive a grade of A.

30. You spent $\dfrac{12}{20} = \dfrac{3}{5}$ of your money, and still have $20 - 12 = 8$ dollars, or $\dfrac{8}{20} = \dfrac{2}{5}$ left over.

31. $36 = 2^2 \cdot 3^2$, $15 = 3 \cdot 5$, and $10 = 2 \cdot 5$, so:

$\text{LCM} = 2^2 \cdot 3^2 \cdot 5 = 180$

32. $25 = 5^2$, $35 = 5 \cdot 7$, and $30 = 2 \cdot 3 \cdot 5$, so:

$\text{LCM} = 2 \cdot 3 \cdot 5^2 \cdot 7 = 1050$

33. $10xy = 2 \cdot 5 \cdot x \cdot y$, $21xz = 3 \cdot 7 \cdot x \cdot z$, and $35yz = 5 \cdot 7 \cdot y \cdot z$, so:

$\text{LCM} = 2 \cdot 3 \cdot 5 \cdot 7 \cdot x \cdot y \cdot z = 210xyz$

34. $24a^2 = 2^3 \cdot 3 \cdot a^2$, $30ab = 2 \cdot 3 \cdot 5 \cdot a \cdot b$, and $40ab^2 = 2^3 \cdot 5 \cdot a \cdot b^2$, so:

$\text{LCM} = 2^3 \cdot 3 \cdot 5 \cdot a^2 \cdot b^2 = 120a^2b^2$

35. $\dfrac{3}{7} + \dfrac{6}{7} = \dfrac{3+6}{7} = \dfrac{9}{7}$

36. $\dfrac{9}{10} + \dfrac{6}{10} = \dfrac{9+6}{10} = \dfrac{15}{10} = \dfrac{3}{2}$

37. $\dfrac{7}{16} - \dfrac{9}{16} = \dfrac{7-9}{16} = \dfrac{-2}{16} = -\dfrac{1}{8}$

38. $\dfrac{9}{32} - \dfrac{15}{32} = \dfrac{9-15}{32} = \dfrac{-6}{32} = -\dfrac{3}{16}$

39. $\dfrac{5}{12} + \dfrac{7}{15} = \dfrac{25}{60} + \dfrac{28}{60} = \dfrac{25+28}{60} = \dfrac{53}{60}$

40. $\dfrac{1}{6} + \dfrac{1}{8} = \dfrac{4}{24} + \dfrac{3}{24} = \dfrac{4+3}{24} = \dfrac{7}{24}$

41. $1 - \dfrac{3}{5} = \dfrac{5}{5} - \dfrac{3}{5} = \dfrac{5-3}{5} = \dfrac{2}{5}$

42. $1 - \dfrac{11}{16} = \dfrac{16}{16} - \dfrac{11}{16} = \dfrac{16-11}{16} = \dfrac{5}{16}$

43. $\dfrac{1}{5} + \dfrac{3}{8} + \dfrac{3}{10} = \dfrac{8}{40} + \dfrac{15}{40} + \dfrac{12}{40} = \dfrac{8+15+12}{40} = \dfrac{35}{40} = \dfrac{7}{8}$

44. $\dfrac{5}{24} - \dfrac{5}{18} + \dfrac{3}{8} = \dfrac{15}{72} - \dfrac{20}{72} + \dfrac{27}{72} = \dfrac{15-20+27}{72} = \dfrac{22}{72} = \dfrac{11}{36}$

45. $\dfrac{7}{6} - \dfrac{9}{10} + \dfrac{4}{15} = \dfrac{35}{30} - \dfrac{27}{30} + \dfrac{8}{30} = \dfrac{35-27+8}{30} = \dfrac{16}{30} = \dfrac{8}{15}$

46. $\dfrac{1}{28} + \dfrac{8}{21} - \dfrac{5}{12} = \dfrac{3}{84} + \dfrac{32}{84} - \dfrac{35}{84} = \dfrac{3+32-35}{84} = \dfrac{0}{84} = 0$

47. $\dfrac{x}{7} - \dfrac{1}{14} = \dfrac{2x}{14} - \dfrac{1}{14} = \dfrac{2x-1}{14}$

48. $\dfrac{y}{4} - \dfrac{1}{3} = \dfrac{3y}{12} - \dfrac{4}{12} = \dfrac{3y-4}{12}$

49. $\dfrac{5}{a} + \dfrac{2}{3} = \dfrac{15}{3a} + \dfrac{2a}{3a} = \dfrac{15+2a}{3a}$

50. $\dfrac{1}{5} - \dfrac{10}{x} = \dfrac{x}{5x} - \dfrac{50}{5x} = \dfrac{x-50}{5x}$

51. $15.2 \cdot 3.5 = 53.2$

52. $8.6 \cdot 9.4 = 80.84$

53. $(1.22)(-5.1) = -6.222$

54. $(-7.5)(-31) = 232.5$

55. $21.2\overline{)76.32} \longrightarrow 212.\overline{)763.2}$

$$\begin{array}{r} 3.6 \\ 212.\overline{)763.2} \\ \underline{636} \\ 1272 \\ \underline{1272} \\ 0 \end{array}$$

56. $0.43\overline{)0.0215} \longrightarrow 43.\overline{)2.15}$

$$\begin{array}{r} 0.05 \\ \hline 215 \\ 0 \end{array}$$

57. $6.12\overline{)495.72} \longrightarrow 612.\overline{)49572.}$

$$\begin{array}{r} 81. \\ \hline 4896 \\ 612 \\ \underline{612} \\ 0 \end{array}$$

58. $15.1\overline{)228.01} \longrightarrow 151.\overline{)2280.1}$

$$\begin{array}{r} 15.1 \\ \hline 151 \\ 770 \\ \underline{755} \\ 151 \\ \underline{151} \\ 0 \end{array}$$

59. $\dfrac{7}{16} = 0.4375$

60. $\dfrac{2}{9} = 0.\overline{2}$

61. $\dfrac{5}{11} = 0.\overline{45}$

62. $\dfrac{1}{6} = 0.1\overline{6}$

63. $\dfrac{4}{13} = 0.\overline{307692}$

64. $\dfrac{33}{8} = 4.125$

In Exercises 65–70, use a calculator to perform the given operation and convert into fraction form.

65. $0.83 + 0.19 = 1.02 = \dfrac{51}{50}$

66. $15.58 - 26.31 = -10.73 = -\dfrac{1073}{100}$

67. $-27.8 - 10.002 = -37.802 = -\dfrac{18,901}{500}$

68. $12.175 - 26.32 = -14.145 = -\dfrac{2829}{200}$

69. $\dfrac{5}{8} + \dfrac{1}{4} + \dfrac{3}{10} = \dfrac{47}{40}$

70. $\dfrac{3}{4} - \dfrac{9}{10} - \dfrac{6}{16} = -\dfrac{21}{40}$

71. $16 \div 2 \cdot 8 + 24 = 8 \cdot 8 + 24$
$$= 64 + 24$$
$$= 88$$

72. $32 - 8 \cdot 14 \div 7 = 32 - 112 \div 7$
$$= 32 - 16$$
$$= 16$$

73. $-36 \div (-2)^2 + 25 - 2(16 - 17) = -36 \div 4 + 25 - 2(-1)$
$$= -9 + 25 + 2$$
$$= 18$$

74. $9(-1 + 2^2) - 7 - 6 \cdot 2^2 = 9(-1 + 4) - 7 - 6 \cdot 2^2$
$$= 9(3) - 7 - 6 \cdot 2^2$$
$$= 9(3) - 7 - 6 \cdot 4$$
$$= 27 - 7 - 24$$
$$= -4$$

75. $2(-1 + 5^2) - 4 - 2 \cdot 3^2 = 2(-1 + 25) - 4 - 2 \cdot 3^2$
$$= 2(24) - 4 - 2 \cdot 3^2$$
$$= 48 - 4 - 2 \cdot 9$$
$$= 48 - 4 - 18$$
$$= 26$$

76. $10 - 2\big[(19 - 14) \div 5 + 3(-4)\big] = 10 - 2\big[5 \div 5 - 12\big]$
$$= 10 - 2\big[1 - 12\big]$$
$$= 10 - 2(-11)$$
$$= 10 + 22$$
$$= 32$$

77. $\dfrac{1}{9} + \dfrac{1}{2} \div \dfrac{1}{6} \cdot \dfrac{5}{12} = \dfrac{1}{9} + \dfrac{1}{2} \cdot \dfrac{6}{1} \cdot \dfrac{5}{12}$

$\qquad\qquad\qquad = \dfrac{1}{9} + 3 \cdot \dfrac{5}{12}$

$\qquad\qquad\qquad = \dfrac{1}{9} + \dfrac{5}{4}$

$\qquad\qquad\qquad = \dfrac{4}{36} + \dfrac{45}{36}$

$\qquad\qquad\qquad = \dfrac{49}{36}$

78. $\left(\dfrac{1}{2} - 2\right) \div \left(1 - \dfrac{2}{3}\right) = \left(\dfrac{1}{2} - \dfrac{4}{2}\right) \div \left(\dfrac{3}{3} - \dfrac{2}{3}\right)$

$\qquad\qquad\qquad = -\dfrac{3}{2} \div \dfrac{1}{3}$

$\qquad\qquad\qquad = -\dfrac{3}{2} \cdot \dfrac{3}{1}$

$\qquad\qquad\qquad = -\dfrac{9}{2}$

79. $\left(\dfrac{1}{15} - \dfrac{1}{12}\right) \div 6 = \left(\dfrac{4}{60} - \dfrac{5}{60}\right) \div 6$

$\qquad\qquad\qquad = -\dfrac{1}{60} \div 6$

$\qquad\qquad\qquad = -\dfrac{1}{60} \cdot \dfrac{1}{6}$

$\qquad\qquad\qquad = -\dfrac{1}{360}$

80. $5\dfrac{1}{2} \div \left(\dfrac{1}{5} - \dfrac{1}{15}\right) = \dfrac{11}{2} \div \left(\dfrac{3}{15} - \dfrac{1}{15}\right)$

$\qquad\qquad\qquad = \dfrac{11}{2} \div \dfrac{2}{15}$

$\qquad\qquad\qquad = \dfrac{11}{2} \cdot \dfrac{15}{2}$

$\qquad\qquad\qquad = \dfrac{165}{4}$

81. $\dfrac{x}{2} - \dfrac{1}{4} = \dfrac{2x}{4} - \dfrac{1}{4}$

$\qquad\qquad = \dfrac{2x - 1}{4}$

82. $\dfrac{4}{7y} - \dfrac{3}{7} \div \dfrac{1}{2} = \dfrac{4}{7y} - \dfrac{3}{7} \cdot \dfrac{2}{1}$

$\qquad\qquad\qquad = \dfrac{4}{7y} - \dfrac{6}{7}$

$\qquad\qquad\qquad = \dfrac{4}{7y} - \dfrac{6y}{7y}$

$\qquad\qquad\qquad = \dfrac{4 - 6y}{7y}$

83. $-5\dfrac{1}{7} \div (2 + 1)^2 = -\dfrac{36}{7} \div 3^2$

$\qquad\qquad\qquad = -\dfrac{36}{7} \div 9$

$\qquad\qquad\qquad = -\dfrac{36}{7} \cdot \dfrac{1}{9}$

$\qquad\qquad\qquad = -\dfrac{4}{7}$

84. $\dfrac{2}{3} \div \dfrac{7}{12} - \dfrac{2}{7} + \left(\dfrac{1}{2}\right)^2 = \dfrac{2}{3} \div \dfrac{7}{12} - \dfrac{2}{7} + \dfrac{1}{4}$

$\qquad\qquad\qquad = \dfrac{2}{3} \cdot \dfrac{12}{7} - \dfrac{2}{7} + \dfrac{1}{4}$

$\qquad\qquad\qquad = \dfrac{8}{7} - \dfrac{2}{7} + \dfrac{1}{4}$

$\qquad\qquad\qquad = \dfrac{6}{7} + \dfrac{1}{4}$

$\qquad\qquad\qquad = \dfrac{24}{28} + \dfrac{7}{28}$

$\qquad\qquad\qquad = \dfrac{31}{28}$

85. $\left(\dfrac{1}{10}\right)^2 + \left(\dfrac{3}{5}\right)^2 = \dfrac{1}{100} + \dfrac{9}{25}$

$\qquad\qquad\qquad = \dfrac{1}{100} + \dfrac{36}{100}$

$\qquad\qquad\qquad = \dfrac{37}{100}$

86. $\left(-\dfrac{5}{9}\right)^2 - \left(\dfrac{2}{3}\right)^2 = \dfrac{25}{81} - \dfrac{4}{9}$

$\qquad\qquad\qquad = \dfrac{25}{81} - \dfrac{36}{81}$

$\qquad\qquad\qquad = -\dfrac{11}{81}$

87. $\left(\dfrac{3}{5}\cdot\dfrac{3}{4}\right)\div\left(\dfrac{4}{5}\cdot\dfrac{-1}{2}\right)=\dfrac{9}{20}\div\dfrac{-2}{5}$

$\qquad\qquad\qquad\qquad =\dfrac{9}{20}\cdot\dfrac{5}{-2}$

$\qquad\qquad\qquad\qquad =-\dfrac{9}{8}$

88. $2-\left(\dfrac{7}{8}+\dfrac{3}{10}\right)=2-\left(\dfrac{35}{40}+\dfrac{12}{40}\right)$

$\qquad\qquad\qquad =2-\dfrac{47}{40}$

$\qquad\qquad\qquad =\dfrac{80}{40}-\dfrac{47}{40}$

$\qquad\qquad\qquad =\dfrac{33}{40}$

89. $\left(\dfrac{5}{16}-\dfrac{1}{6}\right)\left(\dfrac{-9}{16}\right)=\left(\dfrac{15}{48}-\dfrac{8}{48}\right)\cdot\left(\dfrac{-9}{16}\right)$

$\qquad\qquad\qquad =\dfrac{7}{48}\cdot\dfrac{-9}{16}$

$\qquad\qquad\qquad =-\dfrac{63}{768}=-\dfrac{21}{256}$

90. $\left(1-\dfrac{1}{15}\right)\div\left(1+\dfrac{1}{15}\right)=\left(\dfrac{15}{15}-\dfrac{1}{15}\right)\div\left(\dfrac{15}{15}+\dfrac{1}{15}\right)$

$\qquad\qquad\qquad =\dfrac{14}{15}\div\dfrac{16}{15}=\dfrac{14}{15}\cdot\dfrac{15}{16}$

$\qquad\qquad\qquad =\dfrac{14}{16}=\dfrac{7}{8}$

In Exercises 91–100, there are multiple ways to translate the given expression. Keep in mind that you may have a correct translation using different terms than those shown here.

91. $5n+3$ translates to "3 more than 5 times a number."

92. $6x-5$ translates to "5 less than 6 times a number."

93. $8y+3y$ translates to "8 times a number increased by 3 times the same number."

94. $7n+4n$ translates to "7 times a number plus 4 times the same number."

95. $-3(x+2)$ translates to "negative 3 times the sum of a number and 2."

96. $-2(n+6)$ translates to "negative 2 multiplied by 6 more than a number."

97. $5(x-3)$ translates to "5 times the difference between a number and 3."

98. $4(x-10)$ translates to "4 multiplied by 10 less than a number."

99. $\dfrac{7n}{33}$ translates to "7 times a number divided by 33."

100. $\dfrac{50}{6y}$ translates to "50 divided by the product of 6 and a number."

101. "5 times a number increased by 3 times the same number" translates to $5n+3n$.

102. "4 times the sum of a number and 10" translates to $4(x+10)$.

103. "17 more than twice a number" translates to $2y+17$.

104. "A number decreased by 32" translates to $n-32$.

105. "28 decreased by 6 times a number" translates to $28-6n$.

106. "72 plus 8 times the sum of a number and 2" translates to $72+8(x+2)$.

107. "Twice a number plus 3" translates to $2x+3$.

108. "The quotient of a number and 14" translates to $\dfrac{n}{14}$.

109. "22 divided by the sum of a number and 9" translates to $\dfrac{22}{x+9}$.

110. "32 less than the product of a number and 10" translates to $10x-32$.

[End of Chapter 2 Review]

Chapter 2 Test
Solutions to All Exercises

1. **a.** $7x + 8x^2 - 3x^2 = 7x + (8-3)x^2$
    $$= 7x + 5x^2$$
    b. For $x = -2$:
    $$7x + 5x^2 = 7(-2) + 5(-2)^2 = -14 + 5 \cdot 4$$
    $$= -14 + 20 = 6$$

2. **a.** $5y - y - 6 + 2y = (5 - 1 + 2)y - 6$
    $$= 6y - 6$$
    b. For $y = 3$:
    $$6y - 6 = 6(3) - 6 = 18 - 6 = 12$$

3. **a.** $2(x - 5) + 3(x + 4) = 2x - 10 + 3x + 12$
    $$= (2 + 3)x + 2 = 5x + 2$$
    b. For $x = -2$:
    $$5x + 2 = 5(-2) + 2 = -10 + 2 = -8$$

4. The LCM of $\left\{10xy^2, 18x^2y, 15xy\right\}$:
    $$10xy^2 = 2 \cdot 5 \cdot x \cdot y^2$$
    $$18x^2y = 2 \cdot 3^2 \cdot x^2 \cdot y$$
    $$15xy = 3 \cdot 5 \cdot x \cdot y$$
    $$\text{LCM}\left(10xy^2, 18x^2y, 15xy\right) = 2 \cdot 3^2 \cdot 5 \cdot x^2 \cdot y^2$$
    $$= 90x^2y^2$$

5. $(5.61)(3.2) = 17.952$

6. $20.2\overline{)27.27} = 202\overline{)272.70}$

    $$\begin{array}{r} 1.35 \\ 202\overline{)272.70} \\ \underline{-202} \\ 707 \\ \underline{-606} \\ 1010 \\ \underline{-1010} \\ 0 \end{array}$$

7. $\dfrac{-15x}{7} \cdot \dfrac{42}{20x} = \dfrac{-1 \cdot 3 \cdot \cancel{5} \cdot \cancel{x} \cdot \cancel{2} \cdot 3 \cdot \cancel{7}}{\cancel{7} \cdot \cancel{2} \cdot 2 \cdot \cancel{5} \cdot \cancel{x}} = \dfrac{-1 \cdot 3 \cdot 3}{2} = -\dfrac{9}{2}$

8. $\dfrac{6}{7a} \div \dfrac{3a}{28} = \dfrac{6}{7a} \cdot \dfrac{28}{3a} = \dfrac{2 \cdot \cancel{3} \cdot 4 \cdot \cancel{7}}{\cancel{7} \cdot a \cdot \cancel{3} \cdot a} = \dfrac{2 \cdot 4}{a \cdot a} = \dfrac{8}{a^2}$

9. $\dfrac{4}{15} + \dfrac{1}{3} + \dfrac{3}{10} = \dfrac{8}{30} + \dfrac{10}{30} + \dfrac{9}{30} = \dfrac{27}{30} = \dfrac{\cancel{3} \cdot 9}{\cancel{3} \cdot 10} = \dfrac{9}{10}$

10. $\dfrac{7}{2y} - \dfrac{8}{2y} - \dfrac{3}{2y} = \dfrac{7 - 8 - 3}{2y} = \dfrac{-4}{2y} = \dfrac{-1 \cdot \cancel{2} \cdot 2}{\cancel{2} \cdot y} = -\dfrac{2}{y}$

11. $36 \div (-9) \cdot 2 - 16 + 4(-5) = -4 \cdot 2 - 16 + (-20)$
    $$= -8 - 16 - 20$$
    $$= -44$$

12. $8 - 10\left[(2 - 3^2) \div 7 + 5\right] = 8 - 10\left[(2 - 9) \div 7 + 5\right]$
    $$= 8 - 10\left[-7 \div 7 + 5\right]$$
    $$= 8 - 10\left[-1 + 5\right]$$
    $$= 8 - 10 \cdot 4 = 8 - 40$$
    $$= -32$$

13. $\dfrac{7}{8} - \dfrac{1}{3} \div \dfrac{5}{6} + \dfrac{1}{4} = \dfrac{7}{8} - \dfrac{1}{3} \cdot \dfrac{6}{5} + \dfrac{1}{4}$
    $$= \dfrac{7}{8} - \dfrac{2}{5} + \dfrac{1}{4}$$
    $$= \dfrac{35}{40} - \dfrac{16}{40} + \dfrac{10}{40}$$
    $$= \dfrac{35 - 16 + 10}{40}$$
    $$= \dfrac{29}{40}$$

14. $\dfrac{7}{12} \cdot \dfrac{9}{56} \div \dfrac{5}{16} - \left(\dfrac{1}{3}\right)^2 = \dfrac{7}{12} \cdot \dfrac{9}{56} \div \dfrac{5}{16} - \dfrac{1}{9}$
    $$= \dfrac{7}{12} \cdot \dfrac{9}{56} \cdot \dfrac{16}{5} - \dfrac{1}{9}$$
    $$= \dfrac{3}{32} \cdot \dfrac{16}{5} - \dfrac{1}{9}$$
    $$= \dfrac{3}{10} - \dfrac{1}{9}$$
    $$= \dfrac{27}{90} - \dfrac{10}{90}$$
    $$= \dfrac{17}{90}$$

15. $\dfrac{7}{x} + \dfrac{1}{5} = \dfrac{35}{5x} + \dfrac{x}{5x} = \dfrac{35 + x}{5x}$

16. $3.6(7.1) + (6.4)^2 - 82.5 = -15.98$

17. $\dfrac{3}{7} = 0.\overline{428571}$

18. **a.** The product of 5 and a number increased by 18

   b. 3 times the sum of a number and 6

   c. 42 less the product of 7 and a number

19. **a.** $6x - 3$

   b. $2(x + 5)$

   c. $2(x + 15) - 4$

20. $\left(\dfrac{3}{4} + \dfrac{9}{10}\right)\left(\dfrac{2}{3} \div \dfrac{-8}{15}\right) = \left(\dfrac{3 \cdot 5}{4 \cdot 5} + \dfrac{9 \cdot 2}{10 \cdot 2}\right)\left(\dfrac{2}{3} \cdot \dfrac{15}{-8}\right)$

$$= \left(\dfrac{15 + 18}{20}\right)\left(-\dfrac{2 \cdot 3 \cdot 5}{3 \cdot 2 \cdot 2 \cdot 2}\right)$$

$$= \dfrac{33}{20} \cdot \dfrac{5}{-4}$$

$$= \dfrac{-33}{16}$$

21. Original Price $= 60 \div \dfrac{3}{4}$:

$$60 \div \dfrac{3}{4} = 60 \cdot \dfrac{4}{3} = 20 \cdot 4 = \$80$$

22. **a.** $\dfrac{3}{4} \cdot 20 = 15$ gallons.

   b. Cash needed: $\$3.10 \cdot 15 = \$46.50$
   No, you are \$6.50 short.

[End of Chapter Two Test]

Cumulative Review: Chapter 1 – 2
Solutions to All Exercises

1. $18 + 7 - 8 + 3 = 25 - 8 + 3 = 17 + 3 = 20$

2. $9 - 16 - 11 + 4 = -7 - 11 + 4 = -18 + 4 = -14$

3. $26 - 17 = 9$

4. $18 - 33 = -15$

5. $(14)(-9) = -126$

6. $(-27)(-17) = 459$

7. $\dfrac{273}{-7} = -39$

8. $\dfrac{-744}{-6} = 124$

9. **a.** The average of the numbers in the set is
    $$\frac{17 + 15 + 22}{3} = \frac{54}{3} = 18$$
    b. The average of the numbers in the set is
    $$\frac{-18 + (-22) + (-27) + (-37)}{4} = \frac{-104}{4} = -26$$
    c. The average of the numbers in the set is
    $$\frac{23.7 + 38.9 + 47.6}{3} = \frac{110.2}{3} \approx 36.7$$
    d. The average of the numbers in the set is
    $$\frac{9.89 + 7.48 + 6.52 + 5.33}{4} = \frac{29.22}{4} \approx 7.31$$

10. $LCM(18, 27, 36):$
    $$18 = 2 \cdot 3^2$$
    $$27 = 3^3$$
    $$36 = 2^2 \cdot 3^2$$
    $$LCM = 2^2 \cdot 3^3 = 108$$

11. $LCM(12x, 9xy, 24xy^2):$
    $$12x = 2^2 \cdot 3 \cdot x$$
    $$9xy = 3^2 \cdot x \cdot y$$
    $$24xy^2 = 2^3 \cdot 3 \cdot x \cdot y^2$$
    $$LCM = 2^3 \cdot 3^3 \cdot x \cdot y^2 = 72xy^2$$

12. Associative Property of Multiplication

13. Commutative Property of Multiplication

14. Distributive Property of Multiplication over Addition

15. Commutative Property of Addition

16. Associative Property of Addition

17. Multiplicative Identity

18. Multiplicative Inverse

19. Additive Identity

20. Additive Inverse

21. **a.** Since $|-4| = 4$ and $|4| = 4$, the x values satisfying $|x| = 4$ are $x = \pm 4$, or $\{-4, 4\}$.
    b. Since the absolute value of any number is always positive, there is no solution for which $|x| = -5$ is true.

22. $|y| \le 6: \quad \{-6, -5, -4, -3, -2, -1, 0, 1, 2, 3, 4, 5, 6\}$

23. $3 \cdot 2^2 - 5 = 3 \cdot 4 - 5 = 12 - 5 = 7$

24. $(18 + 2 \cdot 3) \div 4 \cdot 3 = (18 + 6) \div 4 \cdot 3$
    $$= 24 \div 4 \cdot 3 = 6 \cdot 3 = 18$$

25. $(13 \cdot 5 - 5) \div 2 \cdot 3 = (65 - 5) \div 2 \cdot 3$
    $$= 60 \div 2 \cdot 3$$
    $$= 30 \cdot 3$$
    $$= 90$$

26. $4\left[6 - (2 \cdot 5 - 2 \cdot 3^2) - 11\right] = 4\left[6 - (2 \cdot 5 - 2 \cdot 9) - 11\right]$
    $$= 4\left[6 - (10 - 18) - 11\right]$$
    $$= 4\left[6 - (-8) - 11\right]$$
    $$= 4(6 + 8 - 11)$$
    $$= 4(14 - 11)$$
    $$= 4 \cdot 3 = 12$$

27. $\dfrac{11}{9} \div \dfrac{22}{15} = \dfrac{11}{9} \cdot \dfrac{15}{22} = \dfrac{\cancel{11} \cdot \cancel{3} \cdot 5}{\cancel{3} \cdot 3 \cdot \cancel{11} \cdot 2} = \dfrac{5}{3 \cdot 2} = \dfrac{5}{6}$

28. $\dfrac{7b}{15} \div \dfrac{2b}{6} = \dfrac{7b}{15} \cdot \dfrac{3}{b} = \dfrac{7 \cdot \cancel{b} \cdot \cancel{3}}{\cancel{3} \cdot 5 \cdot \cancel{b}} = \dfrac{7}{5}$

29. $\dfrac{8x}{21} \cdot \dfrac{7}{12x} = \dfrac{\cancel{2} \cdot \cancel{2} \cdot 2 \cdot \cancel{x} \cdot \cancel{7}}{3 \cdot \cancel{7} \cdot \cancel{2} \cdot \cancel{2} \cdot 3 \cdot \cancel{x}} = \dfrac{2}{3 \cdot 3} = \dfrac{2}{9}$

30. $\dfrac{3}{8} + \dfrac{7}{10} = \dfrac{3 \cdot 5}{8 \cdot 5} + \dfrac{7 \cdot 4}{10 \cdot 4} = \dfrac{15 + 28}{40} = \dfrac{43}{40}$

31. $\dfrac{4}{5y} + \dfrac{2}{5y} = \dfrac{4 + 2}{5y} = \dfrac{6}{5y}$

32. $\dfrac{3}{x} + \dfrac{1}{6} = \dfrac{3 \cdot 6}{x \cdot 6} + \dfrac{1 \cdot x}{6 \cdot x} = \dfrac{18 + x}{6x}$

33. $\dfrac{5}{12} - \dfrac{4}{9} = \dfrac{5 \cdot 3}{12 \cdot 3} - \dfrac{4 \cdot 4}{9 \cdot 4} = \dfrac{15 - 16}{36} = \dfrac{-1}{36}$

34. $\dfrac{7}{4a} - \dfrac{9}{4a} = \dfrac{7 - 9}{4a} = \dfrac{-2}{4a} = -\dfrac{1}{2a}$

35. $\dfrac{1}{2} + \dfrac{3}{8} \div \dfrac{3}{4} - \dfrac{5}{6} = \dfrac{1}{2} + \dfrac{3}{8} \cdot \dfrac{4}{3} - \dfrac{5}{6}$

$\qquad = \dfrac{1}{2} + \dfrac{1}{2} - \dfrac{5}{6}$

$\qquad = \dfrac{3}{6} + \dfrac{3}{6} - \dfrac{5}{6}$

$\qquad = \dfrac{3 + 3 - 5}{6}$

$\qquad = \dfrac{1}{6}$

36. $\left(\dfrac{3}{4} - \dfrac{5}{6}\right) \cdot \left(\dfrac{6}{5} + \dfrac{2}{3}\right) = \left(\dfrac{3 \cdot 3}{4 \cdot 3} - \dfrac{5 \cdot 2}{6 \cdot 2}\right) \cdot \left(\dfrac{6 \cdot 3}{5 \cdot 3} + \dfrac{2 \cdot 5}{3 \cdot 5}\right)$

$\qquad = \left(\dfrac{9 - 10}{12}\right) \cdot \left(\dfrac{18 + 10}{15}\right)$

$\qquad = \dfrac{-1}{12} \cdot \dfrac{28}{15} = \dfrac{-1 \cdot \cancel{4} \cdot 7}{3 \cdot \cancel{4} \cdot 15} = -\dfrac{7}{45}$

37. $\dfrac{7}{18} + \dfrac{5}{24} - \left(\dfrac{3}{8}\right)^2 = \dfrac{7}{18} + \dfrac{5}{24} - \dfrac{9}{64}$

$\qquad = \dfrac{7 \cdot 32}{18 \cdot 32} + \dfrac{5 \cdot 24}{24 \cdot 24} - \dfrac{9 \cdot 9}{64 \cdot 9}$

$\qquad = \dfrac{224 + 120 - 81}{576} = \dfrac{263}{576}$

38. $\left(\dfrac{3}{5}\right)^2 \div \dfrac{8}{5} \cdot \dfrac{8}{3} = \dfrac{9}{25} \cdot \dfrac{5}{8} \cdot \dfrac{8}{3} = \dfrac{\cancel{3} \cdot 3 \cdot \cancel{5} \cdot \cancel{8}}{\cancel{5} \cdot 5 \cdot \cancel{8} \cdot \cancel{3}} = \dfrac{3}{5}$

39. $\dfrac{1}{3} + \dfrac{2}{5} \cdot \dfrac{5}{8} \div \dfrac{3}{2} - \dfrac{1}{2} = \dfrac{1}{3} + \dfrac{2}{5} \cdot \dfrac{5}{8} \cdot \dfrac{2}{3} - \dfrac{1}{2}$

$\qquad = \dfrac{1}{3} + \dfrac{\cancel{2} \cdot \cancel{5} \cdot \cancel{2}}{\cancel{5} \cdot \cancel{2} \cdot \cancel{2} \cdot 2 \cdot 3} - \dfrac{1}{2}$

$\qquad = \dfrac{1}{3} + \dfrac{1}{6} - \dfrac{1}{2} = \dfrac{1 \cdot 2}{3 \cdot 2} + \dfrac{1}{6} - \dfrac{1 \cdot 3}{2 \cdot 3}$

$\qquad = \dfrac{2}{6} + \dfrac{1}{6} - \dfrac{3}{6} = \dfrac{2 + 1 - 3}{6} = 0$

40. $\left(\dfrac{1}{3} - \dfrac{3}{4}\right) \div \left(\dfrac{2}{3} + \dfrac{2}{7}\right) = \left(\dfrac{1 \cdot 4}{3 \cdot 4} - \dfrac{3 \cdot 3}{4 \cdot 3}\right) \div \left(\dfrac{2 \cdot 7}{3 \cdot 7} + \dfrac{2 \cdot 3}{7 \cdot 3}\right)$

$\qquad = \dfrac{4 - 9}{12} \div \dfrac{14 + 6}{21} = \dfrac{-5}{12} \div \dfrac{20}{21}$

$\qquad = \dfrac{-5}{12} \cdot \dfrac{21}{20} = \dfrac{-1 \cdot \cancel{5} \cdot \cancel{3} \cdot 7}{\cancel{3} \cdot 4 \cdot 4 \cdot \cancel{5}}$

$\qquad = \dfrac{-1 \cdot 7}{4 \cdot 4} = -\dfrac{7}{16}$

41. $5.2 \cdot 13.5 = 70.2$

42. $9.6 \cdot 5.4 = 51.84$

43. $(3.23)(-6.1) = -19.703$

44. $31.2\overline{)199.68} \;=\; 312\overline{)1996.8}$ quotient 6.4

$\qquad\qquad \begin{array}{r} 6.4 \\ \hline -1872 \\ \hline 1248 \\ -1248 \\ \hline 0 \end{array}$

45. $0.47\overline{)2.3594} \;=\; 47\overline{)235.94}$ quotient 5.02

$\qquad\qquad \begin{array}{r} 5.02 \\ \hline -235 \\ \hline 094 \\ -94 \\ \hline 0 \end{array}$

46. $6.34\overline{)143.284} \;=\; 634\overline{)14328.4}$ quotient 22.6

$\qquad\qquad \begin{array}{r} 22.6 \\ \hline -1268 \\ \hline 1648 \\ -1268 \\ \hline 3804 \\ -3804 \\ \hline 0 \end{array}$

47. $\dfrac{9}{16} = 0.5625$

48. $\dfrac{7}{9} = 0.\overline{7}$

49. $\dfrac{5}{22} = 0.2\overline{27}$

50. $\dfrac{2}{7} = 0.\overline{285714}$

51. $1.83 + 0.28 = 2.11 = \dfrac{211}{100}$

52. $5.38 - 6.35 = -0.97 = -\dfrac{97}{100}$

53. $-8.8 - 11.003 = -19.803 = -\dfrac{19803}{1000}$

54. $0.15 - 6.45 = -6.3 = -\dfrac{63}{10}$

55. **a.** $|x+y| + |x-y|$ is simplified.
   b. For $x = 5$, $y = -2$,
$$|x+y| + |x-y| = |5+(-2)| + |5-(-2)|$$
$$= |3| + |7|$$
$$= 3 + 7$$
$$= 10$$

56. **a.** $3|x| + 5|y| - 2(x-y) = 3|x| + 5|y| - 2x + 2y$
   b. For $x = 5$, $y = -2$,
$$3|x| + 5|y| - 2x + 2y = 3|5| + 5|-2| - 2(5) + 2(-2)$$
$$= 15 + 10 - 10 - 4$$
$$= 11$$

57. **a.** $3x + 2 - x = 2x + 2$
   b. For $x = 5$,
$$2x + 2 = 2(5) + 2 = 12$$

58. **a.** $3y^2 - 5y + 7 - y^2 + 2y - 10 + y^3$
$$= y^3 + 2y^2 - 3y - 3$$
   b. For $y = -2$,
$$y^3 + 2y^2 - 3y - 3$$
$$= (-2)^3 + 2(-2)^2 - 3(-2) - 3$$
$$= -8 + 2 \cdot 4 - (-6) - 3$$
$$= -8 + 8 + 6 - 3$$
$$= 3$$

59. **a.** $2x - 3$
   b. $\dfrac{x}{6} + 3x$
   c. $2(x + 10) - 13$

60. **a.** The sum of six and four times a number
   b. The product of 18 and the difference between a number and 5

61. $\left(\dfrac{1}{2} - \dfrac{11}{18}\right) \div \left(\dfrac{2}{3} + \dfrac{7}{15}\right) = \left(\dfrac{1 \cdot 9}{2 \cdot 9} - \dfrac{11}{18}\right) \div \left(\dfrac{2 \cdot 5}{3 \cdot 5} + \dfrac{7}{15}\right)$
$$= \dfrac{9 - 11}{18} \div \dfrac{10 + 7}{15} = \dfrac{-2}{18} \cdot \dfrac{15}{17}$$
$$= \dfrac{-1 \cdot \cancel{2} \cdot \cancel{3} \cdot 5}{\cancel{2} \cdot \cancel{3} \cdot 3 \cdot 17} = \dfrac{-1 \cdot 5}{3 \cdot 17}$$
$$= -\dfrac{5}{51}$$

62. $(-2 + (-11)) - (-4)(2) = (-13) + 8 = -5$

63. $5 \cdot \dfrac{2}{3} = \dfrac{10}{3}$ or $3\dfrac{1}{3}$ cups

64.

mph	Frequency	Product
19	4	76
20	5	100
22	7	154
25	3	75
28	1	28
Total	20	433

Mean $= \dfrac{433}{20} \approx 21.7$

65. **a.** $\dfrac{3}{8}$ of $4000 = \dfrac{3}{8} \cdot \dfrac{4000}{1} = 1500$ is the number of

Republicans. $\dfrac{2}{3}$ of these 1500 Republicans favor

Measure XX: $\dfrac{2}{3} \cdot \dfrac{1500}{1} = 1000$.

b. Since the number of other voters is $4000 - 1500$
$= 2500$, the number of non–Republicans who
favor Measure XX is:

$$\dfrac{3}{5} \text{ of } 2500 = \dfrac{3}{5} \cdot \dfrac{2500}{1} = 1500.$$

Then, the total number of voters who favor
Measure XX is $1000 + 1500 = 2500$.

66. $6x + 3x - 5 + 4x = (6 + 3 + 4)x - 5 = 13x - 5$

67. $\dfrac{a}{7} + \dfrac{a}{14} - \dfrac{a}{21} = \dfrac{6a}{42} + \dfrac{3a}{42} - \dfrac{2a}{42}$

$$= \left(\dfrac{6}{42} + \dfrac{3}{42} - \dfrac{2}{42}\right)a$$

$$= \dfrac{7}{42}a = \dfrac{1}{6}a = \dfrac{a}{6}$$

68. $\dfrac{3}{y} + \dfrac{1}{4} = \dfrac{12}{4y} + \dfrac{y}{4y} = \dfrac{12+y}{4y}$

69. **a.** $5x + 5x = (5 + 5)x = 10x$

b. $5x \cdot 5x = 25x^2$

c. $\dfrac{1}{5x} + \dfrac{1}{5x} = \dfrac{1+1}{5x} = \dfrac{2}{5x}$

d. $\dfrac{1}{5x} \cdot \dfrac{1}{5x} = \dfrac{1 \cdot 1}{5x \cdot 5x} = \dfrac{1}{25x^2}$

e. $\dfrac{1}{5x} \div \dfrac{1}{5x} = \dfrac{1}{5x} \cdot \dfrac{5x}{1} = \dfrac{1}{1} = 1$

70. **a.** $6y - 6y = (6 - 6)y = 0y = 0$

b. $6y \cdot (-6y) = (6 \cdot -6)(y \cdot y) = -36y^2$

c. $\dfrac{1}{6y} + \dfrac{1}{6y} = \dfrac{1+1}{6y} = \dfrac{2}{6y} = \dfrac{1}{3y}$

d. $\dfrac{1}{6y}\left(-\dfrac{1}{6y}\right) = -\dfrac{1}{36y^2}$

e. $\dfrac{1}{6y} \div \left(-\dfrac{1}{6y}\right) = \dfrac{1}{6y} \cdot -\dfrac{6y}{1} = \dfrac{1}{1} \cdot -\dfrac{1}{1} = -1$

[End Cumulative Review: Chapters 1 – 2]

Section 3.1
Solutions to Odd Exercises

1. $x - 6 = 1$
           $x - 6 + 6 = 1 + 6$
           $x = 7$

3. $y + 7 = 3$
           $y + 7 - 7 = 3 - 7$
           $y = -4$

5. $x + 15 = -4$
           $x + 15 - 15 = -4 - 15$
           $x = -19$

7. $22 = n - 15$
           $22 + 15 = n - 15 + 15$
           $37 = n$

9. $6 = z + 12$
           $6 - 12 = z + 12 - 12$
           $-6 = z$

11. $x - 20 = -15$
           $x - 20 + 20 = -15 + 20$
           $x = 5$

13. $y + 3.4 = -2.5$
           $y + 3.4 - 3.4 = -2.5 - 3.4$
           $y = -5.9$

15. $x + 3.6 = 2.4$
           $x + 3.6 - 3.6 = 2.4 - 3.6$
           $x = -1.2$

17. $5x = 45$
           $\dfrac{5x}{5} = \dfrac{45}{5}$
           $x = 9$

19. $32 = 4y$
           $\dfrac{32}{4} = \dfrac{4y}{4}$
           $8 = y$

21. $\dfrac{3x}{4} = 15$
           $4\left(\dfrac{3x}{4}\right) = 4(15)$
           $3x = 60$
           $\dfrac{3x}{3} = \dfrac{60}{3}$
           $x = 20$

23. $4 = \dfrac{2y}{5}$
           $5(4) = 5\left(\dfrac{2y}{5}\right)$
           $20 = 2y$
           $\dfrac{20}{2} = \dfrac{2y}{2}$
           $10 = y$

25. $4x - 3x = 10 - 12$
           $x = -2$

27. $7x - 8x = 13 - 25$
           $-x = -12$
           $-1 \cdot (-x) = -1 \cdot (-12)$
           $x = 12$

29. $3n - 2n + 6 = 14$
           $n + 6 = 14$
           $n + 6 - 6 = 14 - 6$
           $n = 8$

31. $1.7y + 1.3y = 6.3$
           $3y = 6.3$
           $\dfrac{3y}{3} = \dfrac{6.3}{3}$
           $y = 2.1$

33. $\dfrac{3}{4}x = \dfrac{5}{3}$
           $4\left(\dfrac{3}{4}x\right) = 4\left(\dfrac{5}{3}\right)$
           $3x = \dfrac{20}{3}$
           $\dfrac{1}{3} \cdot 3x = \dfrac{1}{3} \cdot \dfrac{20}{3}$
           $x = \dfrac{20}{9}$

35. $7.5x = -99.75$

$$\frac{7.5x}{7.5} = \frac{-99.75}{7.5}$$

$$x = -13.3$$

37. $1.5y - 0.5y + 6.7 = -5.3$

$$y + 6.7 = -5.3$$

$$y + 6.7 - 6.7 = -5.3 - 6.7$$

$$y = -12$$

39. $10x - 9x - \dfrac{1}{2} = -\dfrac{9}{10}$

$$x - \frac{1}{2} = -\frac{9}{10}$$

$$x - \frac{1}{2} + \frac{1}{2} = -\frac{9}{10} + \frac{1}{2}$$

$$x = -\frac{9}{10} + \frac{5}{10}$$

$$x = -\frac{4}{10} = -\frac{2}{5}$$

41. $1.4x - 0.4x + 2.7 = -1.3$

$$x + 2.7 = -1.3$$

$$x + 2.7 - 2.7 = -1.3 - 2.7$$

$$x = -4$$

43. $6.1x - 5.1x + 1.3 = 2.5 + 1.7$

$$x + 1.3 = 4.2$$

$$x + 1.3 - 1.3 = 4.2 - 1.3$$

$$x = 2.9$$

45. $\quad 6.2 = -3.5 + 7n - 6n$

$$6.2 = -3.5 + n$$

$$6.2 + 3.5 = -3.5 + n + 3.5$$

$$9.7 = n$$

47. $-8.2 - 1.5 = 2.6y - 1.6y + 3.5$

$$-9.7 = y + 3.5$$

$$-9.7 - 3.5 = y + 3.5 - 3.5$$

$$-13.2 = y$$

49. $1.7x = -5.1 - 1.7$

$$1.7x = -6.8$$

$$\frac{1.7x}{1.7} = \frac{-6.8}{1.7}$$

$$x = -4$$

51. Substitute the given values into an equation based on the perimeter (sum of all side lengths). Solve for x:

$$x + 11.5 + 8.75 = 24$$

$$x + 20.25 = 24$$

$$x + 20.25 - 20.25 = 24 - 20.25$$

$$x = 3.75 \text{ inches}$$

53. Substitute the given value for the perimeter in the formula. Solve for s:

$$4s = 64\frac{1}{2} = \frac{129}{2}$$

$$\frac{1}{4} \cdot 4s = \frac{1}{4} \cdot \frac{129}{2}$$

$$s = \frac{129}{8} = 16\frac{1}{8}\text{m} \ \text{ or } \ 16.125\text{m}$$

55. Substitute the given values for the volume, length, and height in the formula. Solve for w:

$$14 \cdot w \cdot 20 = 1260$$

$$\frac{280w}{280} = \frac{1260}{280} = \frac{9}{2}$$

$$w = 4\frac{1}{2} \text{ cm} \ \text{ or } \ 4.5 \text{ cm}$$

57. Since the sum of the angles must be 180, we can find the third angle using an equation. Solve for x:

$$x + 55 + 20 = 180$$

$$x + 75 - 75 = 180 - 75$$

$$x = 105$$

The triangle is not acute, it is obtuse.

59. From the profit equation $P = R - C$, the price is R (revenue). Substitute all values and solve for R:

$$180 = R - 675$$

$$180 + 675 = R - 675 + 675$$

$$\$855 = R$$

61. Substitute the given values in the distance formula, $d = rt$, and solve for t:

$$350 = 50t$$

$$\frac{350}{50} = t$$

$$t = 7 \text{ hours}$$

63. Substitute the given values in the distance formula, $d = rt$, and solve for t:

$$140 = 40t$$

$$\frac{140}{40} = t$$

$$t = 3.5 \text{ hours}$$

For Exercises 64 – 71, use a calculator to add or subtract the constant terms and coefficients.

65.
$$x - 41.625 = 59.354$$
$$x - 41.625 + 41.625 = 59.354 + 41.625$$
$$x = 100.979$$

67.
$$14.83y - 8.65 - 13.83y = 17.437 + 1.0$$
$$y - 8.65 = 18.437$$
$$y - 8.65 + 8.65 = 18.437 + 8.65$$
$$y = 27.087$$

69.
$$-0.3057y = 316.7052$$
$$\frac{-0.3057y}{-0.3057} = \frac{316.7052}{-0.3057}$$
$$y = -1036$$

71.
$$-y = 143.5 + 178.462 - 200$$
$$-y = 121.962$$
$$-1 \cdot (-y) = -1 \cdot (121.962)$$
$$y = -121.962$$

[End of Section 3.1]

Section 3.2
Solutions to Odd Exercises

1. Write the equation: $3x + 11 = 2$

Add -11 to each side: $3x + 11 - 11 = 2 - 11$

Simplify: $3x = -9$

Divide each side by 3: $\dfrac{3x}{3} = \dfrac{-9}{3}$

Simplify: $x = -3$

3. Write the equation: $5x - 4 = 6$

Add 4 to each side: $5x - 4 + 4 = 6 + 4$

Simplify: $5x = 10$

Divide each side by 5: $\dfrac{5x}{5} = \dfrac{10}{5}$

Simplify: $x = 2$

5. Write the equation: $-5x + 2.9 = 3.5$

Add -2.9 to each side: $-5x + 2.9 - 2.9 = 3.5 - 2.9$

Simplify: $-5x = 0.6$

Divide each side by -5: $\dfrac{-5x}{-5} = \dfrac{0.6}{-5}$

Simplify: $x = -0.12$

7. $5y - 3y + 2 = 2$

$2y + 2 = 2$

$2y + 2 - 2 = 2 - 2$

$2y = 0$

$\dfrac{2y}{2} = \dfrac{0}{2}$

$y = 0$

9. $x - 4x + 25 = 31$

$-3x + 25 = 31$

$-3x + 25 - 25 = 31 - 25$

$-3x = 6$

$\dfrac{-3x}{-3} = \dfrac{6}{-3}$

$x = -2$

11. $-20 = 7y - 3y + 4$

$-20 = 4y + 4$

$-20 - 4 = 4y + 4 - 4$

$-24 = 4y$

$\dfrac{-24}{4} = \dfrac{4y}{4}$

$-6 = y$

13. $4n - 10n + 35 = 1 - 2$

$-6n + 35 = -1$

$-6n + 35 - 35 = -1 - 35$

$-6n = -36$

$\dfrac{-6n}{-6} = \dfrac{-36}{-6}$

$n = 6$

15. $3n - 15 - n = 1$

$2n - 15 = 1$

$2n - 15 + 15 = 1 + 15$

$2n = 16$

$\dfrac{2n}{2} = \dfrac{16}{2}$

$n = 8$

17. $5.4x - 0.2x = 0$

$5.2x = 0$

$\dfrac{5.2x}{5.2} = \dfrac{0}{5.2}$

$x = 0$

19. $\dfrac{5}{8}x - \dfrac{1}{4}x + \dfrac{1}{2} = \dfrac{3}{10}$

$40\left(\dfrac{5}{8}x - \dfrac{1}{4}x + \dfrac{1}{2}\right) = 40\left(\dfrac{3}{10}\right)$

$5 \cdot 5x - 10 \cdot x + 20 = 4 \cdot 3$

$15x + 20 = 12$

$15x + 20 - 20 = 12 - 20$

$15x = -8$

$x = \dfrac{-8}{15}$

21.
$$\frac{y}{2} + \frac{1}{5} = 3$$
$$10\left(\frac{y}{2} + \frac{1}{5}\right) = 10(3)$$
$$5y + 2 = 30$$
$$5y + 2 - 2 = 30 - 2$$
$$5y = 28$$
$$\frac{5y}{5} = \frac{28}{5}$$
$$y = \frac{28}{5}$$

23.
$$\frac{7}{8} = \frac{3}{4}x - \frac{5}{8}$$
$$8\left(\frac{7}{8}\right) = 8\left(\frac{3}{4}x - \frac{5}{8}\right)$$
$$7 = 2 \cdot 3x - 5$$
$$7 + 5 = 6x - 5 + 5$$
$$12 = 6x$$
$$\frac{12}{6} = \frac{6x}{6}$$
$$2 = x$$

25.
$$\frac{y}{7} + \frac{y}{28} + \frac{1}{2} = \frac{3}{4}$$
$$28\left(\frac{y}{7} + \frac{y}{28} + \frac{1}{2}\right) = 28\left(\frac{3}{4}\right)$$
$$4y + y + 14 = 7 \cdot 3$$
$$5y + 14 = 21$$
$$5y + 14 - 14 = 21 - 14$$
$$5y = 7$$
$$\frac{5y}{5} = \frac{7}{5}$$
$$y = \frac{7}{5}$$

27.
$$x + 1.2x + 6.9 = -3.0$$
$$2.2x + 6.9 - 6.9 = -3.0 - 6.9$$
$$2.2x = -9.9$$
$$\frac{2.2x}{2.2} = \frac{-9.9}{2.2}$$
$$x = -4.5$$

29.
$$10 = x - 0.5x + 32$$
$$10 = 0.5x + 32$$
$$10 - 32 = 0.5x + 32 - 32$$
$$-22 = 0.5x$$
$$\frac{-22}{0.5} = \frac{0.5x}{0.5}$$
$$-44 = x$$

31.
$$2.5x + 0.5x - 3.5 = 2.5$$
$$3.0x - 3.5 = 2.5$$
$$3.0x - 3.5 + 3.5 = 2.5 + 3.5$$
$$3.0x = 6.0$$
$$\frac{3.0x}{3.0} = \frac{6.0}{3.0}$$
$$x = 2.0$$

33.
$$6.4 + 1.2x + 0.3x = 0.4$$
$$6.4 + 1.5x = 0.4$$
$$6.4 + 1.5x - 6.4 = 0.4 - 6.4$$
$$1.5x = -6$$
$$\frac{1.5x}{1.5} = \frac{-6}{1.5}$$
$$x = -4$$

35.
$$-12.13 = 2.42y + 0.6y - 13.64$$
$$-12.13 = 3.02y - 13.64$$
$$-12.13 + 13.64 = 3.02y - 13.64 + 13.64$$
$$1.51 = 3.02y$$
$$\frac{1.51}{3.02} = \frac{3.02y}{3.02}$$
$$0.5 = y$$

37.
$$-0.4x + x + 17.2 = 18.1$$
$$0.6x + 17.2 = 18.1$$
$$0.6x + 17.2 - 17.2 = 18.1 - 17.2$$
$$0.6x = 0.9$$
$$\frac{0.6x}{0.6} = \frac{0.9}{0.6}$$
$$x = 1.5$$

39.
$$0 = 17.3x - 15.02x - 0.456$$
$$0 = 2.28x - 0.456$$
$$0 + 0.456 = 2.28x - 0.456 + 0.456$$
$$0.456 = 2.28x$$
$$\frac{0.456}{2.28} = \frac{2.28x}{2.28}$$
$$0.2 = x$$

41. Substitute the given values for perimeter and width into the formula $P = 2l + 2w$ and solve for l:
$$184 = 2l + 2 \cdot 28$$
$$184 = 2l + 56$$
$$184 - 56 = 2l + 56 - 56$$
$$128 = 2l$$
$$\frac{128}{2} = \frac{2l}{2}$$
$$64 = l$$
So the length is 64 feet.

43. Substitute the given values for total cost, monthly payment, and down payment into the formula $C = pt + d$ and solve for t:
$$857.60 = 42.50t + 92.60$$
$$857.60 - 92.60 = 42.50t + 92.60 - 92.60$$
$$\frac{765.00}{42.50} = \frac{42.50t}{42.50}$$
$$18 = t$$
It will take 18 months to pay off the refrigerator.

45.
$$0.15x + 5.23x - 17.815 = 15.003$$
Use your calculator to find the coefficient of x by adding 0.15 and 5.23:
$$5.38x - 17.815 = 15.003$$
Using your calculator, add 17.815 to each side:
$$5.38x - 17.815 + 17.815 = 15.003 + 17.815$$
$$5.38x = 32.818$$
Finish solving by dividing 32.818 by 5.38:
$$\frac{5.38x}{5.38} = \frac{32.818}{5.38}$$
$$x = 6.1$$

47.
$$13.45x - 20x - 17.36 = -24.696$$
Use your calculator to find the coefficient of x by adding 13.45 and -20:
$$-6.55x - 17.36 = -24.696$$
Using your calculator, add 17.36 to each side:
$$-6.55x - 17.36 + 17.36 = -24.696 + 17.36$$
$$-6.55x = -7.336$$
Finish solving by dividing -7.336 by -6.55:
$$\frac{-6.55x}{-6.55} = \frac{-7.336}{-6.55}$$
$$x = 1.12$$

Writing and Thinking About Mathematics
49. Answers will vary.

[End of Section 3.2]

Section 3.3
Solutions to Odd Exercises

1.
$$3x + 2 = x - 8$$
$$3x + 2 - 2 = x - 8 - 2$$
$$3x = x - 10$$
$$3x - x = x - x - 10$$
$$2x = -10$$
$$\frac{2x}{2} = \frac{-10}{2}$$
$$x = -5$$

3.
$$4n - 3 = n + 6$$
$$4n - 3 + 3 = n + 6 + 3$$
$$4n = n + 9$$
$$4n - n = n + 9 - n$$
$$3n = 9$$
$$\frac{3n}{3} = \frac{9}{3}$$
$$n = 3$$

5.
$$3y + 18 = 7y - 6$$
$$3y + 18 - 3y = 7y - 6 - 3y$$
$$18 = 4y - 6$$
$$18 + 6 = 4y - 6 + 6$$
$$24 = 4y$$
$$\frac{24}{4} = \frac{4y}{4}$$
$$6 = y$$

7.
$$14n = 3n$$
$$14n - 3n = 3n - 3n$$
$$11n = 0$$
$$\frac{11n}{11} = \frac{0}{11}$$
$$n = 0$$

9.
$$13x + 5 = 2x + 5$$
$$13x + 5 - 5 = 2x + 5 - 5$$
$$13x = 2x$$
$$13x - 2x = 2x - 2x$$
$$11x = 0$$
$$\frac{11x}{11} = \frac{0}{11}$$
$$x = 0$$

11.
$$2(z + 1) = 3z + 3$$
$$2z + 2 = 3z + 3$$
$$2z + 2 - 2 = 3z + 3 - 2$$
$$2z = 3z + 1$$
$$2z - 3z = 3z - 3z + 1$$
$$-z = 1$$
$$(-1)(-z) = (-1)(1)$$
$$z = -1$$

13.
$$16y + 23y - 5 = 14y$$
$$39y - 5 = 14y$$
$$39y - 5 - 39y = 14y - 39y$$
$$-5 = -25y$$
$$\frac{-5}{-25} = \frac{-25y}{-25}$$
$$\frac{1}{5} = y$$

15.
$$6.5 + 1.2x = 0.5 - 0.3x$$
$$-6.5 + 6.5 + 1.2x = -6.5 + 0.5 - 0.3x$$
$$1.2x = -6.0 - 0.3x$$
$$1.2x + 0.3x = -6.0 - 0.3x + 0.3x$$
$$1.5x = -6.0$$
$$\frac{1.5x}{1.5} = \frac{-6.0}{1.5}$$
$$x = -4$$

17.
$$0.25 + 3x + 6.5 = 0.75x$$
$$3x + 6.75 = 0.75x$$
$$3x + 6.75 - 3x = 0.75x - 3x$$
$$6.75 = -2.25x$$
$$\frac{6.75}{-2.25} = \frac{-2.25x}{-2.25}$$
$$-3 = x$$

19.
$$\frac{2}{3}x + 1 = \frac{1}{3}x - 6$$
$$3\left(\frac{2}{3}x + 1\right) = 3\left(\frac{1}{3}x - 6\right)$$
$$2x + 3 = x - 18$$
$$2x + 3 - 3 = x - 18 - 3$$
$$2x = x - 21$$
$$2x - x = x - x - 21$$
$$x = -21$$

21.

$$\frac{y}{5} + \frac{3}{4} = \frac{y}{2} + \frac{3}{4}$$

$$\frac{y}{5} + \frac{3}{4} - \frac{3}{4} = \frac{y}{2} + \frac{3}{4} - \frac{3}{4}$$

$$\frac{y}{5} = \frac{y}{2}$$

$$10\left(\frac{y}{5}\right) = 10\left(\frac{y}{2}\right)$$

$$2y = 5y$$

$$2y - 5y = 5y - 5y$$

$$-3y = 0$$

$$\frac{-3y}{-3} = \frac{0}{-3}$$

$$y = 0$$

23.

$$\frac{1}{2}(x+1) = \frac{1}{3}(x-1)$$

$$6\left(\frac{1}{2}(x+1)\right) = 6\left(\frac{1}{3}(x-1)\right)$$

$$3(x+1) = 2(x-1)$$

$$3x + 3 = 2x - 2$$

$$3x + 3 - 3 = 2x - 2 - 3$$

$$3x = 2x - 5$$

$$3x - 2x = 2x - 2x - 5$$

$$x = -5$$

25.

$$\frac{5n}{6} + \frac{1}{9} = \frac{3n}{2}$$

$$18\left(\frac{5n}{6} + \frac{1}{9}\right) = 18\left(\frac{3n}{2}\right)$$

$$3 \cdot 5n + 2 = 9 \cdot 3n$$

$$15n + 2 = 27n$$

$$15n - 15n + 2 = 27n - 15n$$

$$2 = 12n$$

$$\frac{2}{12} = \frac{12n}{12}$$

$$\frac{1}{6} = n$$

27.

$$\frac{1}{6}n + \frac{7}{15} - \frac{2}{5}n = 0$$

$$30\left(\frac{1}{6}n + \frac{7}{15} - \frac{2}{5}n\right) = 30(0)$$

$$5n + 2 \cdot 7 - 6 \cdot 2n = 0$$

$$5n + 14 - 12n = 0$$

$$-7n + 14 = 0$$

$$-7n + 14 - 14 = 0 - 14$$

$$-7n = -14$$

$$\frac{-7n}{-7} = \frac{-14}{-7}$$

$$n = 2$$

29.

$$3x + \frac{1}{2}x - \frac{2}{5}x = \frac{x}{10} + \frac{7}{20}$$

$$20\left(3x + \frac{1}{2}x - \frac{2}{5}x\right) = 20\left(\frac{x}{10} + \frac{7}{20}\right)$$

$$20 \cdot 3x + 10x - 4 \cdot 2x = 2x + 7$$

$$60x + 10x - 8x = 2x + 7$$

$$62x = 2x + 7$$

$$62x - 2x = 2x - 2x + 7$$

$$60x = 7$$

$$\frac{60x}{62} = \frac{7}{60}$$

$$x = \frac{7}{60}$$

31.

$$5 - 3(2x+1) = 4(x-5) + 6$$

$$5 - 6x - 3 = 4x - 20 + 6$$

$$2 - 6x = 4x - 14$$

$$2 - 2 - 6x = 4x - 14 - 2$$

$$-6x = 4x - 16$$

$$-6x - 4x = 4x - 4x - 16$$

$$-10x = -16$$

$$\frac{-10x}{-10} = \frac{-16}{-10}$$

$$x = \frac{8}{5}$$

33.

$$8 + 4(2x-3) = 5 - (x+3)$$

$$8 + 8x - 12 = 5 - x - 3$$

$$8x - 4 = -x + 2$$

$$8x - 4 + 4 = -x + 2 + 4$$

$$8x = -x + 6$$

$$8x + x = -x + 6 + x$$

$$9x = 6$$

$$\frac{9x}{9} = \frac{6}{9}$$

$$x = \frac{2}{3}$$

35.
$$4.7 - 0.3x = 0.5x - 0.1$$
$$4.7 - 0.3x + 0.3x = 0.5x - 0.1 + 0.3x$$
$$4.7 = 0.8x - 0.1$$
$$4.7 + 0.1 = 0.8x - 0.1 + 0.1$$
$$4.8 = 0.8x$$
$$\frac{4.8}{0.8} = \frac{0.8x}{0.8}$$
$$6 = x$$

37.
$$0.2(x+3) = 0.1(x-5)$$
$$0.2x + 0.6 = 0.1x - 0.5$$
$$0.2x + 0.6 - 0.1x = 0.1x - 0.5 - 0.1x$$
$$0.1x + 0.6 - 0.6 = -0.5 - 0.6$$
$$0.1x = -1.1$$
$$\frac{0.1x}{0.1} = \frac{-1.1}{0.1}$$
$$x = -11$$

39.
$$0.6x - 22.9 = 1.5x - 18.4$$
$$0.6x - 22.9 - 0.6x = 1.5x - 18.4 - 0.6x$$
$$-22.9 + 18.4 = 0.9x - 18.4 + 18.4$$
$$-4.5 = 0.9x$$
$$\frac{-4.5}{0.9} = \frac{0.9x}{0.9}$$
$$-5 = x$$

41.
$$0.3x + 0.1(x - 15) = 0.116$$
$$0.3x + 0.1x - 1.5 = 0.116$$
$$0.4x - 1.5 = 0.116$$
$$0.4x - 1.5 + 1.5 = 0.116 + 1.5$$
$$0.4x = 1.616$$
$$\frac{0.4x}{0.4} = \frac{1.616}{0.4}$$
$$x = 4.04$$

43.
$$0.12n + 0.25n - 5.895 = 4.3n$$
$$0.37n - 5.895 = 4.3n$$
$$0.37n - 5.895 - 0.37n = 4.3n - 0.37n$$
$$-5.895 = 3.93n$$
$$\frac{-5.895}{3.93} = \frac{3.93n}{3.93}$$
$$-1.5 = n$$

45.
$$0.7(x + 14.1) = 0.3(x + 32.9)$$
$$0.7x + 9.87 = 0.3x + 9.87$$
$$0.7x + 9.87 - 0.3x = 0.3x + 9.87 - 0.3x$$
$$0.4x + 9.87 = 9.87$$
$$0.4x + 9.87 - 9.87 = 9.87 - 9.87$$
$$0.4x = 0$$
$$\frac{0.4x}{0.4} = \frac{0}{0.4}$$
$$x = 0$$

47. Use the formula for the area of a triangle $A = \frac{1}{2}bh$, where b is the length of the base and h is the height. Substitute the given values for area, $A = 180$ in.^{2}, and height, $h = 10$ in., and solve for b:
$$180 = \frac{1}{2}b \cdot 10$$
$$180 = 5b$$
$$\frac{180}{5} = \frac{5b}{5}$$
$$36 = b$$
So the base is 36 inches.

49. Use the trapezoid area formula: $A = \frac{1}{2}h(b+c)$, where h is the height, and b and c are the base lengths. Substitute the given values for area, $A = 108$ cm^2, height, $h = 12$ cm, and base, $b = 8$ cm, and solve for c.
$$108 = \frac{1}{2}(12)(8+c)$$
$$108 = 6(8+c)$$
$$\frac{108}{6} = \frac{6(8+c)}{6}$$
$$18 = 8 + c$$
$$18 - 8 = 8 + c - 8$$
$$10 = c$$
So, the length of the second base is 10 cm.

51. Use the trapezoid area formula: $A = \dfrac{1}{2}h(b+c)$. From the given information, let $c = 3b + 2$. Substitute this, and the given values for area, $A = 250 \text{ ft}^2$, and height, $h = 10$ ft, and solve for b.

$$250 = \frac{1}{2}(10)\big(b + (3b+2)\big)$$
$$250 = 5(4b+2)$$
$$\frac{250}{5} = \frac{5(4b+2)}{5}$$
$$50 = 4b + 2$$
$$50 - 2 = 4b + 2 - 2$$
$$48 = 4b$$
$$\frac{48}{4} = \frac{4b}{4}$$
$$12 = b$$

$b = 12$ ft is the length of one of the bases and $c = 3(12) + 2 = 38$ ft is the length of the other.

53. Let n be the number. Translate the problem into an equation and solve for n:

$$2n + 14.56 = n + 1.46$$
$$2n + 14.56 - n = n + 1.46 - n$$
$$n + 14.56 = 1.46$$
$$n + 14.56 - 14.56 = 1.46 - 14.56$$
$$n = -13.1$$

55. Let x be the fixed price per day. Translate the problem into an equation and solve for x:

$$2x + 0.28(250) = 140$$
$$2x + 70 = 140$$
$$2x + 70 - 70 = 140 - 70$$
$$2x = 70$$
$$\frac{2x}{2} = \frac{70}{2}$$
$$x = 35$$

So the fixed daily rental price is \$35.

57. $0.17x - 23.0138 = 1.35x + 36.234$

Using your calculator, add 23.0138 to each side and subtract $1.35x$ from each side. This action yields:

$$0.17x - 1.35x = 36.234 + 23.0138$$
$$-1.18x = 59.2478$$

Using your calculator, divide both sides by -1.18:

$$\frac{-1.18x}{-1.18} = \frac{59.2478}{-1.18}$$
$$x = -50.21$$

59. $0.32(x + 14.1) = 2.47x + 2.2188$

Use the distributive property to simplify.

$$0.32x + 4.512 = 2.47x + 2.2188$$

Using your calculator, subtract 4.512 and $2.47x$ from both sides:

$$0.32x - 2.47x = 2.2188 - 4.512$$
$$-2.15x = -2.2932$$

Using your calculator, divide both sides by -2.15:

$$\frac{-2.15x}{-2.15} = \frac{-2.2932}{-2.15}$$
$$x \approx 1.066604651$$

[End of Section 3.3]

Section 3.4
Solutions to Odd Exercises

1. False; $3 > -3$

3. True; since $|-5| = 5$ and $5 > -5$.

5. True

7. True

9. True, since $|-4| = 4, |4| = 4$ and $4 \le 4$.

11. Half–open interval;

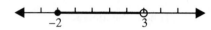

13. Half–open interval;

15. Half–open interval;

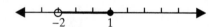

17. Half–open interval;

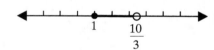

19. Open interval;

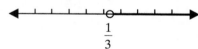

21. $4x + 5 \ge -6$

$4x \ge -11$

$x \ge \dfrac{-11}{4}$

23. $3x + 2 > 2x - 1$

$3x > 2x - 3$

$x > -3$

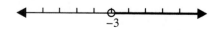

25. $y - 6.1 \le 4.2 - y$

$2y - 6.1 \le 4.2$

$2y \le 10.3$

$y \le 5.15$

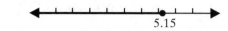

27. $2.9x - 5.7 > -3 - 0.1x$

$2.9x + 0.1x > -3 + 5.7$

$3x > 2.7$

$x > 0.9$

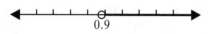

29. $5y + 6 < 2y - 2$

$5y - 2y < -2 - 6$

$3y < -8$

$y < \dfrac{-8}{3}$

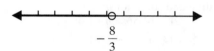

31. $4 + x > 1 - x$

$x + x > 1 - 4$

$2x > -3$

$x > \dfrac{-3}{2}$

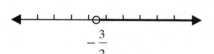

33. $\dfrac{x}{4} + 1 \le 5 - \dfrac{x}{4}$

$4\left(\dfrac{x}{4} + 1\right) \le 4\left(5 - \dfrac{x}{4}\right)$

$x + 4 \le 20 - x$

$x + x \le 20 - 4$

$2x \le 16$

$x \le 8$

35. $\dfrac{x}{3} - 2 > 1 - \dfrac{x}{3}$

$3\left(\dfrac{x}{3} - 2\right) > 3\left(1 - \dfrac{x}{3}\right)$

$x - 6 > 3 - x$

$x + x > 3 + 6$

$2x > 9$

$x > \dfrac{9}{2}$

37. $-(x + 5) \le 2x + 4$

$-x - 5 \le 2x + 4$

$-5 - 4 \le 2x + x$

$-9 \le 3x$

$-3 \le x$

39. $x - (2x + 5) \ge 7 - (4 - x) + 10$

$x - 2x - 5 \ge 7 - 4 + x + 10$

$-x - 5 \ge 13 + x$

$-5 - 13 \ge x + x$

$-18 \ge 2x$

$-9 \ge x$

41. $-5 < 1 - x < 3$

$-5 - 1 < -x < 3 - 1$

$-6 < -x < 2$

$6 > x > -2$

43. $1 < 3x - 2 < 4$

$1 + 2 < 3x < 4 + 2$

$3 < 3x < 6$

$1 < x < 2$

45. $-5 \le 5x + 3 < 9$

$-5 - 3 \le 5x < 9 - 3$

$-8 \le 5x < 6$

$\dfrac{-8}{5} \le x < \dfrac{6}{5}$

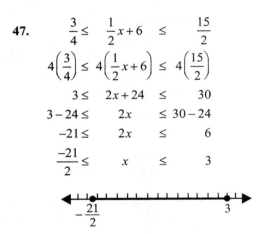

47. $\dfrac{3}{4} \le \dfrac{1}{2}x + 6 \le \dfrac{15}{2}$

$4\left(\dfrac{3}{4}\right) \le 4\left(\dfrac{1}{2}x + 6\right) \le 4\left(\dfrac{15}{2}\right)$

$3 \le 2x + 24 \le 30$

$3 - 24 \le 2x \le 30 - 24$

$-21 \le 2x \le 6$

$\dfrac{-21}{2} \le x \le 3$

49. $-\dfrac{1}{2} < -\dfrac{2}{3}x + 5 \le \dfrac{5}{9}$

$18\left(-\dfrac{1}{2}\right) < 18\left(-\dfrac{2}{3}x + 5\right) \le 18\left(\dfrac{5}{9}\right)$

$-9 < -12x + 90 \le 10$

$-9 - 90 < -12x \le 10 - 90$

$-99 < -12x \le -80$

$\dfrac{-99}{-12} > x \ge \dfrac{-80}{-12}$

$\dfrac{33}{4} > x \ge \dfrac{20}{3}$

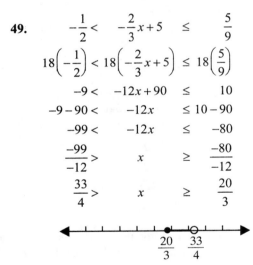

51. Let x be the final exam score. We want the average to be in the half open interval $[80, 90)$:

$$80 \leq \frac{94 + 86 + 78 + 91 + 87 + x}{6} < 90$$

$$80 \leq \frac{436 + x}{6} < 90$$

$$480 \leq 436 + x < 540$$

$$44 \leq x < 104$$

No higher than 100 points is possible, so his grade must between 44 and 100 to receive a B.

53. For C degrees Celsius, we are given that $15 \leq C \leq 65$. Using the formula, this means:

$$15 \leq \frac{5}{9}(F - 32) \leq 65$$

$$135 \leq 5F - 160 \leq 585$$

$$295 \leq 5F \leq 745$$

$$59 \leq F \leq 149$$

Temperature varies from $59°F$ to $149°F$.

55. Let n be the unknown number. Translate the condition that the sum of four times the number and 21 must be greater than 45 and less that 73:

$$45 < 4n + 21 < 73$$

$$24 < 4n < 52$$

$$6 < n < 13$$

The number must be between 6 and 13.

57. Let x be Chuck's score on his final.

$$72 + 68 + 85 + 89 + x + x \geq 540$$

$$314 + 2x \geq 540$$

$$2x \geq 540 - 314$$

$$2x \geq 226$$

$$\frac{2x}{2} \geq \frac{226}{2}$$

$$x \geq 113$$

Chuck must score a 113 on the final exam to make an A. Since Chuck cannot make higher than 100, he cannot make an A in the class.

59. Let x be the number of apples sold at \$2.50 each. Then $200 - x$ is the number of apples sold at \$2 each. The total amount raised is given by the expression:

$$2.50x + 2.00(200 - x)$$

To determine the minimum amount of apples to sell at \$2.50 in order to raise at least \$475.00, set up and solve the following inequality:

$$2.50x + 2.00(200 - x) \geq 475.00$$

$$2.5x + 400 - 2x \geq 475$$

$$0.5x + 400 \geq 475$$

$$0.5x \geq 475 - 400$$

$$0.5x \geq 75$$

$$x \geq 150$$

At least 150 apples must be sold at \$2.50 each in order to raise \$475.00.

[End of Section 3.4]

Section 3.5
Solutions to Odd Exercises

For items $1-5$, apply the formula $I = Prt$. Remember that t is in terms of years.

1. First, convert 90 days to years:
$$t = \frac{90}{360} = \frac{1}{4} \text{ yr}$$
Substitute the given values for P, r and t:
$$I = 4000 \cdot 0.12 \cdot \frac{1}{4}$$
$$I = \$120$$

3. Solve the formula for t: $\quad t = \frac{I}{Pr}$
Substitute the given values for I, P and r:
$$t = \frac{11}{1000(0.055)}$$
$$t = \frac{11}{55} = 0.2$$
$$t = 0.2 \text{ yr, or } 0.2(360) = 72 \text{ days}$$

5. Solve the formula for P: $\quad P = \frac{I}{rt}$
Substitute the given values for I, r and t. Remember that 6 months is $\frac{1}{2} = 0.5$ of a year:
$$P = \frac{450}{0.09(0.5)}$$
$$P = \frac{450}{0.045} = \$10,000$$

7. Use $v = v_0 - 32t$, where t is in seconds, and v_0 is initial velocity. Solve for v_0:
$$v = v_0 - 32t$$
$$v_0 = v + 32t$$
Substitute the given values for v and t:
$$v_0 = 48 + 32(4)$$
$$v_0 = 48 + 128$$
$$v_0 = 176$$
The initial velocity is 176 ft/sec.

9. In the given dosage formula, substitute the values for the adult dosage of 20 ml and the child's age of 3 yrs. Let D be the child's dosage:

$$D = \frac{3}{3+12} \cdot 20$$
$$D = \frac{3}{15} \cdot 20$$
$$D = 4$$
The dosage for a 3 year old child is 4 ml.

11. Use $A = P + Prt$ directly to find A by substituting the given values of P, r and t:
$$A = 1000 + 1000(0.06)(2)$$
$$A = 1000 + 120$$
$$A = \$1120$$

13. Use $N = \frac{l}{d} + 1$ to find N, the number of rafters, by substituting the given values of l and d:
$$N = \frac{26}{2} + 1$$
$$N = 13 + 1$$
$$N = 14$$
There should be 14 rafters.

15. Use $N = \frac{l}{d} + 1$ to find l, the length of the wall:
Solving for l, we get $l = d(N-1)$. Substituting the given values of N and d, we have:
$$l = 16(22 - 1)$$
$$l = 352 - 16$$
$$l = 336$$
Since l and d are given in inches, the length of the wall is 336 inches, or 28 feet.

17. Use $C = ax + k$ to find the fixed cost k: Solving for k, we get $k = C - ax$. Substituting the given values of C, a and x, we have:
$$k = 1097.50 - (9.50)(80)$$
$$k = 1097.50 - 760$$
$$k = \$337.50$$

19. Use $V = C - Crt$, where t is time in years, C is original cost and r is rate of depreciation. Substitute the given values for C, r and t:
$$V = 6000 - 6000(0.10)(6)$$
$$V = 6000 - 3600$$
$$V = \$2400$$

21. Solve $P = a + b + c$ for b:

$$P - a - c = a + b + c - a - c$$
$$P - a - c = b$$
$$\text{or} \quad b = P - a - c$$

23. Solve $F = ma$ for m:

$$\frac{F}{a} = \frac{ma}{a}$$
$$m = \frac{F}{a}$$

25. Solve $A = lw$ for w:

$$\frac{A}{l} = \frac{lw}{l}$$
$$w = \frac{A}{l}$$

27. Solve $R = np$ for n:

$$\frac{R}{p} = \frac{np}{p}$$
$$n = \frac{R}{p}$$

29. Solve $I = A - P$, for P:

$$I - A = A - P - A$$
$$I - A = -P$$
$$-1(I - A) = -1(-P)$$
$$A - I = P \quad \text{or} \quad P = A - I$$

31. Solve $A = \dfrac{m+n}{2}$ for m:

$$2A = m + n$$
$$2A - n = m \quad \text{or} \quad m = 2A - n$$

33. Solve $I = Prt$ for t:

$$\frac{I}{Pr} = \frac{Prt}{Pr}$$
$$\frac{I}{Pr} = t \quad \text{or} \quad t = \frac{I}{Pr}$$

35. Solve $P = a + 2b$ for b:

$$P - a = a - a + 2b$$
$$P - a = 2b$$
$$\frac{P-a}{2} = \frac{2b}{2}$$
$$\frac{P-a}{2} = b \quad \text{or} \quad b = \frac{P-a}{2}$$

37. Solve $\alpha + \beta + \gamma = 180°$, for β:

$$\alpha + \beta + \gamma - \alpha - \gamma = 180° - \alpha - \gamma$$
$$\beta = 180° - \alpha - \gamma$$

39. Solve $V = lwh$ for h:

$$\frac{V}{lw} = \frac{lwh}{lw}$$
$$\frac{V}{lw} = h \quad \text{or} \quad h = \frac{V}{lw}$$

41. Solve $A = \dfrac{1}{2}bh$ for b:

$$A = \frac{1}{2}bh$$
$$2(A) = 2\left(\frac{1}{2}bh\right)$$
$$2A = bh$$
$$\frac{2A}{h} = \frac{bh}{h}$$
$$\frac{2A}{h} = b \quad \text{or} \quad b = \frac{2A}{h}$$

43. Solve $A = \pi r^2$, for π:

$$\frac{A}{r^2} = \frac{\pi r^2}{r^2}$$
$$\frac{A}{r^2} = \pi \quad \text{or} \quad \pi = \frac{A}{r^2}$$

45. Solve $K = \dfrac{mv^2}{2g}$ for g:

$$(g)K = (g)\frac{mv^2}{2g}$$
$$gK = \frac{mv^2}{2}$$
$$\frac{gK}{K} = \frac{mv^2}{2K}$$
$$g = \frac{mv^2}{2K}$$

47. Solve $2x + 3y = 6$ for y:

$$2x - 2x + 3y = 6 - 2x$$
$$3y = 6 - 2x$$
$$\frac{3y}{3} = \frac{6-2x}{3}$$
$$y = \frac{6-2x}{3}$$

49. Solve $5x + 2y = 11$ for x:

$$5x + 2y - 2y = 11 - 2y$$
$$5x = 11 - 2y$$
$$\frac{5x}{5} = \frac{11 - 2y}{5}$$
$$x = \frac{11 - 2y}{5}$$

51. Solve $A = \frac{1}{2}h(b + c)$ for b:

$$2A = h(b + c)$$
$$\frac{2A}{h} = b + c$$
$$\frac{2A}{h} - c = b$$
or $\quad b = \frac{2A}{h} - c = \frac{2A - hc}{h}$

53. Solve $R = \frac{3(x - 12)}{8}$ for x:

$$8R = 3(x - 12)$$
$$\frac{8R}{3} = x - 12$$
$$\frac{8R}{3} + 12 = x \quad \text{or} \quad x = \frac{8R}{3} + 12$$

55. Solve $3y - 2 = x + 4y + 10$ for y:

$$3y - 2 - 4y = x + 4y + 10 - 4y$$
$$-y - 2 = x + 10$$
$$-y - 2 + 2 = x + 10 + 2$$
$$-y = x + 12$$
$$y = -x - 12$$

57. In one day you would pay:
$$C = 25 + 0.12x$$
where C is the total daily rate, and x is the number of miles in a day.

59. Using $P = R - C$, where P is the profit, R is the weekly revenue and C is the weekly cost:
$$R = 683n$$
for n computers made and sold each week.
$C = 325n + 5400$ is the weekly cost of making computers. The weekly profit is:
$$P = R - C$$
$$P = 683n - (325n + 5400)$$
$$P = 683n - 325n - 5400$$
$$P = 358n - 5400$$

Writing and Thinking About Mathematics

61. A z-score, z_x, of a score x from a set of numbers with mean $\overline{x} = 72$ and standard deviation $s = 6$ is computed from the formula:

$$z_x = \frac{x - \overline{x}}{s}$$

a. $z_{78} = \frac{78 - 72}{6} = \frac{6}{6} = 1$

b. $z_{66} = \frac{66 - 72}{6} = \frac{-6}{6} = -1$

c. $z_{81} = \frac{81 - 72}{6} = \frac{9}{6} = 1.5$

d. $z_{60} = \frac{60 - 72}{6} = \frac{-12}{6} = -2$

[End of Section 3.5]

Section 3.6
Solutions to Odd Exercises

1. Let n be the number. Translate the problem into an equation and solve for n:
$$n - 5 = 13 - n$$
$$n - 5 + 5 = 13 - n + 5$$
$$n = 18 - n$$
$$n + n = 18 - n + n$$
$$2n = 18$$
$$\frac{2n}{2} = \frac{18}{2}$$
$$n = 9$$

3. Let n be the number. Translate the problem into an equation and solve for n:
$$36 = 2n + 4$$
$$36 - 4 = 2n + 4 - 4$$
$$32 = 2n$$
$$\frac{32}{2} = \frac{2n}{2}$$
$$16 = n$$

5. Let n be the number. Translate the problem into an equation and solve for n:
$$7n = 2n + 35$$
$$7n - 2n = 2n + 35 - 2n$$
$$5n = 35$$
$$\frac{5n}{5} = \frac{35}{5}$$
$$n = 7$$

7. Let n be the number. Translate the problem into an equation and solve for n:
$$3n + 14 = 6 - n$$
$$3n + 14 + n = 6 - n + n$$
$$4n + 14 = 6$$
$$4n + 14 - 14 = 6 - 14$$
$$4n = -8$$
$$\frac{4n}{4} = \frac{-8}{4}$$
$$n = -2$$

9. Let n be the number. Translate the problem into an equation and solve for n:
$$\frac{2n}{5} = n + 6$$
$$5\left(\frac{2n}{5}\right) = 5(n + 6)$$

$$2n = 5n + 30$$
$$2n - 5n = 5n - 5n + 30$$
$$-3n = 30$$
$$\frac{-3n}{-3} = \frac{30}{-3}$$
$$n = -10$$

11. Let n be the number. Translate the problem into an equation and solve for n:
$$4(n - 5) = n + 4$$
$$4n - 20 = n + 4$$
$$4n - 20 - n = n - n + 4$$
$$3n - 20 = 4$$
$$3n - 20 + 20 = 4 + 20$$
$$3n = 24$$
$$\frac{3n}{3} = \frac{24}{3}$$
$$n = 8$$

13. Let n be the number. Translate the problem into an equation and solve for n:
$$\frac{2n + 5}{11} = 4 - n$$
$$11\left(\frac{2n + 5}{11}\right) = 11(4 - n)$$
$$2n + 5 = 44 - 11n$$
$$2n + 5 + 11n = 44 - 11n + 11n$$
$$13n + 5 = 44$$
$$13n + 5 - 5 = 44 - 5$$
$$13n = 39$$
$$\frac{13n}{13} = \frac{39}{13}$$
$$n = 3$$

15. Let n be the number. Translate the problem into an equation and solve for n:
$$n - 21 = 8n$$
$$n - 21 - 8n = 8n - 8n$$
$$-7n - 21 = 0$$
$$-7n - 21 + 21 = 0 + 21$$
$$-7n = 21$$
$$\frac{-7n}{-7} = \frac{21}{-7}$$
$$n = -3$$

17. Let x be the length of each of the two sides of equal length. Translate the problem into an equation and solve for x:

$$x + x + 4 = 24$$
$$2x + 4 = 24$$
$$2x + 4 - 4 = 24 - 4$$
$$2x = 20$$
$$\frac{2x}{2} = \frac{20}{2}$$
$$x = 10$$

The other two sides are each 10 cm long.

19. Let x be the length of each of the two sides of equal length. The length of the third side is $x + x - 3$, or $2x - 3$. Translate the problem into an equation and solve for x:

$$x + x + (2x - 3) = 45$$
$$4x - 3 = 45$$
$$4x - 3 + 3 = 45 + 3$$
$$4x = 48$$
$$\frac{4x}{4} = \frac{48}{4}$$
$$x = 12$$
$$2x - 3 = 2(12) - 3 = 24 - 3 = 21$$

The lengths are 12 cm, 12 cm, and 21 cm.

21. Let x be the original price. Translate the problem into an equation and solve for x:

$$3x + 1500 = 12,000$$
$$3x + 1500 - 1500 = 12,000 - 1500$$
$$3x = 10,500$$
$$\frac{3x}{3} = \frac{10,500}{3}$$
$$x = 3500$$

The original price was $3500.

23. Let n be the first odd integer. Then $n + 2$ is the next odd integer. Translate the problem to an equation:

$$n + (n + 2) = 60$$
$$2n + 2 = 60$$
$$2n + 2 - 2 = 60 - 2$$
$$2n = 58$$
$$\frac{2n}{2} = \frac{58}{2}$$
$$n = 29$$
$$n + 2 = 31$$

The two numbers are 29 and 31.

25. Let n be the first integer. Then $n + 1$ and $n + 2$ are the second and third integers, respectively. Translate the problem into an equation and solve:

$$n + (n + 1) + (n + 2) = 93$$
$$3n + 3 = 93$$
$$3n + 3 - 3 = 93 - 3$$
$$3n = 90$$
$$\frac{3n}{3} = \frac{90}{3}$$
$$n = 30$$
$$n + 1 = 31$$
$$n + 2 = 32$$

The three numbers are 30, 31, and 32.

27. Let n be the first integer. Then $n + 1$ and $n + 2$ are the second and third integers, respectively. Translate the problem into an equation and solve:

$$171 - n = (n + 1) + (n + 2)$$
$$171 - n = 2n + 3$$
$$171 - n - 3 = 2n + 3 - 3$$
$$168 - n + n = 2n + n$$
$$168 = 3n$$
$$\frac{168}{3} = \frac{3n}{3}$$
$$56 = n$$
$$n + 1 = 57$$
$$n + 2 = 58$$

The integers are 56, 57, and 58.

29. Let n be the first integer. Then $n + 1$, $n + 2$, and $n + 3$, are the second, third, and fourth integers, respectively. Translate the problem into an equation and solve:

$$208 - 3n = (n + 1) + (n + 2) + (n + 3) - 50$$
$$208 - 3n = 3n + 6 - 50$$
$$208 - 3n = 3n - 44$$
$$208 - 3n + 3n = 3n - 44 + 3n$$
$$208 + 44 = 6n - 44 + 44$$
$$252 = 6n$$
$$\frac{252}{6} = \frac{6n}{6}$$
$$n = 42$$
$$n + 1 = 43$$
$$n + 2 = 44$$
$$n + 3 = 45$$

The four integers are 42, 43, 44, and 45.

31. Let n be the first even integer. Then $n + 2$ and $n + 4$ are the second and third even integers, respectively. Translate the problem into an equation and solve:

$$n + 2(n+2) = 4(n+4) - 54$$
$$n + 2n + 4 = 4n + 16 - 54$$
$$3n + 4 = 4n - 38$$
$$3n + 4 + 38 = 4n - 38 + 38$$
$$3n + 42 - 3n = 4n - 3n$$
$$42 = n$$
$$n + 2 = 44$$
$$n + 4 = 46$$

The three integers are 42, 44, and 46.

33. Let n be the first odd integer. Then $n + 2$ and $n + 4$ are the second and third odd integers, respectively. Translate the problem into an equation and solve:

$$4n = 44 + (n+2) + (n+4)$$
$$4n = 44 + 2n + 6$$
$$4n - 2n = 50 + 2n - 2n$$
$$2n = 50$$
$$\frac{2n}{2} = \frac{50}{2}$$
$$n = 25$$
$$n + 2 = 27$$
$$n + 4 = 29$$

The three numbers are 25, 27, and 29.

35. Let n be the first integer. Then $n + 1$, $n + 2$, and $n + 3$, are the second, third, and fourth integers, respectively. Translate the problem into an equation and solve:

$$n + (n+1) + (n+2) + (n+3) = 90$$
$$4n + 6 = 90$$
$$4n + 6 - 6 = 90 - 6$$
$$4n = 84$$
$$\frac{4n}{4} = \frac{84}{4}$$
$$n = 21$$
$$n + 1 = 22$$
$$n + 2 = 23$$
$$n + 3 = 24$$

The integers are 21, 22, 23, and 24.

37. Let n be the number. Translate the problem to an equation and solve for n:

$$24 + n = 2n + 3n$$
$$24 + n = 5n$$
$$24 + n - n = 5n - n$$

$$24 = 4n$$
$$\frac{24}{4} = \frac{4n}{4}$$
$$6 = n$$

The number is 6.

39. Let x be the cost of the lot. Then, $x + 25{,}000$ is the cost of the house. Translate the problem into an equation and solve for x:

$$x + (x + 25{,}000) = 275{,}000$$
$$2x + 25{,}000 = 275{,}000$$
$$2x + 25{,}000 - 25{,}000 = 275{,}000 - 25{,}000$$
$$2x = 250{,}000$$
$$\frac{2x}{2} = \frac{250{,}000}{2}$$
$$x = 125{,}000$$
$$x + 25{,}000 = 150{,}000$$

The cost of the lot is 125,000 and the cost of the house is 150,000.

41. Let x be the cost of one box of golf balls. Translate the problem into an equation and solve for x:

$$x + x = 50 - 10.50$$
$$2x = 39.5$$
$$\frac{2x}{2} = \frac{39.5}{2}$$
$$x = 19.75$$

The cost of one box of golf balls is \$19.75.

43. Since the perimeter is the sum of each of the side lengths, the sum of the given expressions will result in an equation for x:

$$x + (3x - 1) + (2x + 5) = 64$$
$$6x + 4 = 64$$
$$6x + 4 - 4 = 64 - 4$$
$$6x = 60$$
$$\frac{6x}{6} = \frac{60}{6}$$
$$x = 10$$
$$3x - 1 = 3(10) - 1 = 29$$
$$2x + 5 = 2(10) + 5 = 25$$

The lengths of the sides are 10, 25, and 29 in.

45. Let n be the first even integer. Then $n + 2$ and $n + 4$ are
the second and third even integers, respectively.
Translate the problem into an equation and solve:

$$3n + 2(n + 4) = 6(n + 2) - 20$$
$$3n + 2n + 8 = 6n + 12 - 20$$
$$5n + 8 = 6n - 8$$
$$5n + 8 + 8 = 6n - 8 + 8$$
$$5n + 16 - 5n = 6n - 5n$$
$$16 = n$$
$$n + 2 = 18$$
$$n + 4 = 20$$

The three integers are 16, 18, and 20.

47. Answers will vary. One possibility is:
There are two consecutive integers such that 3 times
the second is 53 more than the first. What are the two
integers?

$$3(n + 1) = n + 53$$
$$3n + 3 = n + 53$$
$$3n + 3 - 3 = n + 53 - 3$$
$$3n - n = n + 50 - n$$
$$2n = 50$$
$$\frac{2n}{2} = \frac{50}{2}$$
$$n = 25$$
$$n + 1 = 26$$

The two numbers are 25 and 26.

49. Answers will vary. One possibility is:
The difference between a number and five-sixths of
the same number is equal to five-thirds.

$$x - \frac{5}{6}x = \frac{5}{3}$$
$$6\left(x - \frac{5}{6}x\right) = 6\left(\frac{5}{3}\right)$$
$$6x - 5x = 2 \cdot 5$$
$$x = 10$$

The number is 10.

[End of Section 3.6]

Section 3.7
Solutions to Odd Exercises

1. $0.91 = \dfrac{91}{100} = 91\%$

3. $1.37 = \dfrac{137}{100} = 137\%$

5. $\dfrac{3}{8} = 0.375 = 37.5\%$

7. $1\dfrac{1}{2} = \dfrac{3}{2} = 1.5 = 150\%$

9. $69\% = \dfrac{69}{100} = 0.69$

11. $11.3\% = \dfrac{11.3}{100} = 0.113$

13. $0.5\% = \dfrac{.5}{100} = 0.005$

15. $82\% = \dfrac{82}{100} = 0.82$

17. $35\% = \dfrac{35}{100} = \dfrac{7}{20}$

19. $130\% = \dfrac{130}{100} = \dfrac{13}{10}$

21. **a.** Percentage of income in rent:
$$\dfrac{800}{2500} = \dfrac{8}{25} = 0.32 = 32\%$$

    b. Percentage of income in food and entertainment:
$$\dfrac{400+300}{2500} = \dfrac{700}{2500} = \dfrac{7}{25} = 0.28 = 28\%$$

    c. Percentage of income saved:
$$\dfrac{300}{2500} = \dfrac{3}{25} = \dfrac{12}{100} = 0.12 = 12\%$$

23. **a.** Percentage of population that was married in 2006, to nearest tenth:
$$\dfrac{124.4}{217} \approx 0.5733 \approx 57.3\%$$

    b. Percentage of population that was either widowed or divorced:
$$\dfrac{13.8+23.2}{217} = \dfrac{37}{217} \approx 0.1705 \approx 17.1\%$$

25. Using $R \cdot B = A$,
$$0.81 \cdot 76 = A$$
$$A = 61.56$$

27. Using $R \cdot B = A$,
$$R \cdot 150 = 60$$
$$R = \dfrac{60}{150}$$
$$R = \dfrac{2}{5} = 0.40 = 40\%$$

29. Using $R \cdot B = A$,
$$0.03 \cdot B = 65.4$$
$$B = \dfrac{65.4}{0.03}$$
$$B = 2180$$

31. Using $R \cdot B = A$,
$$R \cdot 32 = 40$$
$$R = \dfrac{40}{32} = 1.25$$
$$R = 125\%$$

33. Using $R \cdot B = A$,
$$1.25 \cdot B = 100$$
$$B = \dfrac{100}{1.25} = 80$$

35. Using Percent Profit $= \dfrac{\text{Profit}}{\text{Investment}}$,

    a. $\dfrac{400}{5000} = \dfrac{2}{25} = 0.08 = 8\%$

    b. $\dfrac{350}{3500} = \dfrac{1}{10} = 0.10 = 10\%$

    c. **b** is the better investment since 10% is greater than 8%.

37. **a.** Carlos paid $100 \cdot \$45 = \$4500$ for the stock. He sold the stock for $5490. His profit is
$$5490 - 4500 = \$990$$

    b. Percent Profit $= \dfrac{\text{Profit}}{\text{Investment}} = \dfrac{990}{4500}$
$$= 0.22 = 22\%$$

39. The commission earned on the sales was
$$0.07(11470 - 5000) = 0.07(6470)$$
$$= \$452.90$$
The monthly income was
$$1500 + 452.90 = \$1952.90$$

41. Using $R \cdot B = A$,
$$R \cdot (170,000) = 2550$$
$$R = \frac{2550}{170,000} = 0.015 = 1.5\%$$
The tax rate was 1.5%.

43. We know that the student got 85% correct, which means that the student missed 15% of the problems. Let B be the total number of problems. Using $R \cdot B = A$, we know that
$$0.15 \cdot B = 6$$
$$B = \frac{6}{0.15} = 40$$
There were 40 problems on the test.

45. a. Using $R \cdot B = A$, let B be the original selling price. If a price is discounted 20%, then the sale price is 80% of the original selling price: Solve for B:
$$0.80 \cdot B = 560$$
$$B = \frac{560}{0.80}$$
$$B = \$700$$

b. The profit was
$$560 - 420 = \$140$$
With base of \$420 the percent of profit, profit/base, was
$$\frac{140}{420} = 0.333... \approx 33.33\%$$

47. a. Using $R \cdot B = A$, let B be the original price. Let A be the dealer discount of \$499.40 and solve for B:
$$0.07 \cdot B = 499.40$$
$$B = \frac{499.40}{0.07} \approx \$7,134.29$$

b. The amount paid, P, is the sum of the final selling price (original price – discount – rebate), the tax, and the license fees. The original price was \$7,134.29. With the discount of \$499.40 and the rebate of \$1000, the final selling price is given by:
$$7134.29 - 499.4 - 1000 = 5634.89$$
Calculate the tax:
$$5\% \cdot (\text{final selling price}) = 0.05(5634.89) \approx 281.74$$

These two values, together with the license fee of 642.60, yields:
$$P = 5634.89 + 281.74 + 642.60$$
$$P = \$6559.23$$

49. a. Percent P of weight lost in first 3 months:
$$P = \frac{20}{200} = 0.10 = 10\%$$

b. Percent P of weight gained back (from 180 pounds):
$$P = \frac{20}{180} \approx 0.111... \approx 11.11\%$$

c. The percents are different because the base amounts are different. The base for the initial loss was 200 pounds, while the base for the later gain was 180 pounds.

Writing and Thinking About Mathematics

51. Use $R \cdot B = A$ as the basis for this problem:
a. From the selling price, they will pay a fee of 6% of the selling price to the realtor. What is left over after subtracting the fee is \$141,000. In other words, the base, B, in the problem is the selling price, not the amount left after the fee is deducted. The couple will get 94% of the selling price because 6% of it is the fee.

b. The algebraic translation of the problem shows this explicitly. Let S be the selling price.
$$S - 0.06S = 141,000$$
$$(1 - 0.06)S = 141,000$$
$$0.94S = 141,000$$
$$S = \frac{141,000}{.94}$$
$$S = 150,000$$
So, the selling price should be \$150,000.

[End of Section 3.7]

Section 3.8
Solutions to Odd Exercises

For Exercises 1–15, the item is matched with the letter of the matching formula.

1. e   3. k   5. c

7. d   9. g   11. l

13. m   15. h

17. Rectangle with $w = 15$ cm and $l = 20$ cm:

$$P = 2l + 2w \qquad A = l \cdot w$$
$$= 2(20) + 2(15) \qquad = 20 \cdot 15$$
$$= 40 + 30 \qquad = 300 \text{ cm}^2$$
$$= 70 \text{ cm}$$

19. Triangle with $h = 12$ mm, $b = 21$ mm, and other sides $a = 13$ mm and $c = 20$ mm:

$$P = a + b + c \qquad A = \frac{1}{2}bh$$
$$= 13 + 21 + 20 \qquad = \frac{1}{2}(21)(12)$$
$$= 54 \text{ mm} \qquad = 21(6)$$
$$= 126 \text{ mm}^2$$

21. Circle with radius 5 in., using $\pi \approx 3.14$:

$$C = 2\pi r \qquad A = \pi r^2$$
$$\approx 2(3.14)(5) \qquad \approx 3.14(5)^2$$
$$\approx 6.28(5) \qquad \approx 3.14(25)$$
$$\approx 31.4 \text{ in.} \qquad \approx 78.5 \text{ in.}^2$$

23. Sphere with radius 7 in., using $\pi \approx 3.14$:

$$V = \frac{4}{3}\pi r^3$$
$$\approx \frac{4}{3}(3.14)(7)^3$$
$$\approx \frac{4}{3}(3.14)(343)$$
$$\approx 1436.02\overline{6}$$
$$\approx 1436.03 \text{ in.}^3$$

25. Cube with sides of 9 m:

$$V = s^3$$
$$= 9^3$$
$$= 729 \text{ m}^3$$

27. $P = 2l + 2w$, with $l = 17$ in. and $w = 11$ in.:

$$P = 2(17) + 2(11)$$
$$= 34 + 22$$
$$= 56 \text{ in.}$$

29. $A = \frac{1}{2}bh$, with $b = 14$ cm and $h = 9$ cm:

$$A = \frac{1}{2}(14)(9)$$
$$= 7(9)$$
$$= 63 \text{ cm}^2$$

31. $V = \pi r^2 h$, with $r = 14$ ft, $h = 10$ ft and $\pi \approx 3.14$:

$$V \approx 3.14(14)^2(10)$$
$$\approx 3.14(196)(10)$$
$$\approx 6154.4 \text{ ft}^3$$

33. $V = s^3$, with $s = 5$ ft:

$$V = 5^3$$
$$= 125 \text{ ft}^3$$

35. $A = bh$, with $b = 20$ cm and $h = 13.6$ cm:

$$A = 20 \cdot 13.6$$
$$= 272 \text{ cm}^2$$

37. $V = \frac{1}{3}\pi r^2 h$, with $r = \frac{d}{2} = \frac{15}{2}$ in., $h = 9$ in., and $\pi \approx 3.14$:

$$V \approx \frac{1}{3}(3.14)\left(\frac{15}{2}\right)^2(9)$$
$$\approx \frac{1}{3}(3.14)\left(\frac{225}{4}\right)(9)$$
$$\approx 529.875 \text{ in.}^3$$

39. The figure is made up of a square and a triangle. The perimeter is the distance around the whole figure. The area is the sum of the area of the triangle and the area of the square:

$$P = 5 + 5 + 6 + 6 + 6 \qquad A = \frac{1}{2}bh + s^2$$
$$= 28 \text{ cm} \qquad = \frac{1}{2}(6)(4) + 6^2$$
$$= 12 + 36$$
$$= 48 \text{ cm}^2$$

Calculator Problems

41. Area of a rectangle with $l = 16.54$ and $w = 12.82$:

$A = lw$

$\quad = (16.54)(12.82)$

$\quad = 212.042$

$\quad \approx 212.04 \text{ m}^2$

43. Area of a trapezoid with $h = 6.53$, $b = 22.36$, and $c = 17.48$:

$A = \dfrac{1}{2}h(b+c)$

$\quad = \dfrac{1}{2}(6.53)(22.36 + 17.48)$

$\quad = \dfrac{1}{2}(6.53)(39.84)$

$\quad = 130.077$

$\quad \approx 130.08 \text{ in.}^2$

45. The volume of a sphere with $r = 114.8$ and $\pi \approx 3.14$:

$V \approx \dfrac{4}{3}\pi r^3$

$\quad \approx \dfrac{4}{3}(3.14)(114.8)^3$

$\quad \approx 6,334,233.21 \text{ cm}^3$

[End of Section 3.8]

Chapter 3 Review
Solutions to All Exercises

1.
$$x - 6 = 1$$
$$x - 6 + 6 = 1 + 6$$
$$x = 7$$

2.
$$x - 4 = 10$$
$$x - 4 + 4 = 10 + 4$$
$$x = 14$$

3.
$$y + 5 = -6$$
$$y + 5 - 5 = -6 - 5$$
$$y = -11$$

4.
$$y + 7 = -1$$
$$y + 7 - 7 = -1 - 7$$
$$y = -8$$

5.
$$0 = n + 8$$
$$0 - 8 = n + 8 - 8$$
$$-8 = n$$

6.
$$0 = n + 13$$
$$0 - 13 = n + 13 - 13$$
$$-13 = n$$

7.
$$4 = x + 9$$
$$4 - 9 = x + 9 - 9$$
$$-5 = x$$

8.
$$13 = x + 17$$
$$13 - 17 = x + 17 - 17$$
$$-4 = x$$

9.
$$6n = 60$$
$$\frac{6n}{6} = \frac{60}{6}$$
$$n = 10$$

10.
$$7n = 42$$
$$\frac{7n}{7} = \frac{42}{7}$$
$$n = 6$$

11.
$$\frac{2}{3} y = -12$$
$$\frac{3}{2}\left(\frac{2}{3} y\right) = \frac{3}{2}(-12)$$
$$y = -18$$

12.
$$\frac{5}{6} y = -25$$
$$\frac{6}{5}\left(\frac{5}{6} y\right) = \frac{6}{5}(-25)$$
$$y = -30$$

13.
$$2.9x - 1.9x = 7 + 11$$
$$x = 18$$

14.
$$3.2y - 2.2y = 1.3 - 0.9$$
$$y = 0.4$$

15.
$$-13.5 = 4.5x$$
$$\frac{-13.5}{4.5} = \frac{4.5x}{4.5}$$
$$-3 = x$$

16.
$$2x + 3 = -7$$
$$2x + 3 - 3 = -7 - 3$$
$$2x = -10$$
$$\frac{2x}{2} = \frac{-10}{2}$$
$$x = -5$$

17.
$$3x + 5 = -19$$
$$3x + 5 - 5 = -19 - 5$$
$$3x = -24$$
$$\frac{3x}{3} = \frac{-24}{3}$$
$$x = -8$$

18.
$$2y + 2y - 6 = 30$$
$$4y - 6 + 6 = 30 + 6$$
$$4y = 36$$
$$\frac{4y}{4} = \frac{36}{4}$$
$$y = 9$$

19.
$$3y + 10 + 2y = -25$$
$$5y + 10 - 10 = -25 - 10$$
$$5y = -35$$
$$\frac{5y}{5} = \frac{-35}{5}$$
$$y = -7$$

20.
$$20 = 5x - 3x - 4$$
$$20 + 4 = 2x - 4 + 4$$
$$24 = 2x$$
$$\frac{24}{2} = \frac{2x}{2}$$
$$12 = x$$

21.
$$-30 = 6x - 4x + 20$$
$$-30 - 20 = 2x + 20 - 20$$
$$-50 = 2x$$
$$\frac{-50}{2} = \frac{2x}{2}$$
$$-25 = x$$

22.
$$0.9y + 0.3y - 5 = 2.2$$
$$1.2y - 5 + 5 = 2.2 + 5$$
$$1.2y = 7.2$$
$$\frac{1.2y}{1.2} = \frac{7.2}{1.2}$$
$$y = 6$$

23.
$$0.5y + 0.6y + 1.7 = -1.6$$
$$1.1y + 1.7 - 1.7 = -1.6 - 1.7$$
$$1.1y = -3.3$$
$$\frac{1.1y}{1.1} = \frac{-3.3}{1.1}$$
$$y = -3$$

24.
$$\frac{1}{5}n + \frac{2}{5}n - \frac{1}{3} = \frac{4}{5}$$
$$\frac{3}{5}n - \frac{1}{3} = \frac{4}{5}$$
$$15\left(\frac{3}{5}n - \frac{1}{3}\right) = 15\left(\frac{4}{5}\right)$$
$$9n - 5 = 12$$
$$9n - 5 + 5 = 12 + 5$$
$$9n = 17$$

$$\frac{9n}{9} = \frac{17}{9}$$
$$n = \frac{17}{9}$$

25.
$$\frac{3}{8}n + \frac{1}{4}n + \frac{1}{2} = \frac{3}{4}$$
$$8\left(\frac{3}{8}n + \frac{1}{4}n + \frac{1}{2}\right) = 8\left(\frac{3}{4}\right)$$
$$3n + 2n + 4 = 6$$
$$5n + 4 - 4 = 6 - 4$$
$$5n = 2$$
$$\frac{5n}{5} = \frac{2}{5}$$
$$n = \frac{2}{5}$$

26.
$$\frac{2}{3}x - \frac{1}{2}x + \frac{5}{6} = 2$$
$$6\left(\frac{2}{3}x - \frac{1}{2}x + \frac{5}{6}\right) = 6(2)$$
$$4x - 3x + 5 = 12$$
$$x + 5 - 5 = 12 - 5$$
$$x = 7$$

27.
$$\frac{3}{4}x - \frac{7}{12}x - \frac{2}{3} = 4$$
$$12\left(\frac{3}{4}x - \frac{7}{12}x - \frac{2}{3}\right) = 12(4)$$
$$9x - 7x - 8 = 48$$
$$2x - 8 + 8 = 48 + 8$$
$$2x = 56$$
$$\frac{2x}{2} = \frac{56}{2}$$
$$x = 28$$

28.
$$4.78 - 0.3x + 0.5x = 2.31$$
$$4.78 + 0.2x - 4.78 = 2.31 - 4.78$$
$$0.2x = -2.47$$
$$\frac{0.2x}{0.2} = \frac{-2.47}{0.2}$$
$$x = -12.35$$

29.
$$5.32 + 0.4x - 0.6x = 7.1$$
$$5.32 - 0.2x - 5.32 = 7.1 - 5.32$$
$$-0.2x = 1.78$$
$$\frac{-0.2x}{-0.2} = \frac{1.78}{-0.2}$$
$$x = -8.9$$

30.
$$8.62 = 0.45x - 0.15x + 5.02$$
$$8.62 - 5.02 = 0.3x + 5.02 - 5.02$$
$$3.6 = 0.3x$$
$$\frac{3.6}{0.3} = \frac{0.3x}{0.3}$$
$$12 = x$$

31.
$$8y - 2.1 = y + 5.04$$
$$8y - 2.1 + 2.1 = y + 5.04 + 2.1$$
$$8y = y + 7.14$$
$$8y - y = y + 7.14 - y$$
$$7y = 7.14$$
$$\frac{7y}{7} = \frac{7.14}{7}$$
$$y = 1.02$$

32.
$$6y - 4.5 = y + 4.5$$
$$6y - 4.5 + 4.5 = y + 4.5 + 4.5$$
$$6y = y + 9$$
$$6y - y = y + 9 - y$$
$$5y = 9$$
$$\frac{5y}{5} = \frac{9}{5}$$
$$y = 1.8$$

33.
$$2.4n = 3.6n$$
$$2.4n - 3.6n = 3.6n - 3.6n$$
$$-1.2n = 0$$
$$\frac{-1.2n}{-1.2} = \frac{0}{-1.2}$$
$$n = 0$$

34.
$$8n = -1.3n$$
$$8n + 1.3n = -1.3n + 1.3n$$
$$9.3n = 0$$
$$\frac{9.3n}{9.3} = \frac{0}{9.3}$$
$$n = 0$$

35.
$$2(x + 3) = 3(x - 7)$$
$$2x + 6 = 3x - 21$$
$$2x + 6 + 21 = 3x - 21 + 21$$
$$2x + 27 = 3x$$
$$2x + 27 - 2x = 3x - 2x$$
$$27 = x$$

36.
$$4(x - 2) = 3(x + 6)$$
$$4x - 8 = 3x + 18$$
$$4x - 8 + 8 = 3x + 18 + 8$$
$$4x = 3x + 26$$
$$4x - 3x = 3x + 26 - 3x$$
$$x = 26$$

37.
$$\frac{1}{3}n + 6 = \frac{1}{4}n + 5$$
$$12\left(\frac{1}{3}n + 6\right) = 12\left(\frac{1}{4}n + 5\right)$$
$$4n + 72 = 3n + 60$$
$$4n + 72 - 72 = 3n + 60 - 72$$
$$4n = 3n - 12$$
$$4n - 3n = 3n - 12 - 3n$$
$$n = -12$$

38.
$$\frac{2}{5}n + 7 = \frac{1}{5}n + 4$$
$$\frac{2}{5}n + 7 - 7 = \frac{1}{5}n + 4 - 7$$
$$\frac{2}{5}n = \frac{1}{5}n - 3$$
$$\frac{2}{5}n - \frac{1}{5}n = \frac{1}{5}n - 3 - \frac{1}{5}n$$
$$\frac{1}{5}n = -3$$
$$5\left(\frac{1}{5}n\right) = 5(-3)$$
$$n = -15$$

39.
$$6(2x-1) = 4(x+3)+14$$
$$12x-6 = 4x+12+14$$
$$12x-6+6 = 4x+26+6$$
$$12x = 4x+32$$
$$12x-4x = 4x+32-4x$$
$$8x = 32$$
$$\frac{8x}{8} = \frac{32}{8}$$
$$x = 4$$

40.
$$8+4(x-4) = 5-(x-14)$$
$$8+4x-16 = 5-x+14$$
$$4x-8+8 = -x+19+8$$
$$4x = -x+27$$
$$4x+x = -x+27+x$$
$$5x = 27$$
$$\frac{5x}{5} = \frac{27}{5}$$
$$x = \frac{27}{5}$$

41.
$$-2(y+2)-4 = 6-(y+12)$$
$$-2y-4-4 = -y+6-12$$
$$-2y-8+2y+6 = -y-6+2y+6$$
$$-2 = y$$

42.
$$7-3(x+5) = -2(x+3)-15$$
$$7-3x-15 = -2x-6-15$$
$$-3x-8+21 = -2x-21+21$$
$$-3x+13 = -2x$$
$$-3x+13+3x = -2x+3x$$
$$13 = x$$

43.
$$\frac{5x}{6}+\frac{1}{18} = \frac{x}{3}$$
$$18\left(\frac{5x}{6}+\frac{1}{18}\right) = 18\left(\frac{x}{3}\right)$$
$$15x+1 = 6x$$
$$15x+1-1 = 6x-1$$
$$15x-6x = 6x-1-6x$$
$$9x = -1$$
$$\frac{9x}{9} = \frac{-1}{9}$$
$$x = -\frac{1}{9}$$

44.
$$\frac{x}{6}+\frac{1}{15}-\frac{2x}{3} = 0$$
$$30\left(\frac{x}{6}+\frac{1}{15}-\frac{2x}{3}\right) = 30(0)$$
$$5x+2-20x = 0$$
$$-15x+2-2 = 0-2$$
$$-15x = -2$$
$$\frac{-15x}{-15} = \frac{-2}{-15}$$
$$x = \frac{2}{15}$$

45.
$$2(y+8.3)+1.2 = y-16.6-1.2$$
$$2y+17.8 = y-17.8$$
$$2y+17.8-17.8 = y-17.8-17.8$$
$$2y = y-35.6$$
$$2y-y = y-35.6-y$$
$$y = -35.6$$

46. $-3 < x < 1$; Open interval

47. $-3 \le x \le 4$; Closed interval

48. $-1 \le x < \dfrac{3}{4}$; Half-open interval

49. $x < \dfrac{2}{3}$; Open interval

50. $x \ge 7.5$; Half-open interval

51.
$$4x+6 \ge 10$$
$$4x+6-6 \ge 10-6$$
$$4x \ge 4$$
$$\frac{4x}{4} \ge \frac{4}{4}$$
$$x \ge 1$$

52.

$$3x - 7 \leq 14$$
$$3x - 7 + 7 \leq 14 + 7$$
$$\frac{3x}{3} \leq \frac{21}{3}$$
$$x \leq 7$$

53.

$$5y + 8 < 2y + 5$$
$$5y + 8 - 8 < 2y + 5 - 8$$
$$5y < 2y - 3$$
$$5y - 2y < 2y - 3 - 2y$$
$$3y < -3$$
$$\frac{3y}{3} < \frac{-3}{3}$$
$$y < -1$$

54.

$$y - 6 > 3y + 4$$
$$y - 6 - 4 > 3y + 4 - 4$$
$$y - 10 > 3y$$
$$y - 10 - y > 3y - y$$
$$-10 > 2y$$
$$\frac{-10}{2} > \frac{2y}{2}$$
$$-5 > y$$

55.

$$\frac{x}{3} - 1 > 2 - \frac{x}{4}$$
$$12\left(\frac{x}{3} - 1\right) > 12\left(2 - \frac{x}{4}\right)$$
$$4x - 12 > 24 - 3x$$
$$4x - 12 + 12 > 24 - 3x + 12$$
$$4x > 36 - 3x$$
$$4x + 3x > 36 - 3x + 3x$$
$$7x > 36$$
$$x > \frac{36}{7}$$

56.

$$\frac{1}{2} + \frac{x}{5} < 1 - \frac{x}{5}$$
$$10\left(\frac{1}{2} + \frac{x}{5}\right) < 10\left(1 - \frac{x}{5}\right)$$
$$5 + 2x < 10 - 2x$$
$$5 + 2x - 5 < 10 - 2x - 5$$
$$2x < 5 - 2x$$
$$2x + 2x < 5 - 2x + 2x$$
$$4x < 5$$
$$\frac{4x}{4} < \frac{5}{4}$$
$$x < \frac{5}{4}$$

57.

$$1 \leq \quad 3x - 5 \quad \leq 10$$
$$1 + 5 \leq 3x - 5 + 5 \leq 10 + 5$$
$$6 \leq \quad 3x \quad \leq 15$$
$$\frac{6}{3} \leq \quad \frac{3x}{3} \quad \leq \frac{15}{3}$$
$$2 \leq \quad x \quad \leq 5$$

58.

$$-4 < \quad 5x + 1 \quad < 16$$
$$-4 - 1 < 5x + 1 - 1 < 16 - 1$$
$$-5 < \quad 5x \quad < 15$$
$$\frac{-5}{5} < \quad \frac{5x}{5} \quad < \frac{15}{5}$$
$$-1 < \quad x \quad < 3$$

59.

$$0 < \quad -\frac{3}{4}x + 1 \quad \leq 4$$
$$0 - 1 < -\frac{3}{4}x + 1 - 1 \leq 4 - 1$$
$$-\frac{4}{3}(-1) > -\frac{4}{3}\left(-\frac{3}{4}x\right) \geq -\frac{4}{3}(3)$$
$$\frac{4}{3} > \quad x \quad \geq -4$$

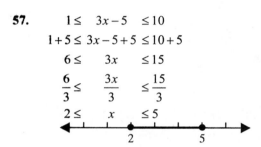

60.
$$0 < -\frac{1}{2}x + 3 \le 2$$

$$0 - 3 < -\frac{1}{2}x + 3 - 3 \le 2 - 3$$

$$-2(-3) > -2\left(-\frac{1}{2}x\right) \ge -2(-1)$$

$$6 > \quad x \quad \ge 2$$

61. Solve for π:
$$L = 2\pi rh$$

$$\frac{L}{2rh} = \frac{2\pi rh}{2rh}$$

$$\frac{L}{2rh} = \pi \text{ or } \pi = \frac{L}{2rh}$$

62. Solve for R:
$$P = R - C$$

$$P + C = R - C + C$$

$$P + C = R, \text{ or } R = P + C$$

63. Solve for b:
$$A = \frac{1}{2}bh$$

$$2A = 2\left(\frac{1}{2}bh\right)$$

$$2A = bh$$

$$\frac{2A}{h} = \frac{bh}{h}$$

$$\frac{2A}{h} = b, \text{ or } b = \frac{2A}{h}$$

64. Solve for α:
$$\alpha + \beta + \gamma = 180$$

$$\alpha + \beta + \gamma - \beta - \gamma = 180 - \beta - \gamma$$

$$\alpha = 180 - \beta - \gamma$$

65. Solve for g:
$$v = v_0 - gt$$

$$v - v_0 = v_0 - gt - v_0$$

$$v - v_0 = -gt$$

$$\frac{v - v_0}{-t} = \frac{-gt}{-t}$$

$$\frac{v - v_0}{-t} = g, \text{ or } g = \frac{v_0 - v}{t}$$

66. Solve for m:
$$K = \frac{mv^2}{2g}$$

$$2g(K) = 2g\left(\frac{mv^2}{2g}\right)$$

$$2gK = mv^2$$

$$\frac{2Kg}{v^2} = \frac{mv^2}{v^2}$$

$$\frac{2gK}{v^2} = m, \text{ or } m = \frac{2gK}{v^2}$$

67. Solve for y:
$$3x + y = 6$$

$$3x + y - 3x = 6 - 3x$$

$$y = 6 - 3x$$

68. Solve for y:
$$6x - y = 3$$

$$6x - y - 3 = 3 - 3$$

$$6x - y - 3 = 0$$

$$6x - y - 3 + y = 0 + y$$

$$6x - 3 = y, \text{ or } y = 6x - 3$$

69. Solve for x:
$$5x - 2y = 10$$

$$5x - 2y + 2y = 10 + 2y$$

$$5x = 10 + 2y$$

$$\frac{5x}{5} = \frac{10 + 2y}{5}$$

$$x = \frac{10 + 2y}{5}$$

70. Solve for x:
$$-3x + 4y = -7$$

$$-3x + 4y - 4y = -7 - 4y$$

$$-3x = -7 - 4y$$

$$\frac{-3x}{-3} = \frac{-7 - 4y}{-3}$$

$$x = \frac{-7 - 4y}{-3}, \text{ or } x = \frac{7 + 4y}{3}$$

71. If $F = 32°$, then $C = \frac{5}{9}(32° - 32) = \frac{5}{9}(0°) = 0°$

72. Using $d = rt$, the average rate of speed is:
$$190 = r \cdot 4$$
$$\frac{190}{4} = \frac{4r}{4}$$
$$r = 47.5 \text{ mph}$$

73. Using $P = 2l + 2w$, the length is:
$$150 = 2l + 2(35)$$
$$150 = 2l + 70$$
$$150 - 70 = 2l + 70 - 70$$
$$80 = 2l$$
$$\frac{80}{2} = \frac{2l}{2}$$
$$40 = l, \quad \text{or} \quad l = 40 \text{ meters}$$

74. Using $V = \dfrac{k}{P}$, the pressure is:
$$20 = \frac{3.2}{P}$$
$$(20)P = \left(\frac{3.2}{P}\right)P$$
$$20P = 3.2$$
$$\frac{20P}{20} = \frac{3.2}{20}$$
$$P = 0.16$$

75. Using $4x + y = 10$, the value of y will be:
$$4(3) + y = 10$$
$$12 + y = 10$$
$$12 + y - 12 = 10 - 12$$
$$y = -2$$

76. Let n be the unknown number. Then:
$$42 = 2n + 4$$
$$42 - 4 = 2n + 4 - 4$$
$$38 = 2n$$
$$\frac{38}{2} = \frac{2n}{2}$$
$$19 = n$$

77. Let n be the unknown number. Then:
$$3n - 7 = 17 - n$$
$$3n - 7 + 7 = 17 - n + 7$$
$$3n = 24 - n$$
$$3n + n = 24 - n + n$$
$$4n = 24$$
$$\frac{4n}{4} = \frac{24}{4}$$
$$n = 6$$

78. Let n be the unknown number. Then:
$$\frac{2n}{9} = n - 7$$
$$9\left(\frac{2n}{9}\right) = 9(n - 7)$$
$$2n = 9n - 63$$
$$2n - 9n = 9n - 63 - 9n$$
$$\frac{-7n}{-7} = \frac{-63}{-7}$$
$$n = 9$$

79. Let n be the unknown number. Then:
$$2(n - 8) = 5n + 11$$
$$2n - 16 - 11 = 5n + 11 - 11$$
$$2n - 27 = 5n$$
$$2n - 27 - 2n = 5n - 2n$$
$$-27 = 3n$$
$$\frac{-27}{3} = \frac{3n}{3}$$
$$-9 = n$$

80. Let n be the first integer. Then $n + 1$ and $n + 2$ are the second and third integers, respectively.
$$n + (n + 1) + (n + 2) = 75$$
$$3n + 3 - 3 = 75 - 3$$
$$3n = 72$$
$$\frac{3n}{3} = \frac{72}{3}$$
$$n = 24$$
$$n + 1 = 25 \quad \text{and} \quad n + 2 = 26$$
The integers are 24, 25, and 26.

81. Let n be the first odd integer. Then $n + 2$ and $n + 4$ are the second and third odd integers, respectively.
$$n + (n + 2) + (n + 4) = 279$$
$$3n + 6 - 6 = 279 - 6$$
$$3n = 273$$
$$\frac{3n}{3} = \frac{273}{3}$$
$$n = 91$$
$$n + 2 = 93 \text{ and } n + 4 = 95$$
The odd integers are 91, 93, and 95.

82. Let n be the first even integer. Then $n + 2$ and $n + 4$ are the second and third even integers, respectively.
$$2n + (n + 2) = (n + 4) + 18$$
$$3n + 2 - 2 = n + 22 - 2$$
$$3n = n + 20$$
$$3n - n = n + 20 - n$$
$$2n = 20$$
$$\frac{2n}{2} = \frac{20}{2}$$
$$n = 10$$
$$n + 2 = 12 \text{ and } n + 4 = 14$$
The even integers are 10, 12, and 14.

83. Let n be the first integer. Then $n + 1$, $n + 2$ and $n + 3$ are the second, third, and fourth integers.
$$n + (n + 1) + (n + 2) + (n + 3) = 190$$
$$4n + 6 - 6 = 190 - 6$$
$$4n = 184$$
$$\frac{4n}{4} = \frac{184}{4}$$
$$n = 46$$
$$n + 1 = 47, \ n + 2 = 48, \text{ and } n + 3 = 49$$
The integers are 46, 47, 48, and 49.

84. Let n be the first integer. Then $n + 1$ and $n + 2$ are the second and third integers, respectively.
$$n + (n + 2) + (n + 4) = -81$$
$$3n + 6 - 6 = -81 - 6$$
$$\frac{3n}{3} = \frac{-87}{3}$$
$$n = -29$$
$$n + 2 = -27 \text{ and } n + 4 = -25$$
The integers are -29, -27, and -25.

85. Let V be the value of the home when it was new. Then:
$$2V + 150,000 = 640,000$$
$$2V + 150,000 - 150,000 = 640,000 - 150,000$$
$$2V = 490,000$$
$$\frac{2V}{2} = \frac{490,000}{2}$$
$$V = 245,000$$
The home was worth \$245,000 when new.

86. 92% of 85 is $0.92 \cdot 85 = 78.2$

87. ___% of 150 is 105:
$$x \cdot 150 = 105$$
$$\frac{150x}{150} = \frac{105}{150}$$
$$x = 0.7 = 70\%$$

88. The tax rate can be found using:
$$370,000r = 4440$$
$$r = \frac{4440}{370,000} = 0.012 = 1.2\%$$

89. Since the player made 85% of her free throws, she missed $100 - 85 = 15\%$ of them. So she missed 15% of $120 = 0.15 \cdot 120 = 18$ free throws.

90. Let p be the sale price. Since the discount is 33%, the sale price will be $100 - 33 = 67\%$ of the original price. Then the sale price is:
$$p = 67\% \text{ of } 90$$
$$p = 0.67 \cdot 90$$
$$p = 60.30$$
The sale price is \$60.30.

91. Let t be the amount in taxes. Then 8.25% of \$260 is $t = 0.0825 \cdot 260 = \$21.45$. The total amount paid will be $260 + t = 260 + 21.45 = \$281.45$.

92. Let x be the percent of weight lost by the dog. Then $x\%$ of $80\text{lb} = x \cdot 80 = 5\text{lb}$. So $x = \frac{5}{80}$, or $x = 0.0625 = 6.25\%$.

93. Using $I = Prt$, solve for t then substitute the given values into the expression:
$$t = \frac{I}{Pr} = \frac{81}{3240 \cdot 0.10} = \frac{81}{324} = 0.25 \text{ years}$$

94. Using $I = Prt$, solve for P then substitute the given values into the expression. Remember to convert the time in months to years:

$$t = \frac{9}{12} = 0.75 \text{ years.}$$

$$P = \frac{I}{rt} = \frac{400}{0.05 \cdot 0.75} = \frac{400}{0.0375} = \$10,666.67$$

95. Using $I = Prt$, solve for r then substitute the given values into the expression. Remember to convert the time in months to years:

$$t = \frac{1}{12} = 0.083 \text{ years}$$

$$r = \frac{I}{Pt} = \frac{40}{2000 \cdot 0.083} = \frac{40}{166} = 0.24 = 24\%$$

96. The perimeter of a parallelogram is: $P = 2a + 2b$, where a and b are the two different side lengths.

97. The area of a circle is $A = \pi r^2$, where r is the radius of the circle.

98. The volume of a sphere is $V = \frac{4}{3}\pi r^3$, where r is the radius of the sphere.

99. The perimeter of the triangle is:
$P = 9.6 + 4.9 + 5.7 = 20.2$ inches.

100. The area of the circle is:
$$A = (3.14)(13)^2 = 3.14 \cdot 169 = 530.66 \text{ cm}^2$$

101. The volume of the cube is:
$$V = s^3 = 3^3 = 27 \text{ ft}^3$$

102. The area of the parallelogram is:
$$A = b \cdot h = 30 \cdot 15.6 = 468 \text{ m}^2$$

103. Remember that the radius is half of the diameter:
$r = \frac{10}{2} = 5$. So, the area of the pizza is:

$$A = \pi r^2$$
$$A = 3.14 \cdot (5)^2$$
$$A = 3.14 \cdot 25$$
$$A = 78.5 \text{ in.}^2$$

104. Remember that the radius is half of the diameter: $r = \frac{8.2}{2} = 4.1$. The volume of the can, to the nearest tenth, is:

$$V = \pi r^2 h$$
$$V = 3.14 \cdot (4.1)^2 \cdot 13$$
$$V = 3.14 \cdot 16.81 \cdot 13$$
$$V = 686.2 \text{ cm}^2$$

105. The figure consists of a triangle and a rectangle. The perimeter is the sum of the sides:
$P = 13 + 13 + 6 + 6 + 10 = 48$ in.
The area is the sum of the areas of the rectangle and the triangle:

$$A = \frac{1}{2}bh + lw$$
$$A = \frac{1}{2} \cdot 10 \cdot 12 + 10 \cdot 6$$
$$A = 60 + 60$$
$$A = 120 \text{ in.}^2$$

[End of Chapter 3 Review]

Chapter 3 Test
Solutions to All Exercises

1.
$$\frac{5}{3}x + 1 = -4$$
$$3\left(\frac{5}{3}x + 1\right) = 3(-4)$$
$$5x + 3 = -12$$
$$5x = -15$$
$$x = -3$$

2.
$$4x - 5 - x = 2x + 5 - x$$
$$3x - 5 = x + 5$$
$$2x - 5 = 5$$
$$2x = 10$$
$$x = 5$$

3.
$$8(3 + x) = -4(2x - 6)$$
$$24 + 8x = -8x + 24$$
$$16x + 24 = 24$$
$$16x = 0$$
$$x = 0$$

4.
$$\frac{3}{2}x + \frac{1}{2}x = -3 + \frac{3}{4}$$
$$\frac{4}{2}x = -2\frac{1}{4}$$
$$4\left(\frac{4}{2}x\right) = 4\left(\frac{-9}{4}\right)$$
$$8x = -9$$
$$x = \frac{-9}{8}$$

5.
$$0.7x + 2 = 0.4x + 8$$
$$0.3x + 2 = 8$$
$$0.3x = 6$$
$$x = 20$$

6.
$$4(2x - 1) + 3 = 2(x - 4) - 5$$
$$8x - 4 + 3 = 2x - 8 - 5$$
$$8x - 1 = 2x - 13$$
$$6x - 1 = -13$$
$$6x = -12$$
$$x = -2$$

7. 62% of 180 is:
$$0.62(180) = 111.6$$

8. 48 is ___ % of 150:
$$R \cdot 150 = 48$$
$$R = \frac{48}{150}$$
$$R = 0.32 \text{ or } 32\%$$

9. Change 9 months to $\frac{3}{4}$ year. Using $I = Prt$,
$$I = 5000(0.09)\left(\frac{3}{4}\right)$$
$$I = \$337.50$$

10. Given that the investment times are the same, $360 profit from $6000 investment is a rate of
$$r = \frac{I}{P} = \frac{360}{6000} = 0.06 \text{, or 6\%, while the \$500 profit}$$
from $10,000 investment is a rate of
$$r = \frac{I}{P} = \frac{500}{10000} = 0.05 \text{, 5\%.}$$
The $6000 investment is better, since 6% is greater than 5%.

11. Using $P = a + b + c$, the perimeter of a triangle is the sum of the lengths of its three sides:
$$(x + 1) + (2x - 1) + x = 12$$
$$4x = 12$$
$$x = 3$$
$$x + 1 = 4$$
$$2x - 1 = 5$$
The three sides have lengths 3 m, 4 m and 5 m.

12. Solve for m:
$$N = mrt + p$$
$$N - p = mrt$$
$$\frac{N - p}{rt} = m \quad \text{or} \quad m = \frac{N - p}{rt}$$

13. Solve for y:
$$5x + 3y - 7 = 0$$
$$3y = -5x + 7$$
$$y = \frac{-5x + 7}{3}$$

14. Substitute $F = -4°$ into the given formula:

$$C = \frac{5}{9}(-4 - 32)$$

$$C = \frac{5}{9}(-36)$$

$$C = -20$$

So $-4°F = -20°C$

15. First, solve for x:

$$3x - 2y = 18$$

$$3x = 2y + 18$$

$$x = \frac{2y + 18}{3} \quad \text{or} \quad x = \frac{2y}{3} + 6$$

Then, substitute $y = 4$.

$$x = \frac{2(4) + 18}{3}$$

$$x = \frac{8 + 18}{3}$$

$$x = \frac{26}{3}$$

16. The figure is a square with side $s = 6$ and a semicircle.
with radius $r = \frac{6}{2} = 3$.

Perimeter: $P = \frac{1}{2}\pi(2r) + 3s$

$$P = \frac{1}{2}(3.14)(2 \cdot 3) + 3(6)$$

$$= 9.42 + 18$$

$$= 27.42 \ \text{ft}$$

Area: $A = \frac{1}{2}\pi r^2 + s^2$

$$A = \frac{1}{2}(3.14)(3)^2 + 6^2$$

$$A = 14.13 + 36$$

$$A = 50.13 \ \text{ft}^2$$

17. The sphere has radius $r = \frac{20}{2} = 10$.

$$V = \frac{4}{3}\pi r^3$$

$$V = \frac{4}{3}(3.14)(10)^3$$

$$V = 4186.\overline{6}$$

To the nearest hundredth, the volume of the sphere is 4186.67 ft^3.

18. $3x + 2 \le -8$

$$3x \le -10$$

$$x \le -\frac{10}{3}$$

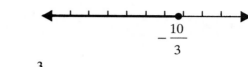

19. $-\frac{3}{4}x - 6 > 9$

$$-\frac{3}{4}x > 15$$

$$4\left(\frac{-3}{4}x\right) > 4(15)$$

$$-3x > 60$$

$$x < -20$$

20. $2(x - 7) < 4(2x + 3)$

$$2x - 14 < 8x + 12$$

$$-6x < 26$$

$$x > \frac{-13}{3}$$

21. $-2 \le 4x + 7 \le 3$

$$-2 - 7 \le 4x \le 3 - 7$$

$$-9 \le 4x \le -4$$

$$-\frac{9}{4} \le x \le -1$$

22. Let x be one of the numbers. Then $2x + 5$ is the other number.

$$(x) + (2x + 5) = -22$$

$$3x + 5 = -22$$

$$3x = -27$$

$$x = -9$$

$$2x + 5 = 2(-9) + 5 = -13$$

The numbers are -9 and -13.

23. Let n be the first integer. Then $n + 1$ is the next integer.

$$2n + 3(n+1) = 83$$
$$2n + 3n + 3 = 83$$
$$5n + 3 = 83$$
$$5n = 80$$
$$n = 16$$
$$n + 1 = 17$$

The integers are 16 and 17.

24. **a.**
$$2W + 2(37) = 122$$
$$2W + 74 = 122$$
$$2W = 48$$
$$W = 24$$
The width is 24 in.

b. $A = lw$
$$A = 37(24)$$
$$A = 888$$
The area is 888 in.^{2}.

25. **a.** Let p be the original price.
$$p - 0.25p = 547.50$$
$$0.75p = 547.50$$
$$p = \frac{547.50}{0.75}$$
$$p = \$730$$

b. Let C be the cost of the suit.
$$C = 547.50 + 0.06(547.50) + 25$$
$$C = 547.50 + 32.85 + 25$$
$$C = \$605.35$$

26. Let n be the first integer. Then the next two consecutive odd integers are $n + 2$ and $n + 4$:
$$3(n+2) = (n) + (n+4) + 27$$
$$3n + 6 = 2n + 31$$
$$n = 25$$
$$n + 2 = 27$$
$$n + 4 = 29$$
The three numbers are 25, 27 and 29.

27. **a.** Use $V = lwh$. Note: 4 in. $= \dfrac{1}{3}$ ft.
$$V = 100(6)\left(\frac{1}{3}\right)$$
$$V = 200$$
So, 200 ft^3 of concrete is needed.

b. Note: 1 yd^3 = 27 ft^3
$$\frac{200}{27} \approx 7.4 \text{ yd}^3$$

28. Use $V = \dfrac{1}{3}\pi r^2 h$.
$$V = \frac{1}{3}(3.14)(4)^2(6)$$
$$V = 100.48$$
The volume of the cone is 100.48 in.^{3}

[End of Chapter 3 Test]

Cumulative Review: Chapter 1 – 3
Solutions to All Exercises

1. True, because $|-5| = 5$.

2. True, because
$$2^3 - 8 = 8 - 8 = 0, \quad 8 - 3 \cdot 4 = 8 - 12 = -4 \quad \text{and}$$
$$0 > -4.$$

3. False. $\dfrac{3}{4} \le |-1|$, since $|-1| = 1$.

4. $|x| \le 6$

5. $|x| = x$

6. Answers may vary. One possible answer is: Division by zero is not defined because if it were, and we had some real number $\dfrac{a}{0} = c$, for $a \ne 0$, then $c \cdot 0 = a$. This would mean that $a = 0$. Since we have chosen a non-zero value for a, there exists a contradiction.

7. **a.** $\text{LCM}(25, 35, 40, 50)$:
$$25 = 5^2$$
$$35 = 5 \cdot 7$$
$$40 = 2^3 \cdot 5$$
$$50 = 2 \cdot 5^2$$
$$\text{LCM} = 2^3 \cdot 5^2 \cdot 7 = 1400$$
b. $\text{LCM}(6x^2, 15xy^2, 20xy)$:
$$6x^2 = 2 \cdot 3 \cdot x^2$$
$$15xy^2 = 3 \cdot 5 \cdot x \cdot y^2$$
$$20xy = 2^2 \cdot 5 \cdot x \cdot y$$
$$\text{LCM} = 2^2 \cdot 3 \cdot 5 \cdot x^2 \cdot y^2 = 60x^2y^2$$

8. **a.** The average of the numbers in the set is
$$\frac{32 + 56 + 70 + 86 + 91}{5} = \frac{335}{5} = 67$$
b. The average of the numbers in the set is
$$\frac{13.62 + 20.15 + 26.24 + 33.19}{4} = \frac{93.2}{4} = 23.3$$

9. $5^2 \cdot 3^2 - 24 \div 2 \cdot 3 = 25 \cdot 9 - 12 \cdot 3$
$$= 225 - 36 = 189$$

10. $-36 \div (-2)^2 + 20 - 2(16 - 17) = -36 \div 4 + 20 - 2(-1)$
$$= -9 + 20 + 2$$
$$= 13$$

11. $(7 - 10)\left[49 \div (-7) + 20 \cdot 3 - 5 \cdot 15 - (-10)\right] - 1$
$$= -3\left[-7 + 60 - 75 + 10\right] - 1$$
$$= -3(-12) - 1$$
$$= 36 - 1$$
$$= 35$$

12. $\left(\dfrac{2}{3}\right)^2 \div \dfrac{5}{18} - \dfrac{3}{8} = \dfrac{4}{9} \cdot \dfrac{18}{5} - \dfrac{3}{8}$
$$= \frac{8}{5} - \frac{3}{8}$$
$$= \frac{64 - 15}{40}$$
$$= \frac{49}{40}$$

13. $\left(\dfrac{9}{10} + \dfrac{2}{15}\right) \div \left(\dfrac{1}{2} - \dfrac{2}{3}\right) = \left(\dfrac{27}{30} + \dfrac{4}{30}\right) \div \left(\dfrac{3}{6} - \dfrac{4}{6}\right)$
$$= \left(\frac{31}{30}\right) \cdot \left(\frac{6}{-1}\right)$$
$$= -\frac{31}{5}$$

14. $15^3 - 160 + 13(-5) = 3375 - 160 - 65 = 3150$

15. **a.** Simplify:
$$-4(y + 3) + 2y = -4y - 12 + 2y$$
$$= -2y - 12$$
b. For $y = 3$:
$$-2(3) - 12 = -6 - 12 = -18$$

16. **a.** Simplify:
$$3(x + 4) - 5 + x = 3x + 12 - 5 + x$$
$$= 4x + 7$$
b. For $x = -2$:
$$4(-2) + 7 = -8 + 7 = -1$$

17. **a.** Simplify:
$$2(x^2 + 4x) - (x^2 - 5x) = 2x^2 + 8x - x^2 + 5x$$
$$= x^2 + 13x$$
b. For $x = -2$:
$$(-2)^2 + 13(-2) = 4 - 26 = -22$$

18. **a.** Simplify:

$$\frac{3(5x-x)}{4} - 2x - 7 = \frac{3 \cdot 4x}{4} - 2x - 7$$
$$= 3x - 2x - 7$$
$$= x - 7$$

b. For $x = -2$:
$$-2 - 7 = -9$$

19. $9 - 2N$

20. $3(x+10)$

21. $24x + 5$

22. $6T + E + 3F$

23. $9x - 8 = 4x - 13$
$$5x = -5$$
$$x = -1$$

24. $6 - (x-2) = 5$
$$6 - x + 2 = 5$$
$$-x + 8 = 5$$
$$-x = -3$$
$$x = 3$$

25. $10 + 4x = 5x - 3(x+4)$
$$10 + 4x = 5x - 3x - 12$$
$$10 + 4x = 2x - 12$$
$$2x = -22$$
$$x = -11$$

26. $4.4 + 0.6x = 1.2 - 0.2x$
$$0.8x = -3.2x$$
$$x = -4$$

27. $-2(5x+12) - 5 = -6(3x-2) - 10$
$$-10x - 24 - 5 = -18x + 12 - 10$$
$$-10x - 29 = -18x + 2$$
$$8x = 31$$
$$x = \frac{31}{8}$$

28. $\dfrac{3}{4}y - \dfrac{1}{2} = \dfrac{5}{8}y + \dfrac{1}{10}$

$$40\left(\frac{3}{4}y - \frac{1}{2}\right) = 40\left(\frac{5}{8}y + \frac{1}{10}\right)$$
$$30y - 20 = 25y + 4$$
$$5y = 24$$
$$y = \frac{24}{5}$$

29. $\dfrac{1}{2} + \dfrac{1}{5}y = \dfrac{2}{15}y + 1$

$$30\left(\frac{1}{2} + \frac{1}{5}y\right) = 30\left(\frac{2}{15}y + 1\right)$$
$$15 + 6y = 4y + 30$$
$$2y = 15$$
$$y = \frac{15}{2}$$

30. $-\dfrac{3}{4}y + \dfrac{3}{8}y = \dfrac{1}{6}y$

$$24\left(-\frac{3}{4}y + \frac{3}{8}y\right) = 24\left(\frac{1}{6}y\right)$$
$$-18y + 9y = 4y$$
$$-9y - 4y = 0$$
$$-13y = 0$$
$$y = 0$$

31. Solve for v:
$$h = vt - 16t^2$$
$$h + 16t^2 = vt$$
$$\frac{h + 16t^2}{t} = v$$

32. Solve for r:
$$A = P + Prt$$
$$A - P = Prt$$
$$\frac{A - P}{Pt} = r$$

33. Solve for y:
$$5x - 3y = 14$$
$$-3y = -5x + 14$$
$$y = \frac{-5x + 14}{-3}$$
$$y = \frac{5x - 14}{3}$$

34. 75% of ___ is 102 translates to:

$$0.75x = 102$$

$$x = \frac{102}{0.75}$$

$$x = 136$$

35. 20.24 is ___% of 92 translates to:

$$20.24 = x(92)$$

$$92x = 20.24$$

$$x = \frac{20.24}{92}$$

$$x = 0.22$$

$$x = 22\%$$

36. 25.2% of 65 is ___ translates to:

$$0.252(65) = x$$

$$16.38 = x$$

37.

$$2x + 5 - 3 \le 6$$

$$2x + 2 \le 6$$

$$2x \le 4$$

$$x \le 2$$

38.

$$-3(7 - 2x) \ge 2 + 3x - 10$$

$$-21 + 6x \ge -8 + 3x$$

$$3x \ge 13$$

$$x \ge \frac{13}{3}$$

39.

$$-\frac{1}{2}x + \frac{7}{8} < \frac{2}{3}x + \frac{1}{2}$$

$$24\left(-\frac{1}{2}x + \frac{7}{8}\right) < 24\left(\frac{2}{3}x + \frac{1}{2}\right)$$

$$-12x + 21 < 16x + 12$$

$$-28x < -9$$

$$x > \frac{9}{28}$$

40.

$$-6.2 \le 2x - 4 \le 10.8$$

$$-2.2 \le 2x \le 14.8$$

$$-1.1 \le x \le 7.4$$

41. Semicircle with radius $r = \frac{10}{2} = 5$ m and a square with side length $s = 10$m:

a. Perimeter:

$$P = \frac{1}{2}(2\pi r) + 3s$$

$$P = \frac{1}{2}(2 \cdot 3.14 \cdot 5) + 3 \cdot 10$$

$$P = 15.7 + 30$$

$$P = 45.7 \text{ m}$$

b. Area:

$$A = \frac{1}{2}\pi r^2 + s^2$$

$$A = \frac{1}{2}(3.14 \cdot 5^2) + 10^2$$

$$A = 39.25 + 100$$

$$A = 139.25 \text{ m}^2$$

42. Volume of a cylinder with base of radius $r = 5$ in. and height $h = 15$ in.:

$$V = \pi r^2 h$$

$$V = 3.14 \cdot 5^2 \cdot 15$$

$$V = 1177.5 \text{ in.}^3$$

43. Use $I = Prt$, where $P = \$2000$, $r = 0.08$ and $t = 0.5$ years:

$$I = \$2000(0.08)(0.5)$$

$$I = \$80$$

44. With a 30% discount, the price before tax is $\$680 - 0.30(680) = \$476$. With the 6.5% tax, the final cost is $\$476 + 0.065(476) = \$506.94$.

45. Let N be the number.

$$2N + 3 = 3N - 8$$

$$3 + 8 = 3N - 2N$$

$$11 = N$$

46. Let N be the first even integer. Then, $N + 2$ and $N + 4$ represent the next two even integers.

$$N + 2(N + 2) = 2(N + 4) + 14$$

$$N + 2N + 4 = 2N + 8 + 14$$

$$3N + 4 = 2N + 22$$

$$N = 18$$

$$N + 2 = 20 \text{ and } N + 4 = 22$$

The three integers are 18, 20, and 22.

47. Let N be the first integer. Then $N + 1$, $N + 2$ and $N + 3$ represent the second, third, and fourth integers, respectively.

$$N + (N + 1) + (N + 3) = (N + 2) - 2$$
$$3N + 4 = N$$
$$2N = -4$$
$$N = -2$$
$$N + 1 = -1 \quad N + 2 = 0 \quad \text{and} \quad N + 3 = 1$$

The four integers are $-2, -1, 0,$ and 1.

48. Let x be the number of miles driven. Set up and solve the following inequality:

ALPHA costs $<$ BETA costs

$$35 + 0.55x < 45 + 0.25x$$
$$0.3x < 10$$
$$x < 33.\overline{3}$$

The cost to drive an ALPHA truck is less than the cost to drive a BETA truck if the number of miles driven is less than 33.33.

49. Let x be the dollar value of the furniture. Set Jay's earnings equal to Kay's for x value of furniture:

$$800 + 0.06(x - 5000) = 0.10x$$
$$800 + 0.06x - 300 = 0.10x$$
$$500 + 0.06x = 0.10x$$
$$500 = 0.04x$$
$$x = 12500$$

They each sold $12,500 in furniture. On $12,500 value in furniture, Jay would earn $800 + 6% of $7,500 (the amount in excess of $5,000), or $800 + $450 = $1250. If Kay sold $12,500 furniture, she would earn 10% of $12,500, which is $1250.

50. Use $V = \dfrac{1}{3}\pi r^2 h$, and solve for h: $3V = \pi r^2 h$

$$\frac{3V}{\pi r^2} = h$$

Substitute the given values for V and r:

$$h = \frac{3(175\pi)}{\pi(7)^2} = \frac{525}{49} \approx 10.71 \text{ in.}$$

51. Let x be the grade on the final exam.

$$75 \le \frac{73 + 65 + 77 + 74 + x}{5} < 85$$
$$375 \le \quad 289 + x \quad < 425$$
$$86 \le \quad\quad x \quad\quad < 136$$

The final exam grade must be between 86 and 100.

52. Use $I = Prt$, where $I = 450$, $r = 0.09$ and $t = 0.5$ yrs.

$$I = Prt \quad \rightarrow \quad P = \frac{I}{rt}$$
$$P = \frac{450}{0.09(.5)}$$
$$P = \$10,000$$

53. A profit of $800 on a $10,000 investment is

$$\frac{800}{10,000} = 0.08 = 8\%.$$

A profit of $1200 on a $15,000 investment is

$$\frac{1200}{15,000} = 0.08 = 8\%.$$

These are equally good investments.

54. It may be helpful to first draw a diagram illustrating how the three sides are related to each other.

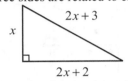

a. Use the given value for the perimeter and the formula for perimeter to find x:

$$x + (2x + 2) + (2x + 3) = 30$$
$$5x + 5 = 30$$
$$5x = 25$$
$$x = 5 \text{ in.}$$
$$2x + 2 = 12 \text{ in. and } 2x + 3 = 13 \text{ in.}$$

b. Use the formula for area to find and the values found in part **a**:

$$A = \frac{1}{2}(5)(12)$$
$$A = 30 \text{ in.}^2$$

[End Cumulative Review: Chapters 1 – 3]

Section 4.1
Solutions to Odd Exercises

1. $\left\{A(-5,1), B(-3,3), C(-1,1), D(1,2), E(2,-2)\right\}$

3. $\left\{A(-3,-2), B(-1,-3), C(-1,3), D(0,0), E(2,1)\right\}$

5. $\left\{A(-4,4), B(-3,-4), C(0,-4), D(0,3), E(4,1)\right\}$

7. $\left\{A(-6,2), B(-1,6), C(0,0), D(1,-6), E(6,3)\right\}$

9. $\left\{A(-5,0), B(-2,2), C(-1,-4), D(0,6), E(2,0)\right\}$

11.

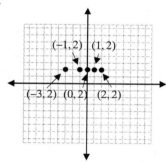

13.

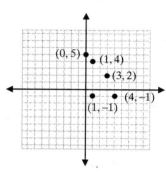

15.

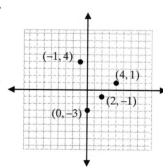

17.

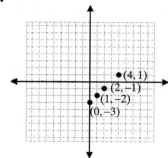

19.

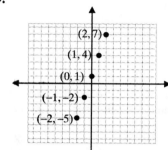

21.

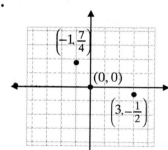

23.

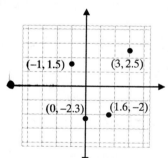

In Exercises 25 – 30, substitute each point into the given equation. If the substitution gives a true statement, then the point satisfies the equation.

25. $2x - y = 4$

  a. $(1,1)$: $2(1) - 1 \overset{?}{=} 4$
  $1 \neq 4$

  b. $(2,0)$: $2(2) - 0 \overset{?}{=} 4$
  $4 = 4$

  c. $(1,-2)$: $2(1) - (-2) \overset{?}{=} 4$
  $4 = 4$

  d. $(3,2)$: $2(3) - 2 \overset{?}{=} 4$
  $4 = 4$

The points **b**, **c**, and **d** satisfy the given equation.

27. $4x + y = 5$

  a. $\left(\frac{3}{4}, 2\right)$: $4\left(\frac{3}{4}\right) + 2 \overset{?}{=} 5$
  $5 = 5$

  b. $(4,0)$: $4(4) + 0 \overset{?}{=} 5$
  $16 \neq 5$

  c. $(1,1)$: $4(1) + 1 \overset{?}{=} 5$
  $5 = 5$

  d. $(0,3)$: $4(0) + 3 \overset{?}{=} 5$
  $3 \neq 5$

The points **a** and **c** satisfy the given equation.

29. $2x + 5y = 8$

  a. $(4,0)$: $2(4) + 5(0) \overset{?}{=} 8$
  $8 = 8$

  b. $(2,1)$: $2(2) + 5(1) \overset{?}{=} 8$
  $9 \neq 8$

  c. $(1,1.2)$: $2(1) + 5(1.2) \overset{?}{=} 8$
  $8 = 8$

  d. $(1.5,1)$: $2(1.5) + 5(1) \overset{?}{=} 8$
  $8 = 8$

The points **a**, **c**, and **d** satisfy the given equation.

For Exercises 31 – 55, substitute the given values into the equation, and solve for the missing coordinate.

31. $x - y = 4$

In $(0, \ )$, we are given $x = 0$, so
  $0 - y = 4$
  $-y = 4$
  $y = -4$

In $(2, \ )$, we are given $x = 2$, so
  $2 - y = 4$
  $-y = 2$
  $y = -2$

In $(\ ,0)$, we are given $y = 0$, so
  $x - 0 = 4$
  $x = 4$

In $(\ ,-3)$, we are given $y = -3$, so
  $x - (-3) = 4$
  $x + 3 = 4$
  $x = 1$

The completed coordinates that satisfy this equation are $(0,-4)$, $(2,-2)$, $(4,0)$, $(1,-3)$.

33. $x + 2y = 6$

In $(0, \ )$, we are given $x = 0$, so
  $0 + 2y = 6$
  $2y = 6$
  $y = 3$

In $(2, \ )$, we are given $x = 2$, so
  $2 + 2y = 6$
  $2y = 4$
  $y = 2$

In $(\ ,0)$, we are given $y = 0$, so
  $x + 2(0) = 6$
  $x + 0 = 6$
  $x = 6$

In $(\ ,4)$, we are given $y = 4$, so
  $x + 2(4) = 6$
  $x + 8 = 6$
  $x = -2$

The completed coordinates that satisfy this equation are $(0,3)$, $(2,2)$, $(6,0)$, $(-2,4)$.

35. $4x - y = 8$

In $(0, \ )$, we are given $x = 0$, so
$$4(0) - y = 8$$
$$0 - y = 8$$
$$-y = 8$$
$$y = -8$$

In $(1, \ )$, we are given $x = 1$, so
$$4(1) - y = 8$$
$$4 - y = 8$$
$$-y = 4$$
$$y = -4$$

In $(\ , 0)$, we are given $y = 0$, so
$$4x - 0 = 8$$
$$4x = 8$$
$$x = 2$$

In $(\ , 4)$, we are given $y = 4$, so
$$4x - 4 = 8$$
$$4x = 12$$
$$x = 3$$

The completed coordinates that satisfy this equation are $(0, -8)$, $(1, -4)$, $(2, 0)$, $(3, 4)$.

37. $2x + 3y = 6$

In $(0, \ )$, we are given $x = 0$, so
$$2(0) + 3y = 6$$
$$0 + 3y = 6$$
$$3y = 6$$
$$y = 2$$

In $(-1, \ )$, we are given $x = -1$, so
$$2(-1) + 3y = 6$$
$$-2 + 3y = 6$$
$$3y = 8$$
$$y = \frac{8}{3}$$

In $(\ , 0)$, we are given $y = 0$, so
$$2x + 3(0) = 6$$
$$2x + 0 = 6$$
$$2x = 6$$
$$x = 3$$

In $(\ , -2)$, we are given $y = -2$, so
$$2x + 3(-2) = 6$$
$$2x - 6 = 6$$
$$2x = 12$$
$$x = 6$$

The completed coordinates that satisfy this equation are $(0, 2)$, $\left(-1, \frac{8}{3}\right)$, $(3, 0)$, $(6, -2)$.

39. $3x - 4y = 7$

In $(0, \ )$, we are given $x = 0$, so,
$$3(0) - 4y = 7$$
$$0 - 4y = 7$$
$$-4y = 7$$
$$y = -\frac{7}{4}$$

In $(1, \ )$, we are given $x = 1$, so,
$$3(1) - 4y = 7$$
$$3 - 4y = 7$$
$$-4y = 4$$
$$y = -1$$

In $(\ , 0)$, we are given $y = 0$, so,
$$3x - 4(0) = 7$$
$$3x - 0 = 7$$
$$3x = 7$$
$$x = \frac{7}{3}$$

In $\left(\ , \frac{1}{2}\right)$, we are given $y = \frac{1}{2}$, so
$$3x - 4\left(\frac{1}{2}\right) = 7$$
$$3x - 2 = 7$$
$$3x = 9$$
$$x = 3$$

The completed coordinates that satisfy this equation are $\left(0, -\frac{7}{4}\right)$, $(1, -1)$, $\left(\frac{7}{3}, 0\right)$, $\left(3, \frac{1}{2}\right)$.

41. $y = 3x$

x	y
0	0
−1	**−3**
−2	−6
2	**6**

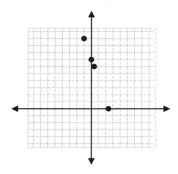

47. $y = \dfrac{3}{4}x + 2$

x	y
0	2
4	**5**
−4	−1
−1	$\dfrac{5}{4}$

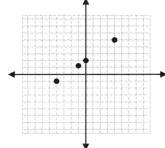

43. $y = 2x - 3$

x	y
0	−3
1	**−1**
−2	−7
$\dfrac{7}{4}$	$\dfrac{1}{2}$

49. $3x - 5y = 9$

x	y
0	$\dfrac{−9}{5}$
3	**0**
−2	−3
$\dfrac{4}{3}$	**−1**

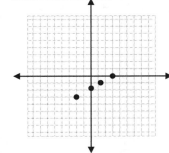

45. $y = 7 - 3x$

x	y
0	7
$\dfrac{7}{3}$	**0**
−1	10
$\dfrac{1}{3}$	6

51. $5x - 2y = 10$

x	y
0	-5
2	**0**
-2	-10
$\dfrac{8}{5}$	**-1**

53. $x - y = 1.5$

x	y
0	-1.5
1.5	0
3.8	**2.3**
-1	**-2.5**

55. $3x + y = -2.4$

x	y
-0.8	**0**
0	-2.4
-1	**0.6**
1.6	-7.2

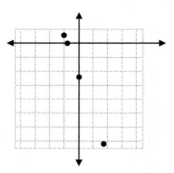

57. Answers will vary. For example, the points
$$(0,4),(2,2),(4,0)$$
are all on the given line.

59. Answers will vary. For example, the points
$$(-5,5),(-3,2),(1,-4)$$
are all on the given line.

61. Answers will vary. For example, the points
$$(-4,1),(0,2),(4,3)$$
are all on the given line.

63. Answers will vary. For example, the points
$$(-4,-3),(-4,0),(-4,2)$$
are all on the given line.

65. Answers will vary. For example, the points
$$(1,5),(3,2),(5,-1)$$
are all on the given line.

67. **a.** $F = \dfrac{9}{5}C + 32$

Substitute each of the given values of C into the formula to find the corresponding value of F.

$$F = \frac{9}{5}(-20) + 32 = -36 + 32 = -4$$

$$F = \frac{9}{5}(-10) + 32 = -18 + 32 = 14$$

$$F = \frac{9}{5}(-5) + 32 \;\; = -9 + 32 \;\; = 23$$

$$F = \frac{9}{5}(0) + 32 \;\;\;\; = 0 + 32 \;\;\;\; = 32$$

$$F = \frac{9}{5}(5) + 32 \;\;\;\; = 9 + 32 \;\;\;\; = 41$$

$$F = \frac{9}{5}(10) + 32 \;\; = 18 + 32 \;\; = 50$$

$$F = \frac{9}{5}(15) + 32 \;\; = 27 + 32 \;\; = 59$$

C	−20	−10	−5	0	5	10	15
F	−4	14	23	32	41	50	59

b.

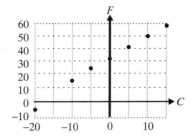

69.   **a.** $V = 9h$

h	V
2	$9 \cdot 2 = 18$
3	$9 \cdot 3 = 27$
5	$9 \cdot 5 = 45$
8	$9 \cdot 8 = 72$
9	$9 \cdot 9 = 81$
10	$9 \cdot 10 = 90$

b.

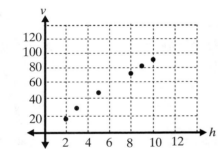

Writing and Thinking About Mathematics

71.   Answers will vary.

[End of Section 4.1]

Section 4.2
Solutions to Odd Exercises

1. $x + y = -12$

   $(6,6)$ does not satisfy the equation:

   $6 + 6 \overset{?}{=} -12$

   $12 \neq -12$

   $(0,12)$ does not satisfy the equation:

   $0 + 12 \overset{?}{=} -12$

   $12 \neq -12$

   $(-14,2)$ satisfies the equation:

   $-14 + 2 \overset{?}{=} -12$

   $12 = -12$

   $(-1,-11)$ satisfies the equation:

   $-1 + (-11) \overset{?}{=} -12$

   $-12 = -12$

3. $y = -2x + 8$

   $(2,4)$ satisfies the equation:

   $4 \overset{?}{=} -2(2) + 8$

   $4 = -4 + 8$

   $(0,8)$ satisfies the equation:

   $8 \overset{?}{=} -2(0) + 8$

   $8 = 0 + 8$

   $(4,0)$ satisfies the equation:

   $0 \overset{?}{=} -2(4) + 8$

   $0 = -8 + 8$

   $(3,-2)$ does not satisfy the equation:

   $-2 \overset{?}{=} -2(3) + 8$

   $-2 \overset{?}{=} -6 + 8$

   $-2 \neq 2$

5. $2x + y = 5$

   $(0,5)$ satisfies the equation:

   $2(0) + 5 \overset{?}{=} 5$

   $0 + 5 \overset{?}{=} 5$

   $5 = 5$

   $(1,3)$ satisfies the equation:

   $2(1) + 3 \overset{?}{=} 5$

   $2 + 3 \overset{?}{=} 5$

   $5 = 5$

   $(2.5,0)$ satisfies the equation:

   $2(2.5) + 0 \overset{?}{=} 5$

   $5 + 0 \overset{?}{=} 5$

   $5 = 5$

   $(-3,1)$ does not satisfy the equation:

   $2(-3) + 1 \overset{?}{=} 5$

   $-6 + 1 \overset{?}{=} 5$

   $-5 \neq 5$

In Exercises 7 – 12, match the given equation with its graph by identifying the x- and y-intercepts. First, let $y = 0$ and solve for x to find the x-intercept. Then, let $x = 0$, and solve for y to find the y-intercept.

7. $4x + 3y = 12$

y-intercept $(x = 0)$:	x-intercept $(y = 0)$:
$4(0) + 3y = 12$	$4x + 3(0) = 12$
$3y = 12$	$4x + 0 = 12$
$y = 4$	$4x = 12$
	$x = 3$

   The intercepts are $(0,4)$ and $(3,0)$, so graph **a** represents this equation.

9. $x + 2y = 8$

y-intercept $(x = 0)$:	x-intercept $(y = 0)$:
$0 + 2y = 8$	$x + 2(0) = 8$
$2y = 8$	$x + 0 = 8$
$y = 4$	$x = 8$

   The intercepts are $(0,4)$ and $(8,0)$, so graph **d** represents this equation.

11. $x + 4y = 0$

y-intercept $(x = 0)$:	x-intercept $(y = 0)$:
$0 + 4y = 0$	$x + 4(0) = 0$
$4y = 0$	$x + 0 = 0$
$y = 0$	$x = 0$

   The only intercept is $(0,0)$, so graph **f** represents this equation.

For Exercises 13 – 46, your points may differ from those given here, but the graph should be the same.

13. $y = 2x$ Points: $(0,0),(1,2),(3,6)$

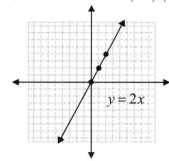

15. $y = -x$ Points: $(-1,1),(0,0),(3,-3)$

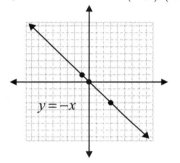

17. $y = x - 4$ Points: $(0,-4),(4,0),(6,2)$

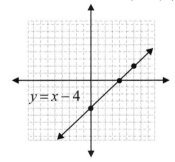

19. $y = x + 2$ Points: $(0,2),(-2,0),(3,5)$

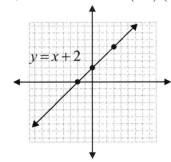

21. $y = 4 - x$ Points: $(0,4),(4,0),(3,1)$

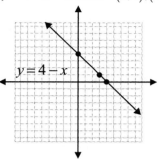

23. $y = 2x - 1$ Points: $(-1,-3),(1,1),(3,5)$

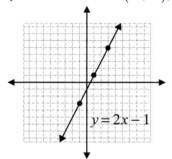

25. $3y = 12$ Points: $(-6,4),(0,4),(3,4)$
A horizontal line:

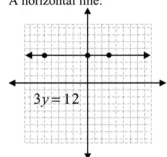

27. $2x - 8 = 0$ Points: $(4,0),(4,-2),(4,6)$
A vertical line:

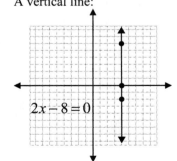

29. $x - 2y = 4$ Points: $(-2, -3), (0, -2), (4, 0)$

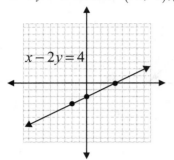

$x - 2y = 4$

31. $2x + y = 0$ Points: $(0, 0), (-1, 2), (-3, 6)$

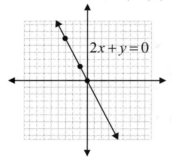

$2x + y = 0$

33. $5y = 0$ Points: $(-1, 0), (0, 0), (3, 0)$
A horizontal line, in particular, the x-axis:

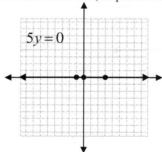

$5y = 0$

35. $4x = 0$ Points: $(0, 0), (0, -2), (0, 5)$
A vertical line, in particular, the y-axis:

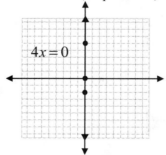

$4x = 0$

37. $2x + 3y = 7$ Points: $(-4, 5), (-1, 3), (2, 1)$

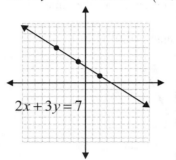

$2x + 3y = 7$

39. $3y - 2x = 4$ Points: $(-2, 0), (1, 2), (4, 4)$

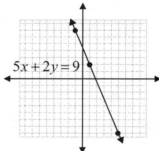

$3y - 2x = 4$

41. $5x + 2y = 9$ Points: $(1, 2), (-1, 7), (5, -8)$

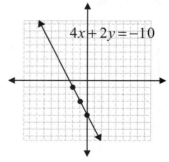

$5x + 2y = 9$

43. $4x + 2y = -10$ Points: $(0, -5), (-1, -3), (-2, -1)$

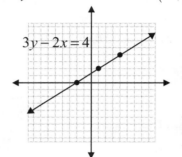

$4x + 2y = -10$

45. $y = \dfrac{1}{3}x - 3$ Points: $(0,-3),(9,0),(3,-2)$

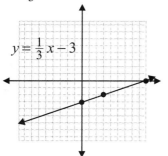

47. $x + y = 4$ Intercepts: $(4,0),(0,4)$

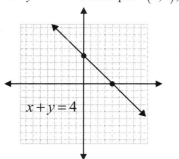

49. $3x - 2y = 6$ Intercepts: $(2,0),(0,-3)$

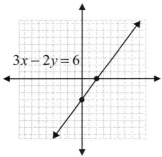

51. $5x + 2y = 10$ Intercepts: $(2,0),(0,5)$

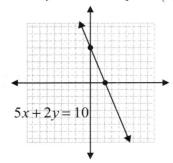

53. $2x - y = 9$ Intercepts: $\left(\dfrac{9}{2},0\right),(0,-9)$

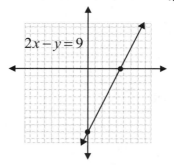

55. $x + 3y = 5$ Intercepts: $(5,0),\left(0,\dfrac{5}{3}\right)$

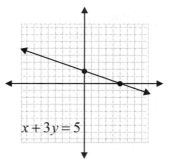

57. $\dfrac{1}{2}x - y = 4$ Intercepts: $(8,0),(0,-4)$

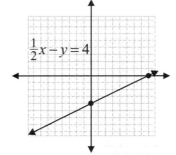

59. $\dfrac{1}{2}x - \dfrac{3}{4}y = 6$ Intercepts: $(12,0),(0,-8)$

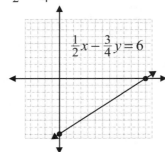

61. $2x + 3y = 5$ Intercepts: $\left(\dfrac{5}{2}, 0\right), \left(0, \dfrac{5}{3}\right)$

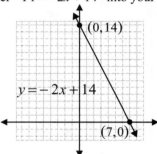

63. $2x + y = 14$

$y = -2x + 14$

Enter "Y1 = −2x + 14" into your calculator.

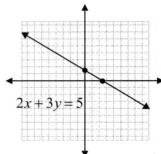

65. $x - 3y = 9$

$-3y = 9 - x$

$y = -3 + \dfrac{x}{3}$

Enter "Y1 = − 3 + x/3" into your calculator.

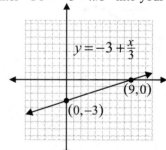

67. $x + y = 5$

$y = 5 - x$

Enter "Y1 = 5 − x" into your calculator.

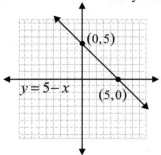

69. $3y - 15 = 0$

$3y = 15$

$y = 5$

Enter "Y1 = 5" into your calculator.

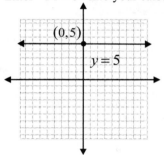

71. $5x - 2y = 10$

$-2y = 10 - 5x$

$y = -5 + \dfrac{5x}{2}$

Enter "Y1 = −5 + 5x/2" into your calculator.

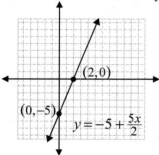

Writing and Thinking About Mathematics

73. $y = x^2$

Points: $(0,0), (1,1), (-1,1), (2,4), (-2,4),$
$(3,9), (-3,9)$

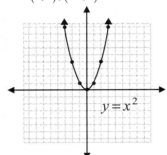

75. $y = x^2 - 5$

Points: $(0,-5), (1,-4), (-1,-4), (2,-1), (-2,-1),$
$(3,4), (-3,4)$

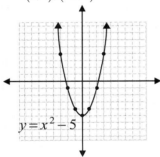

[End of Section 4.2]

Section 4.3
Solutions to Odd Exercises

For Exercises 1 – 20, use $m = \dfrac{y_2 - y_1}{x_2 - x_1}$ to calculate the slope from the given pair of points.

1. $m = \dfrac{-1-4}{2-1} = \dfrac{-5}{1} = -5$

3. $m = \dfrac{-1-7}{-3-4} = \dfrac{-8}{-7} = \dfrac{8}{7}$

5. $m = \dfrac{9-8}{5-(-3)} = \dfrac{1}{8}$

7. $m = \dfrac{8-5}{4-1} = \dfrac{3}{3} = 1$

9. $m = \dfrac{3-0}{-5-0} = \dfrac{3}{-5} = -\dfrac{3}{5}$

11. $m = \dfrac{1-0}{5-(-2)} = \dfrac{1}{7}$

13. $m = \dfrac{4-0}{4-(-4)} = \dfrac{4}{8} = \dfrac{1}{2}$

15. $m = \dfrac{-14-4}{6-(-3)} = \dfrac{-18}{9} = -2$

17. $m = \dfrac{\dfrac{1}{2}-2}{-1-4} = \dfrac{-\dfrac{3}{2}}{-5} = \dfrac{3}{10}$

19. $m = \dfrac{-\dfrac{3}{4}-3}{\dfrac{1}{2}-\dfrac{7}{2}} = \dfrac{-\dfrac{15}{4}}{-\dfrac{6}{2}} = \dfrac{30}{24} = \dfrac{5}{4}$

For Exercises 21 – 40, use the slope-intercept form of a linear equation, $y = mx + b$, substituting the given values of m and b in the equation.

21. $y = \dfrac{2}{3}x$

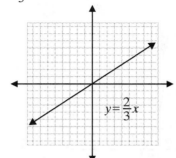

23. $y = -\dfrac{3}{4}x - 3$

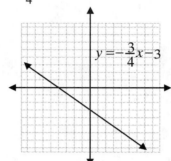

25. $y = -\dfrac{5}{3}x + 3$

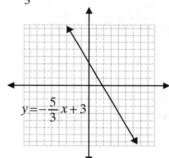

27. $y = 2x - 1$

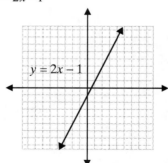

29. $y = -7$

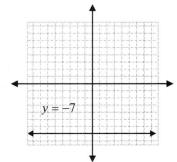

31. $y = \dfrac{3}{2}x - 4$

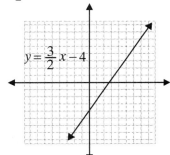

33. $y = x + 5$

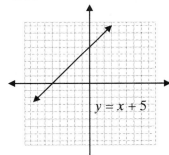

35. $y = \dfrac{2}{5}x - 6$

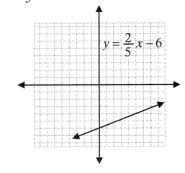

37. $y = \dfrac{4}{3}x - 2$

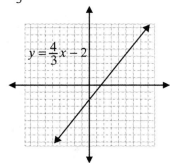

39. $y = -\dfrac{1}{3}x$

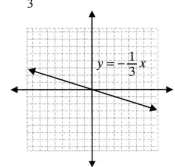

For Exercises 41 − 60, find the slope from the slope-intercept form of the line. First, solve for *y*. The resulting coefficient of *x* is the slope, and the constant term with the sign is the *y*-axis intercept. The line can be graphed by knowing only the *y*-intercept and slope.

41. $y = 2x - 3$

Slope: $m = 2$
y-intercept: $b = -3$

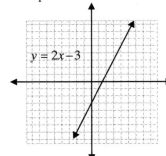

43. $y = 2 - x$
   Slope: $m = -1$
   y-intercept: $b = 2$

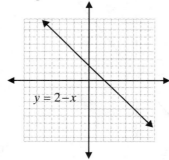

45. $x + y = 5$
   $\quad\quad y = 5 - x$
   Slope: $m = -1$
   y-intercept: $b = 5$

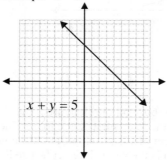

47. $3x - y = 4$
   $\quad\quad y = 3x - 4$
   Slope: $m = 3$
   y-intercept: $b = -4$

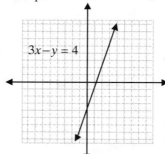

49. $x - 3y = 9$
   $\quad\quad y = \dfrac{1}{3}x - 3$

   Slope: $m = \dfrac{1}{3}$

   y-intercept: $b = -3$

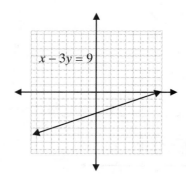

51. $3x + 2y = 6$

   $\quad\quad y = -\dfrac{3}{2}x + 3$

   Slope: $m = -\dfrac{3}{2}$

   y-intercept: $b = 3$

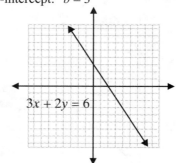

53. $4x - 3y = -3$

   $\quad\quad y = \dfrac{4}{3}x + 1$

   Slope: $m = \dfrac{4}{3}$

   y-intercept: $b = 1$

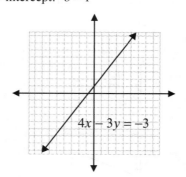

55. $6x + 5y = -15$

$$y = -\frac{6}{5}x - 3$$

Slope: $m = -\dfrac{6}{5}$

y-intercept: $b = -3$

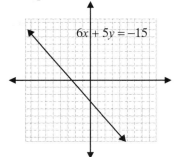

57. $x + 4y = 6$

$$y = -\frac{1}{4}x + \frac{3}{2}$$

Slope: $m = -\dfrac{1}{4}$

y-intercept: $b = \dfrac{3}{2}$

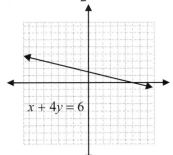

59. $-3x + 6y = 4$

$$y = \frac{1}{2}x + \frac{2}{3}$$

Slope: $m = \dfrac{1}{2}$

y-intercept: $b = \dfrac{2}{3}$

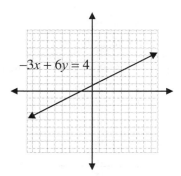

For Exercises 61– 70, find points on the line by substituting any value for one variable into the equation, then solving for the remaining variable. The line can be graphed by knowing only two points. Determine the slope by examining the graph.

61. $y + 5 = 0$, or $y = -5$

This is a horizontal line, so the slope is 0.

$(0, -5)$ and $(-3, -5)$ are two points on the line.

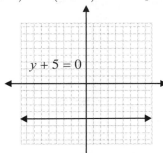

63. $x = -1$

This is a vertical line, so the slope is undefined.

$(-1, 0)$ and $(-1, -2)$ are two points on the line.

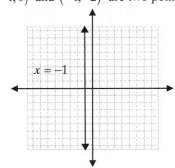

65. $2y = -16$, or $y = -8$

This is a horizontal line, so the slope is 0.

$(7, -8)$ and $\left(\dfrac{2}{3}, -8\right)$ are two points on the line.

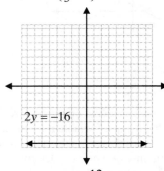

67. $5y - 12 = 0$, or $y = \dfrac{12}{5}$

This is a horizontal line, so the slope is 0. $\left(0, \dfrac{12}{5}\right)$

and $\left(3, \dfrac{12}{5}\right)$ are two points on the line.

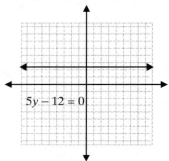

69. $1.5x = -3$, or $x = -2$

This is a vertical line, so the slope is undefined.

$(-2, 0)$ and $(-2, 2)$ are two points on the line.

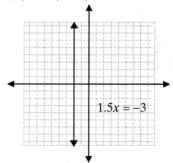

Writing and Thinking About Mathematics

71. Answers will vary. A grade of 12% means the incline, or slope, of the road is 0.12. That is, for every 100 feet of horizontal distance (run), there are 12 feet of vertical distance (rise).

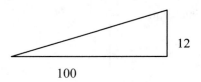

[End of Section 4.3]

108

Section 4.4
Solutions to Odd Exercises

1. Use the slope-intercept form first, with $m = 2$ and $b = 3$:
$$y = 2x + 3$$
$$-2x + y = 3$$
Notice the final equation is in standard form.

3. Use the point-slope form first:
$$y - 3 = -\frac{2}{5}(x - (-1))$$
$$5(y - 3) = -2(x + 1)$$
$$5y - 15 = -2x - 2$$
$$2x + 5y = 13 \quad \text{or} \quad y = -\frac{2}{5}x + \frac{13}{2}$$

5. Use the point-slope form first:
$$y - 2 = \frac{3}{4}(x - (-3))$$
$$4(y - 2) = 3(x + 3)$$
$$4y - 8 = 3x + 9$$
$$-3x + 4y = 17 \quad \text{or} \quad y = \frac{3}{4}x + \frac{17}{4}$$

7. Use the point-slope form first:
$$y - (-1) = 0(x - 3)$$
$$y + 1 = 0$$
$$y = -1$$

9. Since the slope is undefined, the line is vertical with an equation of the form $x = a$. Since $(2, -4)$ is on the line, the equation is $x = 2$.

11. Use the point-slope form first:
$$y - \frac{2}{3} = \frac{3}{2}(x - 1)$$
$$6\left(y - \frac{2}{3}\right) = 6\left(\frac{3}{2}(x - 1)\right)$$
$$6y - 4 = 9x - 9$$
$$-9x + 6y = -5 \quad \text{or} \quad y = \frac{3}{2}x - \frac{5}{6}$$

For Exercises 13 – 24, calculate the slope from the two points, and then find the equation using point-slope form:

13. $m = \dfrac{3 - 2}{-2 - 1} = \dfrac{1}{-3} = -\dfrac{1}{3}$

$$y - 2 = -\frac{1}{3}(x - 1)$$
$$-3(y - 2) = x - 1$$
$$-3y + 6 = x - 1$$
$$7 = x + 3y \quad \text{or} \quad y = -\frac{1}{3}x + \frac{7}{3}$$

15. $m = \dfrac{4 - 2}{3 - 6} = \dfrac{2}{-3} = -\dfrac{2}{3}$
$$y - 4 = -\frac{2}{3}(x - 3)$$
$$3(y - 4) = -2(x - 3)$$
$$3y - 12 = -2x + 6$$
$$2x + 3y = 18 \quad \text{or} \quad y = -\frac{2}{3}x + 6$$

17. $m = \dfrac{2 - 2}{-1 - 8} = \dfrac{0}{-9} = 0$
The line is horizontal: $y = 2$

19. $m = \dfrac{4 - 2}{1 - (-5)} = \dfrac{2}{6} = \dfrac{1}{3}$
$$y - 4 = \frac{1}{3}(x - 1)$$
$$3(y - 4) = x - 1$$
$$3y - 12 = x - 1$$
$$-x + 3y = 11 \quad \text{or} \quad y = \frac{1}{3}x + \frac{11}{3}$$

21. Notice that the x-values for the two given points are the same. So, this line does not have a defined slope. The line is vertical with equation $x = -2$.

23. $m = \dfrac{2 - \dfrac{1}{2}}{-4 - 1} = \dfrac{\dfrac{3}{2}}{-5} = -\dfrac{3}{10}$
$$y - 2 = -\frac{3}{10}(x - (-4))$$
$$10(y - 2) = -3(x + 4)$$
$$10y - 20 = -3x - 12$$
$$3x + 10y = 8 \quad \text{or} \quad y = -\frac{3}{10}x + \frac{4}{5}$$

25. A horizontal line has an equation of the form $y = b$. Since $(-2, 5)$ is on the line, the equation is $y = 5$.

27. A vertical line has an equation of the form $x = a$. Since $(-1, -3)$ is on the line, the equation is $x = -1$.

29. A horizontal line has an equation of the form $y = b$. Since $(-1, -6)$ is on the line, the equation is $y = -6$.

31. The given line is vertical, and any line parallel to it must also be vertical. Since the requested line contains $(0, 3)$, it has the equation $x = 0$.

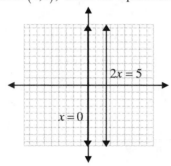

33. The slope of the given line is $\dfrac{3}{4}$, and any line perpendicular to that line must have a slope of $-\dfrac{4}{3}$. Find the equation using point-slope form:

$$y - 5 = -\frac{4}{3}(x - 2) = -\frac{4}{3}x - \frac{8}{3}$$
$$y = -\frac{4}{3}x + \frac{23}{3}$$

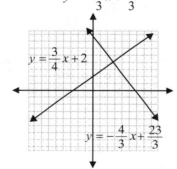

35. The slope of the given line is $-\dfrac{2}{3}$, and any line perpendicular to that line must have a slope of $\dfrac{3}{2}$. Find the equation using point-slope form:

$$y - (-2) = \frac{3}{2}(x - 7) = \frac{3}{2}x - \frac{21}{2}$$
$$y = \frac{3}{2}x - \frac{25}{2}$$

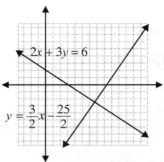

37. The two lines are not parallel, because they have different slopes of 5 and $\dfrac{1}{5}$. They are not perpendicular, either, because the slopes are only reciprocals of each other; they are not *negative* reciprocals.

39. For a line to be parallel to $2x - y = 7$, it must have a slope of 2. To have the same y-intercept as $x - 3y = 6$, it needs a y-intercept at $(0, -2)$. The equation of the line is $y = 2x - 2$.

41. The cost C of the car rental for x miles driven is:
$$C = 0.15x + 30$$

43. **a.** $I = 0.09x + 200$, where I is the weekly income and x is the weekly sales.

  b.

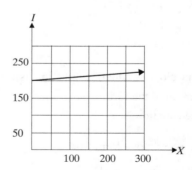

  c. The slope, 0.09, is the rate of weekly earnings with respect to weekly sales. The slope is the percentage of sales that is equal to earnings. For example, for every $100 in sales, the clerk receives an income of $9.

45. a. Use the two given points $(7, 1000)$ and $(6, 1200)$, where price is the first coordinate and attendance is the second coordinate, to find the slope of the line.
$$m = \frac{1200 - 1000}{6 - 7} = -200$$

Use one of the points to write the equation. Let A be attendance and p be price.

$$A - 1200 = -200(p - 6)$$
$$A = -200p + 1200 + 1200$$
$$A = -200p + 2400$$

b. From above, $m = -200$.

c. The slope is the rate of change in attendance relative to the price of admission. Here, the slope indicates a decrease of 200 people for every \$1 increase in the admission price.

d. Since p represents the admission price, negative values for p would not have a practical meaning.

47. a. Find the slope using the given points:
$$m = \frac{2-1}{-1-3} = \frac{1}{-4} = -\frac{1}{4}$$
Use point-slope form to find the equation:
$$y - 1 = -\frac{1}{4}(x - 3)$$
$$4(y - 1) = -(x - 3)$$
$$4y - 4 = -x + 3$$
$$x + 4y = 7$$
Alternatively written in slope intercept form:
$$y = -\frac{1}{4}x + \frac{7}{4}$$

b. Enter "Y1 = $-(1/4)$X$+(7/4)$" into your calculator.

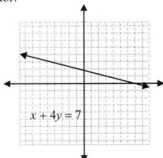

$x + 4y = 7$

49. a. Use the slope-intercept form, where $m = -\frac{1}{2}$ and $b = 0$, to write the equation:
$$y = -\frac{1}{2}x + 0$$
$$y = -\frac{1}{2}x$$

b. Enter "Y1 = $-(1/2)$X" into your calculator.

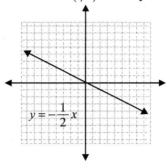

$y = -\frac{1}{2}x$

51. a. Use the point-slope form to write the equation:
$$y - (-4) = \frac{3}{2}(x - (-3))$$
$$2(y + 4) = 3(x + 3)$$
$$2y + 8 = 3x + 9$$
$$-3x + 2y = 1$$
Alternatively written in slope-intercept form:
$$y = \frac{3}{2}x + \frac{1}{2}$$

b. Enter "Y1 = $(3/2)$X$+(1/2)$" into your calculator.

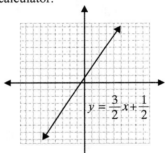

$y = \frac{3}{2}x + \frac{1}{2}$

[End of Section 4.4]

Section 4.5
Solutions to Odd Exercises

1. **a.** $\{(-4,0),(-2,4),(1,2),(2,5),(6,-3)\}$
    b. $D = \{-4,-2,1,2,6\}$
          $R = \{-3,0,2,4,5\}$
    c. It is a function.

3. **a.** $\{(-5,-4),(-4,-2),(-2,-2),(1,-2),(2,1)\}$
    b. $D = \{-5,-4,-2,1,2\}$
          $R = \{-4,-2,1\}$
    c. It is a function.

5. **a.** $\{(-4,1),(-4,-2),(-1,-1),(-1,3),(3,-4)\}$
    b. $D = \{-4,-1,3\}$
          $R = \{-4,-2,-1,1,3\}$
    c. It is not a function.

7. **a.** $\{(-5,-5),(-5,3),(0,5),(1,2),(1,-2)\}$
    b. $D = \{-5,0,1\}$
          $R = \{-5,-2,2,3,5\}$
    c. It is not a function.

9. **a.** $\{(-5,5),(-3,1),(0,-3),(3,-1),(4,2)\}$
    b. $D = \{-5,-3,0,3,4\}$
          $R = \{-3,-1,1,2,5\}$
    c. It is a function.

11. It is a function.

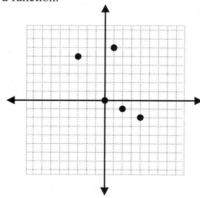

13. It is a function.

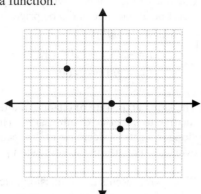

15. It is a function.

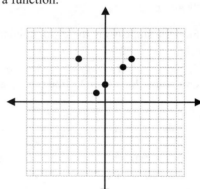

17. It is a function.

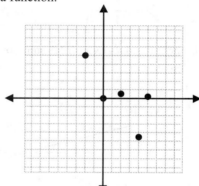

19. It is a function.

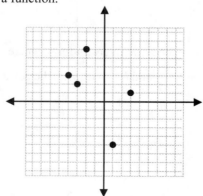

21. The vertical line test indicates that this graph represents a function: $D = -6 \le x \le 6$

$$R = 0 \le y \le 6$$

23. The vertical line test indicates that this graph represents a function: $D = -6 < x \le 6$

$$R = \{-6, -4, -2, 0, 2, 4\}$$

25. The graph does not represent a function, as the domain value $x = 0$ is assigned to $y = 0$, $y = 3$, and $y = 7$. Other x-values also have been assigned multiple values of y.

27. The graph does not represent a function. For example, the domain value $x = 1$ is assigned to two values of y.

29. The vertical line test shows that this graph represents a function: $D =$ All real numbers

$$R = -1.5 \le y \le 1.5$$

31. The graph does not represent a function. For example, the domain value $x = 2$ is assigned to two values of y.

33. Given $f(x) = 2x + 1$:

$$f(-2) = 2(-2) + 1 = -4 + 1 = -3$$
$$f(-1) = 2(-1) + 1 = -2 + 1 = -1$$
$$f(0) = 2(0) + 1 = 0 + 1 = 1$$
$$f(1) = 2(1) + 1 = 2 + 1 = 3$$
$$f(5) = 2(5) + 1 = 10 + 1 = 11$$

35. Given $g(x) = -4x + 2$:

$$g(-2) = -4(-2) + 2 = 8 + 2 = 10$$
$$g(-1) = -4(-1) + 2 = 4 + 2 = 6$$
$$g(0) = -4(0) + 2 = 0 + 2 = 2$$
$$g(1) = -4(1) + 2 = -4 + 2 = -2$$
$$g(5) = -4(5) + 2 = -20 + 2 = -18$$

37. Given $h(x) = x^2$:

$$h(-2) = (-2)^2 = 4$$
$$h(-1) = (-1)^2 = 1$$
$$h(0) = (0)^2 = 0$$
$$h(1) = (1)^2 = 1$$
$$h(5) = (5)^2 = 25$$

39. Given $F(x) = x^2 - 4x + 4$:

$$F(-2) = (-2)^2 - 4(-2) + 4 = 4 + 8 + 4 = 16$$
$$F(-1) = (-1)^2 - 4(-1) + 4 = 1 + 4 + 4 = 9$$
$$F(0) = (0)^2 - 4(0) + 4 = 0 + 0 + 4 = 4$$
$$F(1) = (1)^2 - 4(1) + 4 = 1 - 4 + 4 = 1$$
$$F(5) = (5)^2 - 4(5) + 4 = 25 - 20 + 4 = 9$$

41. Given $H(x) = x^3 - 8x$:

$$H(-2) = (-2)^3 - 8(-2) = -8 + 16 = 8$$
$$H(-1) = (-1)^3 - 8(-1) = -1 + 8 = 7$$
$$H(0) = (0)^3 - 8(0) = 0 - 0 = 0$$
$$H(1) = (1)^3 - 8(1) = 1 - 8 = -7$$
$$H(5) = (5)^3 - 8(5) = 125 - 40 = 85$$

43. Given $P(x) = 0.5x^2 - 2x + 3.5$:

$$P(-2) = 0.5(-2)^2 - 2(-2) + 3.5$$
$$= 2 + 4 + 3.5 = 9.5$$
$$P(-1) = 0.5(-1)^2 - 2(-1) + 3.5$$
$$= 0.5 + 2 + 3.5 = 6$$
$$P(0) = 0.5(0)^2 - 2(0) + 3.5$$
$$= 0 - 0 + 3.5 = 3.5$$
$$P(1) = 0.5(1)^2 - 2(1) + 3.5$$
$$= 0.5 - 2 + 3.5 = 2$$
$$P(5) = 0.5(5)^2 - 2(5) + 3.5$$
$$= 12.5 - 10 + 3.5 = 6$$

45. Given $f(x) = x^3 - 5x^2 + 7x - 2$:

$$f(-2) = (-2)^3 - 5(-2)^2 + 7(-2) - 2$$
$$= -8 - 20 - 14 - 2 = -44$$
$$f(-1) = (-1)^3 - 5(-1)^2 + 7(-1) - 2$$
$$= -1 - 5 - 7 - 2 = -15$$
$$f(0) = (0)^3 - 5(0)^2 + 7(0) - 2$$
$$= 0 - 0 + 0 - 2 = -2$$
$$f(1) = (1)^3 - 5(1)^2 + 7(1) - 2$$
$$= 1 - 5 + 7 - 2 = 1$$
$$f(5) = (5)^3 - 5(5)^2 + 7(5) - 2$$
$$= 125 - 125 + 35 - 2 = 33$$

[End of Section 4.5]

Section 4.6
Solutions to Odd Exercises

1. $y > 3x$

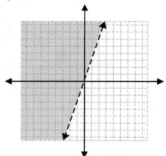

3. $y \leq 2x - 1$

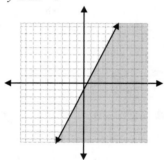

5. $y > -7$

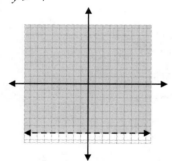

7. $2x - 3 \leq 0$

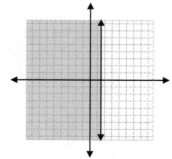

9. $5x - 2y \geq 4$

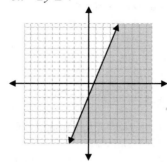

11. $y < -\dfrac{1}{4}x$

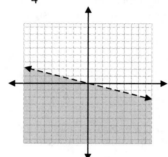

13. $2x - y < 1$

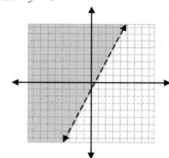

15. $2x - 3y \geq -3$

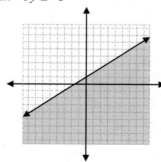

17. $2x + 5y \geq 10$

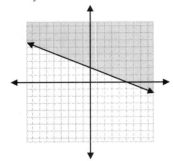

25. $4y \geq 3x - 8$

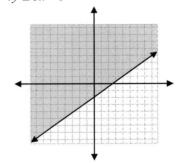

19. $3x < 4 - y$

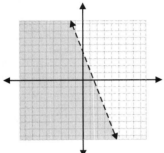

27. $2x + \dfrac{1}{3}y < 2$

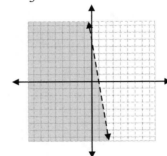

21. $2x - 3y < 9$

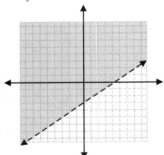

29. $\dfrac{1}{2}x + \dfrac{2}{3}y \geq \dfrac{5}{6}$

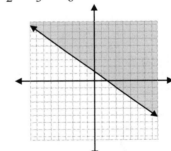

23. $5x - 2y \geq 4$

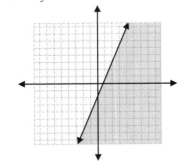

31. $y > 2x$

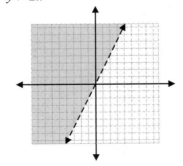

33. $x + 2y > 7$

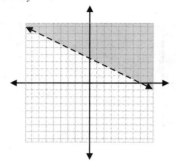

35. $2x - y \leq 6$

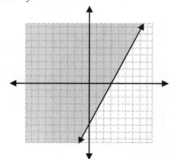

37. $x - 3y \geq 5$

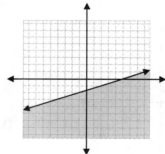

39. $2x + 4y < 9$

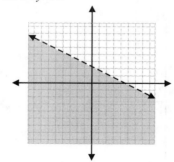

[End of Section 4.6]

Chapter 4 Review
Solutions to All Exercises

1. $\{A(-6,-5), B(-5,6), C(1,-3), D(3,6), E(7,-1)\}$

2. $\{A(-4,-5), B(-3,1), C(1,2), D(1,-1), E(5,3)\}$

3. $\{A(-7,2), B(-1,7), C(-1,-3), D(3,-1), E(4,-6)\}$

4. $\{A(-2,3), B(2,-3), C(3,-4), D(4,-5), E(5,4)\}$

5. $\{A(-3,1), B(-2,-2), C(0,-5), D(1,4)\}$

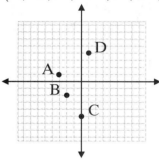

6. $\{E(-3,-4), F(0,0), G(5,-1), H(1.5,-3)\}$

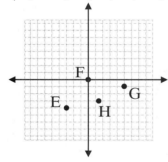

7. $\{M(6,2), N(3,-1), O(0,-2), P(-4,-5)\}$

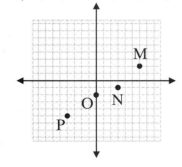

8. $\{Q(-4,-2), R(-4,0), S(-4,3), T(-4,5)\}$

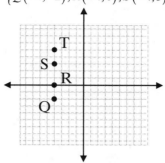

In Exercises 9–12, substitute each point into the given equation. If the substitution gives a true statement, then the point satisfies the equation.

9. $3x - y = -2$

   a. $(1,1)$:　$3(1) - 1 \overset{?}{=} -2$

   $2 \neq -2$

   b. $(-1,1)$:　$3(-1) - 1 \overset{?}{=} -2$

   $-4 \neq -2$

   c. $(2,2)$:　$3(2) - 2 \overset{?}{=} -2$

   $4 \neq -2$

   d. $(-2,2)$:　$3(-2) - 2 \overset{?}{=} -2$

   $-8 \neq -2$

   None of the points satisfy the equation.

10. $x + 2y = 3$

   a. $(-3,1)$:　$-3 + 2(1) \overset{?}{=} 3$

   $-1 \neq 3$

   b. $(2,1)$:　$2 + 2(1) \overset{?}{=} 3$

   $4 \neq 3$

   c. $\left(2, \frac{1}{2}\right)$:　$2 + 2\left(\frac{1}{2}\right) \overset{?}{=} 3$

   $3 = 3$

   d. $(1,-2)$:　$1 + 2(-2) \overset{?}{=} 3$

   $-3 \neq 3$

   The point **c** satisfies the equation.

11. $y = 4x$

    a. $(1, 4)$: $4 \overset{?}{=} 4(1)$

                     $4 = 4$

    b. $(-1, 4)$: $4 \overset{?}{=} 4(-1)$

                    $4 \neq -4$

    c. $\left(\dfrac{1}{4}, 1\right)$: $1 \overset{?}{=} 4\left(\dfrac{1}{4}\right)$

                    $1 = 1$

    d. $(-3, -12)$: $-12 \overset{?}{=} 4(-3)$

                    $-12 = -12$

The points **a**, **c**, and **d** satisfy the equation.

12. $2x - 3y = 5$

    a. $(1, -1)$: $2(1) - 3(-1) \overset{?}{=} 5$

                          $5 = 5$

    b. $\left(2, -\dfrac{1}{3}\right)$: $2(2) - 3\left(-\dfrac{1}{3}\right) \overset{?}{=} 5$

                          $5 = 5$

    c. $(2.5, 0)$: $2(2.5) - 3(0) \overset{?}{=} 5$

                          $5 = 5$

    d. $(0, -2)$: $2(0) - 3(-2) \overset{?}{=} 5$

                          $6 \neq 5$

The points **a**, **b**, and **c** satisfy the equation.

13. $y = 5x$

x	y
0	0
$-\dfrac{1}{5}$	**−1**
−2	−10
1	**5**

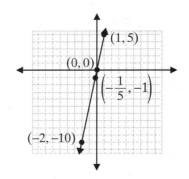

14. $y = 6 - 3x$

x	y
0	6
2	**0**
−1	9
$\dfrac{17}{9}$	$\dfrac{1}{3}$

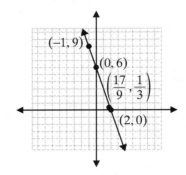

15. $y = 8 - 2x$

x	y
0	8
4	**0**
3	2
$\dfrac{17}{4}$	$-\dfrac{1}{2}$

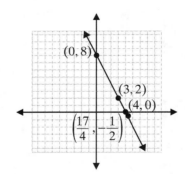

16. $3x + y = -5$

x	y
0	**−5**
$-\dfrac{4}{3}$	**−1**
−3	**4**
−1.5	**−0.5**

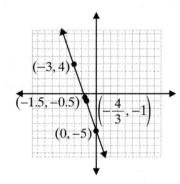

In Exercises 17–20, match the given equation with its graph. Let $y = 0$ and solve for x to find the x-intercept, then let $x = 0$ and solve for y to find the y-intercept.

17. $y = 2x + 4$

y-intercept $(x = 0)$: x-intercept $(y = 0)$:

$$y = 2(0) + 4 \qquad\qquad 0 = 2x + 4$$
$$y = 0 + 4 \qquad\qquad -4 = 2x$$
$$y = 4 \qquad\qquad -2 = x$$

Intercepts: $(0, 4)$ and $(-2, 0)$, so graph **c**.

18. $y = 6 - x$

y-intercept $(x = 0)$: x-intercept $(y = 0)$:

$$y = 6 - 0 \qquad\qquad 0 = 6 - x$$
$$y = 6 \qquad\qquad x = 6$$

Intercepts: $(0, 6)$ and $(6, 0)$, so graph **b**.

19. $y = 3 - 3x$

y-intercept $(x = 0)$: x-intercept $(y = 0)$:

$$y = 3 - 3(0) \qquad\qquad 0 = 3 - 3x$$
$$y = 3 - 0 \qquad\qquad 3x = 3$$
$$y = 3 \qquad\qquad x = 1$$

Intercepts: $(0, 3)$ and $(1, 0)$, so graph **d**.

20. $-5x + y = 10$

y-intercept $(x = 0)$: x-intercept $(y = 0)$:

$$-5(0) + y = 10 \qquad\qquad -5x + 0 = 10$$
$$0 + y = 10 \qquad\qquad -5x = 10$$
$$y = 10 \qquad\qquad x = -2$$

Intercepts: $(0, 10)$ and $(-2, 0)$, so graph **a**.

21. $5x + y = 1.5$

y-intercept $(x = 0)$: x-intercept $(y = 0)$:

$$5(0) + y = 1.5 \qquad\qquad 5x + 0 = 1.5$$
$$0 + y = 1.5 \qquad\qquad 5x = 1.5$$
$$y = 1.5 \qquad\qquad x = 0.3$$

Intercepts: $(0, 1.5)$ and $(0.3, 0)$.

22. $3x - y = 7.5$

y-intercept $(x = 0)$: x-intercept $(y = 0)$:

$$3(0) - y = 7.5 \qquad\qquad 3x - 0 = 7.5$$
$$0 - y = 7.5 \qquad\qquad 3x = 7.5$$
$$y = -7.5 \qquad\qquad x = 2.5$$

Intercepts: $(0, -7.5)$ and $(2.5, 0)$.

23. $x + 2y = 3$

y-intercept $(x = 0)$: x-intercept $(y = 0)$:

$$0 + 2y = 3 \qquad\qquad x + 2(0) = 3$$
$$2y = 3 \qquad\qquad x + 0 = 3$$
$$y = \frac{3}{2} \qquad\qquad x = 3$$

Intercepts: $\left(0, \dfrac{3}{2}\right)$ and $(3, 0)$.

24. $6x - 2y = 7$

y-intercept $(x = 0)$: x-intercept $(y = 0)$:

$$6(0) - 2y = 7 \qquad\qquad 6x - 2(0) = 7$$
$$0 - 2y = 7 \qquad\qquad 6x - 0 = 7$$
$$-2y = 7 \qquad\qquad 6x = 7$$
$$y = -\frac{7}{2} \qquad\qquad x = \frac{7}{6}$$

Intercepts: $\left(0, -\dfrac{7}{2}\right)$ and $\left(\dfrac{7}{6}, 0\right)$.

For Exercises 25–30, your points may differ from those given here, but the graph should be the same.

25. $y = -\dfrac{1}{2}x + 4$ Points: $(0,4),(8,0)$

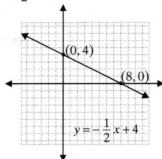

26. $y = \dfrac{2}{3}x$ Points: $(0,0),(3,2)$

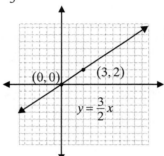

27. $x = 7$ Points: $(7,0),(7,1)$

A vertical line:

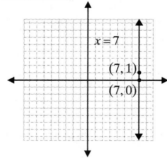

28. $y = -3.5$ Points: $(0,-3.5),(-3,-3.5)$

A horizontal line:

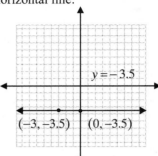

29. $-10 = y$ Points: $(4,-10),(0,-10)$

A horizontal line:

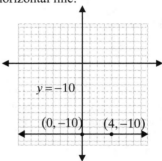

30. $5x + y = 10$ Points: $(0,10),(2,0)$

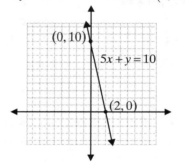

31. $m = \dfrac{5-(-2)}{1-2} = \dfrac{7}{-1} = -7$

32. $m = \dfrac{1-(-2)}{-4-3} = \dfrac{3}{-7} = -\dfrac{3}{7}$

33. $m = \dfrac{-2-3}{-5-1} = \dfrac{-5}{-6} = \dfrac{5}{6}$

34. $m = \dfrac{1-6}{-6-(-1)} = \dfrac{-5}{-5} = 1$

35. $m = \dfrac{4-0}{0-8} = \dfrac{4}{-8} = -\dfrac{1}{2}$

36. $m = \dfrac{3-13}{0-5} = \dfrac{-10}{-5} = 2$

37. $m = \dfrac{\dfrac{1}{2} - \left(-\dfrac{1}{6}\right)}{\dfrac{3}{4} - \dfrac{1}{4}} = \dfrac{\dfrac{3}{6} + \dfrac{1}{6}}{\dfrac{2}{4}} = \dfrac{\dfrac{4}{6}}{\dfrac{1}{2}} = \dfrac{8}{6} = \dfrac{4}{3}$

38. $m = \dfrac{1.8-2.3}{2.6-3.6} = \dfrac{-0.5}{-1} = 0.5$

For Exercises 39–42, use the slope intercept form of a linear equation and the given values of m and b to write the equation.

39. $y = \dfrac{5}{8}x - 1$

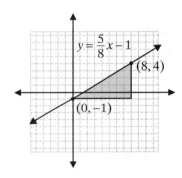

40. $y = -x - 4$

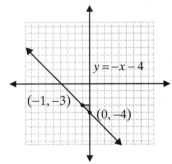

41. $y = 0 \cdot x + 3 = 3$

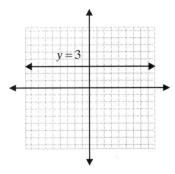

42. $y = \dfrac{2}{3}x + 5$

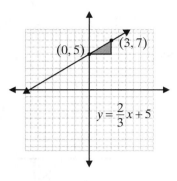

For Exercises 43–50, find the slope from the slope-intercept form of the line. First, solve for y. The resulting coefficient of x is the slope, and the constant term with the sign is the y-axis intercept. The line can be graphed directly by knowing the y-intercept and slope.

43. $y = 2x - 6$

Slope: $\qquad m = 2$

y-intercept: $b = -6$

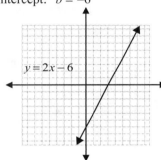

44. $y = -3x + 4$

Slope: $\qquad m = -3$

y-intercept: $b = 4$

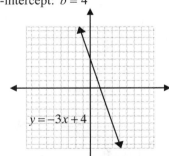

45. $x + y = 1.5$

$\qquad y = -x + 1.5$

Slope: $\qquad m = -1$

y-intercept: $b = 1.5$

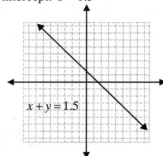

46. $x - 2y = 10$

$\qquad y = \dfrac{1}{2}x - 5$

Slope: $\qquad m = \dfrac{1}{2}$

y-intercept: $b = -5$

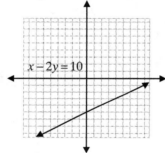

47. $x - y = 8$

$\qquad y = x - 8$

Slope: $\qquad m = 1$

y-intercept: $b = -8$

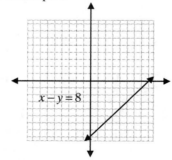

48. $2x - y = 12$

$\qquad y = 2x - 12$

Slope: $\qquad m = 2$

y-intercept: $b = -12$

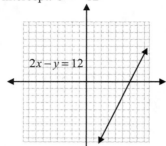

49. $x + 4y = 2$

$\qquad y = -\dfrac{1}{4}x + \dfrac{1}{2}$

Slope: $\qquad m = -\dfrac{1}{4}$

y-intercept: $b = \dfrac{1}{2}$

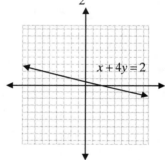

50. $y = 2 - 3x$

Slope: $\qquad m = -3$

y-intercept: $b = 2$

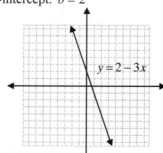

51. Use point-slope form:

$$y - 5 = 3(x - 1)$$

$$y - 5 = 3x - 3$$

$$y = 3x + 2$$

52. Use point-slope form:
$$y - 4 = -1(x - (-1))$$
$$y - 4 = -x - 1$$
$$y = 3 - x$$

53. Use point-slope form:
$$y - 2 = \frac{1}{5}(x - (-3))$$
$$y - 2 = \frac{1}{5}x + \frac{3}{5}$$
$$y = \frac{1}{5}x + \frac{13}{5}$$

54. Use point-slope form:
$$y - (-4) = \frac{2}{3}(x - 2)$$
$$y + 4 = \frac{2}{3}x - \frac{4}{3}$$
$$y = \frac{2}{3}x - \frac{16}{3}$$

55. Undefined slope indicates the line is vertical.
$$x = 5$$

56. Use point-slope form:
$$y - (-7) = 0(x - (-2))$$
$$y + 7 = 0$$
$$y = -7$$

For Exercises 57–62, use the two points to calculate the slope. Then, use one of the two points to find the equation using point-slope form:

57. $m = \dfrac{8 - 3}{1 - 2} = \dfrac{5}{-1} = -5$
$$y - 8 = -5(x - 1)$$
$$y - 8 = 5 - 5x$$
$$y = 13 - 5x$$

58. $m = \dfrac{4 - (-1)}{2 - 5} = \dfrac{5}{-3} = -\dfrac{5}{3}$
$$y - 4 = -\frac{5}{3}(x - 2)$$
$$y - 4 = -\frac{5}{3}x + \frac{10}{3}$$
$$y = -\frac{5}{3}x + \frac{22}{3}$$

59. $m = \dfrac{-2 - 0}{0 - 5} = \dfrac{-2}{-5} = \dfrac{2}{5}$
$$y - (-2) = \frac{2}{5}(x - 0)$$
$$y + 2 = \frac{2}{5}x$$
$$y = \frac{2}{5}x - 2$$

60. $m = \dfrac{1 - 8}{-4 - 0} = \dfrac{-7}{-4} = \dfrac{7}{4}$
$$y - 1 = \frac{7}{4}(x - (-4))$$
$$y - 1 = \frac{7}{4}x + 7$$
$$y = \frac{7}{4}x + 8$$

61. $m = \dfrac{-2 - 1}{-5 - 3} = \dfrac{-3}{-8} = \dfrac{3}{8}$
$$y - (-2) = \frac{3}{8}(x - (-5))$$
$$y + 2 = \frac{3}{8}x + \frac{15}{8}$$
$$y = \frac{3}{8}x - \frac{1}{8}$$

62. $m = \dfrac{6 - (-3)}{-5 - 4} = \dfrac{9}{-9} = -1$
$$y - 6 = -1(x - (-5))$$
$$y - 6 = -x - 5$$
$$y = -x + 1$$

63. The slope of a horizontal line is 0. The requested line contains the point $(-3, 4)$.
The equation for the line is $y = 4$.

64. The x-axis is a horizontal line, so the slope must be 0. The requested line contains the point $(-2, 5)$.
The equation is $y = 5$.

65. The slope of a vertical line is undefined, and the requested line contains the point $(5, -1)$.
The equation is $x = 5$.

66. The slope of a horizontal line is 0. The requested line contains the point $(0, -2)$.
The equation for the line is $y = -2$.

67. The slope of the given line is $\frac{1}{4}$, and any line parallel to that line must have the same slope. Find the equation using point-slope form:

$$y - 2 = \frac{1}{4}\left(x - \frac{5}{2}\right)$$

$$y - 2 = \frac{1}{4}x - \frac{5}{8}$$

$$y = \frac{1}{4}x + \frac{11}{8}$$

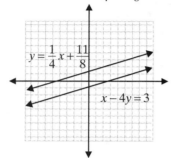

68. The slope of the given line is -3, and any line parallel to it must have the same slope. Find the equation using point-slope form:

$$y - (-1) = -3\left(x - (-6)\right)$$

$$y + 1 = -3x - 18$$

$$y = -3x - 19$$

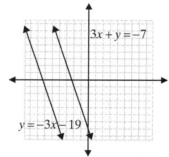

69. The slope of the given line is 2 and any line perpendicular to it must have a slope of $-\frac{1}{2}$. Notice that the given point lies on the y-axis, so you can use slope intercept form to write the equation:

$$y = -\frac{1}{2}x + 1$$

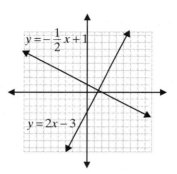

70. The slope of the given line is $-\frac{1}{2}$, and any line perpendicular to it must have a slope of 2. Find the equation using point-slope form.

$$y - 0 = 2\left(x - (-5)\right)$$

$$y = 2x + 10$$

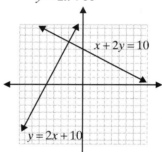

71. The relation is not a function; the domain element 0 has more than one corresponding range element.

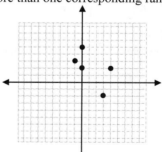

72. The relation is not a function; the domain element -2 has more than one corresponding range element.

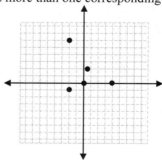

73. The relation is a function.

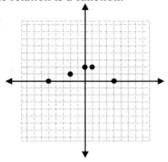

74. The relation is a function.

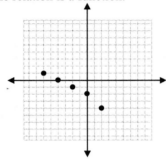

75. The vertical line test indicates that this graph represents a function: $D = -5 \le x \le 5$
$$R = -6 \le y \le 6$$

76. The graph does not represent a function. For example, the domain value $x = 7$ corresponds to $y = 5$ and $y = -5$.

77. The vertical line test indicates that this graph represents a function: $D = -8 \le x \le 8$
$$R = -3 \le y \le 3$$

78. The vertical line test indicates that this graph represents a function: $D = -6 \le x \le 6$
$$R = -6 \le y \le 4$$

79. The graph does not represent a function. For example, the domain value $x = 1$ corresponds to $y = 2$ and $y = 6$.

80. The vertical line test indicates that this graph represents a function:
$$D = -8 \le x \le -2 \text{ and } -1 \le x \le 8$$
$$R = -7 \le y \le 4$$

81. $f(x) = 3x - 10$
$$f(-3) = 3(-3) - 10 = -19$$
$$f(-1) = 3(-1) - 10 = -13$$
$$f(0) = 3(0) - 10 = -10$$

$$f(2) = 3(2) - 10 = -4$$
$$f(3) = 3(3) - 10 = -1$$

82. $g(x) = 6 - 4x$
$$g(-3) = 6 - 4(-3) = 18$$
$$g(-1) = 6 - 4(-1) = 10$$
$$g(0) = 6 - 4(0) = 6$$
$$g(2) = 6 - 4(2) = -2$$
$$g(3) = 6 - 4(3) = -6$$

83. $h(x) = x^2 + 4$
$$h(-3) = (-3)^2 + 4 = 13$$
$$h(-1) = (-1)^2 + 4 = 5$$
$$h(0) = (0)^2 + 4 = 4$$
$$h(2) = (2)^2 + 4 = 8$$
$$h(3) = (3)^2 + 4 = 13$$

84. $f(x) = x^2 - 8x + 7$
$$f(-3) = (-3)^2 - 8(-3) + 7 = 40$$
$$f(-1) = (-1)^2 - 8(-1) + 7 = 16$$
$$f(0) = (0)^2 - 8(0) + 7 = 7$$
$$f(2) = (2)^2 - 8(2) + 7 = -5$$
$$f(3) = (3)^2 - 8(3) + 7 = -8$$

85. $F(x) = 3x^2 + 4x + 1$
$$F(-3) = 3(-3)^2 + 4(-3) + 1 = 16$$
$$F(-1) = 3(-1)^2 + 4(-1) + 1 = 0$$
$$F(0) = 3(0)^2 + 4(0) + 1 = 1$$
$$F(2) = 3(2)^2 + 4(2) + 1 = 21$$
$$F(3) = 3(3)^2 + 4(3) + 1 = 40$$

86. $C(x) = x^2 + 4x$
$$C(-3) = (-3)^2 + 4(-3) = -3$$
$$C(-1) = (-1)^2 + 4(-1) = -3$$
$$C(0) = (0)^2 + 4(0) = 0$$
$$C(2) = (2)^2 + 4(2) = 12$$
$$C(3) = (3)^2 + 4(3) = 21$$

87. $y \leq 2x + 1$

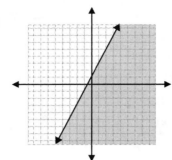

92. $y < \frac{2}{3}x + 6$

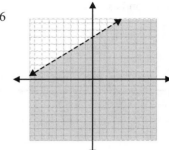

88. $y < -3$

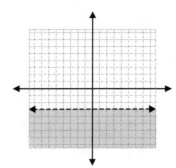

93. $\frac{2}{5}x + y > 5$

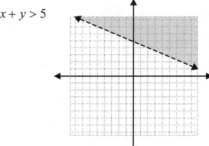

89. $y > -4$

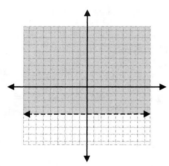

94. $2x + 6 > 2y$

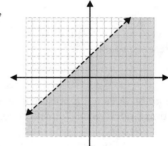

90. $y \geq 3x$

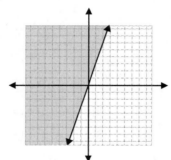

95. $y > 4x$

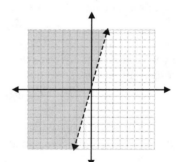

91. $y \geq \frac{1}{4}x - 2$

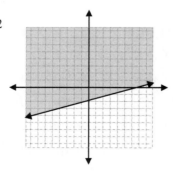

96. $x - y \leq 6$

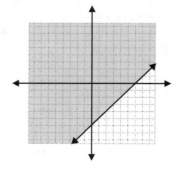

97. $2x + 4y > 0$

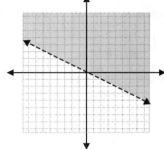

98. $2x + y \leq 3$

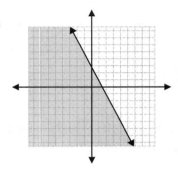

99. $y \geq 4$

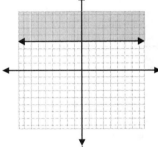

100. $x - 3y < 9$

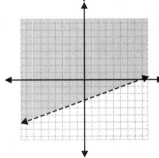

[End of Chapter 4 Review]

Chapter 4 Test
Solutions to All Exercises

1. Points on the line $x - 4y = 7$:

   a. $(2, -2)$: $2 - 4(-2) \overset{?}{=} 7$

                       $10 \neq 7$

   b. $(-1, -2)$: $-1 - 4(-2) \overset{?}{=} 7$

                       $7 = 7$

   c. $\left(5, -\dfrac{1}{2}\right)$: $5 - 4\left(-\dfrac{1}{2}\right) \overset{?}{=} 7$

                       $7 = 7$

   d. $(0, 7)$: $0 - 4(7) \overset{?}{=} 7$

                       $-28 \neq 7$

The points **b** and **c** lie on the given line.

2. $\{A(-3, 2), B(-2, -1), C(0, -3), D(1, 2), E(3, 4),$
$F(5, 0), G(4, -2)\}$

3. $L(0, 2), M(4, -1), N(-3, 2), O(-1, -5), P\left(2, \dfrac{3}{2}\right)$

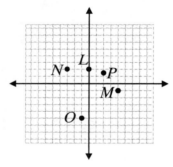

4. $2x - y = 4$

      $y = 2x - 4$

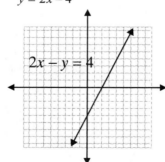

5. $3x + 4y = 9$

   y-intercept $(x = 0)$: x-intercept $(y = 0)$:

     $3(0) + 4y = 9$ $3x + 4(0) = 9$

         $4y = 9$ $3x + 0 = 9$

          $y = \dfrac{9}{4}$ $x = 3$

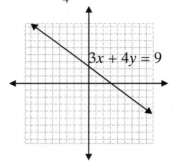

6. Plot the point $(1, -5)$ and use the given slope to graph the rest of the line:

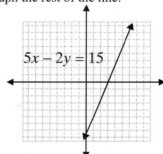

7. $3x + 5y = -15$

   Write in slope-intercept form:

   $y = -\dfrac{3}{5}x - 3$; $m = -\dfrac{3}{5}$ and $b = -3$

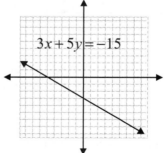

8. $m = \dfrac{-3 - \dfrac{1}{2}}{4 - 2} = \dfrac{-\dfrac{7}{2}}{2} = -\dfrac{7}{4}$

9. Use the point-slope form:
$$y - 2 = \frac{1}{4}\left(x - (-3)\right)$$
$$4(y - 2) = x + 3$$
$$4y - 8 = x + 3$$
$$-x + 4y = 11 \quad \text{or} \quad y = \frac{1}{4}x + \frac{11}{4}$$

10. Use the two points to find the slope, then select one of the points and use the point-slope form of an equation:
$$m = \frac{6 - 1}{-2 - 3} = \frac{5}{-5} = -1$$
$$y - 1 = -1(x - 3)$$
$$y - 1 = -x + 3$$
$$y = -x + 4$$

11. A line parallel to the x-axis will be horizontal, so the slope is 0. Point-slope form gives the equation:
$$y - (-2) = 0(x - 3)$$
$$y + 2 = 0$$
$$y = -2$$

12. **a.** $C = 4t + 9$

   b. C and t only make sense if they are positive because negative values don't have an interpretation as cost and time.

   c. The slope indicates the rate of change in cost per hour. In this situation, the slope indicates that for each additional hour, there is an additional $4 in cost.

13. If the lines are parallel, their slopes are equal. The slope of line $2x + 3y = 6$ is easily determined after solving for y: $y = -\frac{2}{3}x + 2$. The slope is $m = \frac{-2}{3}$.

   The slope of the line $y = -\frac{2}{3}x + 1$ is also $\frac{-2}{3}$.

   Therefore, the lines are parallel.

14. Use the point-slope form. The slope of the given line is 2. Since the slope of the requested line is the same, the point-slope form gives
$$y - (-6) = 2(x - 1)$$
$$y + 6 = 2x - 2$$
$$y = 2x - 8$$

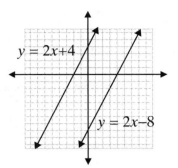

15. The slope of the given line is -1, so the slope of a perpendicular line must be the negative reciprocal of -1 which is 1. Use point-slope form:
$$y - 2 = 1(x - 4)$$
$$y - 2 = x - 4$$
$$y = x - 2$$

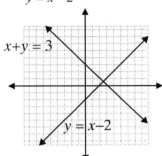

16. $D = \{-1, 0, 2, 5\}$
$R = \{-3, 4, 6\}$

17. The relation is not a function because the domain value 2 is assigned to both -5 and 0 in the range.

18. For $f(x) = 3x - 4$:

   a. $f(-1) = 3(-1) - 4 = -7$

   b. $f(6) = 3(6) - 4 = 14$

19. For $h(x) = 3x^2 - 5$:

   a. $h(-3) = 3(-3)^2 - 5 = 27 - 5 = 22$

   b. $h(2) = 3(2)^2 - 5 = 12 - 5 = 7$

20. It is a function:
$$D = \{-6, -4, -3, 3, 7\}$$
$$R = \{-4, -3, 0, 1, 3\}$$

21. It is a function:

$$D : -2 \leq x \leq 2$$
$$R : 0 \leq y \leq 4$$

22. The graph does not represent a function, as the domain value $x = 0$ is assigned to $y = 5$, $y = 0$ and $y = -5$.

23. It is a function.

$$D : \text{All real numbers}$$
$$R : 0 \leq y$$

24. Graph $y > -2x$.

An open half-plane:

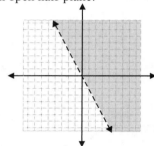

25. Graph $x \leq 5$.

A closed half-plane:

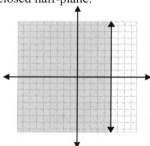

[End of Chapter 4 Test]

Cumulative Review: Chapters 1 – 4
Solutions to All Exercises

1. $3x + 45 = 3(x + 15)$

2. $6x + 16 = 2(3x + 8)$

3. $\text{LCM}(12, 15, 54):$
$12 = 2^2 \cdot 3$
$15 = 3 \cdot 5$
$54 = 2 \cdot 3^3$
$\text{LCM} = 2^2 \cdot 3^3 \cdot 5 = 540$

4. $\text{LCM}(6a^2,\ 24ab^2,\ 30ab,\ 40a^2b):$
$6a^2 = 2 \cdot 3 \cdot a^2$
$24ab^3 = 2^3 \cdot 3 \cdot a \cdot b^2$
$30ab = 2 \cdot 3 \cdot 5 \cdot a \cdot b$
$40a^2b^2 = 2^3 \cdot 5 \cdot a^2 \cdot b$
$\text{LCM} = 2^3 \cdot 3 \cdot 5 \cdot a^2 \cdot b^2 = 120a^2b^2$

5. The average of the set is:
$$\frac{25 + 35 + 42}{3} = \frac{102}{3} = 34$$

6. The average of the set is:
$$\frac{3.6 + 8.9 + 14.7 + 25.3}{4} = \frac{52.5}{4} = 13.125$$

7. The average test scores is:
$$\frac{80 + 88 + 82}{3} = \frac{250}{3} \approx 83.3$$

8. The change in altitude is:
$8000 - 12000 = -4000$ feet

9. False; $-15 < 5$

10. True, since $|-6| = 6$ and $6 \geq 6$.

11. False; $\dfrac{7}{8} \geq \dfrac{7}{10}$

12. True, since all negative numbers are less than 0.

13. $7(4 - x) = 3(x + 4)$
$28 - 7x = 3x + 12$
$-10x = -16$
$x = \dfrac{8}{5}$

14. $-2(5x + 1) + 2x = 4(x + 1)$
$-10x - 2 + 2x = 4x + 4$
$-8x - 2 = 4x + 4$
$-6 = 12x$
$x = -\dfrac{1}{2}$

15. $\dfrac{2}{3}x + \dfrac{1}{2} = \dfrac{3}{4}$
$12\left(\dfrac{2}{3}x + \dfrac{1}{2}\right) = 12\left(\dfrac{3}{4}\right)$
$8x + 6 = 9$
$8x = 3$
$x = \dfrac{3}{8}$

16. $1.5x - 3.7 = 3.6x + 2.6$
$-2.1x = 6.3$
$x = -3$

17. $C = \pi d$
$\dfrac{C}{\pi} = d$ or $d = \dfrac{C}{\pi}$

18. $3x + 5y = 10$
$5y = -3x + 10$
$y = \dfrac{10 - 3x}{5}$

19. $P = 2l + 2w$
$P - 2l = 2w$
$\dfrac{P - 2l}{2} = w$

20. $V = \dfrac{1}{3}\pi r^2 h$
$3V = \pi r^2 h$
$\dfrac{3V}{\pi r^2} = h$

21. $5x + 3 \geq 2x - 15$

$\quad\quad 3x \geq -18$

$\quad\quad\; x \geq -6$

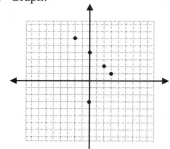

22. $-16 < 3x + 5 < 17$

$\quad -21 < \;\; 3x \;\; < 12$

$\quad\; -7 < \;\;\; x \;\;\; < 4$

23. $\quad\quad \dfrac{1}{2}x + 5 \leq \dfrac{2}{3}x - 7$

$\quad 6\left(\dfrac{1}{2}x + 5\right) \leq 6\left(\dfrac{2}{3}x - 7\right)$

$\quad\quad 3x + 30 \leq 4x - 42$

$\quad\quad\quad\quad 72 \leq x$

24. $\quad -15 < \dfrac{3}{4}x - 9 \leq \dfrac{1}{3}$

$\quad\quad -6 < \;\; \dfrac{3}{4}x \;\; \leq \dfrac{28}{3}$

$\quad\quad -24 < \;\; 3x \;\; \leq \dfrac{112}{3}$

$\quad\quad\; -8 < \;\;\; x \;\;\; \leq \dfrac{112}{9}$

25. $|x| < 5$:

26. $|x| > 6$:

27. This is a function.

$\quad D = \{-3, -2, -1, 1, 3\}$

$\quad R = \{0, 1, 3\}$

28. This is not a function. The domain elements –1 and 4 each have more than one corresponding range element.

$\quad D = \{-3, -1, 1, 4\}$

$\quad R = \{-2, -1, 1, 3, 4\}$

29. This is not a function. The domain element 2 has more than one corresponding range element.

$\quad D = \{-2, 0, 2, 3, 4\}$

$\quad R = \{-1, 0, 1, 2, 3, 4\}$

30. This is a function.

$\quad D : -5 < x \leq -1 \quad \text{and} \quad 0 < x \leq 6$

$\quad R = \{-4, -2, 2, 4, 5\}$

31. a. Graph:

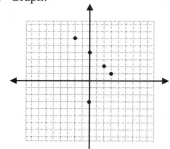

b. $D = \{-2, 0, 2, 3\}$

$\quad R = \{-3, 1, 2, 4, 6\}$

c. This is not a function. The domain element 0 corresponds to more than one range element.

32. a. Graph:

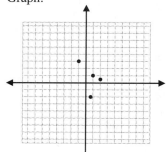

b. $D = \left\{-1, 1, \dfrac{2}{3}, 2\right\}$

$\quad R = \left\{-2, 1, \dfrac{1}{2}, 3\right\}$

c. This is a function.

33. Test each point in $4x - y = 7$:

   a. $(2,1)$: $4(2) - 1 \overset{?}{=} 7$

                       $7 = 7$

   b. $(3,5)$: $4(3) - 5 \overset{?}{=} 7$

                       $7 = 7$

   c. $(1,-3)$: $4(1) - (-3) \overset{?}{=} 7$

                       $7 = 7$

   d. $(0,7)$: $4(0) - 7 \overset{?}{=} 7$

                      $-7 \neq 7$

Points **a**, **b** and **c** lie on the line.

34. Test each point in $2x + 5y = 6$:

   a. $\left(\dfrac{1}{2},1\right)$: $2\left(\dfrac{1}{2}\right) + 5(1) \overset{?}{=} 6$

                       $6 = 6$

   b. $(3,0)$: $2(3) + 5(0) \overset{?}{=} 6$

                       $6 = 6$

   c. $\left(0,\dfrac{6}{5}\right)$: $2(0) + 5\left(\dfrac{6}{5}\right) \overset{?}{=} 6$

                       $6 = 6$

   d. $(-2,2)$: $2(-2) + 5(2) \overset{?}{=} 6$

                       $6 = 6$

All four points lie on the line.

35. Graph: $y = -4x$

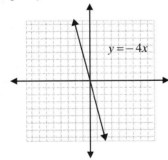

36. Graph: $x + 2y = 4$

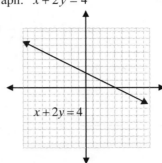

37. Graph: $x + 4 = 0$

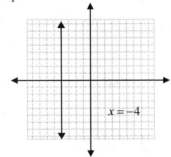

38. Graph: $2y - 6 = 0$

          $2y = 6$

          $y = 3$

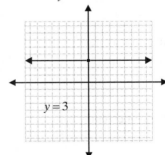

39. Graph: $2x + y = 4$

   y-intercept: $(0,4)$ x-intercept: $(2,0)$

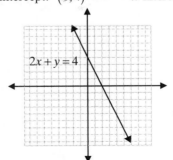

40. Graph: $5x - y = 1$

y-intercept: $(0, -1)$ x-intercept: $\left(\dfrac{1}{5}, 0\right)$

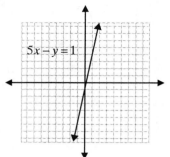

41. Solve the equation for y:
$$y - 2x = 3$$
$$y = 2x + 3$$
$m = 2$ and the y-intercept is 3.

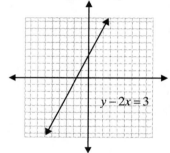

42. Solve the equation for y:
$$2x + 5y = 10$$
$$y = -\frac{2}{5}x + 2$$
$m = \dfrac{-2}{5}$ and the y-intercept is 2.

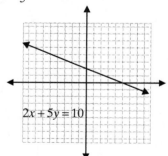

43. Use the two points to find the slope. Then, select one of the two points and use the point-slope form:
$$m = \frac{6-4}{-1-2} = \frac{2}{-3} = -\frac{2}{3}$$
$$y - 4 = -\frac{2}{3}(x-2)$$
$$3(y-4) = -2(x-2)$$
$$3y - 12 = -2x + 4$$
$$2x + 3y = 16 \quad \text{or} \quad y = -\frac{2}{3}x + \frac{16}{3}$$

44. Use the two points to find the slope. Then, select one of the two points and use the point-slope form:
$$m = \frac{2-(-2)}{3-6} = \frac{4}{-3} = -\frac{4}{3}$$
$$y - 2 = -\frac{4}{3}(x-3)$$
$$3(y-2) = -4(x-3)$$
$$3y - 6 = -4x + 12$$
$$4x + 3y = 18 \quad \text{or} \quad y = -\frac{4}{3}x + 6$$

45. Use point-slope form:
$$y - (-1) = 0(x-6)$$
$$y + 1 = 0$$
$$y = -1$$

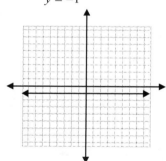

46. Use point-slope form:

$$y - \frac{5}{2} = -\frac{1}{4}(x - 3)$$

$$4\left(y - \frac{5}{2}\right) = 4\left(-\frac{1}{4}(x - 3)\right)$$

$$4y - 10 = -x + 3$$

$$x + 4y = 13 \quad \text{or} \quad y = -\frac{1}{4}x + \frac{13}{4}$$

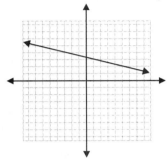

47. Slope is undefined for vertical lines. The line contains $(-3, 4)$, so the equation is $x = -3$.

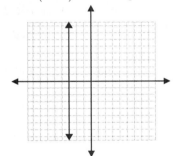

48. Use point-slope form:

$$y - (-5) = \frac{3}{2}(x - 4)$$

$$2y + 10 = 3x - 12$$

$$-3x + 2y = -22 \quad \text{or} \quad y = \frac{3}{2}x - 11$$

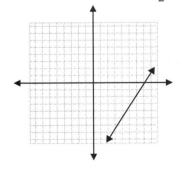

49. Any line parallel to $2x + y = 5$ has slope of –2. Using the point-slope form of a line:

$$y - (-2) = -2(x - 1)$$

$$y + 2 = -2x + 2$$

$$2x + y = 0$$

50. The line has slope equal to the slope of the first line, which is $\frac{1}{2}$. The y-intercept of the second line is $(0, 3)$. So, the new equation is $y = \frac{1}{2}x + 3$, or $x - 2y = -6$.

51. The line has slope equal to the negative reciprocal of the given line $3x + y = 9$, so the slope of the new line must be $\frac{1}{3}$. The given point is actually the y-intercept, so the equation of the line is $y = \frac{1}{3}x + 5$.

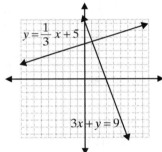

52. The line has slope equal to the negative reciprocal of the given line $y = 3 - x$. So the slope of the new line must be 1. Using the point $(-1, 6)$, the equation of the line is $y - 6 = 1(x + 1)$, or $y = x + 7$.

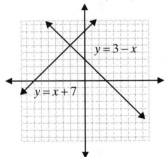

53. The half-plane is closed.

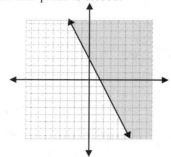

54. The half-plane is open.

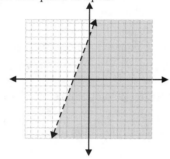

55. For $f(x) = -3x + 4$,

    a. $f(5) = -3(5) + 4 = -11$

    b. $f(-6) = -3(-6) + 4 = 22$

56. For $F(x) = 2x^2 - 5x + 1$,

    a. $F(2) = 2(2)^2 - 5(2) + 1$

           $= 8 - 10 + 1$

           $= -1$

    b. $F(-5) = 2(-5)^2 - 5(-5) + 1$

           $= 50 + 25 + 1$

           $= 76$

57. Let x be the smallest of the three consecutive even integers. Then, $x + 2$ and $x + 4$ are the second and third even integers, respectively.

$$(x + 2) + (x + 4) = 3(x) - 14$$
$$2x + 6 = 3x - 14$$
$$x = 20$$
$$x + 2 = 22$$
$$x + 4 = 24$$

The even integers are 20, 22, and 24.

58. Let w be the width. Then $w + 9$ is the length. Using the formula $P = 2w + 2l$, we have:

$$2w + 2(w + 9) = 82$$
$$2w + 2w + 18 = 82$$
$$4w = 64$$
$$w = 16$$
$$w + 9 = 25$$

The width is 16 cm and length is 25 cm.

59. Let x be the number.

$$2(x - 16) = 4x + 6$$
$$2x - 32 = 4x + 6$$
$$-38 = 2x$$
$$x = -19$$

60. Let x be her final exam score.

$$80 \le \frac{x + 93 + 77 + 90 + 86}{5} < 90$$
$$400 \le \quad x + 346 \quad < 450$$
$$54 \le \quad\quad x \quad\quad < 104$$

So, the range of scores over which she could earn a B grade is 54 to 100, inclusively.

61. Let x be number of miles driven. Acme charges $\$0.25x + \$15$ per day. Zenith charges $\$50$ per day.

$$0.25x + 15 < 50$$
$$0.25x < 35$$
$$x < 140$$

The number of miles must be less than 140.

62. Let x be the first odd integer. Then, $x + 2$ and $x + 4$ are the second and third odd integers.

$$x + 3(x + 2) = 2(x + 4) + 28$$
$$x + 3x + 6 = 2x + 8 + 28$$
$$4x + 6 = 2x + 36$$
$$2x = 30$$
$$x = 15$$
$$x + 2 = 17$$
$$x + 4 = 19$$

The odd integers are 15, 17, and 19.

[End of Cumulative Review: Chapters 1 – 4]

Section 5.1
Solutions to Odd Exercises

1. $\begin{cases} x - y = 6 \\ 2x + y = 0 \end{cases}$

   a. $(1, -2)$: $\quad 1 - (-2) \overset{?}{=} 6 \qquad 2(1) + (-2) \overset{?}{=} 0$
                         $3 \ne 6 \qquad\qquad\qquad 0 = 0$

   b. $(4, -2)$: $\quad 4 - (-2) \overset{?}{=} 6 \qquad 2(4) + (-2) \overset{?}{=} 0$
                         $6 = 6 \qquad\qquad\qquad 2 \ne 0$

   c. $(2, -4)$: $\quad 2 - (-4) \overset{?}{=} 6 \qquad 2(2) + (-4) \overset{?}{=} 0$
                         $6 = 6 \qquad\qquad\qquad 0 = 0$

   d. $(-1, 2)$: $\quad -1 - 2 \overset{?}{=} 6 \qquad 2(-1) + 2 \overset{?}{=} 0$
                         $-3 \ne 6 \qquad\qquad\qquad 0 = 0$

The point **c** lies on both lines.

3. $\begin{cases} 2x + 4y - 6 = 0 \\ 3x + 6y - 9 = 0 \end{cases}$

   a. $(1,1)$: $\quad 2(1) + 4(1) - 6 \overset{?}{=} 0 \qquad 3(1) + 6(1) - 9 \overset{?}{=} 0$
                            $0 = 0 \qquad\qquad\qquad\quad 0 = 0$

   b. $(2,0)$: $\quad 2(2) + 4(0) - 6 \overset{?}{=} 0 \qquad 3(2) + 6(0) - 9 \overset{?}{=} 0$
                            $-2 \ne 0 \qquad\qquad\qquad -3 \ne 0$

   c. $\left(0, \dfrac{3}{2}\right)$: $\quad 2(0) + 4\left(\dfrac{3}{2}\right) - 6 \overset{?}{=} 0 \qquad 3(0) + 6\left(\dfrac{3}{2}\right) - 9 \overset{?}{=} 0$
                            $0 = 0 \qquad\qquad\qquad\quad 0 = 0$

   d. $(-1,3)$: $\quad 2(-1) + 4(3) - 6 \overset{?}{=} 0 \qquad 3(-1) + 6(3) - 9 \overset{?}{=} 0$
                            $4 \ne 0 \qquad\qquad\qquad\quad 6 \ne 0$

The points **a** and **c** lie on both lines.

5. $\begin{cases} x + 2y = 4 \\ x - y = -2 \end{cases}$

   By inspecting the graph, $(0, 2)$ is the solution.

   Check: $\begin{cases} 0 + 2(2) = 4 \\ 0 - 2 = -2 \end{cases}$

7. $\begin{cases} 2x - y = 6 \\ 3x + y = 14 \end{cases}$

   By inspecting the graph, $(4, 2)$ is the solution.

   Check: $\begin{cases} 2(4) - 2 = 6 \\ 3(4) + 2 = 14 \end{cases}$

9. A system is inconsistent when it has no solution and the lines are not the same. This happens when their slopes are equal and their y-intercepts are different. Find the slopes of the lines to determine whether they are parallel.

$$2x + y = 3 \rightarrow y = -2x + 3$$

$$4x + 2y = 5 \rightarrow y = -2x + \frac{5}{2}$$

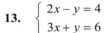

Since $m = -2$ in both equations, the lines are parallel. They are different because $b = 3$ in the first, and $b = \dfrac{5}{2}$ in the second.

11. Find the slopes of the lines to determine whether they are parallel.

$$3x - y = 8 \rightarrow y = 3x - 8$$

$$x - \frac{1}{3}y = 2 \rightarrow y = 3x - 6$$

The lines are parallel, as $m = 3$ in both equations. They are different because $b = -8$ in the first, and $b = -6$ in the second.

13. $\begin{cases} 2x - y = 4 \\ 3x + y = 6 \end{cases}$

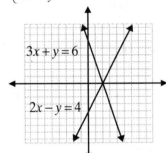

$(2, 0)$ is the solution.
The system is consistent.

15. $\begin{cases} 3x - y = 6 \\ \quad\;\; y = 3x \end{cases}$

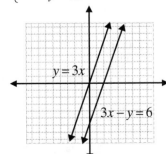

No solution. The lines are parallel.
The system is inconsistent.

17. $\begin{cases} x + 2y = 8 \\ 3x - 2y = 0 \end{cases}$

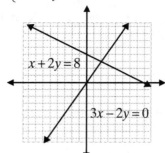

$(2,3)$ is the solution.
The system is consistent.

19. $\begin{cases} 4x - 2y = 10 \\ -6x + 3y = -15 \end{cases}$

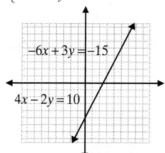

The equations represent the same line.
The system is dependent.

21. $\begin{cases} y = 2x + 5 \\ 4x - 2y = 7 \end{cases}$

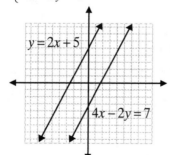

No solution. The lines are parallel.
The system is inconsistent.

23. $\begin{cases} 4x + 3y + 7 = 0 \\ 5x - 2y + 3 = 0 \end{cases}$

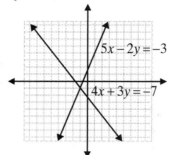

$(-1,-1)$ is the solution.
The system is consistent

25. $\begin{cases} 7x - 2y = 1 \\ y = 3 \end{cases}$

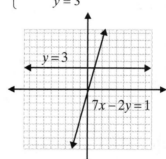

$(1,3)$ is the solution.
The system is consistent.

27. $\begin{cases} y = 4x - 3 \\ x = 2y - 8 \end{cases}$

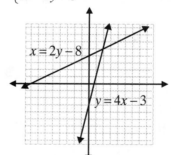

$(2,5)$ is the solution.
The system is consistent.

29. $\begin{cases} y = 7 \\ x = 8 \end{cases}$

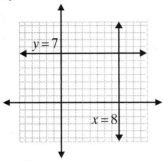

$(8,7)$ is the solution.

The system is consistent.

31. $\begin{cases} \dfrac{2}{3}x + y = 2 \\ x - 4y = 3 \end{cases}$

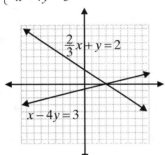

$(3,0)$ is the solution.

The system is consistent.

33. $\begin{cases} x - 2y = 4 \\ x = 4 \end{cases}$

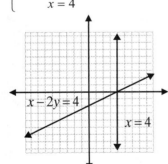

$(4,0)$ is the solution.

The system is consistent.

35. $\begin{cases} x - y = 4 \\ 2y = 2x - 4 \end{cases}$

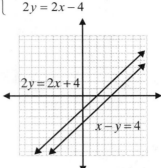

No solution. The lines are parallel.
The system is inconsistent.

37. $\begin{cases} x + y = 4 \\ 2x - 3y = 3 \end{cases}$

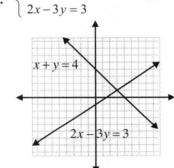

$(3,1)$ is the solution.

The system is consistent.

39. $\begin{cases} 2x + y = 0 \\ 4x + y = -2 \end{cases}$

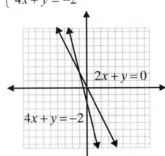

$(-1,2)$ is the solution.

The system is consistent.

41. $\begin{cases} \dfrac{1}{2}x + \dfrac{1}{3}y = \dfrac{1}{6} \\ \dfrac{1}{4}x + \dfrac{1}{4}y = 0 \end{cases}$

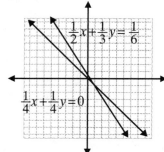

$(1,-1)$ is the solution.

The system is consistent.

43. $\begin{cases} x + y = 25 \\ x - y = 15 \end{cases}$

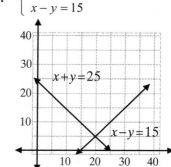

The solution is $(20,5)$.

45. $\begin{cases} x + y = 10 \\ 0.50x + 0.25y = 0.30(10) \end{cases}$

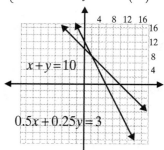

The solution is $(2,8)$.

47. $\begin{cases} x + 2y = 9 \\ x - 2y = -7 \end{cases}$

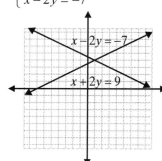

The solution is $(1,4)$.

49. $\begin{cases} y = 2 \\ 2x - 3y = -3 \end{cases}$

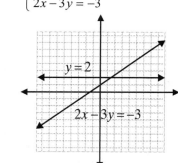

The solution is $\left(\dfrac{3}{2}, 2\right)$.

51. $\begin{cases} y = -3 \\ 2x + y = 0 \end{cases}$

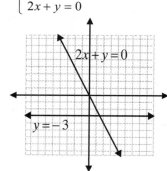

The solution is $\left(\dfrac{3}{2}, -3\right)$.

53. $\begin{cases} x + y = 3.5 \\ -2x + 5y = 7.7 \end{cases}$

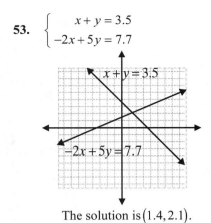

The solution is $(1.4, 2.1)$.

[End of Section 5.1]

Section 5.2
Solutions to Odd Exercises

1. $\begin{cases} x+y=6 \\ \quad y=2x \end{cases}$

$$x+(2x)=6$$
$$3x=6$$
$$x=2$$
$$y=2(2)=4$$

The solution is $(2,4)$.

3. $\begin{cases} x-7=3y \\ \quad y=2x-4 \end{cases}$

$$x-7=3(2x-4)$$
$$x-7=6x-12$$
$$-5x=-5$$
$$x=1$$
$$y=2(1)-4=-2$$

The solution is $(1,-2)$.

5. $\begin{cases} \quad x=3y \\ 3y-2x=6 \end{cases}$

$$3y-2(3y)=6$$
$$3y-6y=6$$
$$-3y=6$$
$$y=-2$$
$$x=3(-2)=-6$$

The solution is $(-6,-2)$.

7. $\begin{cases} x-5y+1=0 \\ \quad x=7-3y \end{cases}$

$$(7-3y)-5y+1=0$$
$$8-8y=0$$
$$-8y=-8$$
$$y=1$$
$$x=7-3(1)=4$$

The solution is $(4,1)$.

9. $\begin{cases} 7x+y=9 \\ \quad y=4-7x \end{cases}$

$$7x+(4-7x)=9$$
$$4=9$$

This statement is never true, so the system is inconsistent. The lines are parallel.

11. $\begin{cases} 3x-y=7 \\ x+y=5 \end{cases}$

Solve the second equation for x:
$$x=5-y$$
Substitute into the first equation:
$$3(5-y)-y=7$$
$$15-3y-y=7$$
$$-4y=-8$$
$$y=2$$
$$x=5-2=3$$

The solution is $(3,2)$.

13. $\begin{cases} 6x+3y=9 \\ \quad y=3-2x \end{cases}$

$$6x+3(3-2x)=9$$
$$6x+9-6x=9$$
$$9=9$$

This statement is always true, so this system is dependent. The equations represent the same line.
The solution is $(x,-2x+3)$.

15. $\begin{cases} \quad x-y=5 \\ 2x+3y=0 \end{cases}$

Solve the first equation for x:
$$x=5+y$$
Substitute into the second equation:
$$2(5+y)+3y=0$$
$$10+2y+3y=0$$
$$10+5y=0$$
$$5y=-10$$
$$y=-2$$
$$x=5+(-2)=3$$

The solution is $(3,-2)$.

17. $\begin{cases} \quad x+y=8 \\ 3x+2y=8 \end{cases}$

Solve the first equation for x:
$$x=8-y$$
Substitute into the second equation:
$$3(8-y)+2y=8$$
$$24-3y+2y=8$$
$$-y=-16$$
$$y=16$$
$$x=8-16=-8$$

The solution is $(-8,16)$.

19. $\begin{cases} 2x + 3y = 5 \\ x - 6y = 0 \end{cases}$

Solve the second equation for x:
$$x = 6y$$
Substitute into the first equation:
$$2(6y) + 3y = 5$$
$$12y + 3y = 5$$
$$15y = 5$$
$$y = \frac{1}{3}$$
$$x = 6\left(\frac{1}{3}\right) = 2$$

The solution is $\left(2, \frac{1}{3}\right)$.

21. $\begin{cases} x + 5y = 1 \\ x - 3y = 5 \end{cases}$

Solve the second equation for x:
$$x = 5 + 3y$$
Substitute into the first equation:
$$(5 + 3y) + 5y = 1$$
$$5 + 8y = 1$$
$$8y = -4$$
$$y = -\frac{1}{2}$$
$$x = 5 + 3\left(-\frac{1}{2}\right)$$
$$x = 5 - \frac{3}{2} = \frac{7}{2}$$

The solution is $\left(\frac{7}{2}, -\frac{1}{2}\right)$.

23. $\begin{cases} 4x - 4y = 9 \\ 3x + y = 2 \end{cases}$

Solve the second equation for y:
$$y = 2 - 3x$$
Substitute into the first equation:
$$4x - 4(2 - 3x) = 9$$
$$4x - 8 + 12x = 9$$
$$16x = 17$$
$$x = \frac{17}{16}$$

$$y = 2 - 3\left(\frac{17}{16}\right)$$
$$y = 2 - \frac{51}{16}$$
$$y = -\frac{19}{16}$$

The solution is $\left(\frac{17}{16}, -\frac{19}{16}\right)$.

25. $\begin{cases} x - 2y = -4 \\ 3x + y = -5 \end{cases}$

Solve the first equation for x:
$$x = 2y - 4$$
Substitute into the second equation:
$$3(2y - 4) + y = -5$$
$$6y - 12 + y = -5$$
$$7y = 7$$
$$y = 1$$
$$x = 2(1) - 4 = -2$$

The solution is $(-2, 1)$.

27. $\begin{cases} 3x - y = -1 \\ 7x - 4y = 0 \end{cases}$

Solve the first equation for y:
$$y = 3x + 1$$
Substitute into the second equation:
$$7x - 4(3x + 1) = 0$$
$$7x - 12x - 4 = 0$$
$$-5x = 4$$
$$x = -\frac{4}{5}$$
$$y = 3\left(-\frac{4}{5}\right) + 1$$
$$y = -\frac{12}{5} + 1 = -\frac{7}{5}$$

The solution is $\left(-\frac{4}{5}, -\frac{7}{5}\right)$.

29. $\begin{cases} x+3y=5 \\ 3x+2y=7 \end{cases}$

Solve the first equation for x:
$$x=5-3y$$
Substitute into the second equation:
$$3(5-3y)+2y=7$$
$$15-9y+2y=7$$
$$-7y=-8$$
$$y=\frac{8}{7}$$
$$x=5-3\left(\frac{8}{7}\right)$$
$$x=5-\frac{24}{7}=\frac{11}{7}$$

The solution is $\left(\dfrac{11}{7},\dfrac{8}{7}\right)$.

31. $\begin{cases} \dfrac{x}{3}+\dfrac{y}{5}=1 \\ x+6y=12 \end{cases}$

Solve the second equation for x:
$$x=12-6y$$
Substitute into the first equation:
$$\frac{(12-6y)}{3}+\frac{y}{5}=1$$
$$4-2y+\frac{y}{5}=1$$
$$-\frac{9}{5}y=-3$$
$$y=-3\left(-\frac{5}{9}\right)=\frac{5}{3}$$
$$x=12-6\left(\frac{5}{3}\right)$$
$$x=12-10=2$$

The solution is $\left(2,\dfrac{5}{3}\right)$.

33. $\begin{cases} 6x-y=15 \\ 0.2x+0.5y=2.1 \end{cases}$

Solve the first equation for y:
$$y=6x-15$$
Substitute into the second equation:
$$0.2x+0.5(6x-15)=2.1$$
$$0.2x+3x-7.5=2.1$$
$$3.2x=9.6$$
$$x=3$$

$$y=6(3)-15$$
$$y=18-15$$
$$y=3$$

The solution is $(3,3)$.

35. $\begin{cases} 0.2x-0.1y=0 \\ y=x+10 \end{cases}$

By substitution:
$$0.2x-0.1(x+10)=0$$
$$0.2x-0.1x-1=0$$
$$0.1x=1$$
$$x=10$$
$$y=10+10$$
$$y=20$$

The solution is $(10,20)$.

37. $\begin{cases} 3x-2y=5 \\ y=\dfrac{3}{2}x+2 \end{cases}$

By substitution:
$$3x-2\left(\frac{3}{2}x+2\right)=5$$
$$3x-3x-4=5$$
$$4=5$$

This statement is never true, so the system is inconsistent. The lines are parallel.

39. $\begin{cases} \dfrac{1}{2}x+\dfrac{1}{3}y=4 \\ 3x+2y=24 \end{cases}$

Solve the first equation for x:
$$\frac{1}{2}x=4-\frac{1}{3}y$$
$$x=2\left(4-\frac{1}{3}y\right)$$
$$x=8-\frac{2}{3}y$$
Substitute into the second equation:
$$3\left(8-\frac{2}{3}y\right)+2y=24$$
$$24-2y+2y=24$$
$$24=24$$

This statement is always true, so this system is dependent. The equations represent the same line.

The solution is $\left(x,-\dfrac{3}{2}x+12\right)$.

41. $\begin{cases} \dfrac{1}{4}x - \dfrac{3}{2}y = 5 \\ -x + 6y = 20 \end{cases}$

Solve the second equation for x:

$$x = 6y - 20$$

Substitute into the first equation:

$$\frac{1}{4}(6y - 20) - \frac{3}{2}y = 5$$

$$\frac{3}{2}y - 5 - \frac{3}{2}y = 5$$

$$-5 = 5$$

This statement is never true, so the system is inconsistent. The lines are parallel.

43. $\begin{cases} x + y = 25 \\ x - y = 15 \end{cases}$

$$x = 25 - y$$
$$25 - y - y = 15$$
$$25 - 2y = 15$$
$$-2y = -10$$
$$y = 5$$
$$x = 25 - 5 = 20$$

The solution is $(20, 5)$. The two numbers are 20 and 5.

45. $\begin{cases} x + y = 10 \\ 0.50x + 0.25y = 0.30(10) \end{cases}$

Solve the first equation for x:

$$x = 10 - y$$

Substitute into the second equation:

$$50(10 - y) + 25y = 300$$
$$500 - 50y + 25y = 300$$
$$-25y = -200$$
$$y = 8$$
$$x = 10 - 8$$
$$x = 2$$

The solution is $(2, 8)$. So, 2 gallons of the 50% solution and 8 gallons of the 25% solution were used.

[End of Section 5.2]

Section 5.3
Solutions to Odd Exercises

1. $\begin{cases} 2x - y = 7 \\ x + y = 2 \end{cases}$

By addition:

$$\begin{array}{rcr} 2x & -y = & 7 \\ x & +y = & 2 \\ \hline 3x & = & 9 \\ x & = & 3 \end{array}$$

Substitute into the second equation:

$$(3) + y = 2$$
$$y = -1$$

Solution is $(3, -1)$.

3. $\begin{cases} 3x + 2y = 0 \\ 5x - 2y = 8 \end{cases}$

By addition:

$$\begin{array}{rcr} 3x & +2y = & 0 \\ 5x & -2y = & 8 \\ \hline 8x & = & 8 \\ x & = & 1 \end{array}$$

Substitute into the first equation:

$$3(1) + 2y = 0$$
$$2y = -3$$
$$y = -\frac{3}{2}$$

Solution is $\left(1, -\frac{3}{2}\right)$.

5. $\begin{cases} 2x + 2y = 5 \\ x + y = 3 \end{cases}$

Solve the second equation for x:

$$x = 3 - y$$

Substitute into the first equation:

$$2(3 - y) + 2y = 5$$
$$6 - 2y + 2y = 5$$
$$6 = 5$$

This statement is never true, so the system is inconsistent. The lines are parallel.

7. $\begin{cases} x = 11 + 2y \\ 2x - 3y = 17 \end{cases}$

By substitution:

$$2(11 + 2y) - 3y = 17$$
$$22 + 4y - 3y = 17$$

$$y = -5$$
$$x = 11 + 2(-5) = 1$$

Solution is $(1, -5)$.

9. $\begin{cases} x - 2y = 4 \\ y = \dfrac{1}{2}x - 2 \end{cases}$

By substitution:

$$x - 2\left(\frac{1}{2}x - 2\right) = 4$$
$$x - x + 4 = 4$$
$$4 = 4$$

This statement is always true, so the system is dependent. The equations represent the same line.

Solution is $\left(x, \dfrac{1}{2}x - 2\right)$.

11. $\begin{cases} x = 3y + 4 \\ y = 6 - 2x \end{cases}$

By substitution:

$$x = 3(6 - 2x) + 4$$
$$x = 18 - 6x + 4$$
$$7x = 22$$
$$x = \frac{22}{7}$$
$$y = 6 - 2\left(\frac{22}{7}\right) = -\frac{2}{7}$$

Solution is $\left(\dfrac{22}{7}, -\dfrac{2}{7}\right)$.

13. $\begin{cases} 7x - y = 16 \\ 2y = 2 - 3x \end{cases}$

Solve the first equation for y:

$$y = 7x - 16$$

Substitute into the second equation:

$$2(7x - 16) = 2 - 3x$$
$$14x - 32 = 2 - 3x$$
$$17x = 34$$
$$x = 2$$
$$y = 7(2) - 16 = -2$$

Solution is $(2, -2)$.

15. $\begin{cases} 4x - 2y = 8 \\ 2x - y = 4 \end{cases}$

Solve the second equation for y:
$$y = 2x - 4$$

Substitute into the first equation:
$$4x - 2(2x - 4) = 8$$
$$4x - 4x + 8 = 8$$
$$8 = 8$$

This statement is always true, so the system is dependent. The equations represent the same line.

Solution is $(x, 2x - 4)$.

17. $\begin{cases} 3x + 2y = 4 \\ x + 5y = -3 \end{cases}$

Solve the second equation for x:
$$x = -5y - 3$$

Substitute into the first equation:
$$3(-5y - 3) + 2y = 4$$
$$-15y - 9 + 2y = 4$$
$$-13y = 13$$
$$y = -1$$
$$x = -5(-1) - 3 = 2$$

Solution is $(2, -1)$.

19. $\begin{cases} \dfrac{1}{2}x + y = -4 \\ 3x - 4y = 6 \end{cases}$

Multiply the first equation by 4, then add:
$$\begin{array}{rr} 2x + 4y = & -16 \\ 3x - 4y = & 6 \\ \hline 5x \quad\;\; = & -10 \\ x \quad\;\; = & -2 \end{array}$$

Substitute into the first equation:
$$\frac{1}{2}(-2) + y = -4$$
$$-1 + y = -4$$
$$y = -3$$

Solution is $(-2, -3)$.

21. $\begin{cases} 4x + 3y = 2 \\ 3x + 2y = 3 \end{cases}$

Solve the second equation for x:
$$x = 1 - \frac{2}{3}y$$

Substitute into the first equation:

$$4\left(1 - \frac{2}{3}y\right) + 3y = 2$$
$$4 - \frac{8}{3}y + 3y = 2$$
$$12 - 8y + 9y = 6$$
$$y = -6$$
$$x = 1 - \frac{2}{3}(-6) = 5$$

The solution is $(5, -6)$.

23. $\begin{cases} 5x - 2y = 17 \\ 2x - 3y = 9 \end{cases}$

Solve the second equation for y:
$$y = \frac{2}{3}x - 3$$

Substitute into the first equation:
$$5x - 2\left(\frac{2}{3}x - 3\right) = 17$$
$$5x - \frac{4}{3}x + 6 = 17$$
$$15x - 4x + 18 = 51$$
$$11x = 33$$
$$x = 3$$
$$y = \frac{2}{3}(3) - 3 = -1$$

The solution is $(3, -1)$.

25. $\begin{cases} 3x + 2y = 14 \\ 7x + 3y = 26 \end{cases}$

Multiply the first equation by 3 and the second by −2, then add.
$$\begin{array}{rr} 9x + 6y = & 42 \\ -14x - 6y = & -52 \\ \hline -5x \quad\;\; = & -10 \\ x \quad\;\; = & 2 \end{array}$$

Substitute into the first equation:
$$3(2) + 2y = 14$$
$$2y = 8$$
$$y = 4$$

Solution is $(2, 4)$.

27. $\begin{cases} 2x+7y=2 \\ 5x+3y=-24 \end{cases}$

Multiply the first equation by 5 and the second by –2, then add:

$$10x+35y=10$$
$$\underline{-10x\ -6y=48}$$
$$29y=58$$
$$y=\ 2$$

Substitute into the first equation:

$$2x+7(2)=2$$
$$2x=-12$$
$$x=-6$$

Solution is $(-6,2)$.

29. $\begin{cases} 3x+3y=9 \\ x+y=3 \end{cases}$

Multiply the second equation by –3, then add:

$$3x+3y=\ 9$$
$$\underline{-3x-3y=-9}$$
$$0=\ 0$$

This statement is always true, so the system is dependent. The equations represent the same line.

Solution is $(x, 3-x)$.

31. $\begin{cases} 10x+4y=7 \\ 5x+2y=15 \end{cases}$

Multiply the second equation by –2, then add:

$$10x+4y=\ \ \ 7$$
$$\underline{-10x-4y=-30}$$
$$0=-23$$

This statement is never true, so the system is inconsistent. The lines are parallel.

33. Remove all fractions by multiplying the first equation by 4 and the second equation by 6.

$$\begin{cases} \dfrac{3}{4}x-\dfrac{1}{2}y=2 \\ \dfrac{1}{3}x-\dfrac{7}{6}y=1 \end{cases} \longrightarrow \begin{cases} 3x-2y=8 \\ 2x-7y=6 \end{cases}$$

Multiply the first equation by –7 and the second by 2, then add:

$$-21x+14y=-56$$
$$\underline{4x-14y=\ \ 12}$$
$$-17x\ \ \ \ \ \ \ =-44$$
$$x\ \ \ \ \ =\dfrac{44}{17}$$

Substitute into the previously modified second equation:

$$2\left(\dfrac{44}{17}\right)-7y=6$$
$$88-17(7y)=17(6)$$
$$-119y=-88+102$$
$$y=-\dfrac{14}{119}=-\dfrac{2}{17}$$

Solution is $\left(\dfrac{44}{17}, \dfrac{-2}{17}\right)$.

35. Remove all decimals by multiplying the first equation by 10 and the second by 100:

$$\begin{cases} x+0.5y=8 \\ 0.1x+0.01y=0.64 \end{cases} \longrightarrow \begin{cases} 10x+5y=80 \\ 10x+y=64 \end{cases}$$

Multiply the second equation by –1, then add:

$$10x+5y=\ \ 80$$
$$\underline{-10x-y=-64}$$
$$4y=\ \ 16$$
$$y=\ \ \ 4$$

Substitute into the original first equation:

$$x+0.5(4)=8$$
$$x+2=8$$
$$x=6$$

Solution is $(6,4)$.

37. Remove all decimals by multiplying each equation by 10:

$$\begin{cases} 0.6x+0.5y=5.9 \\ 0.8x+0.4y=6 \end{cases} \longrightarrow \begin{cases} 6x+5y=59 \\ 8x+4y=60 \end{cases}$$

Multiply the first equation by –4 and the second by 5, then add:

$$-24x-20y=-236$$
$$\underline{40x+20y=\ \ \ 300}$$
$$16x\ \ \ \ \ \ \ =\ \ 64$$
$$x\ \ \ \ \ =\ \ 4$$

Substitute into the previously modified first equation:

$$6(4)+5y=59$$
$$5y=35$$
$$y=7$$

Solution is $(4,7)$.

39. Remove all fractions from the system by multiplying each equation by 6:

$$\begin{cases} \dfrac{2}{3}x + \dfrac{1}{2}y = \dfrac{2}{3} \\ 3x + 2y = \dfrac{17}{6} \end{cases} \longrightarrow \begin{cases} 4x + 3y = 4 \\ 18x + 12y = 17 \end{cases}$$

Multiply the first equation by –4, then add:

$$\begin{array}{r} -16x - 12y = -16 \\ 18x + 12y = 17 \\ \hline 2x \qquad\quad = 1 \\ x \qquad = \dfrac{1}{2} \end{array}$$

Substitute into the previously modified first equation:

$$4\left(\dfrac{1}{2}\right) + 3y = 4$$

$$2 + 3y = 4$$

$$3y = 2$$

$$y = \dfrac{2}{3}$$

Solution is $\left(\dfrac{1}{2}, \dfrac{2}{3}\right)$.

41. Remove all fractions from the system by multiplying the first equation by 12 and the second by 20:

$$\begin{cases} \dfrac{1}{6}x - \dfrac{1}{12}y = -\dfrac{13}{6} \\ \dfrac{1}{5}x + \dfrac{1}{4}y = 2 \end{cases} \longrightarrow \begin{cases} 2x - y = -26 \\ 4x + 5y = 40 \end{cases}$$

Solve the first previously modified equation for y:

$$y = 2x + 26$$

Substitute into the second previously modified equation:

$$4x + 5(2x + 26) = 40$$

$$4x + 10x + 130 = 40$$

$$14x = -90$$

$$x = -\dfrac{45}{7}$$

$$y = 2\left(-\dfrac{45}{7}\right) + 26$$

$$y = \dfrac{92}{7}$$

Solution is $\left(\dfrac{-45}{7}, \dfrac{92}{7}\right)$.

43. Set up the system: $\begin{cases} 3 = 2m + b \\ -2 = 1m + b \end{cases}$

Multiply the second equation by –1, then add:

$$\begin{array}{r} 3 = 2m + b \\ 2 = -m - b \\ \hline 5 = m \end{array}$$

Substitute into the first equation:

$$3 = 2(5) + b$$

$$3 = 10 + b$$

$$-7 = b$$

Use m and b to write the equation of the line in slope intercept form, $y = mx + b$:

$$y = 5x - 7$$

45. Set up the system: $\begin{cases} 1 = -4m + b \\ 2 = 5m + b \end{cases}$

Multiply the second equation by –1, then add:

$$\begin{array}{r} 1 = -4m + b \\ -2 = -5m - b \\ \hline -1 = -9m \\ \dfrac{1}{9} = m \end{array}$$

Substitute into the second equation:

$$2 = 5\left(\dfrac{1}{9}\right) + b$$

$$2 = \dfrac{5}{9} + b$$

$$2 - \dfrac{5}{9} = b$$

$$\dfrac{13}{9} = b$$

Use m and b to write the equation of the line in slope intercept form, $y = mx + b$:

$$y = \dfrac{1}{9}x + \dfrac{13}{9}$$

47. Set up the system: $\begin{cases} -4 = 3m + b \\ 7 = 7m + b \end{cases}$

Multiply the second equation by -1, then add:

$$-4 = \quad 3m + b$$
$$\underline{-7 = -7m - b}$$
$$-11 = -4m$$

$$\frac{11}{4} = \quad m$$

Substitute into the first equation:

$$-4 = 3\left(\frac{11}{4}\right) + b$$

$$-4 = \frac{33}{4} + b$$

$$-4 - \frac{33}{4} = b$$

$$-\frac{49}{4} = b$$

Use m and b to write the equation of the line in slope intercept form, $y = mx + b$:

$$y = \frac{11}{4}x - \frac{49}{4}$$

Calculator Problems

49. $\begin{cases} 0.9x + 1.3y = 1.4 \\ 1.2x - 0.7y = 4.3 \end{cases}$

Enter

"$Y1 = (-9X + 14)/13$" and

"$Y2 = (12X - 43)/7$"

into your calculator.

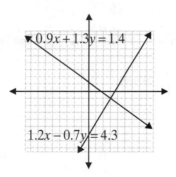

Solution is $(3, -1)$.

51. $\begin{cases} 1.4x + 3.5y = 7.28 \\ 2.4x - 2.1y = -0.48 \end{cases}$

Enter

"$Y1 = (7.28 - 1.4X)/3.5$" and

"$Y2 = (-0.48 - 2.4X)/-2.1$"

into your calculator.

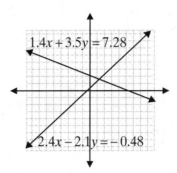

Solution is $(1.2, 1.6)$.

53. $\begin{cases} 1.3x + 4.1y = 9.282 \\ 0.7x - 1.6y = 4.503 \end{cases}$

Enter

"$Y1 = (9.282 - 1.3X)/4.1$" and

"$Y2 = (4.503 - 0.7X)/-1.6$"

into your calculator.

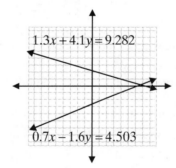

Solution is $(6.726, 0.134)$.

55. Let x be the amount in the 10% account and y be the amount in the 6% account.

$$\begin{cases} x + y = 10{,}000 \\ 0.10x - 0.06y = 40 \end{cases}$$

Multiply the first equation by 10 and the second by -100, then add:

$$\begin{array}{rcl} 10x + 10y &=& 100{,}000 \\ -10x + 6y &=& -4{,}000 \\ \hline 16y &=& 96{,}000 \\ y &=& 6{,}000 \end{array}$$

Substitute into the first equation:

$$x + 6{,}000 = 10{,}000$$
$$x = 4{,}000$$

So, \$6,000 is invested at 6% and \$4,000 at 10%.

57. Let x be the amount in the 30% solution and y be the amount in the 40% solution.

$$\begin{cases} x + y = 100 \\ 0.30x + 0.40y = 0.36(100) \end{cases}$$

Multiply the second equation by 100 to remove all decimals:

$$\begin{cases} x + y = 100 \\ 0.30x + 0.40y = 36 \end{cases} \longrightarrow \begin{cases} x = 100 - y \\ 30x + 40y = 3600 \end{cases}$$

By substitution:

$$30(100 - y) + 40y = 3600$$
$$3000 - 30y + 40y = 3600$$
$$10y = 600$$
$$y = 60$$
$$x = 100 - 60 = 40$$

There should be 60 liters of the 40% solution and 40 liters of the 30% solution.

[End of Section 5.3]

Section 5.4
Solutions to Odd Exercises

1. Let x be one of the numbers and y be the other.
$$\begin{cases} x+y=56 \\ x-y=10 \end{cases}$$
Add the equations to eliminate y:
$$\begin{array}{rcl} x+y &=& 56 \\ x-y &=& 56 \\ \hline 2x &=& 66 \\ x &=& 33 \end{array}$$
Substitute into the first equation:
$$(33)+y=56$$
$$y=23$$
The numbers are 33 and 23.

3. Let x be the larger number and y be the smaller.
$$\begin{cases} x+y=36 \\ 3y+2x=87 \end{cases}$$
Multiply the first equation by -2 and add the equations to eliminate y:
$$\begin{array}{rcl} -2x-2y &=& -72 \\ 2x+3y &=& 87 \\ \hline y &=& 15 \end{array}$$
Substitute into the first equation:
$$x+(15)=36$$
$$x=21$$
The numbers are 15 and 21.

5. Let x be the rate of the boat in still water and y be the rate of the current. Recall that $d=rt$.
$$\begin{cases} \frac{1}{3}(x+y)=4 \\ \frac{1}{2}(x-y)=4 \end{cases}$$
Remove the fractions by multiplying the first equation by 3 and the second by 2. Then use the addition method to eliminate y:
$$\begin{array}{rcl} x+y &=& 12 \\ x-y &=& 8 \\ \hline 2x &=& 20 \\ x &=& 10 \end{array}$$
Substitute into the previously modified first equation:
$$(10)+y=12$$
$$y=2$$
The rate of the current is 2 mph and the rate of the boat in still water is 10 mph.

7. Let x be the time at 52 mph and y be the time at 56 mph.
$$\begin{cases} 52x+56y=190 \\ x+y=3\frac{1}{2} \end{cases}$$
Multiply the second equation by -56 and use the addition method to eliminate y:
$$\begin{array}{rcl} 52x+56y &=& 190 \\ -56x-56y &=& -196 \\ \hline -4x &=& -6 \\ x &=& \frac{3}{2} \end{array}$$
Substitute into the second equation:
$$\left(\frac{3}{2}\right)+y=3\frac{1}{2}$$
$$y=\frac{7}{2}-\frac{3}{2}$$
$$y=2$$
The time at 52 mph was $1\frac{1}{2}$ hours, and the time at 56 mph was 2 hours.

9. Let x be the speed going and y be the speed returning.
$$\begin{cases} x-y=10 \\ 4x+5y=200 \end{cases}$$
Solve first equation for x and use the method of substitution to solve for y:
$$x=10+y$$
$$4(10+y)+5y=400$$
$$40+4y+5y=400$$
$$40+9y=400$$
$$9y=360$$
$$y=40$$
$$x=10+40=50$$
The speed from his home to the city was 50 mph.

11. Let x be Steve's speed and y be Fred's speed.
$$\begin{cases} x=4y \\ 3x+3y=105 \end{cases}$$
Solve by substitution:
$$3(4y)+3y=105$$
$$12y+3y=105$$
$$15y=105$$
$$y=7$$
$$x=4(7)=28$$
Steve's speed was 28 mph and Fred's was 7 mph.

13. Let x be Mary's speed and y be Linda's speed.
$$\begin{cases} x = y + 8 \\ 3x + 3y = 324 \end{cases}$$
Solve by substitution:
$$3(y+8) + 3y = 324$$
$$3y + 24 + 3y = 324$$
$$6y = 300$$
$$y = 50$$
$$x = 50 + 8 = 58$$
Mary's speed was 58 mph and Linda's was 50 mph.

15. Let x be the jogger's time going out and y be the jogger's time returning.
First, note two important facts given in the problem:
$$1 \text{ hour, } 36 \text{ minutes} = \frac{8}{5} \text{ hours}$$
$$\text{distance there} = 10x$$
$$\text{distance back} = 6y$$
$$\begin{cases} x + y = \frac{8}{5} \\ 10x = 6y \end{cases}$$
Multiply first equation by -10 and add $-6y$ to both sides of the second. Solve by the addition method:
$$\begin{array}{r} -10x - 10y = -16 \\ 10x - 6y = 0 \\ \hline -16y = -16 \\ y = 1 \end{array}$$
The jogger ran back for 1 hour at the rate of 6 mph, or for a distance of 6 miles. Then, he ran a total of 12 miles.

17. Let x be the airliner's speed and y be the private plane's speed.
$$\begin{cases} x = 3.5y \\ 2x - 2y = 580 \end{cases}$$
Solve by substitution:
$$2(3.5y) - 2y = 580$$
$$7y - 2y = 580$$
$$5y = 580$$
$$y = 116$$
$$x = 3.5(116) = 406$$
The speed of the airliner is 406 mph and the speed of the private plane is 116 mph.

19. Let x be the number of quarters and y be the number of dimes.
$$\begin{cases} x + y = 27 \\ 0.25x + 0.10y = 5.40 \end{cases}$$
Multiply the second equation by 100 to remove the decimal points and solve the first for y. Solve by substitution:
$$y = 27 - x$$
$$25x + 10(27 - x) = 540$$
$$25x + 270 - 10x = 540$$
$$15x = 270$$
$$x = 18$$
$$y = 27 - (18) = 9$$
The number of quarters is 18 and the number of dimes is 9.

21. Let p be the number of pennies and n be the number of nickels.
$$\begin{cases} p + n = 182 \\ 0.01p + 0.05n = 3.90 \end{cases}$$
Multiply the second equation by -100 to remove the decimal points and to set up the addition method:
$$\begin{array}{r} p + n = 182 \\ -p - 5n = -390 \\ \hline -4n = -208 \\ n = 52 \end{array}$$
Substitute in the first equation:
$$p + (52) = 182$$
$$p = 182 - 52 = 130$$
There are 130 pennies and 52 nickels.

23. Let w be the width and l be the length.
$$\begin{cases} l = 2w - 1 \\ 2(l + 4) + 2(w + 4) = 116 \end{cases}$$
Simplify the second equation and use the substitution method:
$$2l + 8 + 2w + 8 = 116$$
$$2l + 2w = 100$$
$$2(2w - 1) + 2w = 100$$
$$4w - 2 + 2w = 100$$
$$6w = 102$$
$$w = 17$$
$$l = 2(17) - 1 = 33$$
The original width is 17m and the length is 33m.

25. Solve the system.
$$\begin{cases} -1 = -2m + b \\ -7 = 6m + b \end{cases}$$
Multiply the first equation by 3 and use the addition method to eliminate m:
$$\begin{aligned} -3 &= -6m + 3b \\ -7 &= 6m + b \\ \hline -10 &= 4b \\ -\frac{5}{2} &= b \end{aligned}$$
Substitute into the first equation:
$$-1 = -2m + \left(-\frac{5}{2}\right)$$
$$2m = 1 - \frac{5}{2} = -\frac{3}{2}$$
$$m = -\frac{3}{4}$$
The equation is $y = -\dfrac{3}{4}x - \dfrac{5}{2}$.

27. Let A be Anna's age now and B be Beth's age now.
$$\begin{cases} A - 2 = \dfrac{1}{2}(B - 2) \\ A + 8 = \dfrac{2}{3}(B + 8) \end{cases}$$
Multiply the first equation by 2 and the second by 3 to remove the fractions. Then simplify:
$$\begin{cases} 2A - 4 = B - 2 \\ 3A + 24 = 2B + 16 \end{cases}$$
Solve the first equation for B and use substitution:
$$B = 2A - 2$$
$$3A + 24 = 2(2A - 2) + 16$$
$$3A + 24 = 4A + 12$$
$$A = 12$$
$$B = 24 - 2 = 22$$
Anna's age is 12, and Beth's age is 22.

29. Let a be the number of adult tickets and s be the number of student tickets.
$$\begin{cases} a + s = 3500 \\ 3.50a + 2.50s = 9550 \end{cases}$$
Multiply the second equation by 10 to remove the decimal points and solve the first for a to set up the substitution method:
$$\begin{cases} a = 3500 - s \\ 35a + 25s = 95500 \end{cases}$$
Simplify the second by dividing each side of the equation by 5 before substituting:

$$7a + 5s = 19100$$
$$7(3500 - s) + 5s = 19100$$
$$24500 - 7s + 5s = 19100$$
$$-2s = -5400$$
$$s = 2700$$
$$a = 3500 - 2700 = 800$$
There were 800 adults and 2700 students.

31. Let w be the width and l be the length.
$$\begin{cases} l = 3w \\ 2l + 2w = 260 \end{cases}$$
Solve by substitution:
$$2(3w) + 2w = 260$$
$$8w = 260$$
$$w = 32.5$$
$$l = 3(32.5) = 97.5$$
The length is 97.5m and the width is 32.5m.

33. Let g be the number of general admission and r be the number of reserved seats.
$$\begin{cases} g + r = 12500 \\ 2.00g + 3.50r = 36250 \end{cases}$$
Multiply the first equation by -4 and the second by 2 and solve using the addition method:
$$\begin{aligned} -4g - 4r &= -50000 \\ 4g + 7r &= 72500 \\ \hline 3r &= 22500 \\ r &= 7500 \end{aligned}$$
Substitute into the first equation:
$$g + (7500) = 12500$$
$$g = 12500 - 7500$$
$$g = 5000$$
There were 7500 reserved seat tickets and 5000 general admission tickets sold.

35. Let c be the ticket price for a child and a be the ticket price for an adult.
$$\begin{cases} 2c + a = 7 \\ 70c + 160a = 620 \end{cases}$$
Solve the first equation for a and use substitution:
$$a = 7 - 2c$$
$$70c + 160(7 - 2c) = 620$$
$$70c + 1120 - 320c = 620$$
$$c = 2$$
$$a = 7 - 4 = 3$$
Children and adult ticket costs are \$2 and \$3, respectively.

37. Let x be the price of a shirt and y be the price of a pair of slacks.
$$\begin{cases} 2x + y = 55 \\ x + 2y = 68 \end{cases}$$
Solve the second equation for x and use substitution:
$$x = 68 - 2y$$
$$2(68 - 2y) + y = 55$$
$$136 - 4y + y = 55$$
$$-3y = -81$$
$$y = 27$$
$$x = 68 - 2(27) = 68 - 54 = 14$$
A pair of slacks is \$27 and a shirt is \$14.

39. Let x be the number of Model X, and y be the number of Model Y radios to be made.
$$\begin{cases} 4x + 3y = 58 \\ 8x + 7y = 126 \end{cases}$$
Multiply the first equation by –2 and use the addition method:
$$\begin{array}{r} -8x - 6y = -116 \\ \underline{8x + 7y = 126} \\ y = 10 \end{array}$$
Substitute into the first equation:
$$4x + 3(10) = 58$$
$$4x = 28$$
$$x = 7$$
There will be 7 of Model X and 10 of Model Y.

Writing and Thinking About Mathematics

41. Let a be the tens digit of the number, and b be the ones digit.
$$\begin{cases} a + b = 13 \\ 10b + a = 10a + b + 45 \end{cases}$$
To use the addition method, first multiply the first equation by 9 and simplify the second equation:
$$\begin{array}{r} 9a + 9b = 117 \\ \underline{-9a + 9b = 45} \\ 18b = 162 \\ b = 9 \end{array}$$
Substitute into the first equation:
$$a + (9) = 13$$
$$a = 13 - 9 = 4$$
The number is 49.

[End of Section 5.4]

Section 5.5
Solutions to Odd Exercises

1. Let x be the amount invested at 6% and y be the amount invested at 10%.

$$\begin{cases} x + y = 9000 \\ 0.06x + 0.10y = 680 \end{cases}$$

Multiply the second equation by 100, to remove the decimal points, and the first equation by -10, to set up the use of the addition method:

$$-10x - 10y = -90,000$$
$$\underline{6x + 10y = \ 68,000}$$
$$-4x \qquad = -22,000$$
$$x \quad = \quad 5500$$

Substitute into the first equation:

$$(5500) + y = 9000$$
$$y = 9000 - 5500 = 3500$$

$5,500 is invested at 6% and $3,500 is invested at 10%.

3. Let x be the amount invested at 5.5% and y be the amount invested at 6%.

$$\begin{cases} x + y = 10,000 \\ 0.055x - 0.06y = 251 \end{cases}$$

Multiply the second equation by 1000, to remove the decimal points, and multiply the first equation by 60, to set up the use of the addition method:

$$60x + 60y = 600,000$$
$$\underline{55x - 60y = 251,000}$$
$$115x \qquad = 851,000$$
$$x \qquad = \qquad 7400$$

Substitute into the first equation:

$$(7400) + y = 10,000$$
$$y = 10,000 - 7400 = 2600$$

$7,400 is invested at 5.5% and $2,600 is invested at 6%.

5. Let x be the amount invested at 8% and y be the amount invested at 10%.

$$\begin{cases} y - x = 200 \\ 0.08x + 0.10y = 101 \end{cases}$$

Multiply the second equation by 100, to remove the decimal points, and solve for x in the first equation:

$$\begin{cases} x = y - 200 \\ 8x + 10y = 10,100 \end{cases}$$

Solve using the substitution method:

$$8(y - 200) + 10y = 10,100$$
$$8y - 1600 + 10y = 10,100$$

$$18y = 11,700$$
$$y = 650$$
$$x = (650) - 200 = 450$$

$650 is invested at 10% and $450 is invested at 8%.

7. Let x be the amount invested at 11% and y be the amount invested at 13%.

$$\begin{cases} x = y \\ 0.11x + 0.13y = 840 \end{cases}$$

Multiply the second equation by 100, to remove the decimal points. Solve by the substitution method:

$$11x + 13y = 84,000$$
$$11x + 13(x) = 84,000$$
$$24x = 84,000$$
$$x = 3500 = y \ \cdot$$
$$x + y = 3500 + 3500 = 7000$$

The total amount invested is $7,000.

9. Let x be the amount invested at 24% and y be the amount invested at 18%.

$$\begin{cases} x - 9000 = y \\ 0.24x - 0.18y = 2820 \end{cases}$$

Multiply the second equation by 100, to remove the decimal points. Solve by the substitution method:

$$24x - 18y = 282,000$$
$$24x - 18(x - 9000) = 282,000$$
$$6x + 162,000 = 282,000$$
$$6x = 120,000$$
$$x = 20,000$$
$$y = (20,000) - 9000$$
$$y = 11,000$$

$20,000 is invested at 24% and $11,000 is invested at 18%.

11. Let x be the amount invested at 5% and y be the amount invested at 7%.

$$\begin{cases} y = 2x + 500 \\ 0.05x + 0.07y = 187 \end{cases}$$

Multiply the second equation by 100 to get rid of the decimal point. Then, use the substitution method:

$$5x + 7(2x + 500) = 18,700$$
$$19x + 3500 = 18,700$$
$$19x = 15,200$$
$$x = 800$$
$$y = 2(800) + 500 = 2100$$

$800 is invested at 5% and $2,100 is invested at 7%.

13. Let x be the amount invested at 9% and y be the amount invested at 11%.

$$\begin{cases} x + y = 12,000 \\ 0.09x - 0.11y = 380 \end{cases}$$

Multiply the second equation by 100, to remove the decimal points, and solve for x in the first equation:

$$\begin{cases} x = 12,000 - y \\ 9x - 11y = 38,000 \end{cases}$$

Solve using the substitution method:

$$9(12,000 - y) - 11y = 38,000$$
$$108,000 - 9y - 11y = 38,000$$
$$-20y = -70,000$$
$$y = 3500$$
$$x = 12,000 - (3500)$$
$$x = 8500$$

$8,500 is invested at 9% and $3,500 is invested at 11%.

15. Let x be the number of pounds of 20% copper alloy and y be the number of pounds of 70% copper alloy.

$$\begin{cases} x + y = 50 \\ 0.20x + 0.70y = 0.50(50) \end{cases}$$

Multiply the second equation by 10, to remove the decimal points, and solve for x in the first equation:

$$\begin{cases} x = 50 - y \\ 2x + 7y = 250 \end{cases}$$

Solve using the substitution method:

$$2(50 - y) + 7y = 250$$
$$100 - 2y + 7y = 250$$
$$5y = 150$$
$$y = 30$$
$$x = 50 - (30) = 20$$

He must use 30 pounds of the 70% copper alloy and 20 pounds of the 20% alloy.

17. Let x be the number of ounces of 30% blend and y be the number of ounces of 20% blend.

$$\begin{cases} x + y = 50 \\ 0.30x + 0.20y = 0.24(50) \end{cases}$$

Multiply the second equation by 100, to remove the decimal points, and solve for x in the first equation:

$$\begin{cases} x = 50 - y \\ 30x + 20y = 1200 \end{cases}$$

Solve using the substitution method:

$$30(50 - y) + 20y = 1200$$
$$1500 - 30y + 20y = 1200$$
$$-10y = -300$$

$$y = 30$$
$$x = 50 - (30) = 20$$

30 ounces of the 20% blend and 20 ounces of the 30% blend will be needed.

19. Let x be the number of pounds of the 35% protein supplement and y be the number of pounds of the 15% standard ration.

$$\begin{cases} x + y = 1800 \\ 0.35x + 0.15y = 0.20(1800) \end{cases}$$

Multiply the second equation by 100, to remove the decimal points, and solve for x in the first equation:

$$\begin{cases} x = 1800 - y \\ 35x + 15y = 36,000 \end{cases}$$

Simplify the second equation by dividing by 5 and solve using the substitution method:

$$7x + 3y = 7200$$
$$7(1800 - y) + 3y = 7200$$
$$12600 - 7y + 3y = 7200$$
$$-4y = -5,400$$
$$y = 1350$$
$$x = 1800 - (1350) = 450$$

He should use 450 pounds of the 35% supplement and 1350 pounds of the 15% ration.

21. Let x be the number of pounds of 40% fat and y be the number of pounds of 15% fat.

$$\begin{cases} x + y = 50 \\ 0.40x + 0.15y = 0.25(50) \end{cases}$$

Multiply the second equation by 100, to remove the decimal points, and solve for x in the first equation:

$$\begin{cases} x = 50 - y \\ 40x + 15y = 1250 \end{cases}$$

Simplify the second equation by dividing by 5 and solve using the substitution method:

$$8x + 3y = 250$$
$$8(50 - y) + 3y = 250$$
$$400 - 8y + 3y = 250$$
$$-5y = -150$$
$$y = 30$$
$$x = 50 - (30) = 20$$

30 pounds of the 15% fat and 20 pounds of the 40% fat are needed.

23. Let x be the number of grams of pure acid and y be the number of grams of 40% solution.

$$\begin{cases} x + y = 30 \\ 1.00x + 0.40y = 0.60(30) \end{cases}$$

Multiply the second equation by 100, to remove the decimal points, and solve for x in the first equation:

$$\begin{cases} x = 30 - y \\ 100x + 40y = 1800 \end{cases}$$

Simplify the second equation by dividing by 20 and solve using the substitution method:

$$5x + 2y = 90$$
$$5(30 - y) + 2y = 90$$
$$150 - 5y + 2y = 90$$
$$-3y = -60$$
$$y = 20$$
$$x = 30 - (20) = 10$$

10 grams of the pure acid and 20 grams of the 40% solution are needed.

25. Let x be the number of ounces of salt and y be the number of ounces of 4% salt solution.

$$\begin{cases} x + y = 60 \\ 1.00x + 0.04y = 0.20(60) \end{cases}$$

Multiply the second equation by 100, to remove the decimal points, and solve for x in the first equation:

$$\begin{cases} x = 60 - y \\ 100x + 4y = 20(60) \end{cases}$$

Simplify the second equation by dividing by 4 and solve using the substitution method:

$$25x + y = 300$$
$$25(60 - y) + y = 300$$
$$1500 - 25y + y = 300$$
$$-24y = -1200$$
$$y = 50$$
$$x = 60 - (50) = 10$$

10 ounces of the salt and 50 ounces of the 4% salt solution are needed.

[End of Section 5.5]

Solutions to Odd Exercises

1. $\begin{cases} y > 2 \\ x \geq -3 \end{cases}$

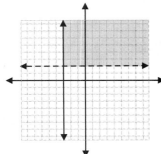

3. $\begin{cases} x < 3 \\ y > -x + 2 \end{cases}$

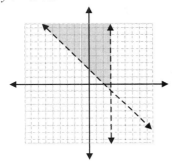

5. $\begin{cases} x \leq 3 \\ 2x + y > 7 \end{cases}$

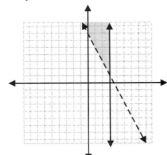

7. $\begin{cases} x - 3y \leq 3 \\ x < 5 \end{cases}$

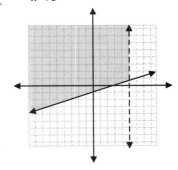

9. $\begin{cases} x - y \geq 0 \\ 3x - 2y \geq 4 \end{cases}$

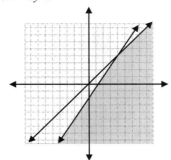

11. $\begin{cases} 3x + y \leq 10 \\ 5x - y \geq 6 \end{cases}$

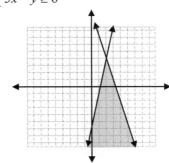

13. $\begin{cases} 3x + 4y \geq -7 \\ y < 2x + 1 \end{cases}$

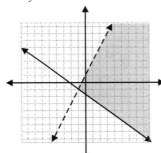

15. $\begin{cases} x + y < 4 \\ 2x - 3y < 3 \end{cases}$

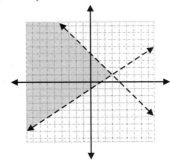

17. $\begin{cases} x + y \geq 0 \\ x - 2y \geq 6 \end{cases}$

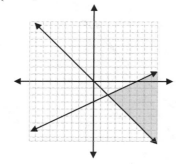

19. $\begin{cases} x + 3y \leq 9 \\ x - y \geq 5 \end{cases}$

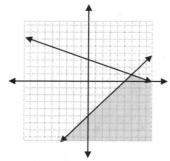

21. $\begin{cases} y \geq 0 \\ 3x - 5y \leq 10 \end{cases}$

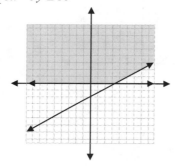

23. $\begin{cases} 4x - 3y \geq 6 \\ 3x - y \leq 3 \end{cases}$

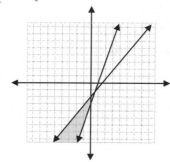

25. $\begin{cases} 3x - 4y \geq -6 \\ 3x + 2y \leq 12 \end{cases}$

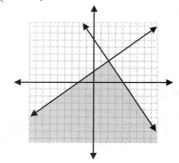

27. $\begin{cases} x + y \leq 8 \\ 3x - 2y \geq -6 \end{cases}$

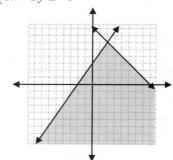

29. $\begin{cases} y \leq x \\ y < 2x + 1 \end{cases}$

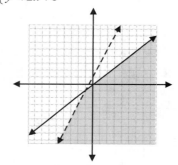

Writing and Thinking About Mathematics

31. **a.** $\begin{cases} y \le 2x - 5 \\ y \ge 2x + 3 \end{cases}$

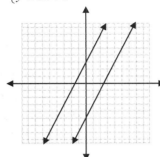

There are no points that satisfy both inequalities.

b. $\begin{cases} y \le -x + 2 \\ y \ge -x - 1 \end{cases}$

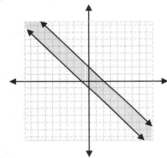

The only points that satisfy both inequalities are those lying on or between the two parallel lines.

c. $\begin{cases} y \le \dfrac{1}{2}x + 3 \\ y \ge \dfrac{1}{2}x - 3 \end{cases}$

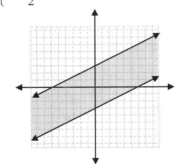

The only points that satisfy both inequalities are those lying on or between the two parallel lines.

[End of Section 5.6]

Chapter 5 Review
Solutions to All Exercises

1. Find the slopes of the lines to determine whether they are parallel. Verify that the y-intercepts are different.
    $$y = 2x \quad \rightarrow \quad y = 2x$$
    $$-2x + y = 5 \quad \rightarrow \quad y = 2x + 5$$
    The lines are parallel with a slope of 2. They are different, as $b = 0$ in the first and $b = 5$ in the second.

2. Find the slopes of the lines to determine whether they are parallel. Verify that the y-intercepts are different.
    $$3x + y = 4 \quad \rightarrow \quad y = -3x + 4$$
    $$6x + 2y = -1 \quad \rightarrow \quad y = -3x - \frac{1}{2}$$
    The lines are parallel with a slope of -3. They are different, as $b = 4$ in the first and $b = -\frac{1}{2}$ in the second.

3. Find the slopes of the lines to determine whether they are parallel. Verify that the y-intercepts are different.
    $$2x - y = -7 \quad \rightarrow \quad y = 2x + 7$$
    $$x - \frac{1}{2}y = 2 \quad \rightarrow \quad y = 2x - 4$$
    The lines are parallel with a slope of 2. They are different, as $b = 7$ in the first and $b = -4$ in the second.

4. Find the slopes of the lines to determine whether they are parallel. Verify that the y-intercepts are different.
    $$5x - 3y = -6 \quad \rightarrow \quad y = \frac{5}{3}x + 2$$
    $$10x - 6y = 3 \quad \rightarrow \quad y = \frac{5}{3}x - \frac{1}{2}$$
    The lines are parallel with a slope of $\frac{5}{3}$. They are different, as $b = 2$ in the first and $b = -\frac{1}{2}$ in the second.

5. $\begin{cases} y = -2x + 5 \\ y = x + 2 \end{cases}$

    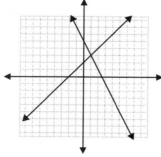

    $(1,3)$ is the solution. The system is consistent.

6. $\begin{cases} y = -\dfrac{1}{2}x \\ 3x - 2y = 8 \end{cases}$

    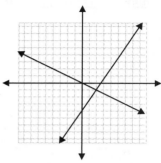

    $(2,-1)$ is the solution. The system is consistent.

7. $\begin{cases} x + 4y = 6 \\ y = -\dfrac{1}{4}x + \dfrac{3}{2} \end{cases}$

    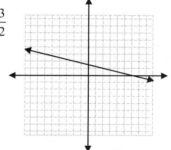

    The equations represent the same line. The system is dependent. $\left(x, \dfrac{-1}{4}x + \dfrac{3}{2}\right)$ is the solution.

8. $\begin{cases} 5x + y = 4 \\ 10x + 2y = 8 \end{cases}$

    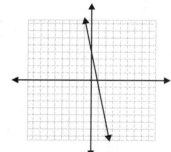

    The equations represent the same line. The system is dependent. $(x, 4 - 5x)$ is the solution.

9. $\begin{cases} -3x + y = 7 \\ -6x + 2y = 9 \end{cases}$

    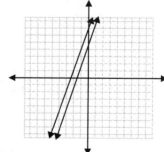

    No solution. The lines are parallel. The system is inconsistent.

10. $\begin{cases} y = -x + 10 \\ x + y = 4 \end{cases}$

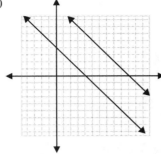

No solution. The lines are parallel. The system is inconsistent.

11. $\begin{cases} x + y = 3 \\ 2x + y = 2 \end{cases}$

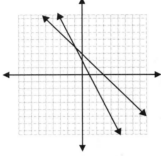

$(-1, 4)$ is the solution. The system is consistent.

12. $\begin{cases} x + 2y = 12 \\ 3x - y = -6 \end{cases}$

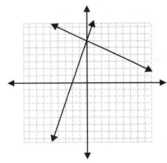

$(0, 6)$ is the solution. The system is consistent.

13. $\begin{cases} 2x + 3y = 10.5 \\ x - y = -1 \end{cases}$

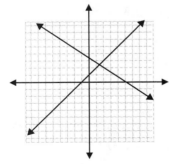

$(1.5, 2.5)$ is the solution. The system is consistent.

14. $\begin{cases} x - 3y = 2.5 \\ 2x + y = -5.5 \end{cases}$

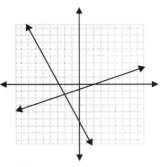

$(-2, -1.5)$ is the solution. The system is consistent.

15. $\begin{cases} y = 7 + x \\ x + 3y = 21 \end{cases}$

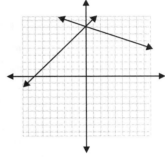

$(0, 7)$ is the solution. The system is consistent.

16. $\begin{cases} x - 5y = -1.5 \\ x - y = 4.3 \end{cases}$

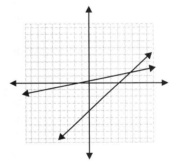

$(5.75, 1.45)$ is the solution. The system is consistent.

17. $\begin{cases} y = 2x \\ x - y = -1 \end{cases}$

$$x - (2x) = -1$$
$$-x = -1$$
$$x = 1$$
$$y = 2(1) = 2$$

The solution is $(1, 2)$.

18. $\begin{cases} y = 4x+1 \\ 8x-2y = -2 \end{cases}$

$$8x - 2(4x+1) = -2$$
$$8x - 8x - 2 = -2$$
$$-2 = -2$$

This statement is always true, so the system is dependent. The equations represent the same line.

The solution is $(x, 4x+1)$.

19. $\begin{cases} 3x-2y = 4 \\ y = \dfrac{3}{2}x+6 \end{cases}$

$$3x - 2\left(\dfrac{3}{2}x+6\right) = 4$$
$$3x - 3x - 12 = 4$$
$$12 = 4$$

This statement is never true, so the system is inconsistent. The lines are parallel.

20. $\begin{cases} y = -2x+5 \\ 4x+2y = -1 \end{cases}$

$$4x + 2(-2x+5) = -1$$
$$4x - 4x + 10 = -1$$
$$10 = -1$$

This statement is never true, so the system is inconsistent. The lines are parallel.

21. $\begin{cases} 5x-y = 6 \\ 2y = 10x-12 \end{cases}$

Solve the first equation for y then substitute:
$$y = 5x - 6$$
$$2(5x-6) = 10x - 12$$
$$10x - 12 = 10x - 12$$
$$-12 = -12$$

This statement is always true, so the system is dependent. The equations represent the same line.

The solution is $(x, 5x-6)$.

22. $\begin{cases} y = -\dfrac{1}{8}x \\ x-4y = 4 \end{cases}$

$$x - 4\left(-\dfrac{1}{8}x\right) = 4$$
$$x + \dfrac{1}{2}x = 4$$
$$2x + x = 8$$

$$3x = 8$$
$$x = \dfrac{8}{3}$$
$$y = -\dfrac{1}{8}\left(\dfrac{8}{3}\right) = -\dfrac{1}{3}$$

The solution is $\left(\dfrac{8}{3}, -\dfrac{1}{3}\right)$.

23. $\begin{cases} 2x-y = 2 \\ 2x+y = -2 \end{cases}$

Solve the second equation for y then substitute:
$$y = -2 - 2x$$
$$2x - (-2-2x) = 2$$
$$4x + 2 = 2$$
$$4x = 0$$
$$x = 0$$
$$y = -2 - 2(0) = -2$$

The solution is $(0, -2)$.

24. $\begin{cases} x+\dfrac{1}{2}y = 8 \\ y = x+1 \end{cases}$

$$x + \dfrac{1}{2}(x+1) = 8$$
$$x + \dfrac{1}{2}x + \dfrac{1}{2} = 8$$
$$2x + x + 1 = 16$$
$$3x = 15$$
$$x = 5$$
$$y = 5 + 1 = 6$$

The solution is $(5, 6)$.

25. $\begin{cases} x+y = 1 \\ x-y = 3 \end{cases}$

$$\begin{array}{r} x + y = 1 \\ \underline{x - y = 3} \\ 2x = 4 \\ x = 2 \end{array}$$

Substitute into the first equation:
$$(2) + y = 1$$
$$y = -1$$

The solution is $(2, -1)$.

26. $\begin{cases} 3x - 2y = 5 \\ x + 3y = 17 \end{cases}$

Multiply the second equation by −3, and then use the addition method:

$$\begin{array}{r} 3x - 2y = 5 \\ \underline{-3x - 9y = -51} \\ -11y = -46 \\ y = \dfrac{46}{11} \end{array}$$

Substitute into the second equation:

$$x + 3\left(\dfrac{46}{11}\right) = 17$$

$$x = 17 - \dfrac{138}{11} = \dfrac{49}{11}$$

The system is consistent; solution is $\left(\dfrac{49}{11}, \dfrac{46}{11}\right)$.

27. $\begin{cases} x + 2y = 1 \\ 2x - 3y = 0 \end{cases}$

Multiply the first equation by −2, and then use the addition method:

$$\begin{array}{r} -2x - 4y = -2 \\ \underline{2x - 3y = 0} \\ -7y = -2 \\ y = \dfrac{2}{7} \end{array}$$

Substitute into the first equation:

$$x + 2\left(\dfrac{2}{7}\right) = 1$$

$$x = 1 - \dfrac{4}{7} = \dfrac{3}{7}$$

The system is consistent; solution is $\left(\dfrac{3}{7}, \dfrac{2}{7}\right)$.

28. $\begin{cases} 5x + y = 1 \\ 3x + y = 3 \end{cases}$

Multiply the second equation by −1, and then use the addition method:

$$\begin{array}{r} 5x + y = 1 \\ \underline{-3x - y = -3} \\ 2x = -2 \\ x = -1 \end{array}$$

Substitute into the first equation:

$$5(-1) + y = 1$$
$$-5 + y = 1$$
$$y = 6$$

The system is consistent; solution is $(-1, 6)$.

29. $\begin{cases} -3x + y = 6 \\ 6x - 2y = -12 \end{cases}$

Multiply the first equation by 2, and then use the addition method:

$$\begin{array}{r} -6x + 2y = 12 \\ \underline{6x - 2y = -12} \\ 0 = 0 \end{array}$$

This statement is always true, so the system is dependent. The equations represent the same line.

The solution is $(x, 3x + 6)$.

30. $\begin{cases} 3x + 4y = 8 \\ y = -\dfrac{3}{4}x + 4 \end{cases}$

Multiply the second equation by −4, and then use the addition method:

$$\begin{array}{r} 3x + 4y = 8 \\ \underline{-4y = 3x - 16} \\ 3x = 3x - 8 \\ 0 = - 8 \end{array}$$

This statement is never true, so the system is inconsistent. The lines are parallel.

31. $\begin{cases} y = -\dfrac{1}{2}x + 10 \\ x + 2y = 6 \end{cases}$

First, multiply the first equation by −2, then use the addition method:

$$\begin{array}{r} -2y = x - 20 \\ \underline{x + 2y = 6} \\ x = x - 14 \\ 0 = -14 \end{array}$$

This statement is never true, so the system is inconsistent. The lines are parallel.

32. $\begin{cases} y = \dfrac{1}{3}x + 10 \\ x - 3y = -30 \end{cases}$

Multiply the first equation by 3, and then use the addition method:

$$\begin{array}{r} 3y = x + 30 \\ \underline{x - 3y = -30} \\ x = x \end{array}$$

This statement is always true, so the system is dependent. The equations represent the same line.

The solution is $\left(x, \dfrac{1}{3}x + 10\right)$.

33. Let A be Alice's age now and J be John's age now. Translate the problem into a system of two linear equations:
$$\begin{cases} A = J + 8 \\ A + 5 = 2(J + 5) \end{cases}$$
Solve by substitution:
$$(J + 8) + 5 = 2(J + 5)$$
$$J + 13 = 2J + 10$$
$$3 = J$$
$$A = (3) + 8 = 11$$
Alice is 11 years old and John is 3 years old.

34. Let q be the number of quarters and d be the number of dimes. Translate the problem into a system of two linear equations:
$$\begin{cases} 0.10 \cdot d + 0.25 \cdot q = 2.10 \\ d = 8q \end{cases}$$
Solve by substitution:
$$0.1(8q) + 0.25q = 2.1$$
$$0.8q + 0.25q = 2.1$$
$$1.05q = 2.1$$
$$q = 2$$
$$d = 8(2) = 16$$
Jorge has 2 quarters and 16 dimes.

35. Let x be one of the numbers and y be the other.
$$\begin{cases} x + y = -12 \\ x = 2y + 3 \end{cases}$$
Solve by substitution:
$$(2y + 3) + y = -12$$
$$3y + 3 = -12$$
$$3y = -15$$
$$y = -5$$
$$x = 2(-5) + 3 = -7$$
The two numbers are -5 and -7.

36. Let x be the smaller number and y be the larger.
$$\begin{cases} y - x = 2 \\ 3x = y + 8 \end{cases}$$
Solve the first equation for x and use substitution:
$$3x = (2 + x) + 8$$
$$3x = x + 10$$
$$2x = 10$$
$$x = 5$$
$$y = 2 + (5) = 7$$
The two numbers are 5 and 7.

37. Let x be the hours hiked at 2 mph. Then $6 - x$ is the number of hours hiked at 3 mph. Translate the problem into an equation and solve:
$$2x + 3(6 - x) = 14.4$$
$$2x + 18 - 3x = 14.4$$
$$18 - x = 14.4$$
$$3.6 = x$$
Karl hiked 3.6 hrs at 2 mph and 2.4 hrs at 3 mph.

38. Let w be the width of the rectangle. Then $w + 5$ is the length. Use the formula for perimeter to find the dimensions of the rectangle:
$$2w + 2(w + 5) = 80$$
$$2w + 2w + 10 = 80$$
$$4w = 70$$
$$w = 17.5$$
$$w + 5 = 22.5$$
So the rectangle is 17.5 meters by 22.5 meters.

39. Let x be the number of shirts sold for $110 each and y be the number of shirts sold for $65 each.
$$\begin{cases} x + y = 50 \\ 110x + 65y = 4600 \end{cases}$$
Solve the first equation for x and use substitution:
$$x = 50 - y$$
$$110(50 - y) + 65y = 4600$$
$$5500 - 110y + 65y = 4600$$
$$-45y = -900$$
$$y = 20$$
$$x = 50 - (20) = 30$$
30 shirts were sold for $110 each and 20 shirts were sold for $65 each.

40. Plug each of the given points into the equation $y = mx + b$:
$$\begin{cases} 5 = 2m + b \\ -1 = -m + b \end{cases}$$
Multiply the second equation by 2. Solve the system for b using the addition method:
$$\begin{array}{rl} 5 = & 2m + b \\ -2 = & -2m + 2b \\ \hline 3 = & 3b \\ 1 = & b \end{array}$$
Substitute into the second equation:
$$-1 = -m + 1$$
$$-2 = -m$$
$$2 = m$$
The equation is $y = 2x + 1$.

41. Let x be the amount invested at 6% and y be the amount invested at 8%.

$$\begin{cases} x + y = 20{,}000 \\ 0.06x + 0.08y = 1300 \end{cases}$$

Multiply the second equation by 100 to remove the decimals. Solve the first equation for x and use substitution to solve the system:

$$x = 20{,}000 - y$$
$$6x + 8y = 130{,}000$$
$$6(20{,}000 - y) + 8y = 130{,}000$$
$$120{,}000 - 6y + 8y = 130{,}000$$
$$2y = 10{,}000$$
$$y = 5000$$
$$x = 20{,}000 - (5000) = 15{,}000$$

So Mr. Smith has $5000 invested at 8% and $15,000 invested at 6%.

42. Let x be the amount invested at 8% and y be the amount invested at 5%.

$$\begin{cases} y = 2x \\ 0.08x + 0.05y = 720 \end{cases}$$

Multiply the second equation by 100 to remove the decimals. Use substitution to solve the system:

$$8x + 5y = 72{,}000$$
$$8x + 5(2x) = 72{,}000$$
$$8x + 10x = 72{,}000$$
$$18x = 72{,}000$$
$$x = 4000$$
$$y = 2(4000) = 8000$$

So Mary Jane has $4000 invested at 8% and $8000 invested at 5%.

43. Let x be the amount invested at 7% and y be the amount invested at 10%.

$$\begin{cases} x = y \\ 0.07x + 0.10y = 1275 \end{cases}$$

Multiply the second equation by 100 to remove the decimals. Use substitution to solve the system:

$$7x + 10y = 127{,}500$$
$$7x + 10(x) = 127{,}500$$
$$17x = 127{,}500$$
$$x = 7500$$
$$y = 7500$$

So there is $7500 invested at each rate.

44. Let x be the amount invested at 4.5% and y be the amount invested at 7%.

$$\begin{cases} x = y \\ 0.045x + 0.07y = 2530 \end{cases}$$

Multiply the second equation by 1000 to remove all decimals. Use substitution to solve the system:

$$45x + 70y = 2{,}530{,}000$$
$$45x + 70x = 2{,}530{,}000$$
$$115x = 2{,}530{,}000$$
$$x = 22{,}000$$
$$y = 22{,}000$$

So there should be $22,000 invested at each rate.

45. Let x be the number of gallons of 25% salt solution and y be the number of gallons of 40% solution.

$$\begin{cases} x + y = 60 \\ 0.25x + 0.40y = 0.35 \cdot 60 \end{cases}$$

Multiply the second equation by 100 to remove the decimals. Solve the first equation for x and use substitution to solve the system:

$$x = 60 - y$$
$$25x + 40y = 2100$$
$$25(60 - y) + 40y = 2100$$
$$1500 - 25y + 40y = 2100$$
$$15y = 600$$
$$y = 40$$
$$x = 60 - (40) = 20$$

So 20 gallons of 25% solution and 40 gallons of 40% solution are required.

46. Let x be the number of pounds that is 22% fat and y be the number of pounds that is 10% fat.

$$\begin{cases} x + y = 80 \\ 0.22x + 0.10y = 0.16 \cdot 80 \end{cases}$$

Multiply the second equation by 100 to remove the decimals. Solve the first equation for x and use substitution to solve the system:

$$x = 80 - y$$
$$22x + 10y = 1280$$
$$22(80 - y) + 10y = 1280$$
$$1760 - 22y + 10y = 1280$$
$$-12y = -480$$
$$y = 40$$
$$x = 80 - (40) = 40$$

So 40 pounds of each are required.

47. Let x be the number of ounces of pure acid and y be the number of ounces of 10% solution.

$$\begin{cases} x + y = 50 \\ 1.00x + 0.10y = 0.64 \cdot 50 \end{cases}$$

Multiply the second equation by 100 to remove all decimals. Solve the first equation for x and use substitution to solve the system:

$$x = 50 - y$$
$$100x + 10y = 32$$
$$100(50 - y) + 10y = 3200$$
$$5000 - 100y + 10y = 3200$$
$$-90y = -1800$$
$$y = 20$$
$$x = 50 - 20 = 30$$

So 30 ounces of pure acid and 20 ounces of 10% solution are required.

48. Let x be the number of tons of 20% alloy and y be the number of tons of 60% solution.

$$\begin{cases} x + y = 100 \\ 0.20x + 0.60y = 0.28 \cdot 100 \end{cases}$$

Multiply the second equation by 100 to remove all decimals. Solve the first equation for x and use substitution to solve the system:

$$x = 100 - y$$
$$20x + 60y = 2800$$
$$20(100 - y) + 60y = 2800$$
$$2000 - 20y + 60y = 2800$$
$$40y = 800$$
$$y = 20$$
$$x = 100 - 20 = 80$$

So 80 tons of 20% alloy and 20 tons of 60% alloy are required.

49. $\begin{cases} y > 2x - 3 \\ y < x + 2 \end{cases}$

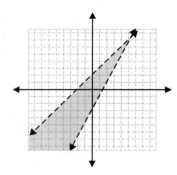

50. $\begin{cases} 2x + y \geq 12 \\ y \leq x + 3 \end{cases}$

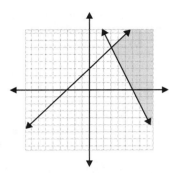

51. $\begin{cases} y \geq x - 2 \\ y \leq 2x + 3 \end{cases}$

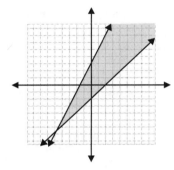

52. $\begin{cases} y < -x \\ y < 2x \end{cases}$

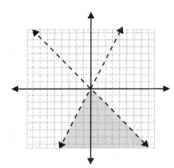

53. $\begin{cases} x \geq -1 \\ y \geq -1 \end{cases}$

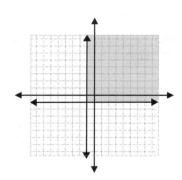

54. $\begin{cases} x < 2 \\ y \geq -2x - 3 \end{cases}$

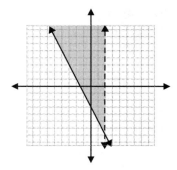

55. $\begin{cases} y \leq 0 \\ y \leq 2x + 2 \end{cases}$

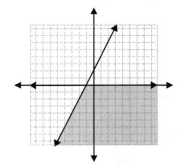

56. $\begin{cases} 2x + y > 10 \\ x - y > 0 \end{cases}$

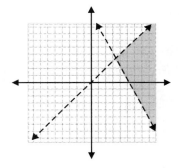

[End of Chapter 5 Review]

Chapter 5 Test
Solutions to All Exercises

1. $\begin{cases} 3x - 7y = 5 \\ 5x - 2y = -11 \end{cases}$

a. $(1,8)$: $\quad 3(1) - 7(8) \overset{?}{=} 5 \qquad 5(1) - 2(8) \overset{?}{=} -11$
$\qquad\qquad\qquad -53 \neq 5 \qquad\qquad -11 = -11$

b. $(4,1)$: $\quad 3(4) - 7(1) \overset{?}{=} 5 \qquad 5(4) - 2(1) \overset{?}{=} -11$
$\qquad\qquad\qquad 5 = 5 \qquad\qquad\quad 18 \neq -11$

c. $(-3,-2)$: $\quad 3(-3) - 7(-2) \overset{?}{=} 5 \quad 5(-3) - 2(-2) \overset{?}{=} -11$
$\qquad\qquad\qquad 5 = 5 \qquad\qquad\quad -11 = -11$

d. $(13,10)$: $\quad 3(13) - 7(10) \overset{?}{=} 5 \quad 5(13) - 2(10) \overset{?}{=} -11$
$\qquad\qquad\qquad -31 \neq 5 \qquad\qquad -45 \neq -11$

The point **c** satisfies the given system of equations.

2. $\begin{cases} x - 2y = 7 \\ 2x - 3y = 5 \end{cases}$

a. $(0,3)$: $\quad 0 - 2(3) \overset{?}{=} 7 \qquad 2(0) - 3(3) \overset{?}{=} 5$
$\qquad\qquad\quad -6 \neq 7 \qquad\qquad\quad -9 \neq 5$

b. $(7,0)$: $\quad 7 - 2(0) \overset{?}{=} 7 \qquad 2(7) - 3(0) \overset{?}{=} 5$
$\qquad\qquad\quad 7 = 7 \qquad\qquad\quad 14 \neq 5$

c. $(1,-1)$: $\quad 1 - 2(-1) \overset{?}{=} 7 \qquad 2(1) - 3(-1) \overset{?}{=} 5$
$\qquad\qquad\quad 3 \neq 7 \qquad\qquad\qquad 5 = 5$

d. $(-11,-9)$: $\quad -11 - 2(-9) \overset{?}{=} 7 \quad 2(-11) - 3(-9) \overset{?}{=} 5$
$\qquad\qquad\qquad 7 = 7 \qquad\qquad\qquad 5 = 5$

The point **d** satisfies the given system of equations.

3. Solve by graphing:
$\begin{cases} y = 2 - 5x \\ x - y = 6 \end{cases}$

The system is consistent: solution is $\left(\dfrac{4}{3}, \dfrac{-14}{3} \right)$.

4. Solve by graphing:
$\begin{cases} x - y = 3 \\ 2x + 3y = 11 \end{cases}$

The system is consistent: solution is $(4,1)$.

5. Solve the system by substitution:
$\begin{cases} 5x - 2y = 0 \\ y = 3x + 4 \end{cases}$

$5x - 2(3x + 4) = 0$

$5x - 6x - 8 = 0$

$-x = 8$

$x = -8$

$y = 3(-8) + 4 = -20$

The system is consistent: solution is $(-8, -20)$.

6. Solve by substitution:
$\begin{cases} x = \dfrac{1}{3}y - 4 \\ 2x + \dfrac{3}{2}y = 5 \end{cases}$

$2\left(\dfrac{1}{3}y - 4 \right) + \dfrac{3}{2}y = 5$

$\dfrac{2}{3}y - 8 + \dfrac{3}{2}y = 5$

$6\left(\dfrac{2}{3}y + \dfrac{3}{2}y \right) = 6(13)$

$4y + 9y = 78$

$13y = 78$

$y = 6$

$x = \dfrac{1}{3}(6) - 4 = 2 - 4 = -2$

The system is consistent: solution is $(-2, 6)$.

7. Solve by the addition method:
$$\begin{cases} -2x + 3y = 6 \\ 4x + y = 1 \end{cases}$$
Multiply the first equation by 2, and then add.
$$-4x + 6y = 12$$
$$\underline{4x + y = 1}$$
$$7y = 13$$
$$y = \frac{13}{7}$$
Substitute into the second equation:
$$4x + \left(\frac{13}{7}\right) = 1$$
$$4x = -\frac{6}{7}$$
$$x = -\frac{3}{14}$$
The system is consistent: solution is $\left(\frac{-3}{14}, \frac{13}{7}\right)$.

8. Solve by the addition method:
$$\begin{cases} 3x + 4y = 10 \\ x + 6y = 1 \end{cases}$$
Multiply the second equation by -3, and then add.
$$3x + 4y = 10$$
$$\underline{-3x - 18y = -3}$$
$$-14y = 7$$
$$y = -\frac{1}{2}$$
Substitute into the second equation:
$$x + 6\left(-\frac{1}{2}\right) = 1$$
$$x - 3 = 1$$
$$x = 4$$
The system is consistent: solution is $\left(4, \frac{-1}{2}\right)$.

9. Solve by the addition method:
$$\begin{cases} x + y = 2 \\ y = -2x - 1 \end{cases}$$
Multiply the second equation by -1, and then add.
$$x + y = 2$$
$$\underline{ - y = 2x + 1}$$
$$x = 2x + 3$$
$$x = -3$$
Substitute into the second equation:
$$y = -2(-3) - 1 = 5$$
The system is consistent: solution is $(-3, 5)$.

10. Solve by substitution:
$$\begin{cases} 6x + 2y - 8 = 0 \\ y = -3x \end{cases}$$
$$6x + 2(-3x) - 8 = 0$$
$$6x - 6x - 8 = 0$$
$$-8 = 0$$
This is never true, so the system is inconsistent.

11. Solve by the addition method:
$$\begin{cases} 7x + 5y = -9 \\ 6x + 2y = 6 \end{cases}$$
Multiply the first equation by 2 and the second by -5, and add to eliminate y.
$$14x + 10y = -18$$
$$\underline{-30x - 10y = -30}$$
$$-16x = -48$$
$$x = 3$$
Substitute into the first equation:
$$7(3) + 5y = -9$$
$$21 + 5y = -9$$
$$5y = -30$$
$$y = -6$$
The system is consistent: solution is $(3, -6)$.

12. Solve by the addition method:
$$\begin{cases} x + 3y = -2 \\ 2x = -6y - 4 \end{cases}$$
Divide the second equation by -2, then add.
$$x + 3y = -2$$
$$\underline{-x = 3y + 2}$$
$$3y = 3y$$
This is always true, so the system is dependent. The equations represent the same line. The solution is $\left(x, \frac{-1}{3}x - \frac{2}{3}\right)$.

13. Solve graphically:
$$\begin{cases} x \geq -3 \\ y \geq 3x \end{cases}$$

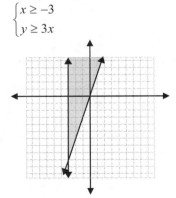

14. Substitute the x and y values into the linear equation $ax + by = 11$ to form a system of equations:

$$\begin{cases} a - 3b = 11 \\ 2a + 5b = 11 \end{cases}$$

Multiply the first equation by -2 and solve by the addition method.

$$\begin{array}{r} -2a + 6b = -22 \\ \underline{2a + 5b = 11} \\ 11b = -11 \\ b = -1 \end{array}$$

Substitute into the first equation:

$$\begin{aligned} a - 3(-1) &= 11 \\ a + 3 &= 11 \\ a &= 8 \end{aligned}$$

The equation is $8x - y = 11$.

15. Let b be the rate of the boat in still water.
Let c be the rate of the current.

$$\begin{cases} 4(b - c) = 48 \\ 3(b + c) = 48 \end{cases} = \begin{cases} 4b - 4c = 48 \\ 3b + 3c = 48 \end{cases}$$

Multiply the first equation by 3 and the second by 4, then solve by the addition method.

$$\begin{array}{r} 12b - 12c = 144 \\ \underline{12b + 12c = 192} \\ 24b = 336 \\ b = 14 \end{array}$$

Substitute into the second equation:

$$\begin{aligned} 3(14 + c) &= 48 \\ 42 + 3c &= 48 \\ 3c &= 6 \\ c &= 2 \end{aligned}$$

The rate of boat in still water is 14 mph, and the rate of the current is 2 mph.

16. Let x be the price of pens and y be the price of pencils.

$$\begin{cases} 2x + 8y = 2.22 \\ 3x + 4y = 2.69 \end{cases}$$

Multiply the second equation by -2 and solve by the addition method.

$$\begin{array}{r} 2x + 8y = 2.22 \\ \underline{-6x - 8y = -5.38} \\ -4x = -3.16 \\ x = 0.79 \end{array}$$

Substitute into the second equation:

$$\begin{aligned} 3(0.79) + 4y &= 2.69 \\ 2.37 + 4y &= 2.69 \\ 4y &= 0.32 \\ y &= 0.08 \end{aligned}$$

Pencils cost \$0.08 each; pens cost \$0.79 each.

17. Let x be the amount invested at 8%, and let y be the amount invested at 6%.

$$\begin{cases} x = 2y - 320 \\ 0.08x + 0.06y = 185.60 \end{cases}$$

Multiply the second equation by 100 to remove the decimals. Convert the first equation to standard form and multiply it by 3, then solve by the addition method.

$$\begin{array}{r} 3x - 6y = -960 \\ \underline{8x + 6y = 18560} \\ 11x = 17600 \\ x = 1600 \end{array}$$

Substitute into the first equation:

$$\begin{aligned} (1600) &= 2y - 320 \\ 1920 &= 2y \\ 960 &= y \end{aligned}$$

\$960 is invested at 6% and \$1600 is invested at 8%.

18. Let x be the number of pounds of 83% copper and y be the number of pounds of 68% copper.

$$\begin{cases} x + y = 2000 \\ 0.83x + 0.68y = 0.80(2000) \end{cases}$$

Multiply the second equation by 100 to remove the decimals. Solve the first equation for x and solve by substitution.

$$\begin{aligned} x &= 2000 - y \\ 83x + 68y &= 160,000 \\ 83(2000 - y) + 68y &= 160,000 \\ 166,000 - 83y + 68y &= 160,000 \\ -15y &= -6,000 \\ y &= 400 \\ x &= 2000 - (400) \\ x &= 1600 \end{aligned}$$

400 lb of 68% copper and 1600 lb of 83% copper must be used.

19. Let W be the width and L be the length, in inches.
$$\begin{cases} 2W + 2L = 60 \\ \qquad L = W + 4 \end{cases}$$
Solve by substitution:
$$2W + 2(W + 4) = 60$$
$$4W + 8 = 60$$
$$4W = 52$$
$$W = 13$$
$$L = (13) + 4 = 17$$
The rectangle is 17 inches by 13 inches.

20. Let n be the number of nickels and let q be the number of quarters.
$$\begin{cases} \qquad\quad n + q = 105 \\ 0.05n + 0.25q = 17.25 \end{cases}$$
Multiply the second equation by 100 to remove the decimals. Solve the first equation for n and solve by substitution.
$$n = 105 - q$$
$$5n + 25q = 1725$$
$$5(105 - q) + 25q = 1725$$
$$525 - 5q + 25q = 1725$$
$$20q = 1200$$
$$q = 60$$
$$n = 105 - (60)$$
$$n = 45$$
There are 45 nickels and 60 quarters.

[End of Chapter 5 Test]

Cumulative Review: Chapters 1 – 5
Solutions to All Exercises

1. $-20 + 15 \div (-5) \cdot 2^3 - 11^2 = -20 + 15 \div (-5) \cdot 8 - 121$
$$= -20 - 3 \cdot 8 - 121$$
$$= -20 - 24 - 121$$
$$= -165$$

2. $24 \div 4 \cdot 6 - 36 \cdot 2 \div 3^2 = 24 \div 4 \cdot 6 - 36 \cdot 2 \div 9$
$$= 6 \cdot 6 - 72 \div 9$$
$$= 36 - 8$$
$$= 28$$

3. $\dfrac{3}{4} \div \dfrac{5}{8} - \dfrac{2}{3} \cdot \dfrac{6}{5} + \left(\dfrac{1}{5}\right)^2 = \dfrac{3}{4} \cdot \dfrac{8}{5} - \dfrac{4}{5} + \dfrac{1}{25}$
$$= \dfrac{6}{5} - \dfrac{4}{5} + \dfrac{1}{25}$$
$$= \dfrac{2}{5} + \dfrac{1}{25}$$
$$= \dfrac{10}{25} + \dfrac{1}{25}$$
$$= \dfrac{11}{25}$$

4. $3\left(4^2 - 16\right) + 5\left(9 - 3^2\right) - \left(6^2 - 9 \cdot 4\right)$
$$= 3(16 - 16) + 5(9 - 9) - (36 - 36)$$
$$= 3 \cdot 0 + 5 \cdot 0 - 0$$
$$= 0$$

5. 110% of $50 = 1.10(50) = 55$

6. $\dfrac{7}{8}$ of $10,000 = \dfrac{7}{8}(10,000) = 8750$

7. $2(x - 7) + 14 = -3x + 1$
$$2x - 14 + 14 = -3x + 1$$
$$5x = 1$$
$$x = \dfrac{1}{5}$$

8. $\dfrac{5x}{6} - \dfrac{2}{3} = \dfrac{x}{2} + \dfrac{1}{4}$
$$10x - 8 = 6x + 3$$
$$4x = 11$$
$$x = \dfrac{11}{4}$$

9. $2.3y - 1.6 = 3(1.2y + 2.5)$
$$2.3y - 1.6 = 3.6y + 7.5$$
$$23y - 16 = 36y + 75$$
$$-91 = 13y$$
$$-7 = y$$

10. $\dfrac{a}{5} + \dfrac{a}{7} = \dfrac{a}{2} - \dfrac{5}{14}$
$$70\left(\dfrac{a}{5} + \dfrac{a}{7}\right) = 70\left(\dfrac{a}{2} - \dfrac{5}{14}\right)$$
$$14a + 10a = 35a - 25$$
$$24a = 35a - 25$$
$$-11a = -25$$
$$a = \dfrac{25}{11}$$

11. $3x - 14 \geq 25$
$$3x \geq 39$$
$$x \geq 13$$

12. $0 \leq 2x + 10 < 1.6$
$$-10 \leq 2x < -8.4$$
$$-5 \leq x < -4.2$$

13. Solve $v = k + gt$ for t:
$$v - k = gt$$
$$t = \dfrac{v - k}{g}$$

14. Solve $3x - 4y = 6$ for y:
$$-4y = 6 - 3x$$
$$y = \dfrac{6 - 3x}{-4} \text{ or } y = \dfrac{3x - 6}{4}$$

15. $2x + 4y = 11$

$4y = -2x + 11$

$y = -\dfrac{1}{2}x + \dfrac{11}{4}$

$m = -\dfrac{1}{2}$ and $b = \dfrac{11}{4}$

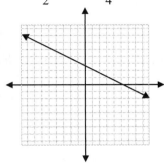

16. First, note that the slope of $x - 3y = -1$ is $\dfrac{1}{3}$. Using the point-slope form of a line, the line parallel to this line through the point $(1, 2)$ is $y - 2 = \dfrac{1}{3}(x - 1)$.

Simplify and put in standard form:

$3y - 6 = x - 1$

$x - 3y = -5$

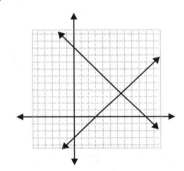

17. $\begin{cases} x - 2y = 6 \\ y = \dfrac{1}{2}x - 3 \end{cases}$

a. $(0, 6)$: $\quad 0 - 2(6) \overset{?}{=} 6 \qquad 6 \overset{?}{=} \dfrac{1}{2}(0) - 3$

$\qquad\qquad\qquad -12 \neq 6 \qquad\quad 6 \neq -3$

b. $(6, 0)$: $\quad 6 - 2(0) \overset{?}{=} 6 \qquad 0 \overset{?}{=} \dfrac{1}{2}(6) - 3$

$\qquad\qquad\qquad\quad 6 = 6 \qquad\qquad 0 = 0$

c. $(2, -2)$: $\quad 2 - 2(-2) \overset{?}{=} 6 \qquad -2 \overset{?}{=} \dfrac{1}{2}(2) - 3$

$\qquad\qquad\qquad\quad 6 = 6 \qquad\qquad -2 = -2$

d. $\left(3, -\dfrac{3}{2}\right)$: $\quad 3 - 2\left(-\dfrac{3}{2}\right) \overset{?}{=} 6 \qquad -\dfrac{3}{2} \overset{?}{=} \dfrac{1}{2}(3) - 3$

$\qquad\qquad\qquad\quad 6 = 6 \qquad\qquad\qquad -\dfrac{3}{2} = -\dfrac{3}{2}$

Points **b**, **c**, and **d** satisfy both equations. In fact, the system is dependent.

18. Graph to estimate the solution:

$\begin{cases} x = y + 3 \\ x + y = 10 \end{cases} = \begin{cases} y = x - 3 \\ y = -x + 10 \end{cases}$

Enter "Y1 = X–3" and "Y2 = –X+10" into your calculator and then use the TRACE button to estimate the point of intersection. Your estimate should be close to $(6.5, 3.5)$. To obtain the exact solution, it is necessary to use one of the algebraic methods for solving systems. This is a consistent system.

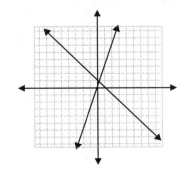

19. Graph to estimate the solution:

$\begin{cases} x + y = 1 \\ 3x - y = 0 \end{cases} = \begin{cases} y = 1 - x \\ y = 3x \end{cases}$

Enter "Y1 = 1–X" and "Y2 = 3X" into your calculator and use the TRACE button to estimate the point of intersection. Your estimate should be close to $(0.25, 0.75)$. To obtain the exact solution, it is necessary to use one of the algebraic methods for solving systems. This is a consistent system.

20. Graph to estimate the solution:
$$\begin{cases} 2x + 3y = 4 \\ 3x - y = 6 \end{cases}$$

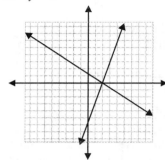

Your estimate should be close to $(2, 0)$. To obtain the exact solution, it is necessary to use one of the algebraic methods for solving systems. This is a consistent system.

21. Graph to estimate the solution:
$$\begin{cases} 5x + 2y = 3 \\ y = 4 \end{cases}$$

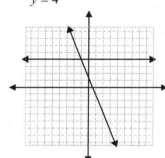

Your estimate should be close to $(-1, 4)$. To obtain the exact solution, it is necessary to use one of the methods shown for solving systems. This is a consistent system.

22. Graph to estimate the solution:
$$\begin{cases} 3x + y = 3 \\ x - 3y = 4 \end{cases}$$

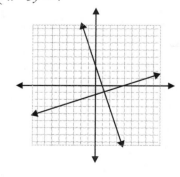

Your estimate should be close to $\left(\dfrac{13}{10}, \dfrac{-9}{10} \right)$. To obtain the exact solution, it is necessary to use one of the methods shown for solving systems. This is a consistent system.

23. Graph to estimate the solution:
$$\begin{cases} y = 4x - 6 \\ 8x - 2y = -4 \end{cases}$$

Notice that these two lines have the same slope:
$$m = 4$$

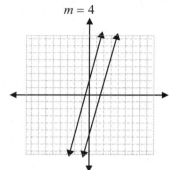

The lines are parallel. The system is inconsistent.

24. Solve by substitution:
$$\begin{cases} x + y = -4 \\ 2x + 7y = 2 \end{cases} = \begin{cases} x = -4 - y \\ 2x + 7y = 2 \end{cases}$$

Substitute:
$$2(-4 - y) + 7y = 2$$
$$-8 - 2y + 7y = 2$$
$$5y = 10$$
$$y = 2$$
$$x = -4 - (2) = -6$$

Consistent system: solution is $(-6, 2)$.

25. Solve by substitution:
$$\begin{cases} x = 2y \\ y = \dfrac{1}{2}x + 9 \end{cases}$$

$$y = \frac{1}{2}(2y) + 9$$
$$y = y + 9$$
$$0 = 9$$

This is never true, so the system is inconsistent.

26. Solve by substitution:

$$\begin{cases} 4x + 3y = 8 \\ x + \dfrac{3}{4}y = 2 \end{cases}$$

$$x = 2 - \frac{3}{4}y$$

$$4\left(2 - \frac{3}{4}y\right) + 3y = 8$$

$$8 - 3y + 3y = 8$$

$$0 = 0$$

This result is always true. The system is dependent.

The solution is $\left(x, \dfrac{-4}{3}x + \dfrac{8}{3}\right)$.

27. Solve by substitution:

$$\begin{cases} 2x + y = 0 \\ 7x + 6y = -10 \end{cases}$$

$$y = -2x$$

$$7x + 6(-2x) = -10$$

$$7x - 12x = -10$$

$$-5x = -10$$

$$x = 2$$

$$y = -2(2) = -4$$

Consistent system: solution is $(2, -4)$.

28. Solve by the addition method:

$$\begin{cases} 2x + y = 7 \\ 2x - y = 1 \end{cases}$$

$$4x = 8$$

$$x = 2$$

$$2(2) - y = 1$$

$$y = 3$$

Consistent system: solution is $(2, 3)$.

29. Solve by the addition method:

$$\begin{cases} 3x - 2y = 9 \\ x - 2y = 11 \end{cases}$$

$$\begin{array}{rcl} 3x - 2y &=& 9 \\ -x + 2y &=& -11 \\ \hline 2x &=& -2 \\ x &=& -1 \end{array}$$

Substitute into the second equation:

$$-1 - 2y = 11$$

$$-2y = 12$$

$$y = -6$$

Consistent system: solution is $(-1, -6)$.

30. Solve by the addition method:

$$\begin{cases} 2x + 4y = 9 \\ 3x + 6y = 8 \end{cases} = \begin{cases} -6x - 12y = -27 \\ 6x + 12y = 16 \end{cases}$$

$$\begin{array}{rcl} -6x - 12y &=& -27 \\ 6x + 12y &=& 16 \\ \hline 0 &=& -11 \end{array}$$

This is never true, so the system is inconsistent.

31. Solve by the addition method:

$$\begin{cases} x + 5y = 10 \\ y = 2 - \dfrac{1}{5}x \end{cases} = \begin{cases} x + 5y = 10 \\ -x - 5y = -10 \end{cases}$$

$$\begin{array}{rcl} x + 5y &=& 10 \\ -x - 5y &=& -10 \\ \hline 0 &=& 0 \end{array}$$

This result is always true. The system is dependent.

The solution is $\left(x, 2 - \dfrac{1}{5}x\right)$.

32. Solve by the addition method:

$$\begin{cases} 3y - x = 7 \\ x - 2y = -2 \end{cases} = \begin{cases} -x + 3y = 7 \\ x - 2y = -2 \end{cases}$$

$$\begin{array}{rcl} -x + 3y &=& 7 \\ x - 2y &=& -2 \\ \hline y &=& 5 \end{array}$$

Substitute into the second equation:

$$x - 2(5) = -2$$

$$x = -2 + 10$$

$$x = 8$$

Consistent system: solution is $(8, 5)$.

33. Multiply the first equation by 15 and the second by -20 to remove the fractions and use addition:

$$\begin{cases} x - \dfrac{2}{5}y = \dfrac{4}{5} \\ \dfrac{3}{4}x + \dfrac{3}{4}y = \dfrac{5}{4} \end{cases} = \begin{cases} 15x - 6y = 12 \\ -15x - 15y = -25 \end{cases}$$

$$\begin{array}{rcl} 15x - 6y &=& 12 \\ -15x - 15y &=& -25 \\ \hline -21y &=& -13 \end{array}$$

$$y = \frac{13}{21}$$

Substitute into the second equation:

$$\frac{3}{4}x + \frac{3}{4}\left(\frac{13}{21}\right) = \frac{5}{4}$$

$$3x = 5 - \frac{13}{7} = \frac{22}{7}$$

$$x = \frac{22}{21}$$

Consistent system: solution is $\left(\dfrac{22}{21}, \dfrac{13}{21}\right)$.

34. Solve by the addition method:
$$\begin{cases} x+3y=7 \\ 5x-2y=1 \end{cases} = \begin{cases} -5x-15y=-35 \\ 5x-2y=1 \end{cases}$$

$$-5x-15y=-35$$
$$\underline{5x-2y=1}$$
$$-17y=-34$$
$$y=2$$

Substitute into the first equation:
$$x+3(2)=7$$
$$x=1$$

Consistent system: solution is $(1,2)$.

35. Solve by the addition method:
$$\begin{cases} 3x-5y=17 \\ x+2y=4 \end{cases} = \begin{cases} 3x-5y=17 \\ -3x-6y=-12 \end{cases}$$

$$3x-5y=17$$
$$\underline{-3x-6y=-12}$$
$$-11y=5$$
$$y=-\frac{5}{11}$$

Substitute into the second equation:
$$x+2\left(\frac{-5}{11}\right)=4$$
$$x=4+\frac{10}{11}$$
$$x=\frac{54}{11}$$

Consistent system: solution is $\left(\frac{54}{11},\frac{-5}{11}\right)$.

36. Substitute the given points in $y=mx+b$:
$$\begin{cases} -1=3m+b \\ 6=2m+b \end{cases} = \begin{cases} 1=-3m-b \\ 6=2m+b \end{cases}$$

Use the addition method:
$$1=-3m-b$$
$$\underline{6=2m+b}$$
$$7=-m$$
$$-7=m$$

Substitute into the second equation:
$$6=2(-7)+b$$
$$6+14=b$$
$$20=b$$

The equation is $y=-7x+20$.

37. Substitute the given points in $y=mx+b$:
$$\begin{cases} 5=-2m+b \\ -3=4m+b \end{cases} = \begin{cases} 5=-2m+b \\ 3=-4m-b \end{cases}$$

Use the addition method:
$$5=-2m+b$$
$$\underline{3=-4m-b}$$
$$8=-6m$$
$$-\frac{4}{3}=m$$

Substitute into the second equation:
$$-3=4\left(\frac{-4}{3}\right)+b$$
$$-3+\frac{16}{3}=b$$
$$\frac{7}{3}=b$$

The equation is $y=-\frac{4}{3}x+\frac{7}{3}$.

38. a. Let x be the first angle. Then $3x+10$ is the second angle. The angles are complimentary, so set up the following equation and solve for x:
$$x+(3x+10)=90$$
$$4x+10=90$$
$$4x=80$$
$$x=\frac{80}{4}=20°$$
$$3x+10=3(20)+10$$
$$=60+10$$
$$=70°$$

So the two angles measure 20° and 70°.

b. Let x be the first angle. Then $\frac{1}{2}x-30$ is the second angle. The angles are supplementary, so set up the following equation and solve for x:
$$x+\left(\frac{1}{2}x-30\right)=180$$
$$\frac{3}{2}x-30=180$$
$$3x-60=360$$
$$3x=420$$
$$x=\frac{420}{3}=140$$
$$\frac{1}{2}x-30=\frac{1}{2}(140)-30=40$$

So the two angles measure 140° and 40°.

39. Let x be the number of 58¢ stamps. Then $15 - x$ is the number of 41¢ stamps.

$$0.58x + 0.41(15 - x) = 7$$
$$58x + 41(15 - x) = 700$$
$$58x + 615 - 41x = 700$$
$$17x = 85$$
$$x = 5$$
$$15 - x = 10$$

There were five 58¢ and ten 41¢ stamps.

40. Let b be the boat's rate in still water and c be the rate of the current.

$$\begin{cases} 2(b+c) = 24 \\ 3(b-c) = 24 \end{cases} = \begin{cases} 6b + 6c = 72 \\ 6b - 6c = 48 \end{cases}$$

Use the addition method:

$$\begin{array}{r} 6b + 6c = 72 \\ 6b - 6c = 48 \\ \hline 12b \quad\;\; = 120 \\ b \quad\;\; = 10 \end{array}$$

Substitute into the first equation:

$$2(10 + c) = 24$$
$$20 + 2c = 24$$
$$2c = 4$$
$$c = 2$$

The rate of boat in still water is 10 mph, and the rate of the current is 2 mph.

41. Let W be the width in yards and $L = W + 3$.

$$2W + 2(W + 3) = 50$$
$$4W + 6 = 50$$
$$4W = 44$$
$$W = 11$$
$$L = 14$$

Width is 11 yards and length is 14 yards.

42. Let A be the number of Model A and B be the number of Model B.

$$\begin{cases} 18A + 7B = 198 \\ 4A + 2B = 52 \end{cases} = \begin{cases} 18A + 7B = 198 \\ -14A - 7B = -182 \end{cases}$$

Use the addition method:

$$\begin{array}{r} 18A + 7B = 198 \\ -14A - 7B = -182 \\ \hline 4A \quad\;\; = 16 \\ A \quad\;\; = 4 \end{array}$$

Substitute into the second equation:

$$4(4) + 2B = 52$$
$$2B = 36$$
$$B = 18$$

There were 4 of Model A and 18 of Model B.

43. Let r be the rate of east bound train. Then, $r + 5$ is the rate of west bound train.

$$6r + 6(r + 5) = 510$$
$$12r + 30 = 510$$
$$12r = 480$$
$$r = 40$$

The east bound train's rate is 40 mph and the west bound train's rate is 45 mph.

44. Let N be the bacteria count at time t. Use the point values $(1, 100)$ and $(3, 2000)$ to find the slope. Then, use the point slope form to write an equation:

$$m = \frac{2000 - 100}{3 - 1} = \frac{1900}{2} = 950$$
$$N - 100 = 950(t - 1)$$
$$N = 950t - 850$$

45. $\{A(-5, -2), B(-3, -2), C(-2, 4), D(1, -1), E(2, 4)\}$

This is a function.

$$D = \{-5, -3, -2, 1, 2\}$$
$$R = \{-2, -1, 4\}$$

46. The graph does not represent a function. It fails the vertical line test; a vertical line would intersect the graph at the two points $(-1, 1)$ and $(-1, -1)$.

47. The graph does represent a function.

48. $F(x) = 3x - 7$

a. $F(-5) = 3(-5) - 7$
$$= -15 - 7 = -22$$

b. $F(10) = 3(10) - 7$
$$= 30 - 7 = 23$$

c. $F(3.5) = 3(3.5) - 7$
$$= 10.5 - 7 = 3.5$$

49. $f(x) = 2x^2 + 3x - 10$

    a. $f(0) = 2(0)^2 + 3(0) - 10$

                $= -10$

    b. $f(-3) = 2(-3)^2 + 3(-3) - 10$

              $= 18 - 9 - 10 = -1$

    c. $f(6) = 2(6)^2 + 3(6) - 10$

              $= 72 + 18 - 10 = 80$

50. Solve the system graphically:

$$\begin{cases} x > 2 \\ y < -x + 5 \end{cases}$$

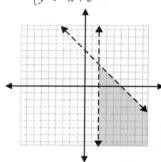

51. Solve the system graphically:

$$\begin{cases} y \geq 1 \\ y \leq 2x + 5 \end{cases}$$

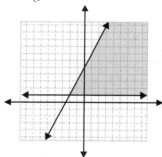

52. Solve the system graphically:

$$\begin{cases} 2x + 5y < 31 \\ x - 2y < -9 \end{cases}$$

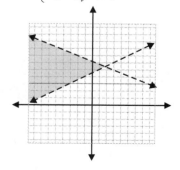

53. Solve the system graphically:

$$\begin{cases} y \leq 4x + 6 \\ y \leq -x + 2 \end{cases}$$

54. Solve the system graphically:

$$\begin{cases} 2x - 3y < 12 \\ x + y < 6 \end{cases}$$

[End of Cumulative Review: Chapters 1 – 5]

Section 6.1
Solutions to Odd Exercises

1. $3^2 \cdot 3 = 3^2 \cdot 3^1 = 3^{2+1} = 3^3 = 27$; product rule.
$3^2 \cdot 3 = 3^2 \cdot 3^1 = 9 \cdot 3 = 27$; 1 exponent rule.

3. $8^3 \cdot 8^0 = 8^{3+0} = 8^3 = 512$; product rule.
$8^3 \cdot 8^0 = 512 \cdot 1 = 512$; 0 exponent rule.

5. $3^{-1} = \dfrac{1}{3^1} = \dfrac{1}{3}$; negative exponent rule.

7. $(-5)^{-2} = \dfrac{1}{(-5)^2} = \dfrac{1}{25}$; negative exponent rule.

9. $(-2)^4 \cdot (-2)^0 = (-2)^{4+0} = (-2)^4 = 16$; product rule.
$(-2)^4 \cdot (-2)^0 = 16 \cdot 1 = 16$; 0 exponent rule.

11. $(-4)^3 \cdot (-4)^0 = (-4)^{3+0} = (-4)^3 = -64$; product rule.
$(-4)^3 \cdot (-4)^0 = -64 \cdot 1 = -64$; 0 exponent rule.

13. $-4 \cdot 5^3 = -4^1 \cdot 5^3 = -4 \cdot 125 = -500$; 1 exponent rule.

15. $3 \cdot 2^{-3} = 3 \cdot \dfrac{1}{2^3} = 3 \cdot \dfrac{1}{8} = \dfrac{3}{8}$; negative exponent rule.
$3 \cdot 2^{-3} = 3^1 \cdot \dfrac{1}{2^3} = 3^1 \cdot \dfrac{1}{8} = \dfrac{3}{8}$; 1 exponent rule.

17. $-3 \cdot 5^{-2} = -3 \cdot \dfrac{1}{5^2} = -\dfrac{3}{25}$; negative exponent rule.
$-3 \cdot 5^{-2} = -3^1 \cdot \dfrac{1}{5^2} = -\dfrac{3^1}{5^2} = -\dfrac{3}{25}$; 1 exponent rule.

19. $x^2 \cdot x^3 = x^{2+3} = x^5$; product rule.

21. $y^2 \cdot y^0 = y^{2+0} = y^2$; product rule.
$y^2 \cdot y^0 = y^2 \cdot 1 = y^2$; 0 exponent rule.

23. $x^{-3} = \dfrac{1}{x^3}$; negative exponent rule.

25. $2x^{-1} = 2 \cdot \dfrac{1}{x^1} = \dfrac{2}{x}$; negative exponent rule.

27. $-8y^{-2} = -8 \cdot \dfrac{1}{y^2} = -\dfrac{8}{y^2}$; negative exponent rule.

29. $5x^6 y^{-4} = 5x^6 \cdot \dfrac{1}{y^4} = \dfrac{5x^6}{y^4}$; negative exponent rule.

31. $3x^0 + y^0 = 3 \cdot 1 + 1 = 4$; 0 exponent rule.

33. $\dfrac{7^3}{7} = \dfrac{7^3}{7^1} = 7^{3-1} = 7^2 = 49$; quotient rule.

35. $\dfrac{10^3}{10^4} = 10^{3-4} = 10^{-1} = \dfrac{1}{10^1} = \dfrac{1}{10}$; quotient rule.
$\dfrac{10^3}{10^4} = 10^3 \cdot 10^{-4} = 10^{3-4} = 10^{-1} = \dfrac{1}{10^1} = \dfrac{1}{10}$;
negative exponent rule.

37. $\dfrac{2^3}{2^6} = 2^{3-6} = 2^{-3} = \dfrac{1}{2^3} = \dfrac{1}{8}$; quotient rule.
$\dfrac{2^3}{2^6} = 2^3 \cdot 2^{-6} = 2^{3-6} = 2^{-3} = \dfrac{1}{2^3} = \dfrac{1}{8}$; negative exponent rule.

39. $\dfrac{x^4}{x^2} = x^{4-2} = x^2$; quotient rule.

41. $\dfrac{x^3}{x} = \dfrac{x^3}{x^1} = x^{3-1} = x^2$; quotient rule.

43. $\dfrac{x^7}{x^3} = x^{7-3} = x^4$; quotient rule.

45. $\dfrac{x^{-2}}{x^2} = x^{-2-2} = x^{-4} = \dfrac{1}{x^4}$; quotient rule.
$\dfrac{x^{-2}}{x^2} = x^{-2} \cdot x^{-2} = x^{-2+(-2)} = x^{-4} = \dfrac{1}{x^4}$; negative exponent rule.

47. $\dfrac{x^4}{x^{-2}} = x^{4-(-2)} = x^6$; quotient rule.

49. $\dfrac{x^{-3}}{x^{-5}} = x^{-3-(-5)} = x^2$; quotient rule.

51. $\dfrac{y^{-2}}{y^{-4}} = y^{-2-(-4)} = y^2$; quotient rule.

53. $3x^3 \cdot x^0 = 3x^{3+0} = 3x^3$; product rule.
$3x^3 \cdot x^0 = 3x^3 \cdot 1 = 3x^3$; 0 exponent rule.

55. $\left(5x^2\right)\left(2x^2\right) = 10x^2 \cdot x^2 = 10x^{2+2} = 10x^4$; product rule.

57. $\left(4x^3\right)\left(9x^0\right) = 36x^3 \cdot x^0 = 36x^{3+0} = 36x^3$; product rule.
$\left(4x^3\right)\left(9x^0\right) = 36x^3 \cdot x^0 = 36x^3 \cdot 1 = 36x^3$; 0 exponent rule.

59. $\left(-2x^2\right)\left(7x^3\right) = -14x^2 \cdot x^3 = -14x^{2+3} = -14x^5$;
product rule.

61. $\left(-4x^5\right)\left(3x\right) = -12x^5 \cdot x^1 = -12x^{5+1} = -12x^6$; product rule.

63. $\dfrac{8y^3}{2y^2} = 4 \cdot \dfrac{y^3}{y^2} = 4y^{3-2} = 4y$; quotient rule.

65. $\dfrac{9y^5}{3y^3} = 3 \cdot \dfrac{y^5}{y^3} = 3y^{5-3} = 3y^2$; quotient rule.

67. $\dfrac{-8y^4}{4y^2} = -2 \cdot \dfrac{y^4}{y^2} = -2y^{4-2} = -2y^2$; quotient rule.

69. $\dfrac{21x^4}{-3x^2} = -7 \cdot \dfrac{x^4}{x^2} = -7x^{4-2} = -7x^2$; quotient rule.

71. $\dfrac{10^4 \cdot 10^{-3}}{10^{-2}} = \dfrac{10^{4+(-3)}}{10^{-2}} = \dfrac{10^1}{10^{-2}} = 10^{1-(-2)} = 10^3 = 1000$;

product and quotient rules.

73. $\left(9x^2\right)^0 = 1$; 0 exponent rule.

75. $\left(-3xy\right)\left(-5x^2y^{-3}\right) = 15x^1y^1 \cdot x^2y^{-3} = 15x^{1+2}y^{1+(-3)}$
$$= 15x^3y^{-2} = \dfrac{15x^3}{y^2};$$
product and negative exponent rules

77. $\dfrac{-8x^{-2}y^4}{4x^2y^{-2}} = -2\dfrac{x^{-2}y^4}{x^2y^{-2}} = -2x^{-2-2}y^{4-(-2)} = -2x^{-4}y^6$
$$= \dfrac{-2y^6}{x^4};$$
quotient and negative exponent rules.

79. $\left(3a^2b^4\right)\left(4ab^5c\right) = 12a^2b^4 \cdot a^1b^5c = 12a^{2+1}b^{4+5}c$
$$= 12a^3b^9c;$$
product rule.

81. $\dfrac{36a^5b^0c}{-9a^{-5}b^{-3}} = -4\dfrac{a^5b^0c}{a^{-5}b^{-3}} = -4a^{5-(-5)}b^{0-(-3)}c$
$$= -4a^{10}b^3c;$$
quotient rule.

Calculator Problems

83. $\left(2 \cdot 16\right)^0 = 1$

85. $1.6^{-2} = 0.390625$

87. $\left(-14.8\right)^2 \cdot 21.3^2 = 99,376.2576$

[End of Section 6.1]

Section 6.2
Solutions to Odd Exercises

1. $-3^4 = -81$

3. $(-3)^4 = 81$

5. $(-10)^6 = 1,000,000$

7. $(6x^3)^2 = 36x^6$

9. $4(-3x^2)^3 = 4(-27x^6) = -108x^6$

11. $-3(7xy^2)^0 = -3(1) = -3$

13. $-2(3x^5y^{-2})^{-3} = -2(3^{-3}x^{-15}y^6) = \dfrac{-2y^6}{27x^{15}}$

15. $\left(\dfrac{-4x}{y^2}\right)^2 = \dfrac{16x^2}{y^4}$

17. $\left(\dfrac{3x^2}{y^3}\right)^2 = \dfrac{9x^4}{y^6}$

19. $\left(\dfrac{x}{y}\right)^{-2} = \left(\dfrac{y}{x}\right)^2 = \dfrac{y^2}{x^2}$

21. $\left(\dfrac{2x}{y^5}\right)^{-2} = \left(\dfrac{y^5}{2x}\right)^2 = \dfrac{y^{10}}{4x^2}$

23. $\left(\dfrac{4a^2}{b^{-3}}\right)^{-3} = \left(\dfrac{4^{-3}a^{-6}}{b^9}\right) = \dfrac{1}{64a^6b^9}$

25. $\left(\dfrac{5xy^3}{y}\right)^2 = \dfrac{25x^2y^6}{y^2} = 25x^2y^4$

27. $\left(\dfrac{2ab^3}{b^2}\right)^4 = 16a^4b^4$

29. $\left(\dfrac{2ab^4}{b^2}\right)^{-3} = \left(\dfrac{b^2}{2ab^4}\right)^3 = \dfrac{1}{8a^3b^6}$

31. $\left(\dfrac{2x^2y}{y^3}\right)^{-4} = \left(\dfrac{y^3}{2x^2y}\right)^4 = \dfrac{y^8}{16x^8}$

33. $\left(\dfrac{2a^2b^{-1}}{b^2}\right)^3 = \dfrac{8a^6}{b^9}$

35. $\dfrac{(7x^{-2}y)^2}{(xy^{-1})^2} = \dfrac{49x^{-4}y^2}{x^2y^{-2}} = \dfrac{49y^4}{x^6}$

37. $\dfrac{(3x^2y^{-1})^{-2}}{(6x^{-1}y)^{-3}} = \dfrac{(6x^{-1}y)^3}{(3x^2y^{-1})^2} = \dfrac{216x^{-3}y^3}{9x^4y^{-2}} = \dfrac{24y^5}{x^7}$

39. $\dfrac{(4x^{-2})(6x^5)}{(9y)(2y^{-1})} = \dfrac{24x^3}{18y^0} = \dfrac{4x^3}{3}$

41. $\left(\dfrac{3xy^3}{4x^2y^{-3}}\right)^{-1}\left(\dfrac{2x^3y^{-1}}{9x^{-3}y^{-1}}\right)^2 = \dfrac{4x^2y^{-3}}{3xy^3} \cdot \dfrac{4x^6y^{-2}}{81x^{-6}y^{-2}}$

$$= \dfrac{16x^8y^{-5}}{243x^{-5}y} = \dfrac{16x^{13}}{243y^6}$$

43. $\left(\dfrac{6x^{-4}yz^{-2}}{4^{-1}x^{-4}y^3z^{-2}}\right)^{-1}\left(\dfrac{2^{-2}xyz^{-3}}{12x^2y^2z^{-1}}\right)^{-2}$

$$= \dfrac{4^{-1}x^{-4}y^3z^{-2}}{6x^{-4}yz^{-2}} \cdot \dfrac{12^2x^4y^4z^{-2}}{2^{-4}x^2y^2z^{-6}}$$

$$= \dfrac{2304x^0y^7z^{-4}}{24x^{-2}y^3z^{-8}} = 96x^2y^4z^4$$

45. $86,000 = 8.6 \times 10^4$

47. $0.0362 = 3.62 \times 10^{-2}$

49. $18,300,000 = 1.83 \times 10^7$

51. $4.2 \times 10^{-2} = 0.042$

53. $7.56 \times 10^6 = 7,560,000$

55. $8.515 \times 10^8 = 851,500,000$

57. $300 \cdot 0.00015 = 3 \times 10^2 \cdot 1.5 \times 10^{-4}$
$$= 4.5 \times 10^{-2}$$

59. $0.0003 \cdot 0.0000025 = 3 \times 10^{-4} \cdot 2.5 \times 10^{-6}$
$$= 7.5 \cdot 10^{-10}$$

61. $\dfrac{3900}{0.003} = \dfrac{3.9 \times 10^3}{3 \times 10^{-3}} = 1.3 \times 10^6$

63. $\dfrac{125}{50,000} = \dfrac{1.25 \times 10^2}{5 \times 10^4} = 2.5 \times 10^{-3}$

65. $\dfrac{0.02 \cdot 3900}{0.013} = \dfrac{2 \times 10^{-2} \cdot 3.9 \times 10^3}{1.3 \times 10^{-2}}$
$$= \dfrac{2 \cdot 3.9 \cdot 10}{1.3 \times 10^{-2}}$$
$$= 2 \cdot 3 \times 10^3 = 6 \times 10^3$$

67. $\dfrac{0.005 \cdot 650 \cdot 3.3}{0.0011 \cdot 2500} = \dfrac{5 \times 10^{-3} \cdot 6.5 \times 10^2 \cdot 3.3 \times 10^0}{1.1 \times 10^{-3} \cdot 2.5 \times 10^3}$
$$= \dfrac{5 \cdot 6.5 \cdot 3.3 \times 10^{-1}}{1.1 \cdot 2.5 \times 10^0}$$
$$= \dfrac{2 \cdot 6.5 \cdot 3 \times 10^{-1}}{1}$$
$$= 39 \times 10^{-1} = 3.9 \times 10^0 = 3.9$$

69. $\dfrac{\left(1.4 \times 10^{-2}\right)(922)}{\left(3.5 \times 10^3\right)\left(2.0 \times 10^6\right)} = \dfrac{1.4 \times 10^{-2} \cdot 9.22 \times 10^2}{3.5 \times 10^3 \cdot 2.0 \times 10^6}$
$$= \dfrac{1.4 \cdot 9.22 \times 10^0}{3.5 \cdot 2 \times 10^9}$$
$$= \dfrac{0.7 \cdot 9.22}{3.5 \times 10^9} = \dfrac{6.454}{3.5 \times 10^9}$$
$$= 1.844 \times 10^{-9}$$

71. $4.3 \cdot 9.46 \times 10^{15} = 40.678 \times 10^{15} = 4.0678 \times 10^{16}$

73. 1.6605×10^{-27} kg

75. $60 \cdot 3 \times 10^{10} = 180 \times 10^{10} = 1.8 \times 10^{12}$ cm/min
$60 \cdot 1.8 \times 10^{12} = 108 \times 10^{12} = 1.08 \times 10^{14}$ cm/hr

77. $2000 \cdot 3.25 \times 10^{-22} = 6500 \times 10^{-22} = 6.5 \times 10^{-19}$

79. $60,000,000,000,000$

81. 8.01×10^8

83. 8.5×10^7

85. 1×10^2

87. 1.864×10^2

[End of Section 6.2]

Section 6.3
Solutions to Odd Exercises

1. $3x^4$ is a monomial.

3. $-2x^{-2}$ is not a polynomial.

5. $14a^7 - 2a - 6$ is a trinomial.

7. $6a^3 + 5a^2 - a^{-3}$ is not a polynomial.

9. $\frac{1}{2}x^3 - \frac{2}{5}x$ is a binomial.

11. $4y$ is a first degree monomial. $a_0 = 0$, $a_1 = 4$

13. $x^3 + 3x^2 - 2x$ is a third degree trinomial.
$a_0 = 0, a_1 = -2, a_2 = 3, a_3 = 1$

15. $-2x^2$ is a second degree monomial.
$a_0 = 0, a_1 = 0, a_1 = 0, a_2 = -2$

17. 0 is a no degree monomial. $a_0 = 0$

19. $6a^5 - 7a^3 - a^2$ is a fifth degree trinomial.
$a_0 = 0, a_1 = 0, a_2 = -1, a_3 = -7, a_4 = 0, a_5 = 6$

21. $2y^3 + 4y$ is a third degree binomial.
$a_0 = 0, a_1 = 4, a_2 = 0, a_3 = 2$

23. 4 is a zero degree monomial.
$a_0 = 4$

25. $2x^3 + 3x^2 - x + 1$ is a third degree polynomial.
$a_0 = 1, a_1 = -1, a_2 = 3, a_3 = 2$

27. $4x^4 - x^2 + 4x - 10$ is a fourth degree polynomial.
$a_0 = -10, a_1 = 4, a_2 = -1, a_3 = 0, a_4 = 4$

29. $9x^3 + 4x^2 + 8x$ is a third degree trinomial.
$a_0 = 0, a_1 = 8, a_2 = 4, a_3 = 9$

31. $p(-1) = (-1)^2 + 14(-1) - 3 = -16$

33. $p(3) = 3(3)^3 - 9(3)^2 - 10(3) - 11$
$= 81 - 81 - 30 - 11 = -41$

35. $p(-2) = 8(-2)^4 + 2(-2)^3 - 6(-2)^2 - 7$
$= 128 - 16 - 24 - 7 = 81$

37. $p(-1) = 2(-1)^4 + 3(-1)^2 - 8(-1)$
$= 2 + 3 + 8 = 13$

39. $p(-2) = (-2)^5 - (-2)^3 + (-2) - 2$
$= -32 + 8 - 2 - 2 = -28$

41. $p(a) = 3a^4 + 5a^3 - 8a^2 - 9a$

43. $f(a+2) = 3(a+2) + 5 = 3a + 11$

45. $g(2a+7) = 5(2a+7) - 10 = 10a + 25$

47. **a.** $p(x) = 2x + 3$

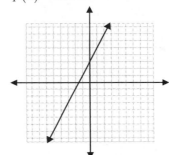

b. $p(x) = -3x + 1$

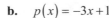
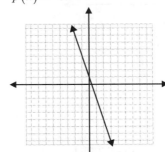

c. $p(x) = \frac{1}{2}x$

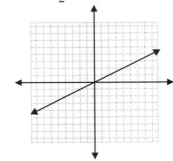

49. **a.** $p(x) = x^3$

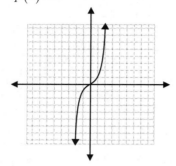

b. $p(x) = x^3 - 4x$

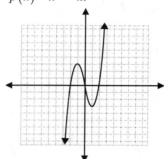

c. $p(x) = x^3 + 2x^2 - 5$

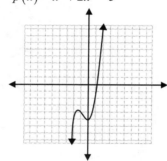

[End of Section 6.3]

Section 6.4
Solutions to Odd Exercises

1. $\left(2x^2 + 5x - 1\right) + \left(x^2 + 2x + 3\right) = 3x^2 + 7x + 2$

3. $\left(x^2 + 7x - 7\right) + \left(x^2 + 4x\right) = 2x^2 + 11x - 7$

5. $\left(2x^2 - x - 1\right) + \left(x^2 + x + 1\right) = 3x^2$

7. $\left(-2x^2 - 3x + 9\right) + \left(3x^2 - 2x + 8\right) = x^2 - 5x + 17$

9. $\left(-4x^2 + 2x - 1\right) + \left(3x^2 - x + 2\right) + \left(x - 8\right) = -x^2 + 2x - 7$

11. $\left(x^2 + 2x - 1\right) + \left(3x^2 - x + 2\right) + \left(x - 8\right) = 4x^2 + 2x - 7$

13.
$$\begin{array}{rrr} x^2 & +4x & -4 \\ -2x^2 & +3x & +1 \\ \hline -x^2 & +7x & -3 \end{array}$$

15.
$$\begin{array}{rrrr} 7x^3 & + 5x^2 & + x - & 6 \\ & - 3x^2 & + 4x + & 11 \\ -3x^3 & - x^2 & - 5x + & 2 \\ \hline 4x^3 & + x^2 & + & 7 \end{array}$$

17.
$$\begin{array}{rrrr} x^3 & + 3x^2 & - & 4 \\ & 7x^2 & + 2x + & 1 \\ x^3 & + x^2 & - 6x & \\ \hline 2x^3 & + 11x^2 & - 4x - & 3 \end{array}$$

19.
$$\begin{array}{rrr} 2x^2 & + 4x - & 3 \\ 3x^2 & - 9x + & 2 \\ \hline 5x^2 & - 5x - & 1 \end{array}$$

21. $\left(2x^2 + 4x + 8\right) - \left(x^2 + 3x + 2\right) = x^2 + x + 6$

23. $\left(x^4 + 8x^3 - 2x^2 - 5\right) - \left(2x^4 + 10x^3 - 2x^2 + 11\right)$
$$= -x^4 - 2x^3 - 16$$

25. $\left(x^2 - 9x + 2\right) - \left(4x^2 - 3x + 4\right) = -3x^2 - 6x - 2$

27. $\left(7x^2 + 4x - 9\right) - \left(-2x^2 + x - 9\right) = 9x^2 + 3x$

29. $\left(3x^4 - 2x^3 - 8x - 1\right) - \left(5x^3 - 3x^2 - 3x - 10\right)$
$$= 3x^4 - 7x^3 + 3x^2 - 5x + 9$$

31. $\left(9x^2 + 6x - 5\right) - \left(13x^2 + 6\right) = -4x^2 + 6x - 11$

33. $\left(x^3 + 4x^2 - 7\right) - \left(3x^3 + x^2 + 2x + 1\right)$
$$= -2x^3 + 3x^2 - 2x - 8$$

35.
$$\begin{array}{rrr} x^3 + 6x^2 & & - 3 \\ -(-x^3 + 2x^2 & - 3x + & 7) \\ \hline 2x^3 + 4x^2 & + 3x - & 10 \end{array}$$

37.
$$\begin{array}{rrr} 5x^4 + 8x^2 & + & 11 \\ -(-3x^4 + 2x^2 & - & 4) \\ \hline 8x^4 + 6x^2 & + & 15 \end{array}$$

39.
$$\begin{array}{rrr} 3x^3 & + 9x & - 17 \\ -(x^3 + 5x^2 & - 2x + & 6) \\ \hline 2x^3 - 5x^2 & + 11x & - 11 \end{array}$$

41. $5x + 2\left(x - 3\right) - \left(3x + 7\right) = 5x + 2x - 6 - 3x - 7$
$$= 4x - 13$$

43. $11 + \left[3x - 2\left(1 + 5x\right)\right] = 11 + 3x - 2 - 10x = -7x + 9$

45. $8x - \left[2x + 4\left(x - 3\right) - 5\right] = 8x - \left[2x + 4x - 12 - 5\right]$
$$= 8x - \left[6x - 17\right]$$
$$= 8x - 6x + 17$$
$$= 2x + 17$$

47. $3x^3 - \left[5 - 7\left(x^2 + 2\right) - 6x^2\right]$
$$= 3x^3 - \left[5 - 7x^2 - 14 - 6x^2\right]$$
$$= 3x^3 - \left[-9 - 13x^2\right]$$
$$= 3x^3 + 13x^2 + 9$$

49. $\left(2x^2 + 4\right) - \left[-8 + 2\left(7 - 3x^2\right) + x\right]$
$$= 2x^2 + 4 - \left[-8 + 14 - 6x^2 + x\right]$$
$$= 2x^2 + 4 - \left[-6x^2 + x + 6\right]$$
$$= 2x^2 + 4 + 6x^2 - x - 6$$
$$= 8x^2 - x - 2$$

51. $2\big[3x+(x-8)-(2x+5)\big]-(x-7)$

$$= 2\big[3x+x-8-2x-5\big]-x+7$$
$$= 2\big[2x-13\big]-x+7$$
$$= 4x-26-x+7$$
$$= 3x-19$$

53. $\left(x^2-1\right)+x\big[4+(3-x)\big]=x^2-1+x\big[4+3-x\big]$

$$= x^2-1+x\big[7-x\big]$$
$$= x^2-1+7x-x^2$$
$$= 7x-1$$

55. $x(2x+1)-\big[5x-x(2x+3)\big]$

$$= 2x^2+x-\big[5x-2x^2-3x\big]$$
$$= 2x^2+x-\big[-2x^2+2x\big]$$
$$= 2x^2+x+2x^2-2x$$
$$= 4x^2-x$$

57. $\big[4x^2+x-1+(6x-5)\big]-\big[3x^2-4x+2\big]$

$$= \big[4x^2+7x-6\big]-\big[3x^2-4x+2\big]$$
$$= 4x^2+7x-6-3x^2+4x-2$$
$$= x^2+11x-8$$

59. $\big[(2x^3+14x-3)-\left(x^2+6x+5\right)\big]+\big[5x^3-8x+1\big]$

$$= \big[2x^3+14x-3-x^2-6x-5\big]+\big[5x^3-8x+1\big]$$
$$= \big[2x^3-x^2+8x-8\big]+5x^3-8x+1$$
$$= 7x^3-x^2-7$$

[End of Section 6.4]

Section 6.5
Solutions to Odd Exercises

1. $-3x^2\left(2x^3+5x\right)=-6x^5-15x^3$

3. $5x^2\left(-4x^2+6\right)=-20x^4+30x^2$

5. $-1\left(y^5-8y+2\right)=-y^5+8y-2$

7. $-4x^3\left(x^5-2x^4+3x\right)=-4x^8+8x^7-12x^4$

9. $7t^3\left(-5t^2+2t+1\right)=-35t^5+14t^4+7t^3$

11. $-x\left(x^3+5x-4\right)=-x^4-5x^2+4x$

13. $3x\left(2x+1\right)-2\left(2x+1\right)=6x^2+3x-4x-2$
$$=6x^2-x-2$$

15. $3a\left(3a-5\right)+5\left(3a-5\right)=9a^2-15a+15a-25$
$$=9a^2-25$$

17. $5x\left(-2x+7\right)-2\left(-2x+7\right)=-10x^2+35x+4x-14$
$$=-10x^2+39x-14$$

19. $x\left(x^2+3x+2\right)+2\left(x^2+3x+2\right)$
$$=x^3+3x^2+2x+2x^2+6x+4$$
$$=x^3+5x^2+8x+4$$

21. $\left(x+4\right)\left(x-3\right)=x\left(x-3\right)+4\left(x-3\right)$
$$=x^2-3x+4x-12$$
$$=x^2+x-12$$

23. $\left(a+6\right)\left(a-8\right)=a\left(a-8\right)+6\left(a-8\right)$
$$=a^2-8a+6a-48$$
$$=a^2-2a-48$$

25. $\left(x-2\right)\left(x-1\right)=x\left(x-1\right)-2\left(x-1\right)$
$$=x^2-x-2x+2$$
$$=x^2-3x+2$$

27. $3\left(t+4\right)\left(t-5\right)=3\left[t\left(t-5\right)+4\left(t-5\right)\right]$
$$=3\left[t^2-5t+4t-20\right]$$
$$=3\left[t^2-t-20\right]$$
$$=3t^2-3t-60$$

29. $x\left(x+3\right)\left(x+8\right)=x\left[x\left(x+8\right)+3\left(x+8\right)\right]$
$$=x\left[x^2+8x+3x+24\right]$$
$$=x\left[x^2+11x+24\right]$$
$$=x^3+11x^2+24x$$

31. $\left(2x+1\right)\left(x-4\right)=2x\left(x-4\right)+1\left(x-4\right)$
$$=2x^2-8x+x-4=2x^2-7x-4$$

33. $\left(6x-1\right)\left(x+3\right)=6x\left(x+3\right)-1\left(x+3\right)$
$$=6x^2+18x-x-3=6x^2+17x-3$$

35. $\left(2x+3\right)\left(2x-3\right)=2x\left(2x-3\right)+3\left(2x-3\right)$
$$=4x^2-6x+6x-9=4x^2-9$$

37. $\left(4x+1\right)\left(4x+1\right)=4x\left(4x+1\right)+1\left(4x+1\right)$
$$=16x^2+4x+4x+1=16x^2+8x+1$$

39. $\left(y+3\right)\left(y^2-y+4\right)=y\left(y^2-y+4\right)+3\left(y^2-y+4\right)$
$$=y^3-y^2+4y+3y^2-3y+12$$
$$=y^3+2y^2+y+12$$

41.
$$
\begin{array}{r}
3x+\ 7\\
\times\quad\underline{x-\ 5}\\
3x^2+\ 7x\\
\underline{-15x-35}\\
3x^2-\ 8x-35
\end{array}
$$

43.
$$
\begin{array}{r}
8x^2+\ 3x-\ 2\\
\times\quad\underline{-\ 2x+\ 7}\\
-16x^3-\ 6x^2+\ 4x\\
\underline{56x^2+21x-14}\\
-16x^3+50x^2+25x-14
\end{array}
$$

45.
$$
\begin{array}{r}
6x^2-\ \ x+\ 8\\
\times\quad\underline{2x^2+\ 5x+\ 6}\\
12x^4-\ 2x^3+16x^2\\
30x^3-\ 5x^2+40x\\
\underline{36x^2-\ 6x+48}\\
12x^4+28x^3+47x^2+34x+48
\end{array}
$$

47. $\left(t+6\right)\left(4t-7\right)=t\left(4t-7\right)+6\left(4t-7\right)$
$$=4t^2-7t+24t-42=4t^2+17t-42$$
For $t=2$: $\left(2+6\right)\left(4\cdot2-7\right)=8\cdot1=8$
$$4\cdot2^2+17\cdot2-42=16+34-42=8$$

49. $(5a-3)(a+4) = 5a(a+4) - 3(a+4)$

$\qquad = 5a^2 + 20a - 3a - 12 = 5a^2 + 17a - 12$

For $a = 2$: $(5 \cdot 2 - 3)(2+4) = 7 \cdot 6 = 42$

$\qquad 5 \cdot 2^2 + 17 \cdot 2 - 12 = 20 + 34 - 12 = 42$

51. $(x-2)(3x+8) = x(3x+8) - 2(3x+8)$

$\qquad = 3x^2 + 8x - 6x - 16 = 3x^2 + 2x - 16$

For $x = 2$: $(2-2)(3 \cdot 2 + 8) = 0 \cdot 14 = 0$

$\qquad 3 \cdot 2^2 + 2 \cdot 2 - 16 = 12 + 4 - 16 = 0$

53. $(3x+7)(2x-5) = 3x(2x-5) + 7(2x-5)$

$\qquad = 6x^2 - 15x + 14x - 35 = 6x^2 - x - 35$

For $x = 2$: $(3 \cdot 2 + 7)(2 \cdot 2 - 5) = 13(-1) = -13$

$\qquad 6 \cdot 2^2 - 2 - 35 = 24 - 2 - 35 = -13$

55. $(5y+2)(5y+2) = 5y(5y+2) + 2(5y+2)$

$\qquad = 25y^2 + 10y + 10y + 4$

$\qquad = 25y^2 + 20y + 4$

For $y = 2$: $(5 \cdot 2 + 2)(5 \cdot 2 + 2) = 12 \cdot 12 = 144$

$\qquad 25 \cdot 2^2 + 20 \cdot 2 + 4 = 100 + 40 + 4 = 144$

57. $(y^2 + 2)(y-4) = y^2(y-4) + 2(y-4)$

$\qquad = y^3 - 4y^2 + 2y - 8$

For $y = 2$: $(2^2 + 2)(2-4) = 6(-2) = -12$

$\qquad 2^3 - 4 \cdot 2^2 + 2 \cdot 2 - 8 = 8 - 16 + 4 - 8 = -12$

59. $(3x-4)(3x+4) = 3x(3x+4) - 4(3x+4)$

$\qquad = 9x^2 + 12x - 12x - 16 = 9x^2 - 16$

For $x = 2$: $(3 \cdot 2 - 4)(3 \cdot 2 + 4) = 2 \cdot 10 = 20$

$\qquad 9 \cdot 2^2 - 16 = 36 - 16 = 20$

61. $(x-2)(x^2 + 2x + 4) = x(x^2 + 2x + 4) - 2(x^2 + 2x + 4)$

$\qquad = x^3 + 2x^2 + 4x - 2x^2 - 4x - 8$

$\qquad = x^3 - 8$

For $x = 2$: $(2-2)(2^2 + 2 \cdot 2 + 4) = 0 \cdot 12$

$\qquad 2^3 - 8 = 8 - 8 = 0$

63. $(5a-6)(5a-6) = 5a(5a-6) - 6(5a-6)$

$\qquad = 25a^2 - 30a - 30a + 36$

$\qquad = 25a^2 - 60a + 36$

For $a = 2$: $(5 \cdot 2 - 6)(5 \cdot 2 - 6) = 4 \cdot 4 = 16$

$\qquad 25 \cdot 2^2 - 60 \cdot 2 + 36 = 100 - 120 + 36 = 16$

65. $(3x+1)(x^2 - x + 9) = 3x(x^2 - x + 9) + 1(x^2 - x + 9)$

$\qquad = 3x^3 - 3x^2 + 27x + x^2 - x + 9$

$\qquad = 3x^3 - 2x^2 + 26x + 9$

For $x = 2$: $(3 \cdot 2 + 1)(2^2 - 2 + 9) = 7 \cdot 11 = 77$

$\qquad 3 \cdot 2^3 - 2 \cdot 2^2 + 26 \cdot 2 + 9 = 24 - 8 + 52 + 9 = 77$

67. $(t-1)(t-2)(t-3) = [t(t-2) - 1(t-2)](t-3)$

$\qquad = [t^2 - 3t + 2](t-3)$

$\qquad = (t^2 - 3t + 2)t - (t^2 - 3t + 2)3$

$\qquad = t^3 - 3t^2 + 2t - 3t^2 + 9t - 6$

$\qquad = t^3 - 6t^2 + 11t - 6$

For $t = 2$: $(2-1)(2-2)(2-3) = (-1)(0)(-1) = 0$

$\qquad 2^3 - 6 \cdot 2^2 + 11 \cdot 2 - 6 = 8 - 24 + 22 - 6 = 0$

69. $(y^2 + y + 2)(y^2 + y - 2)$

$\qquad = y^2(y^2 + y - 2) + y(y^2 + y - 2) + 2(y^2 + y - 2)$

$\qquad = y^4 + y^3 - 2y^2 + y^3 + y^2 - 2y + 2y^2 + 2y - 4$

$\qquad = y^4 + 2y^3 + y^2 - 4$

For $y = 2$: $(2^2 + 2 + 2)(2^2 + 2 - 2) = 8 \cdot 4 = 32$

$\qquad 2^4 + 2 \cdot 2^3 + 2^2 - 4 = 16 + 16 + 4 - 4 = 32$

71. $(x-3)(x+5) - (x+3)(x+2)$

$\qquad = x(x+5) - 3(x+5) - [x(x+2) + 3(x+2)]$

$\qquad = x^2 + 5x - 3x - 15 - [x^2 + 2x + 3x + 6]$

$\qquad = x^2 + 2x - 15 - (x^2 + 5x + 6)$

$\qquad = x^2 + 2x - 15 - x^2 - 5x - 6$

$\qquad = -3x - 21$

73. $(2a+1)(a-5) + (a-4)(a-4)$

$\qquad = 2a(a-5) + 1(a-5) + a(a-4) - 4(a-4)$

$\qquad = 2a^2 - 10a + a - 5 + a^2 - 4a - 4a + 16$

$\qquad = 2a^2 - 9a - 5 + a^2 - 8a + 16$

$\qquad = 3a^2 - 17a + 11$

75. $(y+6)(y-6) + (y+5)(y-5)$

$\qquad = y(y-6) + 6(y-6) + y(y-5) + 5(y-5)$

$\qquad = y^2 - 6y + 6y - 36 + y^2 - 5y + 5y - 25$

$\qquad = 2y^2 - 61$

[End of Section 6.5]

Section 6.6
Solutions to Odd Exercises

1. Difference of two squares:
$(x+3)(x-3) = x^2 - 3x + 3x - 9 = x^2 - 9$

3. Perfect square trinomial:
$(x-5)^2 = x^2 - 10x + 25$

5. Difference of two squares:
$(x-6)(x+6) = x^2 + 6x - 6x - 36 = x^2 - 36$

7. Perfect square trinomial:
$(x+8)(x+8) = x^2 + 16x + 64$

9. $(2x+3)(x-1) = 2x^2 - 2x + 3x - 3 = 2x^2 + x - 3$

11. Perfect square trinomial:
$(3x-4)^2 = 9x^2 - 24x + 16$

13. Difference of two squares:
$(2x+1)(2x-1) = 4x^2 - 2x + 2x - 1 = 4x^2 - 1$

15. Perfect square trinomial:
$(3x-2)(3x-2) = 9x^2 - 12x + 4$

17. Perfect square trinomial:
$(3+x)^2 = 9 + 6x + x^2 = x^2 + 6x + 9$

19. Perfect square trinomial:
$(5-x)(5-x) = 25 - 10x + x^2 = x^2 - 10x + 25$

21. Difference of two squares:
$(5x-9)(5x+9) = 25x^2 + 45x - 45x - 81$
$= 25x^2 - 81$

23. Perfect square trinomial:
$(2x+7)(2x+7) = 4x^2 + 28x + 49$

25. Difference of two squares:
$(9x+2)(9x-2) = 81x^2 - 18x + 18x - 4 = 81x^2 - 4$

27. $(5x^2+2)(2x^2-3) = 10x^4 - 15x^2 + 4x^2 - 6$
$= 10x^4 - 11x^2 - 6$

29. Perfect square trinomial:
$(1+7x)^2 = 1 + 14x + 49x^2 = 49x^2 + 14x + 1$

31. $(x+2)(5x+1) = 5x^2 + x + 10x + 2 = 5x^2 + 11x + 2$

33. $(4x-3)(x+4) = 4x^2 + 16x - 3x - 12 = 4x^2 + 13x - 12$

35. $(3x-7)(x-6) = 3x^2 - 18x - 7x + 42 = 3x^2 - 25x + 42$

37. Perfect square trinomial:
$(5+x)(5+x) = 25 + 10x + x^2 = x^2 + 10x + 25$

39. Difference of two squares:
$(x^2+1)(x^2-1) = x^4 + x^2 - x^2 - 1 = x^4 - 1$

41. Perfect square trinomial:
$(x^2+3)(x^2+3) = x^4 + 3x^2 + 3x^2 + 9 = x^4 + 6x^2 + 9$

43. Perfect square trinomial:
$(x^3-2)^2 = x^6 - 4x^3 + 4$

45. $(x^2-6)(x^2+9) = x^4 + 9x^2 - 6x^2 - 54$
$= x^4 + 3x^2 - 54$

47. Difference of two squares:
$\left(x+\dfrac{2}{3}\right)\left(x-\dfrac{2}{3}\right) = x^2 - \dfrac{2}{3}x + \dfrac{2}{3}x - \dfrac{4}{9} = x^2 - \dfrac{4}{9}$

49. Difference of two squares:
$\left(x+\dfrac{3}{4}\right)\left(x-\dfrac{3}{4}\right) = x^2 - \dfrac{3}{4}x + \dfrac{3}{4}x - \dfrac{9}{16} = x^2 - \dfrac{9}{16}$

51. Perfect square trinomial:
$\left(x+\dfrac{3}{5}\right)\left(x+\dfrac{3}{5}\right) = x^2 + \dfrac{6}{5}x + \dfrac{9}{25}$

53. Perfect square trinomial:
$\left(x-\dfrac{5}{6}\right)^2 = x^2 - \dfrac{10}{6}x + \dfrac{25}{36} = x^2 - \dfrac{5}{3}x + \dfrac{25}{36}$

55. $\left(x+\dfrac{1}{4}\right)\left(x-\dfrac{1}{2}\right) = x^2 - \dfrac{1}{2}x + \dfrac{1}{4}x - \dfrac{1}{8} = x^2 - \dfrac{1}{4}x - \dfrac{1}{8}$

57. $\left(x+\dfrac{1}{3}\right)\left(x+\dfrac{1}{2}\right) = x^2 + \dfrac{1}{2}x + \dfrac{1}{3}x + \dfrac{1}{6} = x^2 + \dfrac{5}{6}x + \dfrac{1}{6}$

Calculator Problems

59. $(x+1.4)(x-1.4) = x^2 - 1.4^2 = x^2 - 1.96$

61. $(x-2.5)^2 = x^2 - 2(2.5)x + 2.5^2 = x^2 - 5x + 6.25$

63. $(x+2.15)(x-2.15) = x^2 - 2.15x + 2.15x - 4.6225$
$$= x^2 - 4.6225$$

65. $(x+1.24)^2 = x^2 + 2(1.24)x + 1.24^2$
$$= x^2 + 2.48x + 1.5376$$

67. $(1.42x + 9.6)^2 = 1.42^2 x^2 + 2(1.42)(9.6)x + 9.6^2$
$$= 2.0164x^2 + 27.264x + 92.16$$

69. $(11.4x + 3.5)(11.4x - 3.5) = 11.4^2 x^2 - 3.5^2$
$$= 129.96x^2 - 12.25$$

71. $(12.6x - 6.8)(7.4x + 15.3)$
$$= 12.6(7.4)x^2 - 15.3(12.6)x$$
$$-6.8(7.4)x - 6.8(15.3)$$
$$= 93.24x^2 + 142.46x - 104.04$$

73. a. Area:
$$A(x) = 20^2 - 4x^2$$
$$= 400 - 4x^2$$

b. Perimeter:
$$P(x) = 4 \cdot (20 - 2x) + 8x$$
$$= 80 - 8x + 8x$$
$$= 80$$

75. Area:
$$A(x) = (x+3)(x+5) - x^2$$
$$= x^2 + 8x + 15 - x^2$$
$$= 8x + 15$$

77. a. Area:
$$A(x) = 10 \cdot 15 - 4x^2$$
$$= 150 - 4x^2$$

b. Perimeter:
$$P(x) = 2 \cdot (15 - 2x) + 2 \cdot (10 - 2x) + 8x$$
$$= 30 - 4x + 20 - 4x + 8x$$
$$= 50$$

c. Volume:
$$V(x) = (15 - 2x) \cdot (10 - 2x) \cdot x$$
$$= (150 - 50x + 4x^2) \cdot x$$
$$= 150x + 50x^2 + 4x^3$$

[End of Section 6.6]

Section 6.7
Solutions to Odd Exercises

1. $\dfrac{4x^2+8x+3}{4}=\dfrac{4x^2}{4}+\dfrac{8x}{4}+\dfrac{3}{4}=x^2+2x+\dfrac{3}{4}$

3. $\dfrac{10x^2-15x-3}{5}=\dfrac{10x^2}{5}-\dfrac{15x}{5}-\dfrac{3}{5}=2x^2-3x-\dfrac{3}{5}$

5. $\dfrac{2x^2+5x}{x}=\dfrac{2x^2}{x}+\dfrac{5x}{x}=2x+5$

7. $\dfrac{x^2+6x-3}{x}=\dfrac{x^2}{x}+\dfrac{6x}{x}-\dfrac{3}{x}=x+6-\dfrac{3}{x}$

9. $\dfrac{4x^2+6x-3}{2x}=\dfrac{4x^2}{2x}+\dfrac{6x}{2x}-\dfrac{3}{2x}=2x+3-\dfrac{3}{2x}$

11. $\dfrac{6x^3-9x^2-3x}{3x^2}=\dfrac{6x^3}{3x^2}-\dfrac{9x^2}{3x^2}-\dfrac{3x}{3x^2}=2x-3-\dfrac{1}{x}$

13. $\dfrac{7x^2y^2+21xy^3-11y^3}{7xy^2}=\dfrac{7x^2y^2}{7xy^2}+\dfrac{21xy^3}{7xy^2}-\dfrac{11y^3}{7xy^2}$

$=x+3y-\dfrac{11y}{7x}$

15. $\dfrac{3x^3y-8xy^2-4y^3}{4xy}=\dfrac{3x^3y}{4xy}-\dfrac{8xy^2}{4xy}-\dfrac{4y^3}{4xy}$

$=\dfrac{3x^2}{4}-2y-\dfrac{y^2}{x}$

17. $\dfrac{5x^2y^2-8xy^2+16xy}{8xy^2}=\dfrac{5x^2y^2}{8xy^2}-\dfrac{8xy^2}{8xy^2}+\dfrac{16xy}{8xy^2}$

$=\dfrac{5x}{8}-1+\dfrac{2}{y}$

19. $\dfrac{8x^3y^2-9x^2y^3+5xy}{9xy^2}=\dfrac{8x^3y^2}{9xy^2}-\dfrac{9x^2y^3}{9xy^2}+\dfrac{5xy}{9xy^2}$

$=\dfrac{8x^2}{9}-xy+\dfrac{5}{9y}$

21. $12+\dfrac{2}{23}$

$$\begin{array}{r} 12 \\ 23\overline{)278} \\ \underline{-23} \\ 48 \\ \underline{-46} \\ 2 \end{array}$$

23. $x+1$

$$\begin{array}{r} x+1 \\ x+2\overline{)x^2+3x+2} \\ \underline{-(x^2+2x)} \\ x+2 \\ \underline{-(x+2)} \\ 0 \end{array}$$

25. $y+4-\dfrac{1}{y+4}$

$$\begin{array}{r} y+4 \\ y+4\overline{)y^2+8y+15} \\ \underline{-(y^2+4y)} \\ 4y+15 \\ \underline{-(4y+16)} \\ -1 \end{array}$$

27. $x-12+\dfrac{42}{x+5}$

$$\begin{array}{r} x-12 \\ x+5\overline{)x^2-7x-18} \\ \underline{-(x^2+5x)} \\ -12x-18 \\ \underline{-(-12x-60)} \\ 42 \end{array}$$

29. $4a+3+\dfrac{20}{a-6}$

$$\begin{array}{r} 4a+3 \\ a-6\overline{)4a^2-21a+2} \\ \underline{-(4a^2-24a)} \\ 3a+2 \\ \underline{-(3a-18)} \\ 20 \end{array}$$

31. $4x-1-\dfrac{1}{2x+3}$

$$\begin{array}{r} 4x-1 \\ 2x+3\overline{)8x^2+10x-4} \\ \underline{-(8x^2+12x)} \\ -2x-4 \\ \underline{-(-2x-3)} \\ -1 \end{array}$$

33. $3x + 2 - \dfrac{2}{2x-1}$

$$
\begin{array}{r}
3x \ + \ 2 \\
2x-1{\overline{\smash{\big)}\,6x^2 + \ x - 4}} \\
\underline{-\left(6x^2 - 3x\right)} \\
4x - 4 \\
\underline{-\left(4x - 2\right)} \\
-2
\end{array}
$$

35. $x - 2 - \dfrac{2}{x+2}$

$$
\begin{array}{r}
x \ - \ 2 \\
x+2{\overline{\smash{\big)}\,x^2 + 0x - 6}} \\
\underline{-\left(x^2 + 2x\right)} \\
-2x - 6 \\
\underline{-\left(-2x - 4\right)} \\
-2
\end{array}
$$

37. $2x^2 + 5x + 2$

$$
\begin{array}{r}
2x^2 + 5x \ + \ 2 \\
x-3{\overline{\smash{\big)}\,2x^3 - \ x^2 - 13x - 6}} \\
\underline{-\left(2x^3 - 6x^2\right)} \\
5x^2 - 13x \\
\underline{-\left(5x^2 - 15x\right)} \\
2x - 6 \\
\underline{-\left(-2x - 6\right)} \\
0
\end{array}
$$

39. $t^2 + 3t + \dfrac{2}{3t+1}$

$$
\begin{array}{r}
t^2 \ + \ 3t \\
3t+1{\overline{\smash{\big)}\,3t^3 + 10t^2 + 3t + 2}} \\
\underline{-\left(3t^3 + \ t^2\right)} \\
9t^2 + 3t \\
\underline{-\left(9t^2 + 3t\right)} \\
0 + 2
\end{array}
$$

41. $2x^2 - 7x + 28 - \dfrac{118}{x+4}$

$$
\begin{array}{r}
2x^2 - 7x + \ 28 \\
x+4{\overline{\smash{\big)}\,2x^3 + \ x^2 + \ 0x - \ 6}} \\
\underline{-\left(2x^3 + 8x^2\right)} \\
-7x^2 + \ 0x \\
\underline{-\left(-7x^2 - 28x\right)} \\
28x - \ 6 \\
\underline{-\left(28x + 112\right)} \\
-118
\end{array}
$$

43. $x^2 + 2x + 4$

$$
\begin{array}{r}
x^2 + 2x \ + \ 4 \\
x-2{\overline{\smash{\big)}\,x^3 + 0x^2 + 0x - 8}} \\
\underline{-\left(x^3 - 2x^2\right)} \\
2x^2 + 0x \\
\underline{-\left(2x^2 - 4x\right)} \\
4x - 8 \\
\underline{-\left(4x - 8\right)} \\
0
\end{array}
$$

45. $2x + \dfrac{3x+3}{x^2-2}$

$$
\begin{array}{r}
2x \\
x^2-2{\overline{\smash{\big)}\,2x^3 + 0x^2 - \ x + 3}} \\
\underline{-\left(2x^3 + 0x^2 - 4x\right)} \\
3x + 3
\end{array}
$$

47. $x + 8 + \dfrac{8x-10}{x^2-x+1}$

$$
\begin{array}{r}
x \ + \ 8 \\
x^2-x+1{\overline{\smash{\big)}\,x^3 + 7x^2 + \ x - \ 2}} \\
\underline{-\left(x^3 - \ x^2 + \ x\right)} \\
8x^2 + 0x - \ 2 \\
\underline{-\left(8x^2 - 8x + \ 8\right)} \\
8x - 10
\end{array}
$$

49. $4x^2 + 6x + 9$

$$
\begin{array}{r}
4x^2 + 6x + 9 \\
2x-3{\overline{\smash{\big)}\,8x^3 + 0x^2 + 0x - 27}} \\
\underline{-\left(8x^3 - 12x^2\right)} \\
12x^2 + 0x \\
\underline{-\left(12x^2 - 18x\right)} \\
18x - 27 \\
\underline{-\left(18x - 27\right)} \\
0
\end{array}
$$

51. $4x + 3 + \dfrac{8}{x-1}$

$$
\begin{array}{r}
4x + 3 \\
x-1{\overline{\smash{\big)}\,4x^2 - x + 5}} \\
\underline{-\left(4x^2 - 4x\right)} \\
3x + 5 \\
\underline{-\left(3x - 3\right)} \\
8
\end{array}
$$

53. $3x^2 + 2x + 5$

$$
\begin{array}{r}
3x^2 + 2x + 5 \\
x-4{\overline{\smash{\big)}\,3x^3 - 10x^2 - 3x - 20}} \\
\underline{-\left(3x^3 - 12x^2\right)} \\
2x^2 - 3x \\
\underline{-\left(2x^2 - 8x\right)} \\
5x - 20 \\
\underline{-\left(5x - 20\right)} \\
0
\end{array}
$$

55. $2a + 3 - \dfrac{4a}{a^2 + 2}$

$$
\begin{array}{r}
2a + 3 \\
a^2+2{\overline{\smash{\big)}\,2a^3 + 3a^2 + 0a + 6}} \\
\underline{-\left(2a^3 + 4a\right)} \\
3a^2 - 4a + 6 \\
\underline{-\left(3a^2 + 6\right)} \\
-4a
\end{array}
$$

57. $4a^2 - 10a + 25$

$$
\begin{array}{r}
4a^2 - 10a + 25 \\
2a+5{\overline{\smash{\big)}\,8a^3 + 0a^2 + 0a + 125}} \\
\underline{-\left(8a^3 + 20a^2\right)} \\
-20a^2 - 0a \\
\underline{-\left(-20a^2 - 50a\right)} \\
50a + 125 \\
\underline{-\left(50a + 125\right)} \\
0
\end{array}
$$

59. $x^3 + x^2 + x + 1$

$$
\begin{array}{r}
x^3 + x^2 + x + 1 \\
x-1{\overline{\smash{\big)}\,x^4 + 0x^3 + 0x^2 + 0x - 1}} \\
\underline{-\left(x^4 - x^3\right)} \\
x^3 + 0x^2 \\
\underline{-\left(x^3 - x^2\right)} \\
x^2 + 0x \\
\underline{-\left(x^2 - x\right)} \\
x - 1 \\
\underline{-\left(x - 1\right)} \\
0
\end{array}
$$

[End of Section 6.7]

Chapter 6 Review
Solutions to All Exercises

1. $5^2 \cdot 5 = 25 \cdot 5 = 125$

2. $4^3 \cdot 4^0 = 64 \cdot 1 = 64$

3. $-3 \cdot 2^{-3} = -3 \cdot \dfrac{1}{2^3} = -\dfrac{3}{8}$

4. $5 \cdot 3^{-2} = 5 \cdot \dfrac{1}{3^2} = \dfrac{5}{9}$

5. $y^{-3} = \dfrac{1}{y^3}$

6. $x^3 \cdot x^4 = x^7$

7. $y^{-2} \cdot y^0 = y^{-2} = \dfrac{1}{y^2}$

8. $4x^{-1} = 4 \cdot \dfrac{1}{x^1} = \dfrac{4}{x}$

9. $\left(6x^2\right)\left(-2x^2\right) = -12x^4$

10. $\dfrac{16y^3}{8y^2} = 2y^{3-2} = 2y$

11. $\dfrac{-12x^7}{2x^3} = -6x^{7-3} = -6x^4$

12. $\left(-3a^3\right)^0 = 1$

13. $\left(3a^3b^2c\right)\left(-2ab^2c\right) = -6a^4b^4c^2$

14. $\dfrac{36a^6b^0}{12a^{-1}b^3} = \dfrac{3a^7}{b^3}$

15. $\dfrac{35y^5 \cdot 2y^{-2}}{14xy^4} = \dfrac{5}{xy}$

16. $\left(5x^3\right)^2 = 25x^6$

17. $\left(-2x^4\right)^3 = -8x^{12}$

18. $17\left(x^{-3}\right)^2 = 17\left(x^{-6}\right) = \dfrac{17}{x^6}$

19. $\left(\dfrac{y}{x}\right)^{-1} = \dfrac{x}{y}$

20. $\left(\dfrac{3x}{y^{-2}}\right)^2 = \dfrac{9x^2}{y^{-4}} = 9x^2y^4$

21. $\left(\dfrac{4a}{b}\right)^{-2} = \left(\dfrac{b}{4a}\right)^2 = \dfrac{b^2}{16a^2}$

22. $\left(\dfrac{3x^3y}{y^3}\right)^2 = \dfrac{9x^6y^2}{y^6} = \dfrac{9x^6}{y^4}$

23. $\left(\dfrac{m^3n^2}{mn^{-1}}\right)^3 = \dfrac{m^9n^6}{m^3n^{-3}} = m^6n^9$

24. $\dfrac{\left(5x^{-2}\right)\left(-2x^6\right)}{\left(y^2\right)^{-3}} = \dfrac{-10x^4}{y^{-6}} = -10x^4y^6$

25. $\dfrac{\left(6x^{-2}\right)\left(5x^{-1}\right)}{\left(3y\right)\left(2y^{-1}\right)} = \dfrac{30x^{-3}}{6y^0} = \dfrac{5}{x^3}$

26. $\left(\dfrac{3xy^3}{4xy^{-2}}\right)^{-2}\left(\dfrac{6x^3y^{-1}}{8x^{-4}y^3}\right)^{-1} = \left(\dfrac{4xy^{-2}}{3xy^3}\right)^2 \dfrac{8x^{-4}y^3}{6x^3y^{-1}}$

$= \dfrac{16x^2y^{-4}}{9x^2y^6} \cdot \dfrac{8x^{-4}y^3}{6x^3y^{-1}}$

$= \dfrac{64x^{-2}y^{-1}}{27x^5y^5} = \dfrac{64}{27x^7y^6}$

27. $\left(\dfrac{15a^3b^{-1}}{20ab^2}\right)^{-1}\left(\dfrac{-3a^2b}{a^{-4}b^{-2}}\right)^{-2} = \dfrac{20ab^2}{15a^3b^{-1}} \cdot \left(\dfrac{a^{-4}b^{-2}}{-3a^2b}\right)^2$

$= \dfrac{20ab^2}{15a^3b^{-1}} \cdot \dfrac{a^{-8}b^{-4}}{9a^4b^2}$

$= \dfrac{4a^{-7}b^{-2}}{27a^7b} = \dfrac{4}{27a^{14}b^3}$

28. $\left(\dfrac{-2xy^3}{12x^{-1}y^3}\right)^2\left(\dfrac{6x^{-4}y^2}{y^{-3}}\right)^2 = \dfrac{4x^2y^6}{144x^{-2}y^6} \cdot \dfrac{36x^{-8}y^4}{y^{-6}}$

$= \dfrac{x^{-6}y^{10}}{x^{-2}} = \dfrac{y^{10}}{x^4}$

29. $\left(\dfrac{3a^4bc^{-1}}{7ab^{-1}c^2}\right)^{-2}\left(\dfrac{9abc^{-1}}{7b^2c^2}\right)^2 = \left(\dfrac{7ab^{-1}c^2}{3a^4bc^{-1}}\right)^2 \cdot \dfrac{81a^2b^2c^{-2}}{49b^4c^4}$

$$= \dfrac{49a^2b^{-2}c^4}{9a^8b^2c^{-2}} \cdot \dfrac{81a^2b^2c^{-2}}{49b^4c^4}$$

$$= \dfrac{9a^4c^2}{a^8b^6c^2} = \dfrac{9}{a^4b^6}$$

30. $97,000 = 9.7 \times 10^4$

31. $29,300,000 = 2.93 \times 10^7$

32. $0.0562 = 5.62 \times 10^{-2}$

33. $0.0075 = 7.5 \times 10^{-3}$

34. $5.3 \times 10^{-2} = 0.053$

35. $7.24 \times 10^{-4} = 0.000724$

36. $8.23 \times 10^5 = 823,000$

37. $9.485 \times 10^7 = 94,850,000$

38. $\dfrac{125}{100,000} = \dfrac{1.25 \times 10^2}{1 \times 10^5} = \dfrac{1.25}{1} \times \dfrac{10^2}{10^5} = 1.25 \times 10^{-3}$

39. $\dfrac{0.0058}{290} = \dfrac{5.8 \times 10^{-3}}{2.9 \times 10^2} = \dfrac{5.8}{2.9} \times \dfrac{10^{-3}}{10^2} = 2 \times 10^{-5}$

40. $\dfrac{2.7 \cdot 0.002 \cdot 25}{54 \cdot 0.0005} = \dfrac{2.7\left(2 \times 10^{-3}\right)\left(2.5 \times 10^1\right)}{\left(5.4 \times 10^1\right)\left(5 \times 10^{-4}\right)}$

$$= \dfrac{2.7 \cdot 2 \cdot 2.5}{5.4 \cdot 5} \times \dfrac{10^{-3} \cdot 10^1}{10^1 \cdot 10^{-4}}$$

$$= 0.5 \times 10^1 = 5 \times 10^0$$

41. $\dfrac{\left(1.5 \times 10^{-3}\right)(822)}{\left(4.11 \times 10^3\right)(300)} = \dfrac{\left(1.5 \times 10^{-3}\right)\left(8.22 \times 10^2\right)}{\left(4.11 \times 10^3\right)\left(3 \times 10^2\right)}$

$$= \dfrac{1.5 \cdot 8.22}{4.11 \cdot 3} \times \dfrac{10^{-3} \cdot 10^2}{10^3 \cdot 10^2}$$

$$= 1 \times 10^{-6}$$

42. $x + 5x = (1+5)x = 6x$
A first degree monomial

43. $5x^2 - x + x^2 = (5+1)x^2 - x = 6x^2 - x$
A second degree binomial

44. $x^3 - 3x^2 + 5x$ is already simplified.
A third degree trinomial

45. $12x^3 - 8x + 4x - 7x^3 = (12-7)x^3 + (-8+4)x$

$$= 5x^3 - 4x$$
A third degree binomial

46. $6x^2 + 4 - 2x^2 + 1 - 4x^2 = (6-2-4)x^2 + 4 + 1$

$$= 0x^2 + 5 = 5$$
A zero degree monomial

47. $11x^3 - 6x^2 - 9x^3 - 5 + 6x^3 = (11-9+6)x^3 - 6x^2 - 5$

$$= 8x^3 - 6x^2 - 5$$
A third degree trinomial

48. $a^3 - 2a^2 - 8a + 3a^3 - 2a^2 + 6a + 9$

$$= (1+3)a^3 + (-2-2)a^2 + (-8+6)a + 9$$

$$= 4a^3 - 4a^2 - 2a + 9$$
A third degree polynomial

49. $3a + 4a^2 + 6a - 10a^3 = -10a^3 + 4a^2 + (3+6)a$

$$= -10a^3 + 4a^2 + 9a$$
A third degree trinomial

50. $p(-1) = (-1)^2 - 13(-1) + 5 = 1 + 13 + 5 = 19$

51. $p(2) = 3(2)^2 + 4(2) - 6 = 12 + 8 - 6 = 14$

52. $p(4) = 3(4)^3 - 3(4)^2 + 10 = 192 - 48 + 10 = 154$

53. $p(-5) = (-5)^2 - 8(-5) + 13 = 25 + 40 + 13 = 78$

54. $p(a) = a^2 + 6a + 9$

55. $p(c) = -3c^3 - 10c + 15$

56. $p(1) = (1)^4 + 3(1)^3 - (1)^2 - 20(1) + 5$

$$= 1 + 3 - 1 - 20 + 5 = -12$$

57. $\left(3x^2 + 6x - 1\right) + \left(x^2 + 3x + 4\right)$

$$= 3x^2 + x^2 + 6x + 3x - 1 + 4$$

$$= 4x^2 + 9x + 3$$

58. $\left(-5x^2 + 2x - 3\right) + \left(4x^2 - x + 6\right)$
$$= -5x^2 + 4x^2 + 2x - x - 3 + 6$$
$$= -x^2 + x + 3$$

59. $\left(x^2 + 7x + 12\right) + \left(2x - 8\right) + \left(x^2 + 10x\right)$
$$= x^2 + x^2 + 7x + 2x + 10x + 12 - 8$$
$$= 2x^2 + 19x + 4$$

60. $\left(-8x^2 + 5x + 2\right) + \left(3x^2 - 9x + 5\right) + \left(x - 11\right)$
$$= -8x^2 + 3x^2 + 5x - 9x + x + 2 + 5 - 11$$
$$= -5x^2 - 3x - 4$$

61. $\quad 2x^2 - 4x + 4$
$$\underline{-x^2 - 3x + 4}$$
$$\quad x^2 - 7x + 8$$

62. $\quad x^3 + 4x^2 + x$
$$\underline{-3x^3 - x^2 + 2x + 7}$$
$$-2x^3 + 3x^2 + 3x + 7$$

63. $\quad x^3 + 4x^2 \quad\quad -6$
$$\quad\quad\quad 8x^2 + 3x - 1$$
$$\underline{\quad x^3 + x^2 - 4x}$$
$$2x^3 + 13x^2 - x - 7$$

64. $\quad 4x^3 - 9x^2 + 7x - 12$
$$\underline{2x^3 - 2x^2 + 2x - 15}$$
$$6x^3 - 11x^2 + 9x - 27$$

65. $\left(3x^2 + 4x - 4\right) - \left(x^2 + 5x + 6\right)$
$$= 3x^2 - x^2 + 4x - 5x - 4 - 6$$
$$= 2x^2 - x - 10$$

66. $\left(x^4 + 7x^3 - x + 13\right) - \left(-x^4 + 6x^2 - x + 3\right)$
$$= x^4 + x^4 + 7x^3 - 6x^2 - x + x + 13 - 3$$
$$= 2x^4 + 7x^3 - 6x^2 + 10$$

67. $\left(5x^2 - x - 10\right) - \left(-x^2 - 10\right) = 5x^2 + x^2 - x - 10 + 10$
$$= 6x^2 - x$$

68. $\left(7x^2 - 14\right) - \left(4x^2 - 2x - 12\right) = 7x^2 - 4x^2 + 2x - 14 + 12$
$$= 3x^2 + 2x - 2$$

69. $\quad 13x^2 - 6x + 7$
$$\underline{-\left(9x^2 + x + 5\right)}$$
$$\quad 4x^2 - 7x + 2$$

70. $\quad 11x^3 \quad\quad + 10x - 20$
$$\underline{-\left(x^3 + 5x^2 - 10x - 22\right)}$$
$$10x^3 - 5x^2 + 20x + 2$$

71. $6x + 2\left(x - 7\right) - \left(3x + 4\right) = 6x + 2x - 14 - 3x - 4$
$$= 5x - 18$$

72. $3x + \left[8x - 4\left(3x + 1\right) - 9\right] = 3x + 8x - 12x - 4 - 9$
$$= -x - 13$$

73. $4x^3 - \left[7 - 5\left(2x - 3\right) + 10x^3\right]$
$$= 4x^3 - \left[7 - 10x + 15 + 10x^3\right]$$
$$= 4x^3 - 7 + 10x - 15 - 10x^3$$
$$= -6x^3 + 10x - 22$$

74. $\left(x^2 + 4\right) - \left[-6 + 2\left(3 - 2x^2\right) + x\right]$
$$= x^2 + 4 - \left[-6 + 6 - 4x^2 + x\right]$$
$$= x^2 + 4 + 6 - 6 + 4x^2 - x$$
$$= 5x^2 - x + 4$$

75. $2\left[4x + 5\left(x - 8\right) - \left(2x + 3\right)\right] + \left(3x - 5\right)$
$$= 2\left[4x + 5x - 40 - 2x - 3\right] + 3x - 5$$
$$= 8x + 10x - 80 - 4x - 6 + 3x - 5$$
$$= 17x - 91$$

76. $-\left[5x^2 + 7\left(x^2 - 30x + 2\right) - x^2 - 14\right]$
$$= -\left[5x^2 + 7x^2 - 210x + 14 - x^2 - 14\right]$$
$$= -5x^2 - 7x^2 + 210x - 14 + x^2 + 14$$
$$= -11x^2 + 210x$$

77. $\left(8x^2 - 10x + 3\right) - \left(3x^2 - 2x + 4\right)$
$$= 8x^2 - 10x + 3 - 3x^2 + 2x - 4$$
$$= 5x^2 - 8x - 1$$

78. $\left[2x^2 - 5x + 2 + x^2 - 14x + 11\right] - \left(4x^2 - 8x + 1\right)$
$$= 3x^2 - 19x + 13 - 4x^2 + 8x - 1$$
$$= -x^2 - 11x + 12$$

79. $-2x^2\left(3x^3 + 4x\right) = -2x^2 \cdot 3x^3 - 2x^2 \cdot 4x = -6x^5 - 8x^3$

80. $7a^3\left(-5a^2 + 3a - 1\right) = 7a^3\left(-5a^2\right) + 7a^3 \cdot 3a + 7a^3\left(-1\right)$
$$= -35a^5 + 21a^4 - 7a^3$$

81. $3a\left(-2a + 6\right) + 5\left(-2a + 6\right)$
$$= 3a\left(-2a\right) + 3a \cdot 6 + 5\left(-2a\right) + 5 \cdot 6$$
$$= -6a^2 + 18a - 10a + 30$$
$$= -6a^2 + 8a + 30$$

82. $x\left(x^2 + 5x + 4\right) - 2\left(x^2 + 5x + 4\right)$
$$= x^3 + 5x^2 + 4x - 2x^2 - 10x - 8$$
$$= x^3 + 3x^3 - 6x - 8$$

83. $x\left(x - 5\right)\left(x - 6\right) = \left(x^2 - 5x\right)\left(x - 6\right)$
$$= x^3 - 6x^2 - 5x^2 + 30x$$
$$= x^3 - 11x^2 + 30x$$

84. $-4\left(x + 3\right)\left(x - 5\right) = \left(-4x - 12\right)\left(x - 5\right)$
$$= -4x^2 + 20x - 12x + 60$$
$$= -4x^2 + 8x + 60$$

85. $\left(2y + 1\right)\left(y^2 - 3y + 2\right) = 2y^3 - 6y^2 + 4y + y^2 - 3y + 2$
$$= 2y^3 - 5y^2 + y + 2$$

86. $\left(x + 2\right)\left(x + 3\right)\left(x + 4\right) = \left(x^2 + 3x + 2x + 6\right)\left(x + 4\right)$
$$= \left(x^2 + 5x + 6\right)\left(x + 4\right)$$
$$= x^3 + 4x^2 + 5x^2 + 20x + 6x + 24$$
$$= x^3 + 9x^2 + 26x + 24$$

87.
$$
\begin{array}{r}
4x + 7 \\
x - 6 \\
\hline
-24x - 42 \\
4x^2 + 7x \\
\hline
4x^2 - 17x - 42
\end{array}
$$

88.
$$
\begin{array}{r}
y^2 + 4y + 1 \\
3y - 8 \\
\hline
-8y^2 - 32y - 8 \\
3y^3 + 12y^2 + 3y \\
\hline
3y^3 + 4y^2 - 29y - 8
\end{array}
$$

89.
$$
\begin{array}{r}
x^2 + 3x + 5 \\
x^2 + 3x - 2 \\
\hline
-2x^2 - 6x - 10 \\
3x^3 + 9x^2 + 15x \\
x^4 + 3x^3 + 5x^2 \\
\hline
x^4 + 6x^3 + 12x^2 + 9x - 10
\end{array}
$$

90.
$$
\begin{array}{r}
2a^2 - a - 3 \\
a^2 + 6a + 5 \\
\hline
10a^2 - 5a - 15 \\
12a^3 - 6a^2 - 18a \\
2a^4 - a^3 - 3a^2 \\
\hline
2a^4 + 11a^3 + a^2 - 23a - 15
\end{array}
$$

91. $\left(2x + 3\right)\left(x + 1\right) = 2x^2 + 2x + 3x + 3 = 2x^2 + 5x + 3$

92. $\left(3x + 5\right)\left(2x + 5\right) = 6x^2 + 15x + 10x + 25$
$$= 6x^2 + 25x + 25$$

93. $\left(x + 13\right)\left(x - 13\right) = x^2 - 169$
A difference of two squares

94. $\left(y + 8\right)\left(y - 8\right) = y^2 - 64$
A difference of two squares

95. $\left(2x + 9\right)^2 = 4x^2 + 2 \cdot 18x + 81 = 4x^2 + 36x + 81$
A perfect square trinomial

96. $\left(3x - 1\right)^2 = 9x^2 - 2 \cdot 3x + 1 = 9x^2 - 6x + 1$
A perfect square trinomial

97. $\left(y^2 + 6\right)\left(y^2 - 4\right) = y^4 - 4y^2 + 6y^2 - 24 = y^4 + 2y - 24$

98. $\left(3x - 7\right)\left(2x + 11\right) = 6x^2 + 33x - 14x - 77$
$$= 6x^2 + 19x - 77$$

99. $\left(x^3 - 10\right)\left(x^3 + 10\right) = x^6 - 100$
A difference of two squares

100. $\left(t+\dfrac{1}{4}\right)\left(t-\dfrac{1}{4}\right)=t^2-\dfrac{1}{16}$

A difference of two squares

101. $\left(y+\dfrac{2}{3}\right)\left(y-\dfrac{1}{2}\right)=y^2-\dfrac{1}{2}y+\dfrac{2}{3}y-\dfrac{1}{3}=y^2+\dfrac{1}{6}y-\dfrac{1}{3}$

102. $(7-3x)^2=49-2\cdot21x+9x^2=49-42x+9x^2$

A perfect square trinomial

103. $\dfrac{9x^2+6x+1}{3}=\dfrac{9x^2}{3}+\dfrac{6x}{3}+\dfrac{1}{3}=3x^2+2x+\dfrac{1}{3}$

104. $\dfrac{-2x^2-8x-3}{-2}=\dfrac{-2x^2}{-2}+\dfrac{-8x}{-2}+\dfrac{-3}{-2}=x^2+4x+\dfrac{3}{2}$

105. $\dfrac{9x^3-12x^2+3x}{x^2}=\dfrac{9x^3}{x^2}+\dfrac{-12x^2}{x^2}+\dfrac{3x}{x^2}=9x-12+\dfrac{3}{x}$

106. $\dfrac{4x^2y-8xy^2+y^3}{2xy}=\dfrac{4x^2y}{2xy}+\dfrac{-8xy^2}{2xy}+\dfrac{y^3}{2xy}$

$=2x-4y+\dfrac{y^2}{2x}$

107. $\dfrac{5a^3-10a^2+4a}{5a}=\dfrac{5a^3}{5a}+\dfrac{-10a^2}{5a}+\dfrac{4a}{5a}=a^2-2a+\dfrac{4}{5}$

108. $\dfrac{4x^2y^2+16xy^2-8y^2}{8y^2}=\dfrac{4x^2y^2}{8y^2}+\dfrac{16xy^2}{8y^2}-\dfrac{8y^2}{8y^2}$

$=\dfrac{x^2}{2}+2x-1$

109. $\dfrac{12a^2b^2+6ab-3ab}{3ab}=\dfrac{12a^2b^2}{3ab}+\dfrac{6ab}{3ab}-\dfrac{3ab}{3ab}$

$=4ab+2-1=4ab+1$

110. $\dfrac{7x^3y^2-8x^2y^3-5xy}{4xy^2}=\dfrac{7x^3y^2}{4xy^2}+\dfrac{-8x^2y^3}{4xy^2}+\dfrac{-5xy}{4xy^2}$

$=\dfrac{7x^2}{4}-2xy-\dfrac{5}{4y}$

111.

$$
\begin{array}{r}
a+3 \\
a-5\overline{\smash{)}a^2-2a-15} \\
-\underline{(a^2-5a)} \\
3a-15 \\
-\underline{(3a-15)} \\
0
\end{array}
$$

112.

$$
\begin{array}{r}
x+5 \\
x+4\overline{\smash{)}x^2+9x+20} \\
-\underline{(x^2+4x)} \\
5x+20 \\
-\underline{(5x+20)} \\
0
\end{array}
$$

113.

$$
\begin{array}{r}
y+7 \\
y-8\overline{\smash{)}y^2-\ y-56} \\
-\underline{(y^2-8y)} \\
7y-56 \\
-\underline{(7y-56)} \\
0
\end{array}
$$

114.

$$
\begin{array}{r}
x-2 \\
x+9\overline{\smash{)}x^2+7x-18} \\
-\underline{(x^2+9x)} \\
-2x-18 \\
-\underline{(-2x-18)} \\
0
\end{array}
$$

115.

$$
\begin{array}{r}
4x-13-\dfrac{23}{x-2} \\
x-2\overline{\smash{)}4x^2-21x+3} \\
-\underline{(4x^2-\ 8x)} \\
-13x+\ 3 \\
-\underline{(-13x-26)} \\
-23
\end{array}
$$

116.

$$2y+3\overline{)8y^2+10y+5} \quad 4y-1+\dfrac{8}{2y+3}$$

$$\underline{-\left(8y^2+12y\right)}$$
$$-2y+5$$
$$\underline{-(-2y-3)}$$
$$8$$

117.

$$2x-1\overline{)6x^2+\ x-6} \quad 3x+2-\dfrac{4}{2x-1}$$

$$\underline{-\left(6x^2-3x\right)}$$
$$4x-6$$
$$\underline{-(4x-2)}$$
$$-4$$

118.

$$x+4\overline{)x^2-0x-\ 9} \quad x-4+\dfrac{7}{x+4}$$

$$\underline{-\left(x^2+4x\right)}$$
$$-4x-\ 9$$
$$\underline{-(-4x-16)}$$
$$7$$

119.

$$x-1\overline{)x^3+0x^2-0x-1} \quad x^2+x+1$$

$$\underline{-\left(x^3-\ x^2\right)}$$
$$x^2-0x$$
$$\underline{-\left(x^2-\ x\right)}$$
$$x-1$$
$$\underline{-(x-1)}$$
$$0$$

120.

$$2x+3\overline{)8x^3+\ 0x^2-\ 0x+27} \quad 4x^2-6x+9$$

$$\underline{-\left(8x^3+12x^2\right)}$$
$$-12x^2-\ 0x$$
$$\underline{-\left(-12x^2-18x\right)}$$
$$18x+27$$
$$\underline{-(18x+27)}$$
$$0$$

121.

$$y+5\overline{)y^3+5y^2} \quad y^2$$

$$\underline{-\left(y^3+5y^2\right)}$$
$$0$$

122.

$$4x-5\overline{)64x^3+\ 0x^2-\ 0x-125} \quad 16x^2+\ 20x+25$$

$$\underline{-\left(64x^3-80x^2\right)}$$
$$80x^2-\ 0x$$
$$\underline{-\left(80x^2-100x\right)}$$
$$100x-125$$
$$\underline{-(100x-125)}$$
$$0$$

[End of Chapter 6 Review]

**Chapter 6 Test
Solutions to All Exercises**

1. $\left(5a^2b^5\right)\left(-2a^3b^{-5}\right) = -10a^5b^0 = -10a^5$

2. $\left(-7x^4y^{-3}\right)^0 = 1$

3. $\dfrac{\left(-8x^2y^{-3}\right)^2}{16xy} = \dfrac{64x^4y^{-6}}{16xy} = \dfrac{4x^3}{y^7}$

4. $\dfrac{3x^{-2}}{9x^{-3}y^2} = \dfrac{x}{3y^2}$

5. $\left(\dfrac{4xy^2}{x^3}\right)^{-1} = \dfrac{x^3}{4xy^2} = \dfrac{x^2}{4y^2}$

6. $\left(\dfrac{2x^0y^3}{x^{-1}y}\right)^2 = \dfrac{4y^6}{x^{-2}y^2} = 4x^2y^4$

7. **a.** $1.35 \times 10^5 = 135,000$
  b. $2.7 \times 10^{-6} = 0.0000027$

8. **a.** $250 \cdot 500,000 = 2.5 \times 10^2 \cdot 5 \times 10^5$
$$= 12.5 \times 10^7$$
$$= 1.25 \times 10^8$$

  b. $\dfrac{65 \cdot 0.012}{1500} = \dfrac{6.5 \times 10 \cdot 1.2 \times 10^{-2}}{1.5 \times 10^3}$
$$= \dfrac{6.5(1.2) \times 10^{-1}}{1.5 \times 10^3}$$
$$= \dfrac{7.8 \times 10^{-1} \cdot 10^{-3}}{1.5}$$
$$= 5.2 \times 10^{-4}$$

9. $3x + 4x^2 - x^3 + 4x^2 + x^3 = 8x^2 + 3x$
A second degree binomial

10. $2x^2 + 3x - x^3 + x^2 - 1 = -x^3 + 3x^2 + 3x - 1$
A third degree polynomial

11. $17 - 14 + 4x^5 - 3x + 2x^4 + x^5 - 8x$
$$= 5x^5 + 2x^4 - 11x + 3$$
A fifth degree polynomial

12. $p(x) = 2x^3 - 5x^2 + 7x + 10$

  a. $p(2) = 2(2)^3 - 5(2)^2 + 7(2) + 10$
$$= 16 - 20 + 14 + 10$$
$$= 20$$

  b. $p(-3) = 2(-3)^3 - 5(-3)^2 + 7(-3) + 10$
$$= -54 - 45 - 21 + 10$$
$$= -110$$

13. $7x + \left[2x - 3(4x + 1) + 5\right] = 7x + \left[2x - 12x - 3 + 5\right]$
$$= 7x - 10x + 2$$
$$= -3x + 2$$

14. $12x - 2\left[5 - (7x + 1) + 3x\right] = 12x - 2\left[5 - 7x - 1 + 3x\right]$
$$= 12x - 2\left[4 - 4x\right]$$
$$= 12x - 8 + 8x$$
$$= 20x - 8$$

15. $\left(5x^3 - 2x + 7\right) + \left(-x^2 + 8x - 2\right)$
$$= 5x^3 - 2x + 7 - x^2 + 8x - 2$$
$$= 5x^3 - x^2 + 6x + 5$$

16. $\left(4x^2 + 3x - 1\right) - \left(6x^2 + 2x + 5\right)$
$$= 4x^2 + 3x - 1 - 6x^2 - 2x - 5$$
$$= -2x^2 + x - 6$$

17. $\left(x^4 + 3x^2 + 9\right) - \left(-6x^4 - 11x^2 + 5\right)$
$$= x^4 + 3x^2 + 9 + 6x^4 + 11x^2 - 5$$
$$= 7x^4 + 14x^2 + 4$$

18. $\left(3x^3 - 2x^2\right) + \left(4x^3 - 7x + 1\right)$
$$= 3x^3 - 2x^2 + 4x^3 - 7x + 1$$
$$= 7x^3 - 2x^2 - 7x + 1$$

19. $5x^2\left(3x^5 - 4x^4 + 3x^3 - 8x^2 - 2\right)$
$$= 15x^7 - 20x^6 + 15x^5 - 40x^4 - 10x^2$$

20. $(7x + 3)(7x - 3) = 49x^2 - 9$
Difference of two squares

21. $(4x + 1)^2 = (4x + 1)(4x + 1)$
$$= 16x^2 + 4x + 4x + 1$$
$$= 16x^2 + 8x + 1$$
Perfect square trinomial

22. $(6x-5)^2 = 36x^2 - 30x - 30x + 25$

$$= 36x^2 - 60x + 25$$

Perfect square trinomial

23. $(2x+5)(6x-3) = 12x^2 - 6x + 30x - 15$

$$= 12x^2 + 24x - 15$$

24. $3x(x-7)(2x-9) = 3x(2x^2 - 9x - 14x + 63)$

$$= 3x(2x^2 - 23x + 63)$$

$$= 6x^3 - 69x^2 + 189x$$

25. $(3x+1)(3x-1) - (2x+3)(x-5)$

$$= 9x^2 - 1 - (2x^2 - 7x - 15)$$

$$= 9x^2 - 1 - 2x^2 + 7x + 15$$

$$= 7x^2 + 7x + 14$$

26.
$$\begin{array}{r}
2x^3 - 3x - 7 \\
\times \quad\quad 5x + 2 \\
\hline
10x^4 \quad\quad -15x^2 - 35x \\
+ \quad 4x^3 \quad\quad\quad - 6x - 14 \\
\hline
10x^4 + 4x^3 - 15x^2 - 41x - 14
\end{array}$$

27. $\dfrac{4x^3 + 3x^2 - 6x}{2x^2} = \dfrac{4x^3}{2x^2} + \dfrac{3x^2}{2x^2} - \dfrac{6x}{2x^2} = 2x + \dfrac{3}{2} - \dfrac{3}{x}$

28. $\dfrac{5a^2b + 6a^2b^2 + 3ab^3}{3a^2b} = \dfrac{5a^2b}{3a^2b} + \dfrac{6a^2b^2}{3a^2b} + \dfrac{3ab^3}{3a^2b}$

$$= \dfrac{5}{3} + 2b + \dfrac{b^2}{a}$$

29. $x - 6 - \dfrac{2}{2x+3}$

$$\begin{array}{r}
x - 6 \\
2x+3 \overline{)2x^2 - 9x - 20} \\
-(2x^2 + 3x) \\
\hline
-12x - 20 \\
-(-12x - 18) \\
\hline
-2
\end{array}$$

30. $x - 9 + \dfrac{15x - 12}{x^2 + x - 3}$

$$\begin{array}{r}
x - 9 \\
x^2 + x - 3 \overline{)x^3 - 8x^2 + 3x + 15} \\
-(x^3 + x^2 - 3x) \\
\hline
-9x^2 + 6x + 15 \\
-(-9x^2 - 9x + 27) \\
\hline
15x - 12
\end{array}$$

31. $(4x^2 + 2x + 1) + (3x^2 - 8x - 10) - (5x^2 - 3x + 4)$

$$= 7x^2 - 6x - 9 - 5x^2 + 3x - 4$$

$$= 2x^2 - 3x - 13$$

32. $6y + 27 + \dfrac{39}{y-3}$

First, note that $\dfrac{(2y+7)(3y-6)}{y-3} = \dfrac{6y^2 + 9y - 42}{y-3}$.

$$\begin{array}{r}
6y + 27 \\
y-3 \overline{)6y^2 + 9y - 42} \\
-(6y^2 - 18y) \\
\hline
27y - 42 \\
-(27y - 81) \\
\hline
39
\end{array}$$

[End of Chapter 6 Test]

Cumulative Review: Chapters 1 – 6
Solutions to All Exercises

1. $3x + 45 = 3(x + 15)$

2. $6x + 16 = 2(3x + 8)$

3. $\text{LCM}(12,15,54):$

   $12 = 2^2 \cdot 3$
   $15 = 3 \cdot 5$
   $54 = 2 \cdot 3^3$
   $\text{LCM} = 2^2 \cdot 3^3 \cdot 5 = 540$

4. $\text{LCM}(6a^2,\ 24ab^3,\ 30ab,\ 40a^2b^2):$

   $6a^2 = 2 \cdot 3 \cdot a^2$
   $24ab^3 = 2^3 \cdot 3 \cdot a \cdot b^3$
   $30ab = 2 \cdot 3 \cdot 5 \cdot a \cdot b$
   $40a^2b^2 = 2^3 \cdot 5 \cdot a^2 \cdot b^2$
   $\text{LCM} = 2^3 \cdot 3 \cdot 5 \cdot a^2 \cdot b^3 = 120a^2b^3$

5. Factors of 32 and their sums:

1, 32	sum = 33
2, 16	sum = 18
4, 8	sum = 12
−1, −32	sum = −33
−2, −16	sum = −18
−4, −8	sum = −12

   2 and 16 are the factors of 32 whose sum is 18.

6. Two factors of 36 whose sum is −13:
   −4 and −9.

7. Two factors of −56 whose sum is 10:
   −4 and 14.

8. Two factors of −48 whose sum is −8:
   −12 and 4.

9. $6x^2 + 10x - (x^2 - 6x) = 6x^2 + 10x - x^2 + 6x$

   $\qquad = 5x^2 + 16x$
   A second degree binomial
   $a_0 = 0,\ a_1 = 16,\ a_2 = 5$

10. $4(2x - 3) - 3(x + 2) = 8x - 12 - 3x - 6$

   $\qquad = 5x - 18$
   A first degree binomial
   $a_0 = -18,\ a_1 = 5$

11. $4x - \left[(6x + 7) - (3x + 4) \right] - 6$

   $\qquad = 4x - (6x + 7 - 3x - 4) - 6$
   $\qquad = 4x - (3x + 3) - 6$
   $\qquad = 4x - 3x - 3 - 6$
   $\qquad = x - 9$
   A first degree binomial
   $a_0 = -9,\ a_1 = 1$

12. $3x^2 - 8(x - 5) + x^4 - 2x^3 + x^2 - 2x$

   $\qquad = 3x^2 - 8x + 40 + x^4 - 2x^3 + x^2 - 2x$
   $\qquad = x^4 - 2x^3 + 4x^2 - 10x + 40$
   A fourth degree polynomial
   $a_0 = 40,\ a_1 = -10,\ a_2 = 4,\ a_3 = -2,\ a_4 = 1$

13. $7(4 - x) = 3(x + 4)$

   $28 - 7x = 3x + 12$
   $-10x = -16$
   $x = \dfrac{8}{5}$

14. $-2(5x + 1) + 2x = 4(x + 1)$

   $-10x - 2 + 2x = 4x + 4$
   $-8x - 2 = 4x + 4$
   $-6 = 12x$
   $x = -\dfrac{1}{2}$

15. $\dfrac{2}{3}x + \dfrac{1}{2} = \dfrac{3}{4}$

   $12\left(\dfrac{2}{3}x + \dfrac{1}{2} \right) = 12\left(\dfrac{3}{4} \right)$
   $8x + 6 = 9$
   $8x = 3$
   $x = \dfrac{3}{8}$

16. $1.5x - 3.7 = 3.6x + 2.6$

   $-2.1x = 6.3$
   $x = -3$

17. $C = \pi d$

   $d = \dfrac{C}{\pi}$

18. $3x + 5y = 10$

$5y = -3x + 10$

$y = \dfrac{10 - 3x}{5}$

19. $5x + 3 \geq 2x - 15$

$3x \geq -18$

$x \geq -6$

20. $-16 < 3x + 5 < 17$

$-21 < \quad 3x \quad < 12$

$-7 < \quad x \quad < 4$

21. Use the point-slope form of a line with the given slope and point:

$y - 2 = \dfrac{2}{3}\left(x - (-1)\right)$

$y - 2 = \dfrac{2}{3}(x + 1)$

$y - 2 = \dfrac{2}{3}x + \dfrac{2}{3}$

$y = \dfrac{2}{3}x + 2 + \dfrac{2}{3} = \dfrac{2}{3}x + \dfrac{8}{3}$

Graph the line:

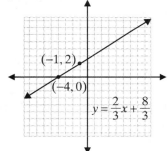

22. First, find the slope of the line from the two given points: $m = \dfrac{1 - (-4)}{-5 - 3} = -\dfrac{5}{8}$. Use the point-slope form of a line with the slope and one of the given points:

$y - 1 = -\dfrac{5}{8}\left(x - (-5)\right)$

$y - 1 = -\dfrac{5}{8}(x + 5)$

$y - 1 = -\dfrac{5}{8}x - \dfrac{25}{8}$

$y = -\dfrac{5}{8}x + 1 - \dfrac{25}{8}$

$y = -\dfrac{5}{8}x - \dfrac{17}{8}$

Graph the line:

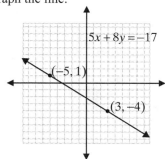

23. The given line has slope $m = -\dfrac{2}{5}$, so the requested line must also have this slope. The given point is actually the y-intercept, so the equation is:

$y = -\dfrac{2}{5}x + 7,$ or $2x + 5y = 35$

24. The lines $2x + 6y = 5$ and $x + 3y = 9$ are parallel because they both have a slope of $-\dfrac{1}{3}$ and their y-intercepts are not equal.

25. The lines $x - 2y = 4$ and $y = -2x + 8$ are perpendicular because their slopes are negative reciprocals of each other: the first has a slope of $\dfrac{1}{2}$ and the second has a slope of -2.

26. $f(x) = 3x^2 - 4x + 2$

a. $f(-2) = 3(-2)^2 - 4(-2) + 2$

$= 12 + 8 + 2 = 22$

b. $f(5) = 3(5)^2 - 4(5) + 2$

$= 75 - 20 + 2 = 57$

27. Graph the inequality:

$$3x + y < 10$$

An open half-plane:

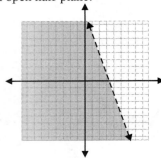

28. Graph the inequality:

$$4x + 2y \geq 9$$

A closed half-plane:

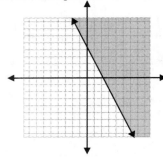

29. Solve (estimate) using a graphing calculator:

$$\begin{cases} 2x + 3y = 18 \\ 3x - 2y = -12 \end{cases}$$

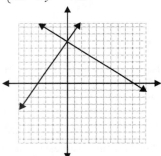

The solution is $(0, 6)$.

30. Solve (estimate) using a graphing calculator:

$$\begin{cases} y = \dfrac{1}{3}x + 5 \\ y = -2x + 1 \end{cases}$$

The solution is $\left(\dfrac{-12}{7}, \dfrac{31}{7} \right)$.

31. Solve the system using algebraic methods:

$$\begin{cases} 3x - 4y = -25 \\ 2x + y = 9 \end{cases}$$

Solve the second equation for y and use substitution:

$$y = -2x + 9$$
$$3x - 4(-2x + 9) = -25$$
$$3x + 8x - 36 = -25$$
$$11x = 11$$
$$x = 1$$

Substitute the results into one of the original equations and solve for y.

$$y = -2(1) + 9$$
$$y = 7$$

The solution is $(1, 7)$.

32. Solve the system using algebraic methods:

$$\begin{cases} x + 8y = -22 \\ 3x - y = -9 \end{cases} = \begin{cases} x + 8y = -22 \\ 24x - 8y = -72 \end{cases}$$

Use the addition method:

$$x + 8y = -22$$
$$\underline{24x - 8y = -72}$$
$$25x \qquad = -94$$
$$x \qquad = -\dfrac{94}{25}$$

206

Substitute into the second equation:

$$3\left(-\frac{94}{25}\right) - y = -9$$

$$-y = -9 + \frac{282}{25} = \frac{57}{25}$$

$$y = -\frac{57}{25}$$

The solution is $\left(\frac{-94}{25}, \frac{-57}{25}\right)$.

33. Solve the system using a graph:
$$\begin{cases} 2x + y < 6 \\ -x + 2y > 4 \end{cases}$$

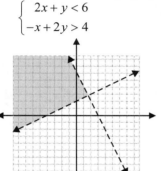

34. Solve the system using a graph:
$$\begin{cases} x \le 3 \\ y \ge -3x + 1 \end{cases}$$

35. $\dfrac{4x^3}{2x^{-2}x^4} = \dfrac{4x^3}{2x^2} = 2x$

36. $\left(4x^2 y\right)^3 = 4^3 x^6 y^3 = 64x^6 y^3$

37. $\left(7x^5 y^{-2}\right)^2 = 7^2 x^{10} y^{-4} = \dfrac{49x^{10}}{y^4}$

38. $\left(\dfrac{6x^2}{y^5}\right)^2 = \dfrac{6^2 x^4}{y^{10}} = \dfrac{36x^4}{y^{10}}$

39. $\left(a^{-3} b^2\right)^{-2} = a^6 b^{-4} = \dfrac{a^6}{b^4}$

40. $\left(\dfrac{3xy^4}{x^2}\right)^{-1} = \dfrac{x^2}{3xy^4} = \dfrac{x}{3y^4}$

41. $\left(\dfrac{8x^{-3} y^2}{xy^{-1}}\right)^0 = 1$

42. $\left(\dfrac{3^{-1} x^3 y^{-1}}{x^{-1} y^2}\right)^2 = \dfrac{3^{-2} x^6 y^{-2}}{x^{-2} y^4} = \dfrac{x^{6-(-2)} y^{-2-4}}{3^2} = \dfrac{x^8}{9 y^6}$

43. $\left(\dfrac{5x^{-2} y^3}{12x^2 y}\right)^{-1} \left(\dfrac{10x^2 y^2}{3x^{-1} y^{-1}}\right)^2 = \dfrac{12x^2 y}{5x^{-2} y^3} \cdot \dfrac{10^2 x^4 y^4}{3^2 x^{-2} y^{-2}}$

$$= \dfrac{12x^4}{5y^2} \cdot \dfrac{100x^6 y^6}{9}$$

$$= \dfrac{80x^{10} y^4}{3}$$

44. $\left(\dfrac{22a^3 b^2 c}{6a^{-3} bc^2}\right)^{-1} \left(\dfrac{14ab^2 c}{7a^{-2} b^2 c^3}\right)^3 = \dfrac{6a^{-3} bc^2}{22a^3 b^2 c} \cdot \dfrac{14^3 a^3 b^6 c^3}{7^3 a^{-6} b^6 c^9}$

$$= \dfrac{3c}{11a^6 b} \cdot \dfrac{8a^9}{c^6}$$

$$= \dfrac{24a^3}{11bc^5}$$

45. **a.** $2.8 \times 10^{-7} = 0.000\,000\,28$

   b. $3.51 \times 10^4 = 35{,}100$

46. $0.0015 \cdot 4200 = 1.5 \times 10^{-3} \cdot 4.2 \times 10^3$

$$= 1.5 \cdot 4.2 \times 10^{-3} \cdot 10^3$$

$$= 6.3 \times 10^0 = 6.3$$

47. $\dfrac{840}{0.00021} = \dfrac{8.4 \times 10^2}{2.1 \times 10^{-4}} = 4 \times 10^6$

48. $\dfrac{0.005 \cdot 77}{0.011 \cdot 3500} = \dfrac{5 \times 10^{-3} \cdot 7.7 \times 10^1}{1.1 \times 10^{-2} \cdot 3.5 \times 10^3}$

$$= \dfrac{5 \cdot 7.7 \times 10^{-2}}{1.1 \cdot 3.5 \times 10^1}$$

$$= 10 \times 10^{-3} = 1 \times 10^{-2}$$

49. **a.** $p(x) = -x^2 + 5x + 2x^2 - x = x^2 + 4x$

    b. A second degree binomial

    c. $p(3) = 3^2 + 4(3) = 9 + 12 = 21$

    d. $p(-5) = (-5)^2 + 4(-5) = 25 - 20 = 5$

50. **a.** $p(x) = 9x - x^3 + 3x^2 - x + x^3 = 3x^2 + 8x$

    b. A second degree binomial

    c. $p(3) = 3(3)^2 + 8(3) = 27 + 24 = 51$

    d. $p(-5) = 3(-5)^2 + 8(-5) = 75 - 40 = 35$

51. **a.** $p(x) = -x^4 + 4x^3 - 3x + x^2 - x^3 + x$

       $= -x^4 + 3x^3 + x^2 - 2x$

    b. A fourth degree polynomial

    c. $p(3) = -(3)^4 + 3(3)^3 + (3)^2 - 2(3)$

       $= -81 + 81 + 9 - 6 = 3$

    d. $p(-5) = -(-5)^4 + 3(-5)^3 + (-5)^2 - 2(-5)$

       $= -625 - 375 + 25 + 10 = -965$

52. **a.** $p(x) = 8x - 7x^2 + x^2 - x^3 - 6x - 4$

       $= -x^3 - 6x^2 + 2x - 4$

    b. A third degree polynomial

    c. $p(3) = -(3)^3 - 6(3)^2 + 2(3) - 4$

       $= -27 - 54 + 6 - 4 = -79$

    d. $p(-5) = -(-5)^3 - 6(-5)^2 + 2(-5) - 4$

       $= 125 - 150 - 10 - 4 = -39$

53. $(-2x^2 - 11x + 1) + (x^3 + 3x - 7) + (2x - 1)$

            $= x^3 - 2x^2 - 6x - 7$

54. $(4x^2 + 2x - 7) - (5x^2 + x - 2)$

        $= 4x^2 + 2x - 7 - 5x^2 - x + 2$

        $= -x^2 + x - 5$

55. $(x^3 + 4x^2 - x) - (-2x^3 + 6x + 3)$

        $= x^3 + 4x^2 - x + 2x^3 - 6x - 3$

        $= 3x^3 + 4x^2 - 7x - 3$

56. $(6x^2 + x - 10) - (x^3 - x^2 + x - 4)$

        $= 6x^2 + x - 10 - x^3 + x^2 - x + 4$

        $= -x^3 + 7x^2 - 6$

57. $(x^2 + 2x + 6) + (5x^2 - x - 2) - (8x + 3)$

        $= x^2 + 2x + 6 + 5x^2 - x - 2 - 8x - 3$

        $= 6x^2 - 7x + 1$

58. $(2x^2 - 5x - 7) - (3x^2 - 4x + 1) + (x^2 - 9)$

        $= 2x^2 - 5x - 7 - 3x^2 + 4x - 1 + x^2 - 9$

        $= -x - 17$

59. $-3x(x^2 - 4x + 1) = -3x^3 + 12x^2 - 3x$

60. $5x^3(x^2 + 2x) = 5x^5 + 10x^4$

61. $(x + 6)(x - 6) = x^2 - 36$

62. $(x + 4)(x - 3) = x^2 - 3x + 4x - 12 = x^2 + x - 12$

63 $(3x + 7)(3x + 7) = 9x^2 + 21x + 21x + 49$

             $= 9x^2 + 42x + 49$

64. $(2x - 1)(2x + 1) = 4x^2 - 1$

65. $(x^2 + 5)(x^2 - 5) = x^4 - 25$

66. $(x^2 - 2)(x^2 - 2) = x^4 - 4x^2 + 4$

67. $-1(x^2 - 5x + 2) = -x^2 + 5x - 2$

68. $(x - 4)^2 = x^2 - 8x + 16$

69. $(2x - 9)(x + 4) = 2x^2 = 8x - 9x - 36$

            $= 2x^2 - x - 36$

70. $(x - 6)(5x + 3) = 5x^2 + 3x - 30x - 18$

            $= 5x^2 - 27x - 18$

71. $(3x - 4)(2x + 3) = 6x^2 + 9x - 8x - 12$

            $= 6x^2 + x - 12$

72. $(3x + 8)(3x - 8) = 9x^2 - 64$

73. $(4x + 1)(x^2 - x) = 4x^3 - 4x^2 + x^2 - x$

            $= 4x^3 - 3x^2 - x$

74. $\dfrac{8x^2 - 14x + 6}{2x} = \dfrac{8x^2}{2x} - \dfrac{14x}{2x} + \dfrac{6}{2x} = 4x - 7 + \dfrac{3}{x}$

75. $\dfrac{13x^3 y + 10x^2 y^2 + 5xy^3}{5x^2 y} = \dfrac{13x}{5} + 2y + \dfrac{y^2}{x}$

76. $\dfrac{x^2 y^2 - 21x^2 y^3 + 7xy - 28x^2 y^4}{7x^2 y^2}$

$= \dfrac{x^2 y^2}{7x^2 y^2} - \dfrac{21x^2 y^3}{7x^2 y^2} + \dfrac{7xy}{7x^2 y^2} - \dfrac{28x^2 y^4}{7x^2 y^2}$

$= \dfrac{1}{7} - 3y + \dfrac{1}{xy} - 4y^2$

77. $x - 2 + \dfrac{0}{x+9}$

$$
\begin{array}{r}
x - 2 \\
x+9 \overline{\smash{\big)}\ x^2 + 7x - 18} \\
\underline{-\left(x^2 + 9x\right)} \\
-2x - 18 \\
\underline{-(-2x - 18)} \\
0
\end{array}
$$

78. $x + 2 + \dfrac{4}{4x-3}$

$$
\begin{array}{r}
x + 2 \\
4x-3 \overline{\smash{\big)}\ 4x^2 + 5x - 2} \\
\underline{-\left(4x^2 - 3x\right)} \\
8x - 2 \\
\underline{-(8x - 6)} \\
4
\end{array}
$$

79. $2x^2 - x + 3 - \dfrac{2}{x+3}$

$$
\begin{array}{r}
2x^2 - x + 3 \\
x+3 \overline{\smash{\big)}\ 2x^3 + 5x^2 + 0 \cdot x + 7} \\
\underline{-\left(2x^3 + 6x^2\right)} \\
-x^2 + 0 \cdot x \\
\underline{-\left(-x^2 - 3x\right)} \\
3x + 7 \\
\underline{-(3x + 9)} \\
-2
\end{array}
$$

80. Let x be the first even integer. Then, $x + 2$ and $x + 4$ are the second and third even integers, respectively.

$x + 2 + x + 4 = 3x - 14$

$2x + 6 = 3x - 14$

$-x = -20$

$x = 20$

$x + 2 = 22$

$x + 4 = 24$

The integers are 20, 22 and 24.

81. Let w be the width of the rectangle. Then, $w + 9$ is the length. Use the formula for perimeter:

$2w + 2(w + 9) = 82$

$2w + 2w + 18 = 82$

$4w = 64$

$w = 16$

$w + 9 = 25$

The rectangle is 16 cm by 25 cm.

82. Let x be the number.

$2(x - 16) = 4x + 6$

$2x - 32 = 4x + 6$

$-2x = 38$

$x = -19$

The number is -19.

83. Let x be the length of one of the congruent sides. Then, $x + 4$ is the length of the third side.

$x + 3 + x + 3 + x + 7 = 127$

$3x + 13 = 127$

$3x = 114$

$x = 38$

$x + 4 = 42$

The lengths of the sides of the triangle are 38, 38 and 42 feet.

84. Substitute the given points in $ax + by = 9$:

$$\begin{cases} -a - 3b = 9 \\ a - 1.5b = 9 \end{cases}$$

Solve by the addition method:

$$
\begin{array}{r}
-a - 3b = 9 \\
\underline{a - 1.5b = 9} \\
-4.5b = 18 \\
b = -4
\end{array}
$$

Substitute into the second equation:

$a - 1.5(-4) = 9$

$a = 3$

The equation of the line is $3x - 4y = 9$.

85. Let x be the amount invested at 6%. Then, $100,000 - x$ is the amount invested at 8%.

$$0.06x + 0.08(100000 - x) = 6700$$
$$0.06x + 8000 - 0.08x = 6700$$
$$-0.02x + 8000 = 6700$$
$$-0.02x = -1300$$
$$x = 65,000$$
$$100,000 - x = 35,000$$

$65,000 is invested at 6% and $35,000 is invested at 8%.

86. Let x be the amount in ounces of the 10% iodine mixture and y be the amount of the 30% iodine mixture. Set up a linear system of equations:

$$\begin{cases} x + y = 8 \\ 0.10x + 0.30y = 0.15(8) \end{cases}$$

Simplify the second equation and multiply the first equation by -10. Then solve by the addition method.

$$-10x - 10y = -80$$
$$\underline{10x + 30y = 120}$$
$$20y = 40$$
$$y = 2$$
$$x = 8 - y = 6$$

The solution is 6 ounces of the 10% mixture and 2 ounces of the 30% mixture.

[End of Cumulative Review: Chapters 1 – 6]

Section 7.1
Solutions to Odd Exercises

1. $\text{GCF}(10, 15, 20) = 5$:
$$10 = 2 \cdot 5$$
$$15 = 3 \cdot 5$$
$$20 = 2^2 \cdot 5$$

3. $\text{GCF}(16, 40, 56) = 2^3 = 8$:
$$16 = 2^4$$
$$40 = 2^3 \cdot 5$$
$$56 = 2^3 \cdot 7$$

5. $\text{GCF}(9, 14, 22) = 1$:
$$9 = 3^2$$
$$14 = 2 \cdot 7$$
$$22 = 2 \cdot 11$$

7. $\text{GCF}(30x^3, 40x^5) = 10x^3$:
$$30x^3 = 2 \cdot 3 \cdot 5 \cdot x^3$$
$$40x^5 = 2^3 \cdot 5 \cdot x^5$$

9. $\text{GCF}(26ab^2, 39a^2b, 52a^2b^2) = 13ab$:
$$26ab^2 = 2 \cdot 13 \cdot a \cdot b^2$$
$$39a^2b = 3 \cdot 13 \cdot a^2 \cdot b$$
$$52a^2b^2 = 2^2 \cdot 13 \cdot a^2 \cdot b^2$$

11. $\text{GCF}(28c^2d^3, 14c^3d^2, 42cd^2) = 14cd^2$:
$$28c^2d^3 = 2^2 \cdot 7 \cdot c^2 \cdot d^3$$
$$14c^3d^2 = 2 \cdot 7 \cdot c^3 \cdot d^2$$
$$42cd^2 = 2 \cdot 3 \cdot 7 \cdot c \cdot d^2$$

13. $\text{GCF}(36xy, 48xy, 60xy) = 12xy$:
$$36xy = 2^2 \cdot 3^2 \cdot x \cdot y$$
$$48xy = 2^4 \cdot 3 \cdot x \cdot y$$
$$60xy = 2^2 \cdot 3 \cdot 5 \cdot x \cdot y$$

15. $\dfrac{x^7}{x^3} = x^{7-3} = x^4$

17. $\dfrac{-8y^3}{2y^2} = -4y^{3-2} = -4y$

19. $\dfrac{9x^5}{3x^2} = 3x^{5-2} = 3x^3$

21. $\dfrac{4x^3y^2}{2xy} = 2x^{3-1}y^{2-1} = 2x^2y$

23. $3m + 27 = 3(m + 9)$

25. $5x^2 - 30x = 5x(x - 6)$

27. $13ab^2 + 13ab = 13ab(b + 1)$

29. $-15xy^2 - 20x^2y - 5xy = -5xy(3y + 4x + 1)$

31. $11x - 121 = 11(x - 11)$

33. $16y^3 + 12y = 4y(4y^2 + 3)$

35. $-8a - 16b = -8(a + 2b)$

37. $-6ax + 9ay = -3a(2x - 3y)$

39. $16x^4y - 14x^2y = 2x^2y(8x^2 - 7)$

41. $-14x^2y^3 - 14x^2y = -14x^2y(y^2 + 1)$

43. $5x^2 - 15x - 5 = 5(x^2 - 3x - 1)$

45. $8m^2x^3 - 12m^2y + 4m^2z = 4m^2(2x^3 - 3y + z)$

47. $34x^4y^6 - 51x^3y^5 + 17x^5y^4$
$$= 17x^3y^4(x^2 - 3y + 2xy^2)$$

49. $15x^4y^2 + 24x^6y^6 - 32x^7y^3$
$$= x^4y^2(15 + 24x^2y^4 - 32x^3y)$$

51. $7y^2(y + 3) + 2(y + 3) = (7y^2 + 2)(y + 3)$

53. $3x(x - 4) + (x - 4) = (3x + 1)(x - 4)$

55. $4x^3(x - 2) - (x - 2) = (4x^3 - 1)(x - 2)$

57. $10y(2y + 3) - 7(2y + 3) = (10y - 7)(2y + 3)$

59. $a(x - 2) - b(x - 2) = (a - b)(x - 2)$

61. $bx + b + cx + c = (bx + b) + (cx + c)$
$$= b(x+1) + c(x+1)$$
$$= (b+c)(x+1)$$

63. $x^3 + 3x^2 + 6x + 18 = (x^3 + 3x^2) + (6x + 18)$
$$= x^2(x+3) + 6(x+3)$$
$$= (x^2 + 6)(x+3)$$

65. $x^2 - 4x + 6xy - 24y = (x^2 - 4x) + (6xy - 24y)$
$$= x(x-4) + 6y(x-4)$$
$$= (x + 6y)(x-4)$$

67. $5xy + yz - 20x - 4z = (5xy + yz) - (20x + 4z)$
$$= y(5x + z) - 4(5x + z)$$
$$= (y - 4)(5x + z)$$

69. $24y - 3yz + 2xz - 16x = (24y - 3yz) + (2xz - 16x)$
$$= 3y(8 - z) + 2x(z - 8)$$
$$= 3y(8 - z) - 2x(8 - z)$$
$$= (3y - 2x)(8 - z)$$
$$\text{or } = (2x - 3y)(z - 8)$$

71. $ax + 5ay + 3x + 15y = (ax + 5ay) + (3x + 15y)$
$$= a(x + 5y) + 3(x + 5y)$$
$$= (a + 3)(x + 5y)$$

73. $4xy + 3x - 4y - 3 = (4xy + 3x) - (4y + 3)$
$$= x(4y + 3) - (4y + 3)$$
$$= (x - 1)(4y + 3)$$

75. $xy + x - y - 1 = (xy + x) - (y + 1)$
$$= x(y + 1) - (y + 1)$$
$$= (x - 1)(y + 1)$$

77. $x^2 - 5 + x^2y + 5y$ is not factorable.

79. $7xy - 3y + 2x^2 - 3x$ is not factorable.

81. $3xy - 4yu - 6xv + 8uv$
$$= (3xy - 6xv) + (8uv - 4yu)$$
$$= 3x(y - 2v) + 4u(2v - y)$$
$$= 3x(y - 2v) - 4u(-2v + y)$$
$$= (3x - 4u)(y - 2v)$$

83. $6ac - 9ad + 2bc - 3bd$
$$= (6ac - 9ad) + (2bc - 3bd)$$
$$= 3a(2c - 3d) + b(2c - 3d)$$
$$= (3a + b)(2c - 3d)$$

[End of Section 7.1]

Section 7.2
Solutions to Odd Exercises

1. Pairs of integer factors of 15:
$\{1, 15\}, \{-1, -15\}, \{3, 5\}, \{-3, -5\}$

3. Pairs of integer factors of 20:
$\{1, 20\}, \{-1, -20\}, \{2, 10\}, \{-2, -10\},$
$\{4, 5\}, \{-4, -5\}$

5. Pairs of integer factors of -6:
$\{1, -6\}, \{-1, 6\}, \{2, -3\}, \{-2, 3\}$

7. Pairs of integer factors of 16:
$\{1, 16\}, \{-1, -16\}, \{2, 8\}, \{-2, -8\}, \{4, 4\},$
$\{-4, -4\}$

9. Pairs of integer factors of -10:
$\{1, -10\}, \{-1, 10\}, \{2, -5\}, \{-2, 5\}$

11. For 12 and 7: 4 and 3
$4 \cdot 3 = 12$
$4 + 3 = 7$

13. For -14 and -5: 2 and -7
$2 \cdot -7 = -14$
$2 + (-7) = -5$

15. For -8 and 7: 8 and -1
$8 \cdot -1 = -8$
$8 + (-1) = 7$

17. For 36 and -12: -6 and -6
$(-6)(-6) = 36$
$(-6) + (-6) = -12$

19. For 20 and -9: -4 and -5
$(-4)(-5) = 20$
$(-4) + (-5) = -9$

21. $x^2 + 6x + 5 = (x + 5)(x + 1)$

23. $p^2 - 9p - 10 = (p + 1)(p - 10)$

25. $a^2 + 12a + 36 = (a + 6)(a + 6)$

27. $x^2 - x - 12 = (x - 4)(x + 3)$

29. $y^2 + y - 30 = (y + 6)(y - 5)$

31. $a^2 + a + 2$ is not factorable.

33. $x^2 + 3x + 5$ is not factorable.

35. $x^2 + 3x - 18 = (x + 6)(x - 3)$

37. $y^2 - 14y + 24 = (y - 12)(y - 2)$

39. $x^2 - 6x - 27 = (x - 9)(x + 3)$

41. $x^3 + 10x^2 + 21x = x(x^2 + 10x + 21)$
$= x(x + 7)(x + 3)$

43. $5x^2 - 5x - 60 = 5(x^2 - x - 12) = 5(x - 4)(x + 3)$

45. $10y^2 - 10y - 60 = 10(y^2 - y - 6)$
$= 10(y - 3)(y + 2)$

47. $4p^4 + 36p^3 + 32p^2 = 4p^2(p^2 + 9p + 8)$
$= 4p^2(p + 8)(p + 1)$

49. $2x^4 - 14x^3 - 36x^2 = 2x^2(x^2 - 7x - 18)$
$= 2x^2(x - 9)(x + 2)$

51. $2x^2 - 2x - 72 = 2(x^2 - x - 36)$
This final trinomial is not factorable.

53. $a^2 - 30a - 216 = (a - 36)(a + 6)$

55. $3y^5 - 21y^4 - 24y^3 = 3y^3(y^2 - 7y - 8)$
$= 3y^3(y - 8)(y + 1)$

57. $x^2 - 2xy - 3y^2 = (x - 3y)(x + y)$

59. $20a^2 + 40ab + 20b^2 = 20(a^2 + 2ab + b^2)$
$= 20(a + b)(a + b)$

61. The area of a rectangle is the product of its length and width. One of the factors of area is the given width, $4x$. You need to find the other factor of the area, which is the length. Complete the factorization:

$$\text{Area} = \text{Width} \times \text{Length}$$

$$4x^2 + 20x = 4x\left(\underline{\hspace{1cm}}\right)$$

$$= 4x(x+5)$$

So, the length is $x + 5$ inches.

63. The volume of a box is the product of the length, width and height. In symbols, $V = lwh$. The height of this box is x. The other factor is the product of length and width. From the figure, length can be represented by $40 - 2x$ and width by $10 - 2x$. The volume can now be written as the product of the length, width and height in factored form:

$$V(x) = (40 - 2x)(10 - 2x)x$$

$$= (400 - 100x + 4x^2)x$$

$$= 4x^3 - 100x^2 + 400x$$

The last line confirms that the given representation is correct.

Writing and Thinking About Mathematics

65. The given polynomial is shown to be factored in two different ways. While both ways are partially correct, neither one is completely factored. If you continue factoring both, you will have the same completely factored form:

$$2x^2 + 10x + 12 = 2(x^2 + 5x + 6)$$

$$= 2(x+3)(x+2)$$

Both ways will *always* result in this factorization, if you factor them completely. Remember, the best first step in factoring is to factor out common monomial factors. Then, proceed to try more complex factoring.

[End of Section 7.2]

Section 7.3
Solutions to Odd Exercises

1. $x^2 + 5x + 6 = (x+3)(x+2)$

3. $2x^2 - 3x - 5 = (2x-5)(x+1)$

5. $6x^2 + 11x + 5 = (6x+5)(x+1)$

7. $-x^2 + 3x - 2 = -(x^2 - 3x + 2) = -(x-2)(x-1)$

9. $x^2 - 3x - 10 = (x-5)(x+2)$

11. $-x^2 + 13x + 14 = -(x^2 - 13x - 14)$
$$= -(x-14)(x+1)$$

13. $x^2 + 8x + 64$ is not factorable.

15. $-2x^3 + x^2 + x = -x(2x^2 - x - 1)$
$$= -x(2x+1)(x-1)$$

17. $4t^2 - 3t - 1 = (4t+1)(t-1)$

19. $5a^2 - a - 6 = (5a-6)(a+1)$

21. $7x^2 + 5x - 2 = (7x-2)(x+1)$

23. $4x^2 + 23x + 15 = (4x+3)(x+5)$

25. $x^2 + 6x - 16 = (x+8)(x-2)$

27. $12x^2 - 38x + 20 = 2(6x^2 - 19x + 10)$
$$= 2(3x-2)(2x-5)$$

29. $3x^2 - 7x + 2 = (3x-1)(x-2)$

31. $9x^2 - 6x + 1 = (3x-1)(3x-1) = (3x-1)^2$

33. $4x^2 + 4x + 1 = (2x+1)(2x+1) = (2x+1)^2$

35. $12y^2 - 7y - 12 = (3y-4)(4y+3)$

37. $3x^2 + 9x + 5$ is not factorable.

39. $8a^2b - 22ab + 12b = 2b(4a^2 - 11a + 6)$
$$= 2b(4a-3)(a-2)$$

41. $x^2 + x + 1$ is not factorable.

43. $16x^2 - 8x + 1 = (4x-1)(4x-1) = (4x-1)^2$

45. $64x^2 - 48x + 9 = (8x-3)(8x-3) = (8x-3)^2$

47. $6x^2 + 2x - 20 = 2(3x^2 + x - 10) = 2(3x-5)(x+2)$

49. $10x^2 + 35x + 30 = 5(2x^2 + 7x + 6)$
$$= 5(2x+3)(x+2)$$

51. $-18x^2 + 72x - 8 = -2(9x^2 - 36x + 4)$
This trinomial cannot be factored any further.

53. $7x^4 - 5x^3 + 3x^2 = x^2(7x^2 - 5x + 3)$
This trinomial cannot be factored any further.

55. $-12m^2 + 22m + 4 = -2(6m^2 - 11m - 2)$
$$= -2(6m+1)(m-2)$$

57. $6x^3 + 9x^2 - 6x = 3x(2x^2 + 3x - 2)$
$$= 3x(2x-1)(x+2)$$

59. $9x^3y^3 + 9x^2y^3 + 9xy^3 = 9xy^3(x^2 + x + 1)$
This trinomial cannot be factored any further.

61. $12x^3 - 108x^2 + 243x = 3x(4x^2 - 36x + 81)$
$$= 3x(2x-9)(2x-9)$$
$$= 3x(2x-9)^2$$

63. $48xy^3 - 100xy^2 + 48xy = 4xy(12y^2 - 25y + 12)$
$$= 4xy(4y-3)(3y-4)$$

65. $21y^4 - 98y^3 + 56y^2 = 7y^2(3y^2 - 14y + 8)$
$$= 7y^2(3y-2)(y-4)$$

67. $ax + ay + 3x + 3y = (ax+ay) + (3x+3y)$
$$= a(x+y) + 3(x+y)$$
$$= (a+3)(x+y)$$

69. $5x^2 + 5y^2 + bx^2 + by^2 = \left(5x^2 + 5y^2\right) + \left(bx^2 + by^2\right)$
$$= 5\left(x^2 + y^2\right) + b\left(x^2 + y^2\right)$$
$$= (5+b)\left(x^2 + y^2\right)$$

Writing and Thinking About Mathematics

71. a. Substitute $t = 3$ and $t = 5$ in the expression
$h(t) = -16t^2 + 784$:

$$h(3) = -16(3)^2 + 784 = 640 \text{ ft}$$
$$h(5) = -16(5)^2 + 784 = 384 \text{ ft}$$

b. The ball started at 784 ft, so after 3 seconds the
ball has dropped $784 - 640 = 144$ ft. After 5
seconds the ball has dropped $784 - 384 = 400$ ft.

c. The ball reaches the ground when
$h(t) = -16t^2 + 784 = 0$.   Factor the polynomial
and set its factors equal to zero:
$$-16t^2 + 784 = 0$$
$$-16\left(t^2 - 49\right) = 0$$
$$-16(t-7)(t+7) = 0$$
Set $t - 7 = 0$ to get $t = 7$.
So, the height is 0 when $t = 7$.
Disregard the other "solution" as it yields a
negative value for t, which does not apply to the
physical aspects of this problem:
$$t + 7 = 0 \rightarrow t = -7$$

[End of Section 7.3]

Section 7.4
Solutions to Odd Exercises

1. $x^2 - 1 = (x+1)(x-1)$

3. $x^2 - 49 = (x+7)(x-7)$

5. $x^2 + 4x + 4 = (x+2)(x+2) = (x+2)^2$

7. $x^2 - 12x + 36 = (x-6)(x-6) = (x-6)^2$

9. $16x^2 - 9 = (4x+3)(4x-3)$

11. $9x^2 - 1 = (3x+1)(3x-1)$

13. $25 - 4x^2 = (5+2x)(5-2x)$

15. $x^2 - 14x + 49 = (x-7)(x-7) = (x-7)^2$

17. $x^2 + 6x + 9 = (x+3)(x+3) = (x+3)^2$

19. $3x^2 - 27y^2 = 3(x^2 - 9y^2) = 3(x+3y)(x-3y)$

21. $x^3 - xy^2 = x(x^2 - y^2) = x(x+y)(x-y)$

23. $x^2 - \dfrac{1}{4} = \left(x + \dfrac{1}{2}\right)\left(x - \dfrac{1}{2}\right)$

25. $x^2 - \dfrac{9}{16} = \left(x + \dfrac{3}{4}\right)\left(x - \dfrac{3}{4}\right)$

27. $x^4 - 1 = (x^2+1)(x^2-1) = (x^2+1)(x+1)(x-1)$

29. $2x^2 - 32x + 128 = 2(x^2 - 16x + 64)$
$= 2(x-8)(x-8) = 2(x-8)^2$

31. $ay^2 + 2ay + a = a(y^2 + 2y + 1)$
$= a(y+1)(y+1) = a(y+1)^2$

33. $4x^2 y^2 - 24xy^2 + 36y^2 = 4y^2(x^2 - 6x + 9)$
$= 4y^2(x-3)(x-3)$
$= 4(x-3)^2$

35. $16x^2 + 81$ is not factorable.

37. $x^2 + x + 1$ is not factorable.

39. $y^2 + 2y + 2$ is not factorable.

41. $16x^2 - 64 = (4x+8)(4x-8)$

43. $ax^2 - 10ax + 25a = a(x-5)(x-5) = a(x-5)^2$

45. $147 - 3y^2 = 3(49 - y^2) = 3(7+y)(7-y)$

47. $x^2 + 5x + \dfrac{25}{4} = \left(x + \dfrac{5}{2}\right)\left(x + \dfrac{5}{2}\right) = \left(x + \dfrac{5}{2}\right)^2$

49. $2x^2 - \dfrac{18}{25} = 2\left(x^2 - \dfrac{9}{25}\right) = 2\left(x + \dfrac{3}{5}\right)\left(x - \dfrac{3}{5}\right)$

51. Since $\left(\dfrac{-6}{2}\right)^2 = 9$, add 9 and factor:
$x^2 - 6x + \underline{9} = (\underline{x-3})^2$

53. Since $\left(\dfrac{-4}{2}\right)^2 = 4$, add 4 and factor:
$x^2 - 4x + \underline{4} = (\underline{x-2})^2$

55. Since $16 = 4^2$ and $2 \cdot 4 = 8$, add $8x$ and factor:
$x^2 + \underline{8x} + 16 = (\underline{x+4})^2$

57. Since $81 = 9^2$ and $-2 \cdot 9 = -18$, add $-18x$ and factor:
$x^2 - \underline{18x} + 81 = (\underline{x-9})^2$

59. Since $\left(\dfrac{1}{2}\right)^2 = \dfrac{1}{4}$, add $\dfrac{1}{4}$ and factor:
$x^2 + x + \underline{\dfrac{1}{4}} = \left(\underline{x + \dfrac{1}{2}}\right)^2$

61. Since $\left(\dfrac{9}{2}\right)^2 = \dfrac{81}{4}$, add $\dfrac{81}{4}$ and factor:
$x^2 - 9x + \underline{\dfrac{81}{4}} = \left(\underline{x - \dfrac{9}{2}}\right)^2$

63. Since $\left(\dfrac{5}{2}\right)^2 = \dfrac{25}{4}$ and $2\left(\dfrac{5}{2}\right) = 5$, add $5x$ and factor:

$$x^2 + \underline{5x} + \frac{25}{4} = \left(x + \frac{5}{2}\right)^2$$

65. Since $\left(\dfrac{3}{2}\right)^2 = \dfrac{9}{4}$ and $2\left(\dfrac{3}{2}\right) = 3$, add $3x$ and factor:

$$x^2 - \underline{3x} + \frac{9}{4} = \left(x - \frac{3}{2}\right)^2$$

[End of Section 7.4]

Section 7.5
Solutions to Odd Exercises

1. $(x-3)(x-2)=0$

Set: $x-3=0$ $x-2=0$

Solve: $x=3$ $x=2$

3. $(2x-9)(x+2)=0$

Set: $2x-9=0$ $x+2=0$

Solve: $2x=9$ $x=-2$

$x=\dfrac{9}{2}$

5. $(x+3)(x+3)=0$

Set: $x+3=0$

Solve: $x=-3$

Note that you only need to set one factor equal to zero in the case of a double root.

7. $(x+5)(x+5)=0$

Set: $x+5=0$

Solve: $x=-5$

9. $2x(x-2)=0$

Set: $2x=0$ $x-2=0$

Solve: $x=0$ $x=2$

11. $x^2-3x=4$

$x^2-3x-4=0$

$(x-4)(x+1)=0$

Set: $x-4=0$ $x+1=0$

Solve: $x=4$ $x=-1$

13. $0=5x^2+15x$

$0=5x(x+3)$

Set: $5x=0$ $x+3=0$

Solve: $x=0$ $x=-3$

15. $2x^2-24=2x$

$2x^2-2x-24=0$

$2(x^2-x-12)=0$

$2(x-4)(x+3)=0$

Set: $x-4=0$ $x+3=0$

Solve: $x=4$ $x=-3$

17. $x^2+8=6x$

$x^2-6x+8=0$

$(x-4)(x-2)=0$

Set: $x-4=0$ $x-2=0$

Solve: $x=4$ $x=2$

19. $2x^2+2x-24=0$

$2(x^2+x-12)=0$

$2(x+4)(x-3)=0$

Set: $x+4=0$ $x-3=0$

Solve: $x=-4$ $x=3$

21. $0=2x^2-5x-3$

$0=(2x+1)(x-3)$

Set: $2x+1=0$ $x-3=0$

Solve: $2x=-1$ $x=3$

$x=-\dfrac{1}{2}$

23. $3x^2-4x-4=0$

$(3x+2)(x-2)=0$

Set: $3x+2=0$ $x-2=0$

Solve: $3x=-2$ $x=2$

$x=-\dfrac{2}{3}$

25. $2x^2-7x-4=0$

$(2x+1)(x-4)=0$

Set: $2x+1=0$ $x-4=0$

Solve: $2x=-1$ $x=4$

$x=-\dfrac{1}{2}$

27. $0=3x^2+2x-8$

$0=(3x-4)(x+2)$

Set: $3x-4=0$ $x+2=0$

Solve: $3x=4$ $x=-2$

$x=\dfrac{4}{3}$

29. $4x^2 - 12x + 9 = 0$

$(2x - 3)^2 = 0$

Set: $\quad 2x - 3 = 0$

Solve: $\quad 2x = 3$

$$x = \frac{3}{2}$$

31. $\quad 8x = 5x^2$

$5x^2 - 8x = 0$

$x(5x - 8) = 0$

Set: $\quad x = 0 \qquad 5x - 8 = 0$

Solve: $\qquad\qquad\qquad 5x = 8$

$$x = \frac{8}{5}$$

33. $\quad 9x^2 = 36$

$9x^2 - 36 = 0$

$9(x^2 - 4) = 0$

$9(x + 2)(x - 2) = 0$

Set: $\quad x + 2 = 0 \qquad x - 2 = 0$

Solve: $\quad x = -2 \qquad x = 2$

35. $5x^2 + 10x + 5 = 0$

$5(x^2 + 2x + 1) = 0$

$5(x + 1)^2 = 0$

Set: $\quad x + 1 = 0$

Solve: $\quad x = -1$

37. $\quad 8x^2 + 32 = 32x$

$8x^2 - 32x + 32 = 0$

$8(x^2 - 4x + 4) = 0$

$8(x - 2)^2 = 0$

Set: $\quad x - 2 = 0$

Solve: $\quad x = 2$

39. $\dfrac{x^2}{3} - 2x + 3 = 0$

$x^2 - 6x + 9 = 0$

$(x - 3)^2 = 0$

Set: $\quad x - 3 = 0$

Solve: $\quad x = 3$

41. $\dfrac{x^2}{5} - x - 10 = 0$

$x^2 - 5x - 50 = 0$

$(x - 10)(x + 5) = 0$

Set: $\quad x - 10 = 0 \qquad x + 5 = 0$

Solve: $\quad x = 10 \qquad x = -5$

43. $\dfrac{x^2}{8} + x + \dfrac{3}{2} = 0$

$x^2 + 8x + 12 = 0$

$(x + 6)(x + 2) = 0$

Set: $\quad x + 6 = 0 \qquad x + 2 = 0$

Solve: $\quad x = -6 \qquad x = -2$

45. $x^2 - x + \dfrac{1}{4} = 0$

$4x^2 - 4x + 1 = 0$

$(2x - 1)^2 = 0$

Set: $\quad 2x - 1 = 0$

Solve: $\quad 2x = 1$

$$x = \frac{1}{2}$$

47. $\quad x^3 + 8x = 6x^2$

$x^3 - 6x^2 + 8x = 0$

$x(x^2 - 6x + 8) = 0$

$x(x - 4)(x - 2) = 0$

Set: $\quad x = 0 \qquad x - 4 = 0 \qquad x - 2 = 0$

Solve: $\qquad\qquad\quad x = 4 \qquad\quad x = 2$

49. $\quad 6x^3 + 7x^2 = -2x$

$6x^3 + 7x^2 + 2x = 0$

$x(6x^2 + 7x + 2) = 0$

$x(3x + 2)(2x + 1) = 0$

Set: $\quad x = 0 \qquad 3x + 2 = 0 \qquad 2x + 1 = 0$

Solve: $\qquad\qquad\quad 3x = -2 \qquad\quad 2x = -1$

$$x = -\frac{2}{3} \qquad\quad x = -\frac{1}{2}$$

51. $0 = x^2 - 100$

$0 = (x + 10)(x - 10)$

Set: $\quad x + 10 = 0 \qquad x - 10 = 0$

Solve: $\quad x = -10 \qquad x = 10$

53.
$$3x^2 - 75 = 0$$
$$3x^2 - 25 = 0$$
$$3(x+5)(x-5) = 0$$
Set: $x + 5 = 0$ $x - 5 = 0$
Solve: $x = -5$ $x = 5$

55. $x^2 + 8x + 16 = 0$
$$(x+4)^2 = 0$$
Set: $x + 4 = 0$
Solve: $x = -4$

57. $3x^2 = 18x - 27$
$$0 = -3x^2 + 18x - 27$$
$$0 = -3(x^2 - 6x + 9)$$
$$0 = -3(x-3)^2$$
Set: $x - 3 = 0$
Solve: $x = 3$

59.
$$(x-1)^2 = 4$$
$$x^2 - 2x + 1 = 4$$
$$x^2 - 2x - 3 = 0$$
$$(x+1)(x-3) = 0$$
Set: $x + 1 = 0$ $x - 3 = 0$
Solve: $x = -1$ $x = 3$

61.
$$(x+5)^2 = 9$$
$$x^2 + 10x + 25 - 9 = 0$$
$$x^2 + 10x + 16 = 0$$
$$(x+2)(x+8) = 0$$
Set: $x + 2 = 0$ $x + 8 = 0$
Solve: $x = -2$ $x = -8$

63. $(x+4)(x-1) = 6$
$$x^2 + 3x - 4 - 6 = 0$$
$$x^2 + 3x - 10 = 0$$
$$(x+5)(x-2) = 0$$
Set: $x + 5 = 0$ $x - 2 = 0$
Solve: $x = -5$ $x = 2$

65. $27 = (x+2)(x-4)$
$$27 = x^2 - 2x - 8$$
$$0 = -27 + x^2 - 2x - 8$$
$$0 = x^2 - 2x - 35$$
$$0 = (x+5)(x-7)$$
Set: $x + 5 = 0$ $x - 7 = 0$
Solve: $x = -5$ $x = 7$

67. $x(x+7) = 3(x+4)$
$$x^2 + 7x = 3x + 12$$
$$x^2 + 4x - 12 = 0$$
$$(x+6)(x-2) = 0$$
Set: $x + 6 = 0$ $x - 2 = 0$
Solve: $x = -6$ $x = 2$

69.
$$3x(x+1) = 2(x+1)$$
$$3x^2 + 3x = 2x + 2$$
$$3x^2 + x - 2 = 0$$
$$(3x-2)(x+1) = 0$$
Set: $3x - 2 = 0$ $x + 1 = 0$
Solve: $3x = 2$ $x = -1$
$$x = \frac{2}{3}$$

71.
$$x(2x+1) = 6(x+2)$$
$$2x^2 + x = 6x + 12$$
$$2x^2 - 5x - 12 = 0$$
$$(2x+3)(x-4) = 0$$
Set: $2x + 3 = 0$ $x - 4 = 0$
Solve: $2x = -3$ $x = 4$
$$x = -\frac{3}{2}$$

[End of Section 7.5]

Section 7.6
Solutions to Odd Exercises

1. Let x be the integer.
$$x^2 = 7x$$
$$x^2 - 7x = 0$$
$$x(x-7) = 0$$
Set: $\qquad x = 0 \qquad\quad x - 7 = 0$
Solve: $\qquad\qquad\qquad\quad x = 7$
There are two solutions: 0 and 7.

3. Let x be the positive integer.
$$x^2 = x + 12$$
$$x^2 - x - 12 = 0$$
$$(x+3)(x-4) = 0$$
Set: $\qquad x + 3 = 0 \qquad x - 4 = 0$
Solve: $\qquad x = -3 \qquad\quad x = 4$
Disregard $x = -3 < 0$. The integer is 4.

5. Let x be the first number. Then, $x + 7$ is the second number.
$$x(x+7) = 78$$
$$x^2 + 7x - 78 = 0$$
$$(x-6)(x+13) = 0$$
Set: $\qquad x - 6 = 0 \qquad x + 13 = 0$
Solve: $\qquad x = 6 \qquad\qquad x = -13$
$$6 + 7 = 13 \qquad -13 + 7 = -6$$
There are two solutions: 6 and 13, or −13 and −6.

7. Let x be the positive integer.
$$x^2 + 3x = 54$$
$$x^2 + 3x - 54 = 0$$
$$(x+9)(x-6) = 0$$
Set: $\qquad x + 9 = 0 \qquad x - 6 = 0$
Solve: $\qquad x = -9 \qquad\quad x = 6$
Disregard $x = -9 < 0$. The number is 6.

9. Let x be the first positive integer. Then, $x - 8$ is the second integer.
$$x - 8 + x^2 = 124$$
$$x^2 + x - 132 = 0$$
$$(x+12)(x-11) = 0$$
Set: $\qquad x + 12 = 0 \qquad x - 11 = 0$
Solve: $\qquad x = -12 \qquad\quad x = 11$
$$11 - 8 = 3$$
Disregard $x = -12 < 0$. The integers are 11 and 3.

11. Let x be the first number. Then, $x - 5$ is the second number.
$$(x-5)^2 + x^2 = 97$$
$$x^2 - 10x + 25 + x^2 - 97 = 0$$
$$2x^2 - 10x - 72 = 0$$
$$2(x^2 - 5x - 36) = 0$$
$$2(x+4)(x-9) = 0$$
Set: $\qquad x + 4 = 0 \qquad x - 9 = 0$
Solve: $\qquad x = -4 \qquad\quad x = 9$
$$-4 - 5 = -9 \qquad 9 - 5 = 4$$
There are two solutions: −4 and −9, or 4 and 9.

13. Let x be the first integer. Then, $x + 1$ is the next integer.
$$x(x+1) = 72$$
$$x^2 + x - 72 = 0$$
$$(x+9)(x-8) = 0$$
Set: $\qquad x + 9 = 0 \qquad x - 8 = 0$
Solve: $\qquad x = -9 \qquad\quad x = 8$
$$8 + 1 = 9$$
Disregard $x = -9 < 0$. The two integers are 8 and 9.

15. Let x be the first positive integer. Then, $x + 1$ is the next positive integer.
$$x^2 + (x+1)^2 = 85$$
$$x^2 + x^2 + 2x + 1 - 85 = 0$$
$$2x^2 + 2x - 84 = 0$$
$$2(x^2 + x - 42) = 0$$
$$2(x+7)(x-6) = 0$$
Set: $\qquad x + 7 = 0 \qquad x - 6 = 0$
Solve: $\qquad x = -7 \qquad\quad x = 6$
$$6 + 1 = 7$$
Disregard $x = -7 < 0$. The integers are 6 and 7.

17. Let x be the first positive integer. Then, $x + 1$ is the next positive integer.
$$(x+1)^2 - x^2 = 17$$
$$x^2 + 2x + 1 - x^2 - 17 = 0$$
$$2x - 16 = 0$$
$$2(x-8) = 0$$
Set: $\qquad x - 8 = 0$
Solve: $\qquad x = 8$
$$8 + 1 = 9$$
The two integers are 8 and 9.

19. Let x be the first even integer. Then, $x + 2$ is the next even integer.
$$x(x+2) = 120$$
$$x^2 + 2x - 120 = 0$$
$$(x+12)(x-10) = 0$$
Set: $\quad x+12 = 0 \qquad x-10 = 0$
Solve: $\qquad x = -12 \qquad\quad x = 10$
$$-12 + 2 = -10 \qquad 10 + 2 = 12$$
There are two possible solutions: -12 and -10, or 10 and 12.

21. Let w be the width of the rectangle. Then, $2w$ is the length. Use the formula for area, $A = lw,$ to write the equation.
$$(2w)w = 72$$
$$2w^2 = 72$$
$$w^2 = 36$$
$$w^2 - 36 = 0$$
$$(w+6)(w-6) = 0$$
Set: $\quad w+6 = 0 \qquad w-6 = 0$
Solve: $\qquad w = -6 \qquad\quad w = 6$
$$2 \cdot 6 = 12$$
Disregard $w = -6$, since width must be positive. The width is 6 in. and the length is 12 in.

23. Let w be the width of the rectangle. Then, $w + 12$ is the length. Use the formula for area, $A = lw$, to write the equation.
$$(w+12)w = 85$$
$$w^2 + 12w = 85$$
$$w^2 + 12w - 85 = 0$$
$$(w-5)(w+17) = 0$$
Set: $\quad w+17 = 0 \qquad w-5 = 0$
Solve: $\qquad w = -17 \qquad\quad w = 5$
$$5 + 12 = 17$$
Disregard $w = -17$, since width must be positive. The width is 5 m and the length is 17 m.

25. Let w be the width of the rectangle. Then, $w + 4$ is the length. Use the formula for area, $A = lw$, to write the equation.
$$(w+4)w = 117$$
$$w^2 + 4w = 117$$
$$w^2 + 4w - 117 = 0$$
$$(w+13)(w-9) = 0$$

Set: $\quad w+13 = 0 \qquad w-9 = 0$
Solve: $\qquad w = -13 \qquad\quad w = 9$
$$9 + 4 = 13$$
Disregard $w = -13$, since width must be positive. The width is 9 ft and the length is 13 ft.

27. Let h be the height of the triangle. Then, $h + 5$ is the base of the triangle. Use the formula for area, $A = \frac{1}{2}bh,$ to write the equation.
$$\frac{1}{2}(h+5)h = 42$$
$$(h+5)h = 84$$
$$h^2 + 5h - 84 = 0$$
$$(h+12)(h-7) = 0$$
Set: $\quad h+12 = 0 \qquad h-7 = 0$
Solve: $\qquad h = -12 \qquad\quad h = 7$
$$7 + 5 = 12$$
Disregard $h = -12$, since height must be positive. The height is 7 m and the base is 12 m.

29. Let h be the height of the triangle. Then, $h - 6$ is the base of the triangle. Use the formula for area, $A = \frac{1}{2}bh,$ to write the equation.
$$\frac{1}{2}(h-6)h = 56$$
$$(h-6)h = 112$$
$$h^2 - 6h - 112 = 0$$
$$(h+8)(h-14) = 0$$
Set: $\quad h+8 = 0 \qquad h-14 = 0$
Solve: $\qquad h = -8 \qquad\quad h = 14$
$$14 - 6 = 8$$
Disregard $h = -8$, since height must be a positive number. The height is 14 ft and the base is 8 ft.

31. Let w be the width of the rectangle. Use the formula for the perimeter, $2l + 2w = P$, to find an expression for the length in terms of width.
$$2w + 2l = 20$$
$$2l = 20 - 2w$$
$$l = \frac{20 - 2w}{2}$$
$$l = 10 - w$$
Then, use the formula for area, $A = lw$, to write the equation.

$$24 = (10 - w)w$$
$$24 = 10w - w^2$$
$$w^2 - 10w + 24 = 0$$
$$(w - 6)(w - 4) = 0$$

Set: $w - 6 = 0$ $w - 4 = 0$

Solve: $w = 6$ $w = 4$

$10 - 6 = 4$ $10 - 4 = 6$

The solution is width is 4 cm and length is 6 cm. The second "solution" is not really different, since it is immaterial which side is called "width" and "length."

33. Let R be the number of rows. Then $\dfrac{140}{R}$ is the number of trees in each row. Translate the problem to an equation.

$$\frac{140}{R} + 13 = R$$
$$140 + 13R = R^2$$
$$R^2 - 13R - 140 = 0$$
$$(R - 20)(R + 7) = 0$$

Set: $R - 20 = 0$ $R + 7 = 0$

Solve: $R = 20$ $R = -7$

$$\frac{140}{20} = 7$$

Disregard $R = -7$, since R must be positive. There are 7 trees in each row.

35. Let n be the number of rows of seats. Then, $n + 7$ is the number of seats in each row.

$$n(n + 7) = 144$$
$$n^2 + 7n - 144 = 0$$
$$(n + 16)(n - 9) = 0$$

Set: $n + 16 = 0$ $n - 9 = 0$

Solve: $n = -16$ $n = 9$

$9 + 7 = 16$

Disregard $n = -16$, since n must be positive. There are 9 rows of 16 seats each.

37. Let l be the length of the rectangle. Then, $l - 5$ is the width in meters. Use the formula $A = lw$, to write the equation.

$$(l + 6)(l - 5 + 6) = 300$$
$$(l + 6)(l + 1) = 300$$
$$l^2 + 7l + 6 = 300$$
$$l^2 + 7l - 294 = 0$$
$$(l + 21)(l - 14) = 0$$

Set: $l + 21 = 0$ $l - 14 = 0$

Solve: $l = -21$ $l = 14$

$14 - 5 = 9$

Disregard $l = -21$, since length must be positive. The length is 14 m and the width is 9 m.

39. Assume the corral is rectangular. The amount of fencing is 2 widths, or $2w$, and 1 length. Since 52 is the total amount, set up the following and solve for l: $2w + l = 52$

$$l = 52 - 2w$$

Substitute this expression into the formula: $A = lw$.

$$(52 - 2w)w = 320$$
$$52w - 2w^2 = 320$$
$$2w^2 - 52w + 320 = 0$$
$$2(w^2 - 26w + 160) = 0$$
$$2(w - 16)(w - 10) = 0$$

Set: $w - 16 = 0$ $w - 10 = 0$

Solve: $w = 16$ $w = 10$

$52 - 2 \cdot 16 = 20$ $52 - 2 \cdot 10 = 32$

The corral may have dimensions 10 yd by 32 yd, or 16 yd by 20 yd.

41. Let w be the width of the rectangle. Then, $3w$ is the length in feet.

$$3w(w) = 48$$
$$3w^2 = 48$$
$$w^2 = 16$$
$$w = \pm 4$$

Disregard w = -4, since width must be positive.

$$3w = 3(4) = 12$$

The width is 4 ft and the length is 12 ft.

43. Let x be one of the numbers. Then, the other number is $x + 8$.

$$x(x + 8) = -16$$
$$x^2 + 8x + 16 = 0$$
$$(x + 4)^2 = 0$$

Set: $x + 4 = 0$

Solve: $n = -x$

$-4 + 8 = 4$

The numbers are –4 and 4.

[End of Section 7.6]

224

Section 7.7
Solutions to Odd Exercises

1. Use your calculator and the Body-Mass-Index formula
$BMI = \dfrac{W \times 704.5}{H^2}$ with $W = 130$lb and $H = 5$ ft 7 in.,
or 67 in., to find the BMI.
$$BMI = \frac{130 \cdot 704.5}{67^2}$$
$$= \frac{91585}{4489}$$
$$= 20.4021$$
The BMI is 20.4.

3. Use your calculator and the Body-Mass-Index formula
$BMI = \dfrac{W \times 704.5}{H^2}$ with $BMI = 24.37$ and $W = 170$lb
to find the height, H.
$$24.37 = \frac{170 \cdot 704.5}{H^2}$$
$$H^2 = \frac{119765}{24.37}$$
$$H^2 = 4914.44$$
$$H = 70.1031$$
The height is 70 in., or 5 ft 10 in.

5. Use the safe load formula, $L = \dfrac{kbd^2}{l}$, with $l = 180$ in.,
$b = 3$ in., $d = 4$ in. and $k = 3000$ lb/ in.^{2} to find the safe
load L.
$$L = \frac{3000(3)4^2}{180}$$
$$L = \frac{144,000}{180}$$
$$L = 800$$
The safe load for this beam is 800 lb.

7. Use the safe load formula, $L = \dfrac{kbd^2}{l}$, with $l = 192$ in.,
$d = 8$ in., $k = 6000$ lb/in.^{2} and $L = 12,000$ lb to find
how wide, in inches, the beam should be. First, solve
the equation for b. Then, substitute the given values.
$$L = \frac{kbd^2}{l}$$
$$L \cdot l = kbd^2$$
$$b = \frac{L \cdot l}{kd^2}$$
$$b = \frac{12,000 \cdot 192}{6,000 \cdot 8^2}$$

$$b = \frac{2,304,000}{384,000}$$
$$b = 6$$
The width needed for the beam is 6 in.

9. Use the safe load formula, $L = \dfrac{kbd^2}{l}$, with $l = 144$
in., $b = 6$ in., $k = 4800$ lb/in.^{2} and $L = 20,000$ lb to
find how thick, in inches, the beam should be.
First, solve the equation for d^2.
$$L = \frac{kbd^2}{l}$$
$$L \cdot l = kbd^2$$
$$d^2 = \frac{L \cdot l}{kb}$$
$$d^2 = \frac{20,000 \cdot 144}{4800 \cdot 6}$$
$$d^2 = \frac{2,880,000}{28,800}$$
$$d^2 = 100$$
$$d = 10$$
The thickness needed to support a safe load of
20,000 lb on this beam is 10 in.

11. Use the height formula $h = -16t^2 + v_0 t$ to find h
with $t = 5$ sec., and $v_0 = 120$ ft/sec.
$$h = -16(5)^2 + 120(5)$$
$$h = -400 + 600$$
$$h = 200$$
The height of the object at $t = 5$ sec is 200 ft.

13. Use the height formula $h = -16t^2 + v_0 t$ to find t
with $h = 384$ ft, and $v_0 = 160$ ft/sec. Substitute the
given values for the variables, and solve for t:
$$384 = -16t^2 + 160t$$
$$16t^2 - 160t + 384 = 0$$
$$16(t^2 - 10t + 24) = 0$$
$$16(t - 6)(t - 4) = 0$$
Set: $\quad t - 6 = 0 \quad\quad t - 4 = 0$
Solve: $\quad\quad t = 6 \quad\quad\quad t = 4$
There are two solutions: the height is 384 ft at $t = 4$
seconds and again at $t = 6$ seconds.

15. Use the power output formula $P_o = E_g I - r_g I^2$ to find I

with $P_o = 120$ kw, $E_g = 16$ volts, and $r_g = \dfrac{1}{2}$ ohm.

Substitute the given values for the variables in the equation, and solve for I:

$$120 = 16I - \frac{1}{2}I^2$$
$$240 = 32I - I^2$$
$$I^2 - 32I + 240 = 0$$
$$(I - 20)(I - 12) = 0$$

Set: $\quad I - 20 = 0 \qquad I - 12 = 0$

Solve: $\quad\quad I = 20 \qquad\quad I = 12$

There are two solutions: $I = 12$ or 20 amps.

17. Use the power output formula $P_o = E_g I - r_g I^2$ to find I

with $P_o = 210$ kw, $E_g = 20$ volts, and $r_g = \dfrac{2}{5}$ ohm.

Substitute the given values for the variables in the equation, and solve for I.

$$210 = 20I - \frac{2}{5}I^2$$
$$1050 = 100I - 2I^2$$
$$2I^2 - 100I + 1050 = 0$$
$$I^2 - 50I + 525 = 0$$
$$(I - 35)(I - 15) = 0$$

Set: $\quad I - 35 = 0 \qquad I - 15 = 0$

Solve: $\quad\quad I = 35 \qquad\quad I = 15$

There are two solutions: $I = 15$ or 35 amps.

19. Total sales is $T = \text{price} \cdot \text{demand} = p(36 - p)$.

Given that $T = \$320$, find the price p.

$$320 = p(36 - p)$$
$$320 = 36p - p^2$$
$$p^2 - 36p + 320 = 0$$
$$(p - 16)(p - 20) = 0$$

Set: $\quad p - 16 = 0 \qquad p - 20 = 0$

Solve: $\quad\quad p = 16 \qquad\quad p = 20$

There are two solutions: the price is $16 or $20.

21. Total sales $T = \text{price} \cdot \text{demand} = p(10 - 5p)$.

Given that $T = \$4.95$, find the price p.

$$4.95 = p(10 - 5p)$$
$$4.95 = 10p - 5p^2$$
$$5p^2 - 10p + 4.95 = 0$$
$$p^2 - 2p + .99 = 0$$

$$100p^2 - 200p + 99 = 0$$
$$(10p - 11)(10p - 9) = 0$$

Set: $\quad 10p - 11 = 0 \qquad 10p - 9 = 0$

Solve: $\quad\quad 10p = 11 \qquad\quad 10p = 9$

$$p = \frac{11}{10} \qquad\qquad p = \frac{9}{10}$$

The selling price is $1.10 or $0.90.

23. Given that the demand formula in number of units per month is $D = -20p^2 + ap + 1200$, find the selling price p when $D = 1403$ and $a = 128$:

$$1403 = -20p^2 + 128p + 1200$$
$$0 = -20p^2 + 128p - 203$$
$$0 = -1(20p^2 - 128p + 203)$$
$$0 = -1(2p - 7)(10p - 29)$$

Set: $\quad 2p - 7 = 0 \qquad 10p - 29 = 0$

Solve: $\quad\quad 2p = 7 \qquad\quad 10p = 29$

$$p = \frac{7}{2} = 3.5 \qquad p = \frac{29}{10} = 2.9$$

The selling price is $3.50 or $2.90.

25. Given that the demand formula in number of units per month is $D = -20p^2 + ap + 1200$, find the selling price p when $D = 1120$ and $a = 60$:

$$1120 = -20p^2 + 60p + 1200$$
$$\frac{1120}{-20} = \frac{-20p^2 + 60p + 1200}{-20}$$
$$-56 = p^2 - 3p - 60$$
$$p^2 - 3p - 4 = 0$$
$$(p - 4)(p + 1) = 0$$

Set: $\quad p - 4 = 0 \qquad p + 1 = 0$

Solve: $\quad\quad p = 4 \qquad\quad p = -1$

Disregard the negative solution, as it has no meaning in this context. The selling price is $4.

27. Use $V = \pi r^2 h$, with $r = 6$ in., $h = 20$ in. to find V. Use the approximation $\pi \approx 3.14$.

$$V = 3.14(6)^2(20)$$
$$V = 2260.8$$

The volume is 2260.8 sq. in.

29. Use $V = \pi r^2 h$, with $V = 1099$ in.^{3}, $h = 14$ in. to find the radius r of the cylinder.

$$1099 = 3.14 r^2 (14)$$
$$1099 = 43.96 r^2$$
$$r^2 = 25$$
$$r^2 - 25 = 0$$
$$(r+5)(r-5) = 0$$
$$r = \pm 5$$

The radius is 5 in. Disregard $r = -5$, as a negative solution has no meaning in this context.

[End of Section 7.7]

Chapter 7 Review
Solutions to All Exercises

1. $11x - 22 = 11(x - 2)$

2. $3y + 24 = 3(y + 8)$

3. $-7a - 14b = -7(a + 2b)$

4. $-4y^2 + 28y = -4y(y - 7)$

5. $16x^3 y - 12x^2 y = 4x^2 y(4x - 3)$

6. $am^2 + 11am + 25a = a(m^2 + 11m + 25)$

7. $-7x^4 t^3 + 98t^4 x^3 + 35t^5 x = -7xt^3 (x^3 - 14x^2 t - 5t^2)$

8. $-6x^2 y^4 - 6x^3 y^4 + 24x^2 y^3 = -6x^2 y^3 (y + xy - 4)$

9. $8y(y + 5) + 3(y + 5) = (8y + 3)(y + 5)$

10. $3a(a + 7) - 2(a + 7) = (3a - 2)(a + 7)$

11. $a(x + 4) + b(x + 4) = (a + b)(x + 4)$

12. $2a(x - 10) + 3b(x - 10) = (2a + 3b)(x - 10)$

13. $ax - a + cx - c = a(x - 1) + c(x - 1) = (a + c)(x - 1)$

14. $4x + 4y + bx + by = 4(x + y) + b(x + y)$
$= (4 + b)(x + y)$

15. $x - 4xy + 2z - 8zy = x(1 - 4y) + 2z(1 - 4y)$
$= (x + 2z)(1 - 4y)$

16. $z^2 + 5 + cz^2 + 5c = (z^2 + 5) + c(z^2 + 5)$
$= (1 + c)(z^2 + 5)$

17. $x^2 - x^2 y + 6 + 6y$ is not factorable.

18. $7ab - 3b + 2a^2 - 3a$ is not factorable.

19. $10ab - 2b^2 + 7ac - 35bc$ is not factorable.

20. $x - 3xy - z + 3zy = x(1 - 3y) - z(1 - 3y)$
$= (x - z)(1 - 3y)$

21. $m^2 + 7m + 6 = (m + 1)(m + 6)$

22. $a^2 - 4a + 3 = (a - 3)(a - 1)$

23. $x^2 + 11x + 18 = (x + 9)(x + 2)$

24. $y^2 + 8y + 15 = (y + 5)(y + 3)$

25. $n^2 - 8n + 12 = (n - 6)(n - 2)$

26. $x^2 - 10x + 24 = (x - 4)(x - 6)$

27. $x^2 + 3x - 10 = (x - 2)(x + 5)$

28. $y^2 + 13y + 36 = (y + 9)(y + 4)$

29. $y^2 + 2y + 24$ is not factorable.

30. $c^2 + 3c + 10$ is not factorable.

31. $x^2 + 12x + 35 = (x + 5)(x + 7)$

32. $x^2 + 17x + 72 = (x + 9)(x + 8)$

33. $a^2 - 5a - 50 = (a - 10)(a + 5)$

34. $x^2 - 4x - 21 = (x - 7)(x + 3)$

35. $x^2 + 29x + 100 = (x + 4)(x + 25)$

36. $2x^2 + 7x + 3 = (2x + 1)(x + 3)$

37. $-2x^2 + 3x - 1 = (2x - 1)(-x + 1) = (2x - 1)(1 - x)$

38. $-12x^2 + 32x - 5 = (6x - 1)(5 - 2x)$

39. $12x^2 + x - 6 = (3x - 2)(4x + 3)$

40. $6y^2 - 11y + 4 = (2y - 1)(3y - 4)$

41. $12y^2 - 11y + 4$ is not factorable.

42. $63x^2 - 3x - 30 = 3(21x^2 - x - 10) = 3(3x + 2)(7x - 5)$

43. $16x^2 + 12x - 70 = 2(8x^2 + 6x - 35) = 2(2x + 5)(4x - 7)$

44. $24 + x - 3x^2 = -3x^2 + x + 24 = -1 \cdot (3x^2 - x - 24)$
$$= -1 \cdot (3x + 8)(x - 3)$$

45. $14 + 11x - 15x^2 = -15x^2 + 11x + 14$
$$= -1 \cdot (15x^2 - 11x - 14)$$
$$= -1 \cdot (5x - 7)(3x + 2)$$

46. $2y^2 - 13y + 5$ is not factorable.

47. $3y^2 + 10y + 6$ is not factorable.

48. $200 + 20x - 4x^2 = -4x^2 + 20x + 200$
$$= -4(x^2 - 5x - 50)$$
$$= -4(x - 10)(x + 5)$$

49. $7y^2 + 14y - 168 = 7(y^2 + 2y - 24) = 7(y - 4)(y + 6)$

50. $18x^2 - 15x + 2 = (3x - 2)(6x - 1)$

51. $y^2 - 49 = (y + 7)(y - 7)$

52. $x^2 - 100 = (x + 10)(x - 10)$

53. $64 + 49x^2$ is not factorable.

54. $36 + 25y^2$ is not factorable.

55. $25x^2 - 36 = (5x + 6)(5x - 6)$

56. $2x^2 - 98 = 2(x^2 - 49) = 2(x + 7)(x - 7)$

57. $3x^2 - 147 = 3(x^2 - 49) = 3(x + 7)(x - 7)$

58. $64a^2 - 1 = (8a + 1)(8a - 1)$

59. $12x^2 - 60x - 75 = 3(4x^2 - 20x - 25)$

60. $21x^3 - 13x^2 - 2x = x(21x^2 - 13x - 2)$

61. $x^2 + 12x + 36 = (x + 6)^2$

62. $6y^2 + 120y + 600 = 6(y^2 + 20y + 100) = 6(y + 10)^2$

63. $5x^2 - 150x + 1125 = 5(x^2 - 30x + 225) = 5(x - 15)^2$

64. $7a^2 - 42a + 63 = 7(a^2 - 6a + 9) = 7(a - 3)^2$

65. $-2ax^2 - 44ax - 242a = -2a(x^2 + 22x + 121)$
$$= -2a(x + 11)^2$$

66. $y^2 + 22y + \underline{121} = \left(y + 11\right)^2$

67. $x^2 - 12x + \underline{36} = \left(x - 6\right)^2$

68. $x^2 - 7x + \dfrac{49}{\underline{4}} = \left(x - \dfrac{7}{2}\right)^2$

69. $y^2 + \underline{24y} + 144 = \left(y + 12\right)^2$

70. $x^2 - \underline{14x} + 49 = \left(x - 7\right)^2$

71. $3x^2 + 15x = 0$
$3x(x + 5) = 0$
Set: $3x = 0$ $x + 5 = 0$
Solve: $x = 0$ $x = -5$

72. $4x^2 - 24x = 0$
$4x(x - 6) = 0$
Set: $4x = 0$ $x - 6 = 0$
Solve: $x = 0$ $x = 6$

73. $3x^2 - 17x + 10 = 0$
$(3x - 2)(x - 5) = 0$
Set: $3x - 2 = 0$ $x - 5 = 0$
Solve: $3x = 2$ $x = 5$
$$x = \frac{2}{3}$$

74. $6y^2 + y - 35 = 0$

$(3y - 7)(2y + 5) = 0$

Set: $3y - 7 = 0$ $2y + 5 = 0$

Solve: $3y = 7$ $2y = -5$

$y = \dfrac{7}{3}$ $y = \dfrac{-5}{2}$

75. $16x^3 - 100x = 0$

$4x(4x^2 - 25) = 0$

$4x(2x + 5)(2x - 5) = 0$

Set: $4x = 0$ $2x + 5 = 0$ $2x - 5 = 0$

Solve: $x = 0$ $2x = -5$ $2x = 5$

$x = \dfrac{-5}{2}$ $x = \dfrac{5}{2}$

76. $x^3 - 4x^2 - 12x = 0$

$x(x^2 - 4x - 12) = 0$

$x(x - 6)(x + 2) = 0$

Set: $x = 0$ $x - 6 = 0$ $x + 2 = 0$

Solve: $x = 0$ $x = 6$ $x = -2$

77. $25y^3 = 36y$

$25y^3 - 36y = 0$

$y(5y + 6)(5y - 6) = 0$

Set: $y = 0$ $5y + 6 = 0$ $5y - 6 = 0$

Solve: $y = 0$ $5y = -6$ $5y = 6$

$y = \dfrac{-6}{5}$ $y = \dfrac{6}{5}$

78. $x^3 = 81x$

$x^3 - 81x = 0$

$x(x + 9)(x - 9) = 0$

Set: $x = 0$ $x + 9 = 0$ $x - 9 = 0$

Solve: $x = 0$ $x = -9$ $x = 9$

79. $2a^2 + 28a + 98 = 0$

$2(a^2 + 14a + 49) = 0$

$2(a + 7)^2 = 0$

Set: $a + 7 = 0$

Solve: $a = -7$

80. $3y^2 - 28y + 64 = 0$

$(3y - 16)(y - 4) = 0$

Set: $3y - 16 = 0$ $y - 4 = 0$

Solve: $3y = 16$ $y = 4$

$y = \dfrac{16}{3}$

81. $x^2 = x + 12$

$x^2 - x - 12 = 0$

$(x + 3)(x - 4) = 0$

Set: $x + 3 = 0$ $x - 4 = 0$

Solve: $x = -3$ $x = 4$

82. $x^2 = 2x + 35$

$x^2 - 2x - 35 = 0$

$(x - 7)(x + 5) = 0$

Set: $x - 7 = 0$ $x + 5 = 0$

Solve: $x = 7$ $x = -5$

83. $(x + 3)^2 = 9$

$x^2 + 6x + 9 = 9$

$x^2 + 6x = 0$

$x(x + 6) = 0$

Set: $x = 0$ $x + 6 = 0$

Solve: $x = -6$

84. $(x - 5)^2 = 64$

$x^2 - 10x + 25 = 64$

$x^2 - 10x - 39 = 0$

$(x - 13)(x + 3) = 0$

Set: $x - 13 = 0$ $x + 3 = 0$

Solve: $x = 13$ $x = -3$

85. $5x^2 - 65x = -200$

$5x^2 - 65x + 200 = 0$

$5(x^2 - 13x + 40) = 0$

$5(x - 8)(x - 5) = 0$

Set: $x - 8 = 0$ $x - 5 = 0$

Solve: $x = 8$ $x = 5$

86. Let n be the smaller number. Then $n + 3$ is the larger number. Translate the problem into an equation and solve by factoring:

$$n^2 + (n+3)^2 = 269$$

$$n^2 + n^2 + 6n + 9 = 269$$

$$2n^2 + 6n - 260 = 0$$

$$2(n-10)(n+13) = 0$$

Set: $\quad n - 10 = 0 \qquad n + 13 = 0$

Solve: $\qquad n = 10 \qquad n = -13$

The numbers are positive, so disregard $n = -13$. The numbers are 10 and $n + 3 = 10 + 3 = 13$.

87. Let x be the number of streets. Then there are $x(x+5)$ houses in the tract. Translate the problem into an equation and solve by factoring:

$$x(x+5) = 126$$

$$x^2 + 5x - 126 = 0$$

$$(x-9)(x+14) = 0$$

Set: $\quad x - 9 = 0 \qquad x + 14 = 0$

Solve: $\qquad x = 9 \qquad x = -14$

Disregard the negative solution, as it has no meaning in this context. There are 9 streets.

88. Use the formulas for area and perimeter to set up a system of equations.

$$\begin{cases} 450 = l \cdot w \\ 90 = 2l + 2w \end{cases}$$

Solve the second equation for l, then substitute:

$$l = \frac{90 - 2w}{2} = 45 - w$$

$$450 = w(45 - w)$$

$$450 = 45w - w^2$$

$$w^2 - 45w + 450 = 0$$

$$(w-15)(w-30) = 0$$

Set: $\quad w - 15 = 0 \qquad w - 30 = 0$

Solve: $\qquad w = 15 \qquad w = 30$

Plug $w = 15$ into the area equation and solve for l:

$$450 = l \cdot 15$$

$$30 = l$$

Note that using $w = 30$ will only reverse the values of the width and length. So the rectangle is 15 yards by 30 yards.

89. Let n be the first odd integer. Then $n + 2$ is the second odd integer. Translate to problem into an equation and solve by factoring:

$$n^2 + (n+2)^2 = 290$$

$$n^2 + n^2 + 4n + 4 = 290$$

$$2n^2 + 4n - 286 = 0$$

$$2(n-11)(n+13) = 0$$

Set: $\quad n - 11 = 0 \qquad n + 13 = 0$

Solve: $\qquad n = 11 \qquad n = -13$

The numbers are positive, so disregard $n = -13$. The numbers are 11 and $n + 2 = 11 + 2 = 13$

90. Let n be the first integer. Then $n + 1$ is the second integer. Translate the problem into an equation and solve by factoring:

$$n(n+1) = 156$$

$$n^2 + n - 156 = 0$$

$$(n-12)(n+13) = 0$$

Set: $\quad n - 12 = 0 \qquad n + 13 = 0$

Solve: $\qquad n = 12 \qquad n = -13$

The numbers are negative, so disregard $n = 12$. The numbers are -13 and $n + 1 = -13 + 1 = -12$

91. Let b be the length of the base. Then $b - 12$ is the height. Translate the problem into an equation and solve by factoring:

$$\frac{1}{2} b(b - 12) = 80$$

$$\frac{1}{2} b^2 - 6b = 80$$

$$b^2 - 12b = 160$$

$$b^2 - 12b - 160 = 0$$

$$(b-20)(b+8) = 0$$

Set: $\quad b - 20 = 0 \qquad b + 8 = 0$

Solve: $\qquad b = 20 \qquad b = -8$

Disregard the negative solution, as it has no meaning in this context. So the base is 20 in.

92. Use the formula for area of a rectangle to translate the problem into an equation and solve by factoring. Use the given fact that $l = w + 10$:

$$(l+3)(w+3) = 299$$

$$((w+10)+3)(w+3) = 299$$

$$(w+13)(w+3) = 299$$

$$w^2 + 16w + 39 = 299$$

$$w^2 + 16w - 260 = 0$$

$$(w-10)(w+26) = 0$$

Set: $\quad w - 10 = 0 \qquad w + 26 = 0$

Solve: $\qquad w = 10 \qquad w = -26$

Disregard the negative solution, as it has no meaning in this context. The width is 10 inches and the length is $w + 10 = 10 + 10 = 20$ inches.

93. Let s be the length of one side of the square. Use the formulas for area and perimeter of a square to translate the problem into an equation and solve by factoring:

$$s^2 = 4s$$
$$s^2 - 4s = 0$$
$$s(s - 4) = 0$$

Set: $\quad s = 0 \quad s - 4 = 0$

Solve: $\qquad\qquad s = 4$

Disregard $s = 0$ as it has no meaning here. So the sides are all 4 inches long.

94. Let n be the smaller number. Then $n + 20$ is the larger number. Translate the problem into an equation and solve by factoring:

$$n(n + 20) = -84$$
$$n^2 + 20n = -84$$
$$n^2 + 20n + 84 = 0$$
$$(n + 6)(n + 14) = 0$$

Set: $\quad n + 6 = 0 \quad n + 14 = 0$

Solve: $\quad n = -6 \quad n = -14$

So the numbers are either -6 and $-6 + 20 = 14$ or -14 and $-14 + 20 = 6$.

95. Let r be the number of rows. Then $r + 10$ is the number of seats in each row. Translate the problem into an equation and solve by factoring.

$$r(r + 10) = 600$$
$$r^2 + 10r = 600$$
$$r^2 + 10r - 600 = 0$$
$$(r + 30)(r - 20) = 0$$

Set: $\quad r + 30 = 0 \quad r - 20 = 0$

Solve: $\quad r = -30 \quad r = 20$

Disregard the negative solution, as it has no meaning in this context. There are 20 rows of seats and $20 + 10 = 30$ seats in each row.

96. Use the formula for volume of a cylinder with $\pi = 3.14$, $r = 3$, and $h = 15$:

$$V = \pi r^2 h = (3.14)(3)^2(15) = 423.9 \text{ cm}^3$$

97. Use the formula for height of a projectile with $h = 24$ to solve for time t:

$$24 = -16t^2 + 56t$$

$$0 = -16t^2 + 56t - 24$$
$$0 = -8(2t - 1)(t - 3)$$

Set: $\quad 2t - 1 = 0 \qquad t - 3 = 0$

Solve: $\quad 2t = 1 \qquad\qquad t = 3$

$$t = \frac{1}{2} = 0.5$$

The projectile will be 24 ft above the ground after 0.5 seconds and again after 3 seconds.

98. Use the formula for height of a projectile with $h = 200$ to solve for time t:

$$200 = -16t^2 + 120t$$
$$0 = -16t^2 + 120t - 200$$
$$0 = -8(2t - 5)(t - 5)$$

Set: $\quad 2t - 5 = 0 \qquad t - 5 = 0$

Solve: $\quad 2t = 5 \qquad\qquad t = 5$

$$t = \frac{5}{2} = 2.5$$

The projectile will be 200 ft above the ground after 2.5 seconds and again after 5 seconds.

99. Use the formula for demand with $D = 5000$ to solve for price p:

$$5000 = -20p^2 + 500p + 2000$$
$$0 = -20p^2 + 500p - 3000$$
$$0 = -20(p - 10)(p - 15)$$

Set: $\quad p - 10 = 0 \quad p - 15 = 0$

Solve: $\quad p = 10 \qquad\quad p = 15$

The price will be either $10 or $15.

100. Use the formula for demand with $D = 2500$ to solve for price p:

$$2500 = -20p^2 + 200p + 2000$$
$$0 = -20p^2 + 200p - 500$$
$$0 = -20(p - 5)^2$$

Set: $\quad p - 5 = 0$

Solve: $\quad p = 5$

The price will be $5.

[End of Chapter 7 Review]

Chapter 7 Test
Solutions to All Exercises

1. $\text{GCF}(30, 75, 90) = 3 \cdot 5 = 15$
$$30 = 2 \cdot 3 \cdot 5$$
$$75 = 3 \cdot 5^2$$
$$90 = 2 \cdot 3^2 \cdot 5$$

2. $\text{GCF}\left(40x^2y, 48x^2y^3, 56x^2y^2\right) = 2^3 \cdot x^2 \cdot y = 8x^2y$
$$40x^2y = 2^3 \cdot 5 \cdot x^2y$$
$$48x^2y^3 = 2^4 \cdot 3 \cdot x^2y^3$$
$$56x^2y^2 = 2^3 \cdot 7 \cdot x^2y^2$$

3. $\dfrac{16x^4}{8x} = \dfrac{\cancel{8x} \cdot 2x^3}{\cancel{8x}} = 2x^3$

4. $\dfrac{42x^3y^3}{-6x^3y} = -\dfrac{\cancel{6x^3y} \cdot 7y^2}{\cancel{6x^3y}} = -7y^2$

5. $20x^3y^2 + 30x^3y + 10x^3 = 10x^3\left(2y^2 + 3y + 1\right)$
$$= 10x^3(2y+1)(y+1)$$

6. $x^2 - 9x + 20 = (x-4)(x-5)$

7. $-x^2 - 14x - 49 = -\left(x^2 + 14x + 49\right) = -(x+7)^2$

8. $6x^2 - 6 = 6\left(x^2 - 1\right) = 6(x+1)(x-1)$

9. $12x^2 + 2x - 10 = 2\left(6x^2 + x - 5\right) = 2(6x-5)(x+1)$

10. $3x^2 + x - 24 = (3x-8)(x+3)$

11. $16x^2 - 25y^2 = (4x+5y)(4x-5y)$

12. $2x^3 - x^2 - 3x = x\left(2x^2 - x - 3\right) = x(2x-3)(x+1)$

13. $6x^2 - 13x + 6 = (3x-2)(2x-3)$

14. $2xy - 3y + 14x - 21 = (2xy - 3y) + (14x - 21)$
$$= y(2x-3) + 7(2x-3)$$
$$= (y+7)(2x-3)$$

15. $4x^2 + 25$ is not factorable. In general, a sum of squares is not factorable.

16. $-3x^3 + 6x^2 - 6x = -3x\left(x^2 - 2x + 2\right)$

17. $\left(\dfrac{-10}{2}\right)^2 = (-5)^2 = 25,$ so add 25 and factor:
$$x^2 - 10x + \underline{25} = (x-5)^2$$

18. $\left(\dfrac{16}{2}\right)^2 = (8)^2 = 64,$ so add 64 and factor:
$$x^2 + 16x + \underline{64} = (x+8)^2$$

19. $(x+2)(3x-5) = 0$
Set: $\quad x + 2 = 0 \qquad 3x - 5 = 0$
Solve: $\quad x = -2 \qquad x = \dfrac{5}{3}$
The solutions are $x = -2$ and $x = \dfrac{5}{3}$.

20. $\quad x^2 - 7x - 8 = 0$
$(x+1)(x-8) = 0$
Set: $\quad x + 1 = 0 \qquad x - 8 = 0$
Solve: $\quad x = -1 \qquad x = 8$
The solutions are $x = -1$ and $x = 8$.

21. $\quad -3x^2 = 18x$
$3x^2 + 18x = 0$
$3x(x+6) = 0$
Set: $\quad 3x = 0 \qquad x + 6 = 0$
Solve: $\quad x = 0 \qquad x = -6$
The solutions are $x = 0$ and $x = -6$.

22. $\dfrac{2x^2}{5} - 6 = \dfrac{4x}{5}$
$2x^2 - 30 = 4x$
$2x^2 - 4x - 30 = 0$
$2\left(x^2 - 2x - 15\right) = 0$
$2(x-5)(x+3) = 0$
Set: $\quad x - 5 = 0 \qquad x + 3 = 0$
Solve: $\quad x = 5 \qquad x = -3$
The solutions are $x = 5$ and $x = -3$.

23. $4x^2 - 17x - 15 = 0$

$(4x + 3)(x - 5) = 0$

Set: $4x + 3 = 0$ $x - 5 = 0$

Solve: $x = -\dfrac{3}{4}$ $x = 5$

The solutions are $x = -\dfrac{3}{4}$ and $x = 5$.

24. $8x^2 - 2x - 15 = 0$

$(4x + 5)(2x - 3) = 0$

Set: $4x + 5 = 0$ $2x - 3 = 0$

Solve: $4x = -5$ $2x = 3$

$x = -\dfrac{5}{4}$ $x = \dfrac{3}{2}$

The solutions are $x = -\dfrac{5}{4}$ and $x = \dfrac{3}{2}$.

25.

$(2x - 7)(x + 1) = 6x - 19$

$2x^2 - 5x - 7 - 6x + 19 = 0$

$2x^2 - 11x + 12 = 0$

$(2x - 3)(x - 4) = 0$

Set: $2x - 3 = 0$ $x - 4 = 0$

Solve: $2x = 3$ $x = 4$

$x = \dfrac{3}{2}$

The solutions are $x = \dfrac{3}{2}$ and $x = 4$.

26. Let x be one number. Then, $5x - 10$ is the other number. Translate the problem into an equation and solve:

$(5x - 10)x = 120$

$5x^2 - 10x - 120 = 0$

$5(x^2 - 2x - 24) = 0$

$5(x - 6)(x + 4) = 0$

Set: $x - 6 = 0$ $x + 4 = 0$

Solve: $x = 6$ $x = -4$

$5x - 10 = 20$ $5x - 10 = -30$

The numbers are either 6 and 20 or –4 and –30.

27. Let w be the width of the rectangle. Then, $2w - 7$ is the length. Using the area formula for a rectangle:

$(2w - 7)w = 165$

$2w^2 - 7w - 165 = 0$

$(2w + 15)(w - 11) = 0$

Set: $2w + 15 = 0$ $w - 11 = 0$

Solve: $w = -\dfrac{15}{2}$ $w = 11$

$2w - 7 = 2(11) - 7 = 15$

Disregard the negative solution, since width must be positive. The width is 11 cm and length is 15 cm.

28. Let x be the first of two consecutive positive integers. Then, $x + 1$ is the next integer.

$x(x + 1) = 342$

$x^2 + x - 342 = 0$

$(x - 18)(x + 19) = 0$

Set: $x - 18 = 0$ $x + 19 = 0$

Solve: $x = 18$ $x = -19$

$x + 1 = 18 + 1 = 19$

Disregard the negative solution, since the integers must be positive. The two integers are 18 and 19.

29. Let x be the larger of the two positive numbers. Then, $x - 9$ is the smaller one.

$x - 9 + x^2 = 147$

$x^2 + x - 156 = 0$

$(x - 12)(x + 13) = 0$

Set: $x - 12 = 0$ $x + 13 = 0$

Solve: $x = 12$ $x = -13$

$x - 9 = 12 - 9 = 3$

Disregard the negative solution. The two numbers are 12 and 3.

30. **a.** Area of the given figure:

$A(x) = 20(12) - x(x + 3)$

$= 240 - x^2 - 3x$

$= -(x^2 + 3x - 240)$

b. Perimeter of the remaining figure:

$P(x) = 20 + 12 + 20 - (x + 3)$

$+ x + (x + 3) + 12 - x$

$= 64 \text{ in.}$

[End of Chapter 7 Test]

Cumulative Review: Chapters 1 – 7
Solutions to All Exercises

1. $\text{LCM}(20,12,24) = 120$

$$20 = 2^2 \cdot 5$$
$$12 = 2^2 \cdot 3$$
$$24 = 2^3 \cdot 3$$
$$\text{LCM} = 2^3 \cdot 3 \cdot 5 = 120$$

2. $\text{LCM}\{8x^2, 14x^2 y, 21xy\} = 168x^2 y$

$$8x^2 = 2^3 \cdot x^2$$
$$14x^2 y = 2 \cdot 7 \cdot x^2 \cdot y$$
$$21xy = 3 \cdot 7 \cdot x \cdot y$$
$$\text{LCM} = 2^3 \cdot 3 \cdot 7 \cdot x^2 \cdot y = 168x^2 y$$

3. $\dfrac{7}{12} + \dfrac{9}{16} = \dfrac{7 \cdot 4}{12 \cdot 4} + \dfrac{9 \cdot 3}{16 \cdot 3} = \dfrac{28 + 27}{48} = \dfrac{55}{48}$

4. $\dfrac{11}{15a} - \dfrac{5}{12a} = \dfrac{11 \cdot 4}{15a \cdot 4} - \dfrac{5 \cdot 5}{12a \cdot 5} = \dfrac{44 - 25}{60a} = \dfrac{19}{60a}$

5. $\dfrac{6x}{25} \cdot \dfrac{5}{4x} = \dfrac{\cancel{2} \cdot 3 \cdot \cancel{x} \cdot \cancel{5}}{\cancel{5} \cdot 5 \cdot \cancel{2} \cdot 2 \cdot \cancel{x}} = \dfrac{3}{10}$

6. $\dfrac{40}{92} \div \dfrac{2}{15x} = \dfrac{40}{92} \cdot \dfrac{15x}{2} = \dfrac{\cancel{2} \cdot \cancel{4} \cdot 5 \cdot 15 \cdot x}{\cancel{4} \cdot 23 \cdot \cancel{2}} = \dfrac{75x}{23}$

7. $\dfrac{-24x^4 y^2}{3x^2 y} = \dfrac{-24}{3} \cdot x^{4-2} \cdot y^{2-1} = -8x^2 y$

8. $\dfrac{36x^3 y^5}{9xy^4} = \dfrac{36}{9} \cdot x^{3-1} \cdot y^{5-4} = 4x^2 y$

9. $\dfrac{-4x^2 y}{12xy^{-3}} = \dfrac{-4}{12} \cdot x^{2-1} \cdot y^{1-(-3)} = \dfrac{-xy^4}{3}$

10. $\dfrac{21xy^2}{3x^{-1}y} = \dfrac{21}{3} \cdot x^{1-(-1)} y^{2-1} = 7x^2 y$

11. $r = \{(2,-3),(3,-2),(5,0),(7.1,3.2)\}$

    a. $D = \{2,3,5,7.1\}$

    b. $R = \{-3,-2,0,3.2\}$

    c. r is a function because, for each element of the domain, there is exactly one element of the range assigned to it.

12. For $f(x) = x^2 - 3x + 4$:

    a. $f(-6) = (-6)^2 - 3(-6) + 4$
    $$= 36 + 18 + 4$$
    $$= 58$$

    b. $f(0) = (0)^2 - 3(0) + 4$
    $$= 4$$

    c. $f\left(\dfrac{1}{2}\right) = \left(\dfrac{1}{2}\right)^2 - 3\left(\dfrac{1}{2}\right) + 4$
    $$= \dfrac{1}{4} - \dfrac{3}{2} + 4$$
    $$= \dfrac{1}{4} - \dfrac{6}{4} + \dfrac{16}{4} = \dfrac{11}{4}$$

13. First, find the slope of the line by using the two given points:
    $$m = \dfrac{3 - (-4)}{-5 - 2} = \dfrac{7}{-7} = -1$$
    Use the point-slope form to find the equation:
    $$y - 3 = -1(x - (-5))$$
    $$y - 3 = -x - 5$$
    $$y = -x - 2$$

    Graph:

    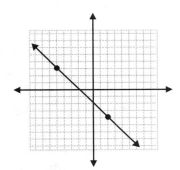

14. Solve the given equation for y to find the slope: $y = 2x - 7$. The slope is 2. Use the point-slope form with the given point to find the parallel line.
    $$y - 6 = 2(x - (-1))$$
    $$y = 2x + 8$$

    Graph:

    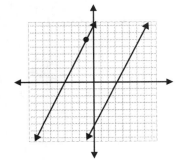

15. The given line has a slope of 3. Use the point-slope form with the given point to find the parallel line.

$$y - 1 = 3(x - (-1))$$
$$y = 3x + 4$$

Graph:

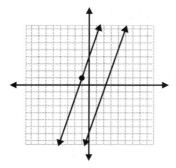

16. The given line has a slope of 3, so a perpendicular line must have a slope of $-\dfrac{1}{3}$. Use the point-slope form to find the equation of the perpendicular line that contains the point $(-1, 1)$.

$$y - 1 = -\frac{1}{3}(x - (-1))$$
$$y = -\frac{1}{3}x + \frac{2}{3}$$

Graph:

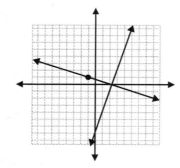

17. The given line has a slope of 2, so a perpendicular line must have a slope of $-\dfrac{1}{2}$. The given point is actually the y-intercept. Write the equation for the line in slope-intercept form:

$$y = -\frac{1}{2}x + 4$$

Graph:

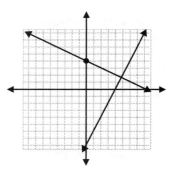

18. The graph represents a function.

19. The vertical line test indicates that the graph does not represent a function. For example, a line at $x = 0$ will intersect the graph at two distinct points.

20. $\begin{cases} 3x - y = 12 \\ x + 2y = -3 \end{cases}$

Multiply the first equation by 2 and solve:

$$6x - 2y = 24$$
$$\underline{x + 2y = -3}$$
$$7x \quad\ = 21$$
$$x \quad\ = 3$$

Substitute into the first equation:

$$3(3) - y = 12$$
$$-y = 12 - 9$$
$$y = -3$$

The system is consistent. The solution is $(3, -3)$.

21. $\begin{cases} y = 4x - 1 \\ 5x + 2y = -2 \end{cases}$

Solve by the substitution method:

$$5x + 2(4x - 1) = -2$$
$$5x + 8x - 2 = -2$$
$$13x = 0$$
$$x = 0$$
$$y = 4x - 1 = 4(0) - 1 = -1$$

The system is consistent. The solution is $(0, -1)$.

22. $\begin{cases} y = 2x - 7 \\ 4x - 2y = 3 \end{cases}$

Solve by the substitution method:

$$4x - 2(2x - 7) = 3$$
$$4x - 4x + 14 = 3$$
$$14 = 3$$

This statement is <u>never</u> true. The system is inconsistent.

23. Use a calculator to solve the system:
$$\begin{cases} 5x - y = 5 \\ 3x + y = 5.4 \end{cases}$$
$$x = \frac{13}{10}, \ y = \frac{3}{2}$$

24. Use a calculator to solve the system:
$$\begin{cases} -3x + y = -8.9 \\ 2x - y = 6.5 \end{cases}$$
$$x = 2.4, \ y = -1.7$$

25. Use a graphing calculator to estimate the points of intersection of the two functions $f(x) = x$ and $g(x) = 20x^2 - 1$.

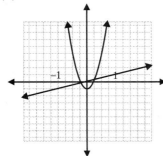

The points of intersection are $\left(\frac{-1}{5}, \frac{-1}{5}\right)$ and $\left(\frac{1}{4}, \frac{1}{4}\right)$.

26. Graph $y \le 4x - 5$:

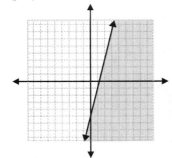

27. Graph $y > 6x + 2$:

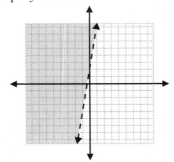

28. Graph $3x + 2y \ge 10$:

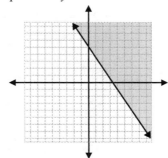

29. Graph $\begin{cases} 2x + y > 8 \\ x + 2y < 2 \end{cases}$:

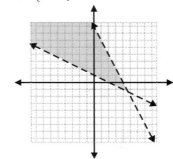

30. Graph $\begin{cases} y \le 7 \\ x \le 3 \end{cases}$:

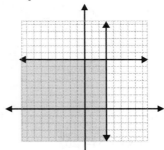

31. $2(4x + 3) + 5(x - 1) = 8x + 6 + 5x - 5$
$$= 13x + 1$$

32. $2x(x + 5) - (x + 1)(x - 3) = 2x^2 + 10x - (x^2 - 2x - 3)$
$$= 2x^2 + 10x - x^2 + 2x + 3$$
$$= x^2 + 12x + 3$$

33. $(x^2 + 3x - 1) + (2x^2 - 8x + 4) = 3x^2 - 5x + 3$

34. $(2x^2 + 6x - 7) + (2x^2 - x - 1) = 4x^2 + 5x - 8$

35. $\left(2x^2 - 5x + 3\right) - \left(x^2 - 2x + 8\right)$

$$= 2x^2 - 5x + 3 - x^2 + 2x - 8$$
$$= x^2 - 3x - 5$$

36. $\left(x + 1\right) - \left(4x^2 + 3x - 2\right) = x + 1 - 4x^2 - 3x + 2$

$$= -4x^2 - 2x + 3$$

37. $\left(2x - 7\right)\left(x + 4\right) = 2x^2 + 8x - 7x - 28$

$$= 2x^2 + x - 28$$

38. $-\left(x + 6\right)\left(3x - 1\right) = -\left(3x^2 - x + 18x - 6\right)$

$$= -\left(3x^2 + 17x - 6\right)$$
$$= -3x^2 - 17x + 6$$

39. $\left(x + 6\right)^2 = x^2 + 12x + 36$

40. $\left(2x - 7\right)^2 = 4x^2 - 28x + 49$

41. $\dfrac{15x^3 - 10x^2 + 3x}{5x^2} = \dfrac{15x^3}{5x^2} - \dfrac{10x^2}{5x^2} + \dfrac{3x}{5x^2}$

$$= 3x - 2 + \dfrac{3}{5x}$$

42. $\dfrac{8x^2y^2 - 5xy^2 + 4xy}{4xy^2} = \dfrac{8x^2y^2}{4xy^2} - \dfrac{5xy^2}{4xy^2} + \dfrac{4xy}{4xy^2}$

$$= 2x - \dfrac{5}{4} + \dfrac{1}{y}$$

43. $3x + 1 + \dfrac{22}{x - 2}$

$$
\begin{array}{r}
3x + 1 \\
x - 2 \overline{) 3x^2 - 5x + 20} \\
-\left(3x^2 - 6x\right) \\
\hline
x + 20 \\
-\left(x - 2\right) \\
\hline
22
\end{array}
$$

44. $x^2 - 8x + 38 - \dfrac{142}{x + 4}$

$$
\begin{array}{r}
x^2 - 8x + 38 \\
x + 4 \overline{) x^3 - 4x^2 + 6x + 10} \\
-\left(x^3 + 4x^2\right) \\
\hline
-8x^2 + 6x + 10 \\
-\left(-8x^2 - 32x\right) \\
\hline
38x + 10 \\
-\left(38x + 152\right) \\
\hline
-142
\end{array}
$$

45. $8x - 20 = 4\left(2x - 5\right)$

46. $6x - 96 = 6\left(x - 16\right)$

47. $2x^2 - 15x + 18 = \left(x - 6\right)\left(2x - 3\right)$

48. $6x^2 - x - 12 = \left(2x - 3\right)\left(3x + 4\right)$

49. $5x - 10 = 5\left(x - 2\right)$

50. $-12x^2 - 16x = -4x\left(3x + 4\right)$

51. $16x^2y - 24xy = 8xy\left(2x - 3\right)$

52. $10x^4 - 25x^3 + 5x^2 = 5x^2\left(2x^2 - 5x + 1\right)$

53. $4x^2 - 1 = \left(2x + 1\right)\left(2x - 1\right)$

54. $y^2 - 20y + 100 = \left(y - 10\right)\left(y - 10\right) = \left(y - 10\right)^2$

55. $3x^2 - 48y^2 = 3\left(x^2 - 16y^2\right) = 3\left(x + 4y\right)\left(x - 4y\right)$

56. $x^2 - 7x - 18 = \left(x - 9\right)\left(x + 2\right)$

57. $5x^2 + 40x + 80 = 5\left(x^2 + 8x + 16\right) = 5\left(x + 4\right)\left(x + 4\right)$

$$= 5\left(x + 4\right)^2$$

58. $x^2 + x + 3$ is not factorable.

59. $25x^2 + 20x + 4 = \left(5x + 2\right)\left(5x + 2\right) = \left(5x + 2\right)^2$

60. $3x^2 + 5x + 2 = (3x+2)(x+1)$

61. $2x^3 - 20x^2 + 50x = 2x(x^2 - 10x + 25)$
$$= 2x(x-5)(x-5) = 2x(x-5)^2$$

62. $4x^3 + 100x = 4x(x^2 + 25)$

63. $xy + 3x + 2y + 6 = (xy + 3x) + (2y + 6)$
$$= x(y+3) + 2(y+3)$$
$$= (x+2)(y+3)$$

64. $ax - 2a - 2b + bx = (ax - 2a) - (2b - bx)$
$$= a(x-2) - b(2-x)$$
$$= a(x-2) + b(x-2)$$
$$= (a+b)(x-2)$$

65. $\left(\dfrac{-4}{2}\right)^2 = (-2)^2 = 4$, so add 4 and factor:
$$x^2 - 4x + \underline{4} = (x-2)^2$$

66. $\left(\dfrac{18}{2}\right)^2 = (9)^2 = 81$, so add 81 and factor:
$$x^2 + 18x + \underline{81} = (x+9)^2$$

67. $\left(\dfrac{-8}{2}\right)^2 = (-4)^2 = 16$, so add 16 and factor:
$$x^2 - 8x + \underline{16} = (x-4)^2$$

68. $x^2 - \underline{10x} + 25 = (x-5)^2$

69. $x^2 - \underline{5x} + \dfrac{25}{4} = \left(x - \dfrac{5}{\underline{2}}\right)^2$

70. $(x-7)(x+1) = 0$
Set: $\quad x - 7 = 0 \qquad x + 1 = 0$
Solve: $\quad x = 7 \qquad\quad x = -1$

71. $x(3x+5) = 0$
Set: $\qquad x = 0 \qquad 3x + 5 = 0$
Solve: $\qquad\qquad\qquad 3x = -5$
$$x = -\dfrac{5}{3}$$

72. $4x^2 + 9x - 9 = 0$
$(x+3)(4x-3) = 0$
Set: $\quad x + 3 = 0 \qquad 4x - 3 = 0$
Solve: $\quad x = -3 \qquad\quad 4x = 3$
$$x = \dfrac{3}{4}$$

73. $21x - 3x^2 = 0$
$3x(7 - x) = 0$
Set: $\qquad 3x = 0 \qquad 7 - x = 0$
Solve: $\qquad x = 0 \qquad\quad 7 = x$

74. $x^2 + 8x + 12 = 0$
$(x+2)(x+6) = 0$
Set: $\quad x + 2 = 0 \qquad x + 6 = 0$
Solve: $\quad x = -2 \qquad\quad x = -6$

75. $\qquad x^2 = 3x + 28$
$\quad x^2 - 3x - 28 = 0$
$(x-7)(x+4) = 0$
Set: $\quad x - 7 = 0 \qquad x + 4 = 0$
Solve: $\quad x = 7 \qquad\qquad x = -4$

76. $x^3 + 5x^2 - 6x = 0$
$x(x^2 + 5x - 6) = 0$
$x(x+6)(x-1) = 0$
Set: $\quad x = 0 \qquad x + 6 = 0 \qquad x - 1 = 0$
Solve: $\qquad\qquad\qquad x = -6 \qquad\quad x = 1$

77. $\dfrac{1}{4}x^2 + x - 15 = 0$
$\quad x^2 + 4x - 60 = 0$
$(x+10)(x-6) = 0$
Set: $\quad x + 10 = 0 \qquad x - 6 = 0$
Solve: $\quad x = -10 \qquad\quad x = 6$

78. $15 - 12x - 3x^2 = 0$

$3x^2 + 12x - 15 = 0$

$3(x^2 + 4x - 5) = 0$

$3(x - 1)(x + 5) = 0$

Set: $x - 1 = 0$ $x + 5 = 0$

Solve: $x = 1$ $x = -5$

79. $x^3 + 14x^2 + 49x = 0$

$x(x^2 + 14x + 49) = 0$

$x(x + 7)^2 = 0$

Set: $x = 0$ $x + 7 = 0$

Solve: $x = -7$

80. $8x = 12x + 2x^2$

$2x^2 + 4x = 0$

$2x(x + 2) = 0$

Set: $2x = 0$ $x + 2 = 0$

Solve: $x = 0$ $x = -2$

81. $6x^2 = 24x$

$6x^2 - 24x = 0$

$6x(x - 4) = 0$

Set: $6x = 0$ $x - 4 = 0$

Solve: $x = 0$ $x = 4$

82. $2x(x - 2) = x + 25$

$2x^2 - 4x - x - 25 = 0$

$2x^2 - 5x - 25 = 0$

$(2x + 5)(x - 5) = 0$

Set: $2x + 5 = 0$ $x - 5 = 0$

Solve: $2x = -5$ $x = 5$

$$x = -\frac{5}{2}$$

83. $(4x - 3)(x + 1) = 2$

$4x^2 + x - 3 - 2 = 0$

$4x^2 + x - 5 = 0$

$(4x + 5)(x - 1) = 0$

Set: $4x + 5 = 0$ $x - 1 = 0$

Solve: $4x = -5$ $x = 1$

$$x = -\frac{5}{4}$$

84. $2x(x + 5)(x - 2) = 0$

Set: $2x = 0$ $x + 5 = 0$ $x - 2 = 0$

Solve: $x = 0$ $x = -5$ $x = 2$

85. Let b be the rate of the boat in still water. Let c be the rate of current. Recall the formula $d = rt$. Set up a system of equations based on the trip downstream and the trip upstream:

$$\begin{cases}(b + c)3 = 21 \\ (b - c)7 = 21\end{cases} = \begin{cases}3b + 3c = 21 \\ 7b - 7c = 21\end{cases}$$

Multiply the first equation by 7 and the second by 3. Solve by the addition method.

$$21b + 21c = 147$$
$$\underline{21b - 21c = 63}$$
$$42b \qquad = 210$$
$$b \qquad = 5$$

Substitute into the first equation:

$$3(5 + c) = 21$$
$$15 + 3c = 21$$
$$3c = 6$$
$$c = 2$$

The rate of the boat is 5 mph and the rate of the stream is 2 mph.

86. Let x be the amount invested at 6%. Then, $10,000 - x$ is the amount invested at 8%.

$$0.06x + 0.08(10,000 - x) = 650$$
$$6x + 8(10,000 - x) = 65,000$$
$$6x + 80,000 - 8x = 65,000$$
$$-2x + 80,000 = 65,000$$
$$-2x = -15,000$$
$$x = 7,500$$
$$10,000 - 7,500 = 2,500$$

Karl invested \$7,500 at 6% and \$2,500 at 8%.

87. Let l be the length of the rectangle and w be the width. Set up a system of two equations based on perimeter and area.

$$\begin{cases}2l + 2w = 60 \\ lw = 221\end{cases}$$

Solve the first equation for l and use substitution.

$$l = 30 - w$$
$$(30 - w)w = 221$$
$$30w - w^2 = 221$$
$$0 = w^2 - 30w + 221$$
$$0 = (w - 17)(w - 13)$$

Set: $w - 17 = 0$ $w - 13 = 0$ [End of Cumulative Review: Chapters 1 – 7]

Solve: $w = 17$ $w = 13$

$l = 30 - 17 = 13$ $l = 30 - 13 = 17$

Notice the two solutions are actually just switching the values for the width and the length. So, the dimensions of the rectangle are 13 in. by 17 in.

88. Let x be the smaller of the two positive numbers. Then, $x + 9$ is the larger.

$$x + (x + 9)^2 = 147$$
$$x^2 + 18x + 81 + x - 147 = 0$$
$$x^2 + 19x - 66 = 0$$
$$(x + 22)(x - 3) = 0$$

Set: $x + 22 = 0$ $x - 3 = 0$

Solve: $x = -22$ $x = 3$

Disregard the negative solution, since the numbers must be positive. The two numbers are 3 and 12.

89. Let n be the first of the two consecutive integers. Then, $n + 1$ is the next one.

$$n^2 + (n + 1)^2 = 145$$
$$n^2 + n^2 + 2n + 1 - 145 = 0$$
$$2n^2 + 2n - 144 = 0$$
$$2(n^2 + n - 72) = 0$$
$$2(n + 9)(n - 8) = 0$$

Set: $n + 9 = 0$ $n - 8 = 0$

Solve: $n = -9$ $n = 8$

$n + 1 = -8$ $n + 1 = 9$

There are two pairs of numbers that satisfy the conditions of the problem: $-9, -8$ and $8, 9$.

90. Using $h = -16t^2 + v_0 t$, where $v_0 = 88$ and $h = 120$ feet above the ground for some value of t, find t.

$$120 = -16t^2 + 88t$$
$$16t^2 - 88t + 120 = 0$$
$$8(2t^2 - 11t + 15) = 0$$
$$8(2t - 5)(t - 3) = 0$$

Set: $2t - 5 = 0$ $t - 3 = 0$

Solve: $2t = 5$ $t = 3$

$$t = \frac{5}{2} = 2.5$$

So at 2.5 seconds and again at 3 seconds, the ball is 120 feet from the ground.

Section 8.1
Solutions to Odd Exercises

For Exercises 1 – 20, the denominators must not be equal to zero. You must determine which values of the variable would cause the denominators to equal zero.

1. For $\dfrac{-3}{5x}$, $5x \neq 0$, so $x \neq 0$.

3. For $\dfrac{5y^2}{3y-4}$, $3y - 4 \neq 0$, so $y \neq \dfrac{4}{3}$.

5. For $\dfrac{x+1}{x^2-4x}$, $x^2 - 4x \neq 0$. This means that $x(x-4) \neq 0$. By the Zero-Product Principle, if either factor is equal to 0, then the product is 0. So, $x \neq 0$ and $x - 4 \neq 0$, or $x \neq 4$.

7. For $\dfrac{y^2-2y+3}{y^2-3y-10}$, $y^2 - 3y - 10 \neq 0$.
   Since $y^2 - 3y - 10 = (y-5)(y+2)$, $y \neq 5$ and $y \neq -2$.

9. For $\dfrac{x^2+2x}{x^2+4}$, the denominator $x^2 + 4$ is never equal to zero for any real number x. So there are no restrictions for x in the set of real numbers.

11. For $\dfrac{4-y}{y-4}$, $y - 4 \neq 0$, so $y \neq 4$.

13. For $\dfrac{5x+20}{6x+24}$, $6x + 24 \neq 0$.
    Since $6x + 24 = 6(x+4)$, $x \neq -4$.

15. For $\dfrac{x^2}{x^2+5x}$, $x^2 + 5x \neq 0$.
    Since $x^2 + 5x = x(x+5)$, $x \neq 0$ and $x \neq -5$.

17. For $\dfrac{a^2-7a}{a^2-49}$, $a^2 - 49 \neq 0$.
    Since $a^2 - 49 = (a-7)(a+7)$, $a \neq 7$ and $a \neq -7$.

19. For $\dfrac{m^2+3}{m^2+9}$, the denominator $m^2 + 9$ is never equal to zero for any real number m. So there are no restrictions for m in the set of real numbers.

21. For $x = 5$, evaluate $\dfrac{x-3}{3x^2}$:
    $$\dfrac{5-3}{3 \cdot 5^2} = \dfrac{2}{75}$$

23. For $x = -3$, evaluate $\dfrac{5x^2}{x^2-4}$:
    $$\dfrac{5(-3)^2}{(-3)^2-4} = \dfrac{5(9)}{9-4} = \dfrac{45}{5} = 9$$

25. For $x = 0$, evaluate $\dfrac{2x^2+5x}{x^2-1}$:
    $$\dfrac{2(0)^2+5(0)}{(0)^2-1} = \dfrac{0}{-1} = 0$$

27. For $m = 2$, evaluate $\dfrac{2m-7}{m^2+8m+12}$:
    $$\dfrac{2 \cdot 2 - 7}{2^2 + 8 \cdot 2 + 12} = \dfrac{4-7}{4+16+12} = \dfrac{-3}{32}$$

29. For $x = 1000$, evaluate $-\dfrac{15-x}{x-15}$. Notice that simplifying first yields $-\dfrac{15-x}{x-15} = \dfrac{x-15}{x-15} = 1$ for all values of x.

For Exercises 31 – 64, reduce to lowest terms and state any restrictions on the variable. <u>Always determine the restriction for the variable from the original form of the expression—not from the lowest terms.</u>

31. $\dfrac{4x-12}{4x} = \dfrac{4(x-3)}{4x} = \dfrac{x-3}{x}$ and $x \neq 0$.

33. $\dfrac{x^2+2x}{2x} = \dfrac{x(x+2)}{2x} = \dfrac{x+2}{2}$ and $x \neq 0$.

35. $\dfrac{2y+3}{4y+6} = \dfrac{2y+3}{2(2y+3)} = \dfrac{1}{2}$
    Since $4y + 6 = 2(2y+3) \neq 0$, $y \neq -\dfrac{3}{2}$.

37. $\dfrac{3x-6}{6x+3} = \dfrac{3(x-2)}{3(2x+1)} = \dfrac{x-2}{2x+1}$
    Since $6x + 3 = 3(2x+1) \neq 0$, $x \neq -\dfrac{1}{2}$.

39. $\dfrac{x}{x^2-4x} = \dfrac{x}{x(x-4)} = \dfrac{1}{x-4}$
    Since $x^2 - 4x = x(x-4) \neq 0$, $x \neq 0$ and $x \neq 4$.

41. $\dfrac{7x-14}{2-x} = \dfrac{7(x-2)}{-(x-2)} = \dfrac{7}{-1} = -7$

Since $2-x \neq 0$, $x \neq 2$.

43. $\dfrac{5+3x}{3x+5} = \dfrac{3x+5}{3x+5} = 1$

Since $3x+5 \neq 0$, $x \neq -\dfrac{5}{3}$.

45. $\dfrac{4-4x^2}{4x^2-4} = \dfrac{-4(x^2-1)}{4(x^2-1)} = \dfrac{-1}{1} = -1$

Since $4x^2-4 = 4(x+1)(x-1) \neq 0$, $x \neq -1$ and $x \neq 1$.

47. $\dfrac{x^2+2x}{x^2+4x+4} = \dfrac{x(x+2)}{(x+2)(x+2)} = \dfrac{x}{x+2}$

Since $x^2+4x+4 = (x+2)^2 \neq 0$, $x \neq -2$.

49. $\dfrac{x^2+7x+10}{x^2-25} = \dfrac{(x+5)(x+2)}{(x+5)(x-5)} = \dfrac{x+2}{x-5}$

Since $x^2-25 = (x+5)(x-5) \neq 0$, $x \neq -5$ and $x \neq 5$.

51. $\dfrac{x^2-3x-18}{x^2+6x+9} = \dfrac{(x-6)(x+3)}{(x+3)(x+3)} = \dfrac{x-6}{x+3}$

Since $x^2+6x+9 = (x+3)^2 \neq 0$, $x \neq -3$.

53. $\dfrac{x^2-5x+6}{8x-2x^3} = \dfrac{(x-2)(x-3)}{-2x(-4+x^2)} = \dfrac{(x-2)(x-3)}{-2x(x-2)(x+2)}$

$\qquad = \dfrac{x-3}{-2x(x+2)}$

$\qquad = \dfrac{3-x}{2x(x+2)}$

Since $8x-2x^2 = -2x(x-2)(x+2) \neq 0$, $x \neq 0$,

$x \neq -2$ and $x \neq 2$.

55. $\dfrac{16x^2+40x+25}{4x^2+13x+10} = \dfrac{(4x+5)(4x+5)}{(4x+5)(x+2)} = \dfrac{4x+5}{x+2}$

Since $4x^2+13x+10 = (4x+5)(x+2) \neq 0$, $x \neq -\dfrac{5}{4}$

and $x \neq -2$.

57. $\dfrac{3x-6}{8-4x} = \dfrac{3(x-2)}{-4(x-2)} = -\dfrac{3}{4}$

Since $8-4x = -4(x-2) \neq 0$, $x \neq 2$.

59. $\dfrac{a^2+6a+8}{a^2+4a+4} = \dfrac{(a+2)(a+4)}{(a+2)(a+2)} = \dfrac{a+4}{a+2}$

Since $a^2+4a+4 = (a+2)^2 \neq 0$, $a \neq -2$.

61. $\dfrac{3y^2-3y-18}{2y^2-10y+12} = \dfrac{3(y-3)(y+2)}{2(y-3)(y-2)} = \dfrac{3(y+2)}{2(y-2)}$

Since $2y^2-10y+12 = 2(y-3)(y-2) \neq 0$,

$y \neq 3$ and $y \neq 2$

63. $\dfrac{5y^2+17y+6}{5y^2+16y+3} = \dfrac{(y+3)(5y+2)}{(y+3)(5y+1)} = \dfrac{5y+2}{5y+1}$

Since $5y^2+16y+3 = (y+3)(5y+1) \neq 0$, $y \neq -3$

and $y \neq -\dfrac{1}{5}$.

Writing and Thinking About Mathematics

65. Answers will vary. One possible answer is:
The first line states that $a = b$. This would mean that $a - b = 0$. Then, notice in the fifth line of the work, both sides of the equation are divided by $a - b$, which equals 0. Division by 0 is not defined. Therefore, it is not surprising that this work led to an erroneous conclusion.

[End of Section 8.1]

Section 8.2
Solutions to Odd Exercises

1. $\dfrac{2x-4}{3x+6} \cdot \dfrac{3x}{x-2} = \dfrac{2(x-2)}{3(x+2)} \cdot \dfrac{3x}{x-2} = \dfrac{2x}{x+2}$

3. $\dfrac{4x^2}{x^2+3x} \cdot \dfrac{x^2-9}{2x-2} = \dfrac{4x^2}{x(x+3)} \cdot \dfrac{(x+3)(x-3)}{2(x-1)}$

$\qquad = \dfrac{2x(x-3)}{x-1}$

$\qquad \text{or} = \dfrac{2x^2-6x}{x-1}$

5. $\dfrac{x^2-4}{x} \cdot \dfrac{3x}{x+2} = \dfrac{(x+2)(x-2)}{x} \cdot \dfrac{3x}{x+2} = 3x-6$

7. $\dfrac{x^2-1}{5} \div \dfrac{x^2+2x+1}{10} = \dfrac{(x+1)(x-1)}{5} \cdot \dfrac{10}{(x+1)(x+1)}$

$\qquad = \dfrac{2(x-1)}{x+1}$

$\qquad \text{or} = \dfrac{2x-2}{x+1}$

9. $\dfrac{10x^2-5x}{6x^2+12x} \div \dfrac{2x-1}{x^2+2x} = \dfrac{5x(2x-1)}{6x(x+2)} \cdot \dfrac{x(x+2)}{2x-1} = \dfrac{5x}{6}$

11. $\dfrac{x^2+x}{x^2+2x+1} \cdot \dfrac{x^2-x-2}{x^2-1} = \dfrac{x(x+1)}{(x+1)(x+1)} \cdot \dfrac{(x-2)(x+1)}{(x+1)(x-1)}$

$\qquad = \dfrac{x(x-2)}{(x+1)(x-1)}$

$\qquad \text{or} = \dfrac{x^2-2x}{(x+1)(x-1)} = \dfrac{x^2-2x}{x^2-1}$

13. $\dfrac{x+3}{x^2+3x-4} \cdot \dfrac{x^2+x-2}{x+2} = \dfrac{x+3}{(x+4)(x-1)} \cdot \dfrac{(x+2)(x-1)}{x+2}$

$\qquad = \dfrac{x+3}{x+4}$

15. $\dfrac{6x^2-7x-3}{x^2-1} \cdot \dfrac{x-1}{2x-3} = \dfrac{(2x-3)(3x+1)}{(x+1)(x-1)} \cdot \dfrac{x-1}{2x-3}$

$\qquad = \dfrac{3x+1}{x+1}$

17. $\dfrac{x^2-8x+15}{x^2-9x+14} \cdot \dfrac{7-x}{x^2+4x-21}$

$\qquad = \dfrac{(x-5)(x-3)}{(x-7)(x-2)} \cdot \dfrac{-(x-7)}{(x+7)(x-3)}$

$\qquad = \dfrac{-(x-5)}{(x-2)(x+7)}$

$\qquad \text{or} \quad = \dfrac{5-x}{x^2+5x-14}$

19. $\dfrac{3x^2-7x+2}{1-9x^2} \cdot \dfrac{3x+1}{x-2} = \dfrac{(3x-1)(x-2)}{-(3x-1)(3x+1)} \cdot \dfrac{3x+1}{x-2}$

$\qquad = \dfrac{x-2}{-(x-2)}$

$\qquad = -1$

21. $\dfrac{4x^2-1}{x^2-16} \div \dfrac{2x+1}{x^2-4x} = \dfrac{(2x+1)(2x-1)}{(x+4)(x-4)} \cdot \dfrac{x(x-4)}{2x+1}$

$\qquad = \dfrac{x(2x-1)}{x+4}$

$\qquad \text{or} \quad = \dfrac{2x^2-x}{x+4}$

23. $\dfrac{x^2+x-6}{2x^2+6x} \div \dfrac{x^2-5x+6}{8x^2}$

$\qquad = \dfrac{(x+3)(x-2)}{2x(x+3)} \cdot \dfrac{8x^2}{(x-3)(x-2)}$

$\qquad = \dfrac{4x}{x-3}$

25. $\dfrac{2x^2-5x-12}{x^2-10x+24} \div \dfrac{4x^2-9}{x^2-9x+18}$

$\qquad = \dfrac{(2x+3)(x-4)}{(x-6)(x-4)} \cdot \dfrac{(x-6)(x-3)}{(2x+3)(2x-3)}$

$\qquad = \dfrac{x-3}{2x-3}$

27. $\dfrac{6x^2-x-2}{12x^2+5x-2} \div \dfrac{4x^2-1}{8x^2-6x+1}$

$\qquad = \dfrac{(2x+1)(3x-2)}{(3x+2)(4x-1)} \cdot \dfrac{(4x-1)(2x-1)}{(2x+1)(2x-1)}$

$\qquad = \dfrac{3x-2}{3x+2}$

29. $\dfrac{3x^2+11x+6}{4x^2+16x+7} \div \dfrac{9x^2+12x+4}{2x^2-x-28}$

$= \dfrac{(3x+2)(x+3)}{(2x+1)(2x+7)} \cdot \dfrac{(2x+7)(x-4)}{(3x+2)(3x+2)}$

$= \dfrac{(x+3)(x-4)}{(2x+1)(3x+2)}$

or $= \dfrac{x^2-x-12}{6x^2+7x+2}$

31. $\dfrac{x^2-16}{x^3-8x^2+16x} \cdot \dfrac{x^2}{x^2+4x} \cdot \dfrac{2x^2-2x}{x^2-2x+1}$

$= \dfrac{(x-4)(x+4)}{x(x-4)(x-4)} \cdot \dfrac{x^2}{x(x+4)} \cdot \dfrac{2x(x-1)}{(x-1)(x-1)}$

$= \dfrac{2x}{(x-4)(x-1)}$

or $= \dfrac{2x}{x^2-5x+4}$

33. $\dfrac{x^2-3x}{x^2+2x+1} \cdot \dfrac{x-3}{x^2-9} \cdot \dfrac{x^2-3x-18}{x-6}$

$= \dfrac{x(x-3)}{(x+1)(x+1)} \cdot \dfrac{x-3}{(x-3)(x+3)} \cdot \dfrac{(x-6)(x+3)}{(x-6)}$

$= \dfrac{x(x-3)}{(x+1)^2}$

or $= \dfrac{x^2-3x}{x^2+2x+1}$

35. $\dfrac{2x^2+5x-3}{x^2+2x-3} \cdot \dfrac{x^2+3x-4}{4x^2-1} \div \dfrac{x^2+4x}{x^3+5x^2}$

$= \dfrac{(2x-1)(x+3)}{(x+3)(x-1)} \cdot \dfrac{(x+4)(x-1)}{(2x-1)(2x+1)} \cdot \dfrac{x^2(x+5)}{x(x+4)}$

$= \dfrac{x(x+5)}{2x+1}$

or $= \dfrac{x^2+5x}{2x+1}$

37. $\dfrac{x^2+4x+3}{x^2+8x+7} \div \dfrac{x^2-7x-8}{35+12x+x^2} \cdot \dfrac{x^2+8x+15}{x^2-9x+8}$

$= \dfrac{(x+1)(x+3)}{(x+1)(x+7)} \cdot \dfrac{(x+7)(x+5)}{(x-8)(x+1)} \cdot \dfrac{(x+3)(x+5)}{(x-8)(x-1)}$

$= \dfrac{(x+3)^2(x+5)^2}{(x-8)^2(x+1)(x-1)}$

or $= \dfrac{(x+3)^2(x+5)^2}{(x-8)^2(x^2-1)}$

39. $\dfrac{2x^2-7x+3}{36x^2-1} \div \dfrac{6x^2+5x+1}{2x+1} \div \dfrac{2x^2-5x-3}{18x^2+3x-1}$

$= \dfrac{(2x-1)(x-3)}{(6x-1)(6x+1)} \cdot \dfrac{2x+1}{(3x+1)(2x+1)} \cdot \dfrac{(6x-1)(3x+1)}{(2x+1)(x-3)}$

$= \dfrac{2x-1}{(6x+1)(2x+1)}$ or $= \dfrac{2x-1}{12x^2+8x+1}$

41. Since $A = LW$, for the area of a rectangle, $W = \dfrac{A}{L}$.

$W = \dfrac{4x^2-4x-15}{2x+3}$

$W = \dfrac{(2x-5)(2x+3)}{2x+3}$

$W = 2x-5$

The width is $2x-5$ feet.

[End of Section 8.2]

Section 8.3
Solutions to Odd Exercises

1. $\dfrac{4x}{x+2} + \dfrac{8}{x+2} = \dfrac{4x+8}{x+2} = \dfrac{4(x+2)}{x+2} = 4$

3. $\dfrac{3x-1}{2x-6} + \dfrac{x-11}{2x-6} = \dfrac{3x-1+x-11}{2x-6}$

$= \dfrac{4x-12}{2x-6}$

$= \dfrac{4(x-3)}{2(x-3)}$

$= 2$

5. $\dfrac{x^2}{x^2+2x+1} + \dfrac{-2x-3}{x^2+2x+1} = \dfrac{x^2-2x-3}{x^2+2x+1}$

$= \dfrac{(x-3)(x+1)}{(x+1)(x+1)}$

$= \dfrac{x-3}{x+1}$

7. $\dfrac{-(3x+2)}{x^2-7x+6} + \dfrac{4x-4}{x^2-7x+6} = \dfrac{-3x-2+4x-4}{x^2-7x+6}$

$= \dfrac{x-6}{(x-6)(x-1)}$

$= \dfrac{1}{x-1}$

9. $\dfrac{x^2+2}{x^2-4} + \dfrac{-(4x-2)}{x^2-4} = \dfrac{x^2+2-4x+2}{x^2-4}$

$= \dfrac{x^2-4x+4}{(x-2)(x+2)}$

$= \dfrac{(x-2)(x-2)}{(x-2)(x+2)}$

$= \dfrac{x-2}{x+2}$

11. $\dfrac{3x+2}{4x-2} - \dfrac{x+2}{4x-2} = \dfrac{3x+2-(x+2)}{4x-2}$

$= \dfrac{3x+2-x-2}{4x-2}$

$= \dfrac{2x}{2(2x-1)}$

$= \dfrac{x}{2x-1}$

13. $\dfrac{2x^2-3x+1}{x^2-3x-4} - \dfrac{x^2+x+6}{x^2-3x-4}$

$= \dfrac{2x^2-3x+1-(x^2+x+6)}{x^2-3x-4}$

$= \dfrac{2x^2-3x+1-x^2-x-6}{x^2-3x-4}$

$= \dfrac{x^2-4x-5}{x^2-3x-4}$

$= \dfrac{(x-5)(x+1)}{(x-4)(x+1)}$

$= \dfrac{x-5}{x-4}$

15. $2x-6 = 2(x-3)$, so the LCM of the denominators

is $2(x-3)$.

$\dfrac{3}{x-3} - \dfrac{2}{2x-6} = \dfrac{3(2)-2}{2(x-3)}$

$= \dfrac{4}{2(x-3)}$

$= \dfrac{2}{x-3}$

17. $5x-10 = 5(x-2)$, and $3x-6 = 3(x-2)$, so the

LCM of the denominators is $5 \cdot 3(x-2)$.

$\dfrac{8}{5x-10} - \dfrac{6}{3x-6} = \dfrac{8(3)-6(5)}{5 \cdot 3(x-2)}$

$= \dfrac{24-30}{5 \cdot 3(x-2)}$

$= \dfrac{-6}{5 \cdot 3(x-2)}$

$= \dfrac{-2}{5(x-2)}$ or $\dfrac{-2}{5x-10}$

19. LCM of the denominators is $x(x+4)$.

$\dfrac{2}{x} + \dfrac{1}{x+4} = \dfrac{2(x+4)+1(x)}{x(x+4)}$

$= \dfrac{2x+8+x}{x(x+4)}$

$= \dfrac{3x+8}{x(x+4)}$ or $\dfrac{3x+8}{x^2+4x}$

21. LCM of the denominators is $(x+4)(x-4)$.

$$\frac{x}{x+4}+\frac{2}{x-4}=\frac{x(x-4)+2(x+4)}{(x+4)(x-4)}$$

$$=\frac{x^2-4x+2x+8}{(x+4)(x-4)}$$

$$=\frac{x^2-2x+8}{(x+4)(x-4)}$$

$$\text{or }\frac{x^2-2x+8}{x^2-16}$$

23. LCM of the denominators is $x-2$.

Notice that $2-x=-(x-2)$.

$$\frac{3}{2-x}+\frac{6}{x-2}=\frac{3}{-(x-2)}+\frac{6}{x-2}$$

$$=\frac{-3+6}{x-2}$$

$$=\frac{3}{x-2}$$

25. LCM of the denominators is $x-5$.

Notice that $5-x=-(x-5)$.

$$\frac{x}{5-x}+\frac{2x+3}{x-5}=\frac{x}{-(x-5)}+\frac{2x+3}{x-5}$$

$$=\frac{-x+2x+3}{x-5}$$

$$=\frac{x+3}{x-5}$$

27. LCM of the denominators is $(x-1)(x+2)$.

$$\frac{x}{x-1}-\frac{1}{x+2}=\frac{x(x+2)-1(x-1)}{(x-1)(x+2)}$$

$$=\frac{x^2+2x-x+1}{(x-1)(x+2)}$$

$$=\frac{x^2+x+1}{(x-1)(x+2)}$$

$$\text{or }\frac{x^2+x+1}{x^2+x-2}$$

29. LCM of the denominators is $(x-4)(x+4)$.

Notice that $4-x=-(x-4)$.

$$\frac{x+1}{x+4}+\frac{2x}{4-x}=\frac{x+1}{x+4}-\frac{2x}{x-4}$$

$$=\frac{(x+1)(x-4)-2x(x+4)}{(x+4)(x-4)}$$

$$=\frac{x^2-3x-4-2x^2-8x}{(x+4)(x-4)}$$

$$=\frac{-x^2-11x-4}{(x+4)(x-4)}$$

$$=\frac{x^2+11x+4}{(4+x)(4-x)}\quad\text{or}\quad\frac{x^2+11x+4}{16-x^2}$$

31. LCM of the denominators is $(x+2)(x-2)$.

$$\frac{8}{x-2}-\frac{4}{x+2}=\frac{8(x+2)-4(x-2)}{(x-2)(x+2)}$$

$$=\frac{8x+16-4x+8}{(x-2)(x+2)}$$

$$=\frac{4x+24}{(x-2)(x+2)}\quad\text{or}\quad\frac{4x+24}{x^2-4}$$

33. LCM of the denominators is $4\cdot3(x-2)$.

$$\frac{x}{4x-8}-\frac{3x+1}{3x-6}=\frac{x(3)-(3x+1)(4)}{12(x-2)}$$

$$=\frac{3x-12x-4}{12(x-2)}$$

$$=\frac{-9x-4}{12(x-2)}$$

$$=\frac{9x+4}{12(2-x)}\quad\text{or}\quad\frac{9x+4}{24-12x}$$

35. Factor first: $x^2+4x-5=(x+5)(x-1)$.

LCM of the denominators is $(x+5)(x-1)$.

$$\frac{2x+3}{x^2+4x-5}-\frac{4}{x-1}=\frac{(2x+3)-4(x+5)}{(x+5)(x-1)}$$

$$=\frac{2x+3-4x-20}{(x+5)(x-1)}$$

$$=\frac{-2x-17}{(x+5)(x-1)}$$

$$\text{or }\frac{-2x-17}{x^2+4x-5}$$

37. Factor first: $x^2 - 36 = (x+6)(x-6)$.

LCM of the denominators is $(x+6)(x-6)$.

$$\frac{3x}{6+x} - \frac{2x}{x^2-36} = \frac{3x(x-6)-2x}{(x+6)(x-6)}$$

$$= \frac{3x^2-18x-2x}{(x+6)(x-6)}$$

$$= \frac{3x^2-20x}{(x+6)(x-6)}$$

$$\text{or} \quad \frac{3x^2-20x}{x^2-36}$$

39. Factor first: $x^2 - 8x + 7 = (x-7)(x-1)$.

LCM of the denominators is $(x-7)(x-1)$.

$$\frac{2x+1}{x-7} + \frac{3x}{x^2-8x+7} = \frac{(2x+1)(x-1)+3x}{(x-7)(x-1)}$$

$$= \frac{2x^2-x-1+3x}{(x-7)(x-1)}$$

$$= \frac{2x^2+2x-1}{(x-7)(x-1)}$$

$$\text{or} \quad \frac{2x^2+2x-1}{x^2-8x+7}$$

41. LCM of the denominators is $2(x+2)(x+3)$.

$$\frac{x+1}{x+2} - \frac{x+2}{2x+6} = \frac{2(x+3)(x+1)-(x+2)(x+2)}{2(x+2)(x+3)}$$

$$= \frac{(2x+6)(x+1)-(x^2+4x+4)}{2(x+2)(x+3)}$$

$$= \frac{2x^2+8x+6-x^2-4x-4}{2(x+2)(x+3)}$$

$$= \frac{x^2+4x+2}{2(x+2)(x+3)}$$

$$\text{or} \quad \frac{x^2+4x+2}{2x^2+10x+12}$$

43. LCM of the denominators is $2(x-4)(x+4)$.

$$\frac{x+2}{x^2-16} - \frac{x+1}{2x-8} = \frac{(x+2)(2)-(x+1)(x+4)}{(x-4)(x+4)(2)}$$

$$= \frac{2x+4-(x^2+5x+4)}{2(x^2-16)}$$

$$= \frac{2x+4-x^2-5x-4}{2(x^2-16)}$$

$$= \frac{-x^2-3x}{2(x^2-16)}$$

$$= \frac{x^2+3x}{-2(x^2-16)} \quad \text{or} \quad \frac{x^2+3x}{-2x^2+32}$$

45. First, factor the two denominators:

$$x^2 + x - 2 = (x+2)(x-1)$$

$$x^2 - 1 = (x+1)(x-1)$$

LCM of the denominators is $(x+2)(x+1)(x-1)$.

$$\frac{1}{x^2+x-2} - \frac{1}{x^2-1} = \frac{1(x+1)-1(x+2)}{(x+2)(x+1)(x-1)}$$

$$= \frac{x+1-x-2}{(x+2)(x+1)(x-1)}$$

$$= \frac{-1}{(x+2)(x+1)(x-1)}$$

$$\text{or} \quad \frac{-1}{x^3+2x^2-x-2}$$

47. First, factor the two denominators:

$$x^2 + x - 12 = (x+4)(x-3)$$

$$x^2 - 9 = (x+3)(x-3)$$

LCM of the denominators is $(x+4)(x+3)(x-3)$.

$$\frac{2x}{x^2+x-12} + \frac{3x}{x^2-9} = \frac{2x(x+3)+3x(x+4)}{(x+4)(x+3)(x-3)}$$

$$= \frac{2x^2+6x+3x^2+12x}{(x+4)(x+3)(x-3)}$$

$$= \frac{5x^2+18x}{(x+4)(x+3)(x-3)}$$

$$\text{or} \quad \frac{5x^2+18x}{x^3+4x^2-9x-36}$$

49. First, factor the two denominators:
$$x^2 + 4x + 4 = (x+2)(x+2)$$
$$x^2 + 3x + 2 = (x+2)(x+1)$$

LCM of the denominators is $(x+2)(x+2)(x+1)$.

$$\frac{x-3}{x^2+4x+4} - \frac{3x}{x^2+3x+2}$$
$$= \frac{(x-3)(x+1) - 3x(x+2)}{(x+2)(x+2)(x+1)}$$
$$= \frac{x^2 - 2x - 3 - 3x^2 - 6x}{(x+2)(x+2)(x+1)}$$
$$= \frac{-2x^2 - 8x - 3}{(x+2)(x+2)(x+1)}$$
$$\text{or} \quad \frac{-2x^2 - 8x - 3}{x^3 + 5x^2 + 8x + 4}$$

51. First, factor the two denominators:
$$x^2 - 16 = (x+4)(x-4)$$
$$x^2 + 5x + 4 = (x+4)(x+1)$$

LCM of the denominators is $(x+4)(x-4)(x+1)$.

$$\frac{x}{x^2-16} - \frac{3x}{x^2+5x+4} = \frac{x(x+1) - 3x(x-4)}{(x+4)(x-4)(x+1)}$$
$$= \frac{x^2 + x - 3x^2 + 12x}{(x+4)(x-4)(x+1)}$$
$$= \frac{-2x^2 + 13x}{(x+4)(x-4)(x+1)}$$
$$\text{or} \quad \frac{-2x^2 + 13x}{x^3 + x^2 - 16x - 16}$$

53. First, factor the two denominators:
$$2x^2 + 5x - 3 = (2x-1)(x+3)$$
$$2x^2 - 9x + 4 = (2x-1)(x-4)$$

LCM of the denominators is $(2x-1)(x+3)(x-4)$.

$$\frac{x-2}{2x^2+5x-3} + \frac{2x-5}{2x^2-9x+4}$$
$$= \frac{(x-2)(x-4) + (2x-5)(x+3)}{(2x-1)(x+3)(x-4)}$$
$$= \frac{x^2 - 6x + 8 + 2x^2 + x - 15}{(2x-1)(x+3)(x-4)}$$
$$= \frac{3x^2 - 5x - 7}{(2x-1)(x+3)(x-4)}$$
$$\text{or} \quad \frac{3x^2 - 5x - 7}{2x^3 - 3x^2 - 23x + 12}$$

55. First, factor the two denominators:
$$x^2 + 3x - 10 = (x+5)(x-2)$$
$$x^2 - 4 = (x+2)(x-2)$$

LCM of the denominators is $(x+5)(x-2)(x+2)$.

$$\frac{x}{x^2+3x-10} + \frac{3x}{x^2-4} = \frac{x(x+2) + 3x(x+5)}{(x+5)(x-2)(x+2)}$$
$$= \frac{x^2 + 2x + 3x^2 + 15x}{(x+5)(x-2)(x+2)}$$
$$= \frac{4x^2 + 17x}{(x+5)(x-2)(x+2)}$$
$$\text{or} \quad \frac{4x^2 + 17x}{x^3 + 5x^2 - 4x - 20}$$

[End of Section 8.3]

Section 8.4
Solutions to Odd Exercises

1. Multiply the numerator and denominator by the LCM of the fractions: 10.

$$\frac{\dfrac{4}{5}}{\dfrac{7}{10}} = \left(\frac{\dfrac{4}{5}}{\dfrac{7}{10}}\right)\cdot\frac{10}{10} = \frac{8}{7}$$

3. Multiply the numerator and denominator by the LCM of the fractions: 12.

$$\frac{\dfrac{1}{2}+\dfrac{1}{3}}{\dfrac{5}{6}-\dfrac{1}{4}} = \left(\frac{\dfrac{1}{2}+\dfrac{1}{3}}{\dfrac{5}{6}-\dfrac{1}{4}}\right)\cdot\frac{12}{12} = \frac{6+4}{10-3} = \frac{10}{7}$$

5. Multiply the numerator and denominator by the LCM of the fractions: 3.

$$\frac{1+\dfrac{1}{3}}{2+\dfrac{2}{3}} = \left(\frac{1+\dfrac{1}{3}}{2+\dfrac{2}{3}}\right)\cdot\frac{3}{3} = \frac{3+1}{6+2} = \frac{4}{8} = \frac{1}{2}$$

7. Invert the divisor (denominator) and multiply.

$$\frac{\dfrac{3x}{5}}{\dfrac{x}{5}} = \frac{3x}{5}\cdot\frac{5}{x} = 3$$

9. Invert the divisor (denominator) and multiply.

$$\frac{\dfrac{4}{xy^2}}{\dfrac{6}{x^2 y}} = \frac{4}{xy^2}\cdot\frac{x^2 y}{6} = \frac{2x}{3y}$$

11. Invert the divisor (denominator) and multiply.

$$\frac{\dfrac{8x^2}{5y}}{\dfrac{x}{10y^2}} = \frac{8x^2}{5y}\cdot\frac{10y^2}{x} = 16xy$$

13. Invert the divisor (denominator) and multiply.

$$\frac{\dfrac{12x^3}{7y^4}}{\dfrac{3x^5}{2y}} = \frac{12x^3}{7y^4}\cdot\frac{2y}{3x^5} = \frac{8}{7y^3 x^2}$$

15. Invert the divisor (denominator) and multiply.

$$\frac{\dfrac{3x}{x+1}}{\dfrac{x-2}{x+1}} = \frac{3x}{x+1}\cdot\frac{x+1}{x-2} = \frac{3x}{x-2}$$

17. Invert the divisor (denominator) and multiply.

$$\frac{\dfrac{x+3}{2x}}{\dfrac{2x-1}{4x^2}} = \frac{x+3}{2x}\cdot\frac{4x^2}{2x-1} = \frac{2x(x+3)}{2x-1} = \frac{2x^2+6x}{2x-1}$$

19. Invert the divisor (denominator) and multiply.

$$\frac{\dfrac{x^2-9}{x}}{x-3} = \frac{(x+3)(x-3)}{x}\cdot\frac{1}{x-3} = \frac{x+3}{x}$$

21. Invert the divisor (denominator), factor and multiply.

$$\frac{\dfrac{x+2}{x-2}}{\dfrac{4x^2+8x}{x^2-4}} = \frac{x+2}{x-2}\cdot\frac{(x+2)(x-2)}{4x(x+2)} = \frac{x+2}{4x}$$

23. Multiply numerator and denominator by the LCM of the denominators of the fractions. LCM of the denominators is $3x^2$.

$$\frac{\dfrac{1}{x}-\dfrac{1}{3x}}{\dfrac{x+6}{x^2}} = \left(\frac{\dfrac{1}{x}-\dfrac{1}{3x}}{\dfrac{x+6}{x^2}}\right)\cdot\frac{3x^2}{3x^2}$$

$$= \frac{3x-x}{3(x+6)} = \frac{2x}{3(x+6)} \quad\text{or}\quad \frac{2x}{3x+18}$$

25. Multiply numerator and denominator by the LCM of the denominators of the fractions. LCM of the denominators is x^2.

$$\frac{1+\dfrac{1}{x}}{1-\dfrac{1}{x^2}} = \left(\frac{1+\dfrac{1}{x}}{1-\dfrac{1}{x^2}}\right)\cdot\frac{x^2}{x^2} = \frac{x^2+x}{x^2-1}$$

$$= \frac{x(x+1)}{(x-1)(x+1)} = \frac{x}{x-1}$$

27. Multiply numerator and denominator by the LCM of the denominators of the fractions. LCM of the denominators is $x^2 y$.

$$\frac{\dfrac{1}{x}-\dfrac{1}{y}}{\dfrac{y}{x^2}-\dfrac{1}{y}}=\left(\frac{\dfrac{1}{x}-\dfrac{1}{y}}{\dfrac{y}{x^2}-\dfrac{1}{y}}\right)\cdot\frac{x^2 y}{x^2 y}=\frac{xy-x^2}{y^2-x^2}$$

$$=\frac{x(y-x)}{(y+x)(y-x)}=\frac{x}{y+x}$$

29. Multiply numerator and denominator by the LCM of the denominators of the fractions. LCM of the denominators is $x(x+1)=x^2+x$.

$$\frac{2-\dfrac{4}{x}}{\dfrac{x^2-4}{x^2+x}}=\left(\frac{2-\dfrac{4}{x}}{\dfrac{x^2-4}{x^2+x}}\right)\cdot\frac{x^2+x}{x^2+x}$$

$$=\frac{2x^2+2x-4x-4}{x^2-4}$$

$$=\frac{2x^2-2x-4}{(x+2)(x-2)}$$

$$=\frac{2(x-2)(x+1)}{(x+2)(x-2)}$$

$$=\frac{2(x+1)}{x+2}\quad\text{or}\quad\frac{2x+2}{x+2}$$

31. Multiply numerator and denominator by the LCM of the denominators of the fractions. LCM of the denominators is $x+1$.

$$\frac{x-\dfrac{2}{x+1}}{x+\dfrac{x-3}{x+1}}=\left(\frac{x-\dfrac{2}{x+1}}{x+\dfrac{x-3}{x+1}}\right)\cdot\frac{x+1}{x+1}$$

$$=\frac{x(x+1)-2}{x(x+1)+x-3}$$

$$=\frac{x^2+x-2}{x^2+x+x-3}$$

$$=\frac{x^2+x-2}{x^2+2x-3}$$

$$=\frac{(x+2)(x-1)}{(x+3)(x-1)}$$

$$=\frac{x+2}{x+3}$$

33. Perform the multiplication first, then simplify.

$$\frac{1}{x+1}-\frac{3}{2x}\cdot\frac{4x}{x+1}=\frac{1}{x+1}-\frac{12x}{2x(x+1)}$$

$$=\frac{1}{x+1}-\frac{6}{x+1}$$

$$=\frac{1-6}{x+1}$$

$$=\frac{-5}{x+1}$$

35. Use the LCM of the first two fractions and simplify the subtraction. Then, invert the divisor and multiply.

$$\left(\frac{8}{x}-\frac{3}{4x}\right)\div\frac{4x+5}{x}=\frac{8(4)-3}{4x}\cdot\frac{x}{4x+5}$$

$$=\frac{32-3}{4x}\cdot\frac{x}{4x+5}$$

$$=\frac{29}{4(4x+5)}\quad\text{or}\quad\frac{29}{16x+20}$$

37. Following the rules for order of operations, multiply first.

$$\frac{x}{x-1}-\frac{3}{x-1}\cdot\frac{x+2}{x}=\frac{x}{x-1}-\frac{3x+6}{x(x-1)}$$

$$=\frac{x^2-(3x+6)}{x(x-1)}$$

$$=\frac{x^2-3x-6}{x(x-1)}$$

$$\text{or}\quad\frac{x^2-3x-6}{x^2-x}$$

39. Following the rules for order of operations, carry out the division first.

$$\frac{x-1}{x+4}+\frac{x-6}{x^2+3x-4}\div\frac{x-4}{x-1}$$

$$=\frac{x-1}{x+4}+\frac{x-6}{(x+4)(x-1)}\cdot\frac{x-1}{x-4}$$

$$=\frac{x-1}{x+4}+\frac{x-6}{(x+4)(x-4)}$$

$$=\frac{(x-1)(x-4)+x-6}{(x+4)(x-4)}$$

$$=\frac{x^2-5x+4+x-6}{(x+4)(x-4)}$$

$$=\frac{x^2-4x-2}{(x+4)(x-4)}\quad\text{or}\quad\frac{x^2-4x-2}{x^2-16}$$

41. **a.** Simplify, beginning with the "farthest" denominator and simplifying in steps:

$$1+\cfrac{1}{1+\cfrac{1}{1+\cfrac{1}{1+1}}}=1+\cfrac{1}{1+\cfrac{1}{1+\cfrac{1}{2}}}$$

$$=1+\cfrac{1}{1+\cfrac{1}{\cfrac{3}{2}}}$$

$$=1+\cfrac{1}{1+\cfrac{2}{3}}$$

$$=1+\cfrac{1}{\cfrac{5}{3}}$$

$$=1+\cfrac{3}{5}$$

$$=\cfrac{8}{5}$$

b. Simplify, beginning with the "farthest" denominator and simplifying in steps:

$$2-\cfrac{1}{2-\cfrac{1}{2-\cfrac{1}{2-1}}}=2-\cfrac{1}{2-\cfrac{1}{2-1}}$$

$$=2-\cfrac{1}{2-1}$$

$$=2-1$$

$$=1$$

c. Simplify, beginning with the "farthest" denominator and simplifying in steps:

$$x+\cfrac{1}{x+\cfrac{1}{x+\cfrac{1}{x+1}}}=x+\cfrac{1}{x+\cfrac{1}{\cfrac{x(x+1)+1}{x+1}}}$$

$$=x+\cfrac{1}{x+\cfrac{x+1}{x^2+x+1}}$$

$$=x+\cfrac{1}{\cfrac{x(x^2+x+1)+x+1}{x^2+x+1}}$$

$$=x+\cfrac{1}{\cfrac{x^3+x^2+2x+1}{x^2+x+1}}$$

$$=x+\cfrac{x^2+x+1}{x^3+x^2+2x+1}$$

$$=\cfrac{x(x^3+x^2+2x+1)+x^2+x+1}{x^3+x^2+2x+1}$$

$$=\cfrac{x^4+x^3+3x^2+2x+1}{x^3+x^2+2x+1}$$

[End of Section 8.4]

Section 8.5
Solutions to Odd Exercises

1. $x \neq 0$. Use the rule:
   "product of means = product of extremes."
   $$\frac{4}{5x} = \frac{2}{25}$$
   $$4(25) = 2(5x)$$
   $$100 = 10x$$
   $$x = 10$$

3. $x \neq -4$. Use the rule:
   "product of means = product of extremes."
   $$\frac{21}{x+4} = \frac{7}{8}$$
   $$8(21) = 7(x+4)$$
   $$168 = 7x + 28$$
   $$140 = 7x$$
   $$x = 20$$

5. $x \neq 1$. Use the rule:
   "product of means = product of extremes."
   $$\frac{2x}{x-1} = \frac{3}{2}$$
   $$2(2x) = 3(x-1)$$
   $$4x = 3x - 3$$
   $$x = -3$$

7. $x \neq 0, 2$. Use the rule:
   "product of means = product of extremes."
   $$\frac{10}{x} = \frac{5}{x-2}$$
   $$10(x-2) = x(5)$$
   $$10x - 20 = 5x$$
   $$5x = 20$$
   $$x = 4$$

9. $x \neq 4, -7$. Use the rule:
   "product of means = product of extremes."
   $$\frac{6}{x-4} = \frac{5}{x+7}$$
   $$6(x+7) = 5(x-4)$$
   $$6x + 42 = 5x - 20$$
   $$x = -62$$

11. No restrictions on x. Multiply by the LCM: 12.
    $$\frac{4x}{3} - \frac{3}{4} = \frac{5x}{6}$$
    $$12\left(\frac{4x}{3}\right) - 12\left(\frac{3}{4}\right) = 12\left(\frac{5x}{6}\right)$$
    $$16x - 9 = 10x$$
    $$6x = 9$$
    $$x = \frac{3}{2}$$

13. No restrictions on x. Multiply by the LCM: 15.
    $$\frac{x-2}{3} - \frac{x-3}{5} = \frac{13}{15}$$
    $$15\left(\frac{x-2}{3}\right) - 15\left(\frac{x-3}{5}\right) = 15\left(\frac{13}{15}\right)$$
    $$5x - 10 - 3x + 9 = 13$$
    $$2x - 1 = 13$$
    $$2x = 14$$
    $$x = 7$$

15. $x \neq 0$. Reduce fractions to lowest terms and
    multiply by the LCM: $12x$.
    $$\frac{9}{3x} = \frac{1}{4} - \frac{1}{6x}$$
    $$12x\left(\frac{3}{x}\right) = 12x\left(\frac{1}{4}\right) - 12x\left(\frac{1}{6x}\right)$$
    $$36 = 3x - 2$$
    $$38 = 3x$$
    $$x = \frac{38}{3}$$

17. $x \neq 0$. Multiply by the LCM: $21x$.
    $$\frac{1}{x} - \frac{8}{21} = \frac{3}{7x}$$
    $$21x\left(\frac{1}{x}\right) - 21x\left(\frac{8}{21}\right) = 21x\left(\frac{3}{7x}\right)$$
    $$21 - 8x = 9$$
    $$12 = 8x$$
    $$x = \frac{3}{2}$$

19. $x \neq 6$. Multiply by the LCM: $x - 6$.

$$\frac{5}{x-6} - 5 = \frac{2x}{x-6}$$

$$(x-6)\left(\frac{5}{x-6}\right) - (x-6)(5) = (x-6)\left(\frac{2x}{x-6}\right)$$

$$5 - 5x + 30 = 2x$$

$$35 = 7x$$

$$x = 5$$

21. $x \neq 3$. Multiply by the LCM: $x - 3$.

$$3 + \frac{2x}{x-3} = \frac{4}{x-3}$$

$$(x-3)(3) + (x-3)\left(\frac{2x}{x-3}\right) = (x-3)\left(\frac{4}{x-3}\right)$$

$$3x - 9 + 2x = 4$$

$$5x = 13$$

$$x = \frac{13}{5}$$

23. $x \neq 0, -4$. Multiply by the LCM: $3x(x+4)$.

$$\frac{1}{3} + \frac{1}{x+4} = \frac{1}{x}$$

$$3x(x+4)\left(\frac{1}{3} + \frac{1}{x+4}\right) = 3x(x+4)\left(\frac{1}{x}\right)$$

$$x^2 + 4x + 3x = 3x + 12$$

$$x^2 + 4x - 12 = 0$$

$$(x+6)(x-2) = 0$$

Set each factor equal to 0 and solve:
$x = -6$ and $x = 2$

25. $x \neq 0, -1$. Multiply by the LCM: $3x(x+1)$.

$$\frac{4}{x} - \frac{2}{x+1} = \frac{4}{3}$$

$$3x(x+1)\left(\frac{4}{x} - \frac{2}{x+1}\right) = 3x(x+1)\left(\frac{4}{3}\right)$$

$$12x + 12 - 6x = 4x^2 + 4x$$

$$-4x^2 + 2x + 12 = 0$$

$$-2(2x^2 - x - 6) = 0$$

$$2x^2 - x - 6 = 0$$

$$(2x+3)(x-2) = 0$$

Set each factor equal to 0 and solve:

$$x = -\frac{3}{2} \text{ and } x = 2$$

27. $x \neq -4, 3$. Multiply by the LCM: $(x+4)(x-3)$.

$$\frac{6}{x+4} + \frac{2}{x-3} = -1$$

$$(x+4)(x-3)\left(\frac{6}{x+4} + \frac{2}{x-3}\right) = -1(x+4)(x-3)$$

$$6x - 18 + 2x + 8 = -x^2 - x + 12$$

$$x^2 + 9x - 22 = 0$$

$$(x+11)(x-2) = 0$$

Set each factor equal to 0 and solve:
$x = -11$ and $x = 2$

29. $x \neq 3$. Multiply by the LCM: $x - 3$.

$$\frac{6}{x-3} + 1 = \frac{2x}{x-3}$$

$$(x-3)\left(\frac{6}{x-3} + 1\right) = (x-3)\left(\frac{2x}{x-3}\right)$$

$$6 + x - 3 = 2x$$

$$3 = x$$

But, $x = 3$ is not valid in the original equation, so there is no solution.

31. $x \neq -3, -2$. Multiply by the LCM: $(x+3)(x+2)$.

$$\frac{x}{x+3} + \frac{1}{x+2} = 1$$

$$\left(\frac{x}{x+3} + \frac{1}{x+2}\right)(x+3)(x+2) = 1(x+3)(x+2)$$

$$x^2 + 2x + x + 3 = x^2 + 5x + 6$$

$$-3 = 2x$$

$$x = -\frac{3}{2}$$

33. $x \neq 4, \frac{1}{2}$. Multiply by the LCM: $(x-4)(2x-1)$.

$$\frac{x}{x-4} - \frac{4}{2x-1} = 1$$

$$\left(\frac{x}{x-4} - \frac{4}{2x-1}\right)(x-4)(2x-1) = 1(x-4)(2x-1)$$

$$2x^2 - x - 4x + 16 = 2x^2 - 9x + 4$$

$$4x = -12$$

$$x = -3$$

35. $x \neq \dfrac{1}{4}, -1$. Multiply by the LCM: $(4x-1)(x+1)$.

$$\frac{2}{4x-1} + \frac{1}{x+1} = \frac{3}{x+1}$$

$$(4x-1)(x+1)\left(\frac{2}{4x-1} + \frac{1}{x+1}\right) = (4x-1)(x+1)\left(\frac{3}{x+1}\right)$$

$$2x+2+4x-1 = 12x-3$$

$$4 = 6x$$

$$x = \frac{2}{3}$$

37. $x \neq 4, -5$. Multiply by the LCM: $(x-4)(x+5)$.

$$\frac{x}{x-4} - \frac{12x}{x^2+x-20} = \frac{x-1}{x+5}$$

$$(x-4)(x+5)\left(\frac{x}{x-4} - \frac{12x}{(x-4)(x+5)}\right) = (x-4)(x+5)\left(\frac{x-1}{x+5}\right)$$

$$x^2+5x-12x = (x-4)(x-1)$$

$$x^2+5x-12x = x^2-5x+4$$

$$-2x = 4$$

$$x = -2$$

39. $x \neq 1, -1$. Multiply by the LCM: $(x+1)(x-1)$.

$$\frac{12}{x+1} + 1 = \frac{5}{x-1}$$

$$(x+1)(x-1)\left(\frac{12}{x+1} + 1\right) = (x+1)(x-1)\left(\frac{5}{x-1}\right)$$

$$12x-12+x^2-1 = 5x+5$$

$$x^2+7x-18 = 0$$

$$(x+9)(x-2) = 0$$

Set each factor equal to 0 and solve:

$x = -9$ and $x = 2$

41. $x \neq 3, -1$. Multiply by the LCM: $(x-3)(x+1)$.

$$\frac{x}{x-3} + \frac{8}{x^2-2x-3} = \frac{-2}{x+1}$$

$$(x-3)(x+1)\left(\frac{x}{x-3} + \frac{8}{x^2-2x-3}\right) = (x-3)(x+1)\left(\frac{-2}{x+1}\right)$$

$$x^2+x+8 = -2x+6$$

$$x^2+3x+2 = 0$$

$$(x+2)(x+1) = 0$$

Set each factor equal to 0 and solve:

$x = -2, -1$.

However, $x \neq -1$, because the original equation is not defined for that value. The only solution is $x = -2$.

43. $x \neq -5, -4$. Multiply by the LCM: $(x+5)(x+4)$.

$$\frac{x}{x+5} - \frac{5}{x^2+9x+20} = \frac{2}{x+4}$$

$$(x+5)(x+4)\left(\frac{x}{x+5} - \frac{5}{x^2+9x+20}\right) = (x+5)(x+4)\left(\frac{2}{x+4}\right)$$

$$x^2+4x-5 = 2x+10$$

$$x^2+2x-15 = 0$$

$$(x+5)(x-3) = 0$$

Set each factor equal to 0 and solve:

$x = -5$ and $x = 3$.

However, $x \neq -5$, since the original equation is not defined for that value. The only solution is $x = 3$.

45. Use proportions:

$$\frac{6}{1}\frac{\text{cents}}{\text{dollars}} = \frac{51}{x}\frac{\text{cents}}{\text{dollars}}$$

$$6x = 1(51)$$

$$x = \frac{51}{6}$$

$$x = \$8.50$$

47. Use proportions:

$$\frac{3}{100}\frac{\text{defective}}{\text{produced}} = \frac{x}{4800}\frac{\text{defective}}{\text{produced}}$$

$$100x = 3(4800)$$

$$x = 144$$

There were 144 defective bulbs.

49. Use proportions:

$$\frac{1}{24}\frac{\text{teachers}}{\text{students}} = \frac{21}{x}\frac{\text{teachers}}{\text{students}}$$

$$x = 24(21)$$

$$x = 504$$

There are 504 students.

51. Use proportions:

$$\frac{17}{50}\frac{\text{hits}}{\text{\# at bat}} = \frac{204}{x}\frac{\text{hits}}{\text{\# at bat}}$$

$$17x = 50(204)$$

$$x = 600$$

He needs 600 "times at bat" to get 204 hits.

53. Use proportions:

$$\frac{4}{10} \cdot \frac{\text{needed}}{\text{capacity}} = \frac{x}{22} \cdot \frac{\text{needed}}{\text{capacity}}$$

$$4(22) = 10x$$

$$x = \frac{88}{10}$$

$$x = 8.8$$

So, 8.8 quarts are needed.

55. Use proportions:

$$\frac{30}{x} \cdot \frac{\text{shadow}}{\text{height}} = \frac{4\frac{1}{2}}{6} \cdot \frac{\text{shadow}}{\text{height}}$$

$$4.5x = 180$$

$$x = \frac{180}{4.5}$$

$$x = 40$$

The height of the flagpole is 40 feet.

Writing and Thinking About Mathematics

57. **a.** Simplify the rational expression.

$$\frac{5}{x} + \frac{2}{x-1} + \frac{2x}{x-1}$$

Find the LCM of the denominators:

LCM is $x(x-1)$.

Multiply each fraction by 1 in the form of

$$\frac{x(x-1)}{x(x-1)}$$ so the fractions will have the same

denominator, but their values will be unchanged:

$$\frac{5}{x} \cdot \frac{x(x-1)}{x(x-1)} + \frac{2}{x-1} \cdot \frac{x(x-1)}{x(x-1)} + \frac{2x}{x-1} \cdot \frac{x(x-1)}{x(x-1)}$$

$$= \frac{5x - 5 + 2x + 2x^2}{x(x-1)}$$

$$= \frac{2x^2 + 7x - 5}{x^2 - x}$$

b. Solve the given equation.

$$\frac{5}{x} + \frac{2}{x-1} = \frac{2x}{x-1}$$

Notice that $x \neq 0$ and $x \neq 1$. To solve the equation, first simplify by multiplying both sides by the LCM of the denominators. The LCM is $x(x-1)$.

This process "divides out" the denominators by using the principle of "multiplying both sides of an equation by the same non-zero expression to create an equivalent equation."

HOWEVER, answers must always be checked, since multiplying by a variable expression can introduce extraneous solutions.

$$x(x-1)\left(\frac{5}{x} + \frac{2}{x-1}\right) = x(x-1)\left(\frac{2x}{x-1}\right)$$

$$5(x-1) + 2x = x(2x)$$

$$5x - 5 + 2x = 2x^2$$

$$-2x^2 + 7x - 5 = 0$$

$$2x^2 - 7x + 5 = 0$$

$$(2x-5)(x-1) = 0$$

Set: $2x - 5 = 0$ $x - 1 = 0$

Solve: $2x = 5$ $x = 1$

$$x = \frac{5}{2}$$

Checking these, you will find that $x \neq 1$, because the equation is not defined for that value.

A substitution of $x = \frac{5}{2}$ will yield a true statement, so this value is the only solution.

[End of Section 8.5]

Section 8.6
Solutions to Odd Exercises

1. Let n be the smaller of the two numbers.
Then $48 - n$ is the other number.
$$\frac{7}{9} = \frac{n}{48 - n}$$
$$7(48 - n) = 9n$$
$$336 - 7n = 9n$$
$$336 = 16n$$
$$n = 21 \text{ and } 48 - n = 27$$
The two numbers are 21 and 27.

3. Let n be the number of women.
Then $44 - n$ is the number of men.
$$\frac{n}{44 - n} = \frac{6}{5}$$
$$5n = 6(44 - n)$$
$$5n = 264 - 6n$$
$$11n = 264$$
$$n = 24 \text{ and } 44 - n = 20$$
There are 24 women and 20 men.

5. Let n be the numerator.
Then $n + 2$ is the denominator.
$$\frac{n + 3}{(n + 2) + 3} = \frac{5}{6}$$
$$6(n + 3) = 5(n + 5)$$
$$6n + 18 = 5n + 25$$
$$n = 7 \text{ and } n + 2 = 9$$
The numerator is 7 and the denominator is 9. The
original fraction is $\frac{7}{9}$.

7. One minute is the basic time unit in the problem. Let t
be the number of minutes for both copiers working
together to make 675 copies. Creating a "work table"
you would have:

	Minutes to Make 675 Copies	Part of Work in One Min.
Copier 1	$\frac{675}{8}$	$\frac{8}{675}$
Copier 2	$\frac{675}{10}$	$\frac{10}{675}$
Together	t	$\frac{1}{t}$

$$\frac{8}{675} + \frac{10}{675} = \frac{1}{t}$$
$$\frac{18}{675} = \frac{1}{t}$$
$$t = \frac{675}{18} = 37.5$$
Copier 1: $8 \cdot 37.5 = 300$ copies
Copier 2: $10 \cdot 37.5 = 375$ copies

9. From the uniform motion formula, $d = rt$, use the
form $t = \frac{d}{r}$. Let r be Derek's rate in mph.
Then $r + 15$ is Julie's rate.

	Rate	Time	Distance
Derek	r	$\frac{135}{r}$	135
Julie	$r + 15$	$\frac{180}{r + 15}$	180

Using the fact that the times for the distances are the
same, you have the equation:
$$\frac{135}{r} = \frac{180}{r + 15}$$
$$135(r + 15) = 180r$$
$$135r + 2025 = 180r$$
$$2025 = 45r$$
$$r = 45 \text{ and } r + 15 = 60$$
Derek's and Julie's rates were, respectively, 45 mph
and 60 mph.

11. Let r be the speed of the wind.

	Rate	Time	Distance
With Wind	$320 + r$	$\frac{690}{320 + r}$	690
Against Wind	$320 - r$	$\frac{510}{320 - r}$	510

The times are the same, so the equation is
$$\frac{690}{320 + r} = \frac{510}{320 - r}$$
$$690(320 - r) = 510(320 + r)$$
$$220,800 - 690r = 163,200 + 510r$$
$$57,600 = 1200r$$
$$r = \frac{57,600}{1200} = 48$$
The speed of the wind is 48 mph.

13. Let r be the rate for first 750 miles.

	Rate	Time	Distance
1st Part	r	$\dfrac{750}{r}$	750
2nd Part	$r+20$	$\dfrac{780}{r+20}$	780

The times are the same, so the equation is
$$\frac{750}{r} = \frac{780}{r+20}$$
$$750(r+20) = 780r$$
$$750r + 15,000 = 780r$$
$$15,000 = 30r$$
$$r = \frac{15,000}{30}$$
$$r = 500 \quad \text{and} \quad r+20 = 520$$
The rate for the first 750 miles was 500 mph and for the second part 520 mph.

15. Let r be the rate for second airliner.

	Rate	Time	Distance
1st Airliner	420	$\dfrac{1680}{420} = 4$	1680
2nd Airliner	r	$4 - \dfrac{1}{2} = 3.5$	1680

The second airliner travels $\dfrac{1}{2}$ hour less time than the first airliner, or 3.5 hours. Using $r \cdot t = d$,
$$r(3.5) = 1680$$
$$3.5r = 1680$$
$$r = \frac{1680}{3.5}$$
$$r = 480$$
The rate of the second airliner was 480 mph.

17. Let r be the rate of the boat in still water.

	Rate	Time	Distance
Downstream	$r+3$	$\dfrac{9}{r+3}$	9
Upstream	$r-3$	$\dfrac{9}{r-3}$	9

The total trip took $2\dfrac{1}{4}$, or $\dfrac{9}{4}$, hours, so the equation is:
$$\frac{9}{r+3} + \frac{9}{r-3} = \frac{9}{4}.$$

$$4(r+3)(r-3)\left(\frac{9}{r+3} + \frac{9}{r-3}\right) = 4(r+3)(r-3)\left(\frac{9}{4}\right)$$
$$36(r-3) + 36(r+3) = 9(r^2 - 9)$$
$$36r - 108 + 36r + 108 = 9r^2 - 81$$
$$9r^2 - 72r - 81 = 0$$
$$9(r^2 - 8r - 9) = 0$$
$$9(r-9)(r+1) = 0$$
Set each factor to 0 and solve:
$$r = 9 \text{ and } r = -1$$
Disregard $r = -1$, since a negative rate does not apply in this context. The rate of the boat in still water is 9 mph.

19. Let r be the rate that he skis. He skis 5 times faster than the lift rate, so the rate of the lift is one-fifth his ski rate, or $\dfrac{r}{5}$.

	Rate	Time	Distance
Lift	$\dfrac{r}{5}$	$\dfrac{1\frac{3}{4}}{\frac{r}{5}}$	$1\dfrac{3}{4}$
Downhill	r	$\dfrac{1\frac{3}{4}}{r}$	$1\dfrac{3}{4}$

The total time is 45 minutes, or $\dfrac{3}{4}$ hour, so the equation is:

$$\frac{1\frac{3}{4}}{\frac{r}{5}} + \frac{1\frac{3}{4}}{r} = \frac{3}{4}$$

To simplify, convert the mixed fractions to improper fractions and carry out the "divisions" by inverting the denominators and multiplying:
$$\frac{7}{4} \cdot \frac{5}{r} + \frac{7}{4} \cdot \frac{1}{r} = \frac{3}{4}$$
$$\frac{35}{4r} + \frac{7}{4r} = \frac{3}{4}$$
$$4r\left(\frac{35}{4r}\right) + 4r\left(\frac{7}{4r}\right) = 4r\left(\frac{3}{4}\right)$$
$$35 + 7 = 3r$$
$$3r = 42$$
$$r = 14$$

He skis at a rate of 14 mph.

21. Let x be the number of hours it takes Tom and Ellen working together.

	Hours to finish	Part in One Hour
Tom	6	$\dfrac{1}{6}$
Ellen	4	$\dfrac{1}{4}$
Together	x	$\dfrac{1}{x}$

The sum of the part each can do in one hour is the part they can do together:

$$\frac{1}{6}+\frac{1}{4}=\frac{1}{x}$$
$$12x\left(\frac{1}{6}+\frac{1}{4}\right)=12x\left(\frac{1}{x}\right)$$
$$2x+3x=12$$
$$5x=12$$
$$x=2.4$$

Together they can fix the car in 2.4 hrs, or 2 hours and 24 minutes.

23. Let x be the number of hours it takes with both pipes working.

	Hours to finish	Part in One Hour
Fill Pipe	$\dfrac{1}{4}$	$\dfrac{1}{\frac{1}{4}}=4$
Drain Pipe	$\dfrac{3}{4}$	$\dfrac{1}{\frac{3}{4}}=\dfrac{4}{3}$
Together	x	$\dfrac{1}{x}$

The difference between the parts that each pipe completes in one hour toward filling the tank is the part of the tank that is filled in one hour when both pipes are working at once. The drain pipe's part is subtracted from the fill pipe's part because the former is taking out while the latter is filling.

$$4-\frac{4}{3}=\frac{1}{x}$$
$$3x\left(\frac{8}{3}\right)=3x\left(\frac{1}{x}\right)$$
$$8x=3$$
$$x=\frac{3}{8} \text{ or } x=0.375$$

It takes 0.375 hours, or 22.5 minutes, to fill the tank.

25. Let x be the number of hours it takes Rod working alone. Convert minutes to hours.

	Hours to finish	Part in One Hour
Rick	$\dfrac{36}{60}=\dfrac{3}{5}$	$\dfrac{1}{\frac{3}{5}}=\dfrac{5}{3}$
Rod	x	$\dfrac{1}{x}$
Together	$\dfrac{20}{60}=\dfrac{1}{3}$	$\dfrac{1}{\frac{1}{3}}=3$

The sum of the part each can do in one hour equals the part they can do together:

$$\frac{5}{3}+\frac{1}{x}=3$$
$$3x\left(\frac{5}{3}+\frac{1}{x}\right)=3x(3)$$
$$5x+3=9x$$
$$3=4x$$
$$x=\frac{3}{4}\text{ hr}=45\text{ min}$$

Rod can finish the work in 45 minutes.

27. Let x be the number of hours for Judy to finish alone. Then Maria would finish in $2x$ hours working alone.

	Hours to finish	Part in One Hour
Judy	x	$\dfrac{1}{x}$
Maria	$2x$	$\dfrac{1}{2x}$
Together	1	1

The sum of the part each can do in one hour equals the part they can do together:

$$\frac{1}{x}+\frac{1}{2x}=1$$
$$2x\left(\frac{1}{x}+\frac{1}{2x}\right)=2x(1)$$
$$2+1=2x$$
$$x=\frac{3}{2}\text{ and }2x=3$$

Judy would finish in $\dfrac{3}{2}$ hours, while Maria would finish in 3 hours.

29. Let x be the number of days for the old boiler.
Then $x + 5$ is the number of days for new boiler.

	Days to Burn Load	Amount of Load for One Day
New Boiler	$x + 5$	$\dfrac{1}{x+5}$
Old Boiler	x	$\dfrac{1}{x}$
Together	6	$\dfrac{1}{6}$

The sum of the amount each can do in one hour is the amount they can do together:

$$\frac{1}{x+5} + \frac{1}{x} = \frac{1}{6}$$

$$6x(x+5)\left(\frac{1}{x+5} + \frac{1}{x}\right) = 6x(x+5)\left(\frac{1}{6}\right)$$

$$6x + 6x + 30 = x^2 + 5x$$

$$12x + 30 = x^2 + 5x$$

$$x^2 - 7x - 30 = 0$$

$$(x - 10)(x + 3) = 0$$

Set each factor equal to 0 and solve:

$$x = 10 \quad \text{and} \quad x = -3.$$

$$x + 5 = 15$$

Disregard $x = -3$, because negative values do not fit the context of the problem. So, the old boiler would take 10 days and the new boiler would take 15 days.

[End of Section 8.6]

Section 8.7
Solutions to Odd Exercises

1. y varies as x varies means that for some number k,

$y = kx$, or that $\dfrac{y}{x} = k$, $x \neq 0$.

Using $y = 2$, $x = 10$, solve for k:

$$k = \frac{2}{10} = \frac{1}{5}$$

Then $y = \dfrac{1}{5}x$. For $x = 7$, $y = \dfrac{1}{5} \cdot 7 = \dfrac{7}{5}$.

3. $\dfrac{y}{x^2} = k$, for some number k. Since $y = 9$ when $x = 2$, solve for k:

$$k = \frac{9}{2^2} = \frac{9}{4}$$

Then $y = \dfrac{9}{4}x^2$. For $x = 4$, $y = \dfrac{9}{4} \cdot 4^2 = 36$.

5. y inversely proportional to x means that $xy = k$, for some number k, or $y = \dfrac{k}{x}$, $x \neq 0$. Since $y = 5$ when $x = 4$, solve for k:

$$k = 4(5) = 20$$

Then $y = \dfrac{20}{x}$. For $x = 2$, $y = \dfrac{20}{2} = 10$.

7. $P = kG$ where P is the total price and G is the total number of gallons. Solve for k using $P = 25$ and $G = 20$:

$$k = \frac{P}{G} = \frac{25}{20} = \frac{5}{4}$$

Then $P = \dfrac{5}{4}G$. For $G = 30$, $P = \dfrac{5}{4}(30) = \dfrac{75}{2}$.

So, $P = \$37.50$.

9. "Gravitational force F is proportional to the square of the distance D" means that there is a number k for which $F = \dfrac{k}{D^2}$, or that $k = FD^2$.

In this problem, we are interested in the weight W (i.e., gravitational force) of the astronaut when he is 100 miles from Earth's surface. Note that the surface is 4000 miles from the center of the Earth. When the astronaut is on the surface of the Earth, the force F is 200 lb (his weight). Therefore, we can find k in the proportion:

$$k = 200(4000)^2 = 3,200,000,000$$

$$k = 3.2 \times 10^9$$

Then, for $D = 4100$, which is the distance the astronaut is from the center of the Earth,

$$F = \frac{3.2 \times 10^9}{D^2}$$

$$F = \frac{3.2 \times 10^9}{4100^2}$$

$$F = 190.36$$

The astronaut's gravitational force 100 miles from the Earth's surface is approximately 190.4 lb.

11. Pressure and volume of gas in a container are inversely related. This means that $V = \dfrac{k}{P}$, for some number k. Given that, when $P = 15$ grams per cm^2, $V = 300$ cm^3, determine the value of k:

$$k = VP = 300(15) = 4500$$

Then, for $P = 20$,

$$V = \frac{4500}{P}$$

$$V = \frac{4500}{20} = 225 \text{ cm}^3$$

13. "Circumference C varies directly with diameter d" means that $C = kd$ for some number k. For $C = 3.14$, $d = 1$, so $k = 3.14$. Note that these values are rounded to the nearest hundredth. Then, for $d = 2$,

$$C = 3.14(2) = 6.28 \text{ ft.}$$

15. Given that the bases B of the triangles are inversely proportional to the heights H, for some number k,

$H = \dfrac{k}{B}$, or $k = HB$. For $H = 6$, $B = 10$ solve for k:

$k = 6(10) = 60$. Then, for $B = 15$ in.,

$$H = \frac{k}{B} = \frac{60}{15} = 4 \text{ in.}$$

17. Use the given formula $P = kAv^2$, where P is the lifting force, A is the wing area, and v is the velocity. Given values $P = 12{,}000$, $A = 110$ and $v = 90$, solve for k:

$$k = \frac{12{,}000}{110(90)^2} \approx 0.013468$$

For $P = 12{,}000$ and $v = 80$

$$P = kAv^2 \text{ or } A = \frac{P}{kv^2}$$

$$A \approx \frac{12{,}000}{0.013468(80)^2} \approx 139.219 \text{ ft}^2$$

19. For constant temperature, it is given that $P = \dfrac{k}{V}$. For $v = 300$, it is known that $P = 16$, so find k:
$$k = PV = 16(300) = 4800$$
Then, for $V = 25$,
$$P = \dfrac{4800}{V}$$
$$P = \dfrac{4800}{25} = 192$$
So, the pressure is 192 lb/ft^2.

21. For constant temperature, it is given that $P = \dfrac{k}{V}$.
When $P = 140$ lb/ft^2, it is given that $V = 1000$ ft^3, so solve for k:
$$k = PV = 140(1000) = 140,000$$
Then, for $V = 200$,
$$P = \dfrac{140,000}{V}$$
$$P = \dfrac{140,000}{200} = 700$$
So, the pressure is 700 lb/ft^2.

23. $R = \dfrac{kL}{d^2}$, where R is the resistance in ohms, L is the wire length, d is the diameter of the wire, and k is the constant of proportionality. Notice that the linear measurements are given in mixed units, and you have to change these to the same units:
$$L = 100 \text{ ft} = 1200 \text{ in.}$$
For $L = 1200$ in., $d = 0.01$ in. and $R = 8$ ohms, find k:
$$k = \dfrac{Rd^2}{L} = \dfrac{8(0.01)^2}{1200} = 6.67 \times 10^{-7}$$
Find R for $d = 0.015$ in. and $L = 150$ ft $= 1800$ in.:
$$R = 6.67 \times 10^{-7} \left(\dfrac{1800}{0.015^2} \right)$$
$$R = 5.\overline{3} \text{ or } 5\dfrac{1}{3} \text{ ohms}$$

25. $\dfrac{T_1}{T_2} = \dfrac{R_2}{R_1}$, where T is the number of teeth in the gear and R is the number of revolutions per minute of the gear. For $T_1 = 96$, $T_2 = 27$, and $R_1 = 108$ revolutions, find R_2:
$$\dfrac{96}{27} = \dfrac{R_2}{108}$$
$$R_2 = \dfrac{96(108)}{27} = 384 \text{ rpm}$$

27. $\dfrac{T_1}{T_2} = \dfrac{R_2}{R_1}$, where T is the number of teeth in the gear and R is the number of revolutions per minute. For $T_1 = 12$, $R_1 = 60$, and $R_2 = 20$, find T_2:
$$\dfrac{12}{T_2} = \dfrac{20}{60}$$
$$T_2 = \dfrac{12(60)}{20} = 36 \text{ teeth}$$

29. $\dfrac{W_1}{W_2} = \dfrac{L_2}{L_1}$, where W is weight and L is distance from the fulcrum of the lever. $L_1 + L_2 = 5$ can be rewritten as $L_1 = 5 - L_2$. Given $W_1 = 40$ and $W_2 = 160$, substitute these values and the previous expression into the first equation:
$$\dfrac{40}{160} = \dfrac{L_2}{5 - L_2}$$
$$40(5 - L_2) = 160L_2$$
$$200 - 40L_2 = 160L_2$$
$$200 = 200L_2$$
$$L_2 = 1$$
$$L_1 = 5 - L_2 = 5 - 1 = 4$$
The 40 lb weight is to be placed 4 ft from the fulcrum for a balanced lever.

[End of Section 8.7]

Chapter 8 Review
Solutions to All Exercises

For Exercises 1–20, recall that the denominator can never be zero. Use this fact to determine any necessary restrictions on the variable.

1. $\dfrac{4}{9x}$ $\longrightarrow$ $9x \neq 0$

$x \neq 0$

2. $\dfrac{23}{7y}$ $\longrightarrow$ $7y \neq 0$

$y \neq 0$

3. $\dfrac{5x}{x-7}$ $\longrightarrow$ $x - 7 \neq 0$

$x \neq 7$

4. $\dfrac{3x}{x+4}$ $\longrightarrow$ $x + 4 \neq 0$

$x \neq -4$

5. $\dfrac{14}{2x+1}$ $\longrightarrow$ $2x + 1 \neq 0$

$x \neq -\dfrac{1}{2}$

6. $\dfrac{-10}{3x-1}$ $\longrightarrow$ $3x - 1 \neq 0$

$x \neq \dfrac{1}{3}$

7. $\dfrac{12}{a^2+25}$ $\longrightarrow$ $a^2 + 25 \neq 0$

$a^2 \neq -25$

Since the square of a number is never negative, this is always true. There is no restriction on a.

8. $\dfrac{3a}{a^2+a-2}$ $\longrightarrow$ $a^2 + a - 2 \neq 0$

$(a-1)(a+2) \neq 0$

$a - 1 \neq 0 \qquad a + 2 \neq 0$

$a \neq 1 \quad$ and $\quad a \neq -2$

9. $\dfrac{y^2+2y}{y^2+6y+5}$ $\longrightarrow$ $y^2 + 6y + 5 \neq 0$

$(y+5)(y+1) \neq 0$

$y + 5 \neq 0 \qquad\qquad y + 1 \neq 0$

$y \neq -5 \quad$ and $\quad y \neq -1$

10. $\dfrac{5a^2+5a}{a^2+2a+1}$ $\longrightarrow$ $a^2 + 2a + 1 \neq 0$

$(a+1)^2 \neq 0$

$a + 1 \neq 0$

$a \neq -1$

11. $\dfrac{x+2}{x^2+4x+4} = \dfrac{x+2}{(x+2)^2} = \dfrac{1}{x+2}$

$x^2 + 4x + 4 \neq 0$

$(x+2)^2 \neq 0$

$x + 2 \neq 0$

$x \neq -2$

12. $\dfrac{y-3}{y^2-6y+9} = \dfrac{y-3}{(y-3)^2} = \dfrac{1}{y-3}$

$y^2 - 6y + 9 \neq 0$

$(y-3)^2 \neq 0$

$y - 3 \neq 0$

$y \neq 3$

13. $\dfrac{x^2-2x-3}{-x^2+2x+3} = \dfrac{x^2-2x-3}{-(x^2-2x-3)} = \dfrac{1}{-1} = -1$

$-x^2 + 2x + 3 \neq 0$

$-(x-3)(x+1) \neq 0$

$x - 3 \neq 0 \quad$ and $\quad x + 1 \neq 0$

$x \neq 3 \qquad\qquad x \neq -1$

14. $\dfrac{3-y}{y-3} = \dfrac{-(y-3)}{y-3} = \dfrac{-1}{1} = -1$

$y - 3 \neq 0$

$y \neq 3$

15. $\dfrac{4+x}{x^2-16} = \dfrac{4+x}{(x+4)(x-4)} = \dfrac{1}{x-4}$

$x^2 - 16 \neq 0$

$(x+4)(x-4) \neq 0$

$x + 4 \neq 0 \quad$ and $\quad x - 4 \neq 0$

$x \neq -4 \qquad\qquad x \neq 4$

16. $\dfrac{-5+x}{2x^2-50}=\dfrac{x-5}{2(x+5)(x-5)}=\dfrac{1}{2(x+5)}$

$$2x^2-50\neq 0$$
$$2(x+5)(x-5)\neq 0$$
$$x+5\neq 0 \quad \text{and} \quad x-5\neq 0$$
$$x\neq -5 \qquad\qquad x\neq 5$$

17. $\dfrac{4x-8}{x^2-2x}=\dfrac{4(x-2)}{x(x-2)}=\dfrac{4}{x}$

$$x^2-2x\neq 0$$
$$x(x-2)\neq 0$$
$$x-2\neq 0 \quad \text{and} \quad x\neq 0$$
$$x\neq 2$$

18. $\dfrac{2x+3}{4x^2+12x+9}=\dfrac{2x+3}{(2x+3)^2}=\dfrac{1}{2x+3}$

$$4x^2+12x+9\neq 0$$
$$(2x+3)^2\neq 0$$
$$2x+3\neq 0$$
$$2x\neq -3$$
$$x\neq \dfrac{-3}{2}$$

19. $\dfrac{x^2-25}{x^2-10x+25}=\dfrac{(x+5)(x-5)}{(x-5)^2}=\dfrac{x+5}{x-5}$

$$x^2-10x+25\neq 0$$
$$(x-5)^2\neq 0$$
$$x-5\neq 0$$
$$x\neq 5$$

20. $\dfrac{6x^2+7x-5}{3x^2-4x-15}=\dfrac{(2x-1)(3x+5)}{(x-3)(3x+5)}=\dfrac{2x-1}{x-3}$

$$3x^2-4x-15\neq 0$$
$$(3x+5)(x-3)\neq 0$$
$$3x+5\neq 0 \quad \text{and} \quad x-3\neq 0$$
$$3x\neq -5 \qquad\qquad x\neq 3$$
$$x\neq \dfrac{-5}{3}$$

21. $\dfrac{x}{x-2}\cdot\dfrac{x^2-4}{x^2}=\dfrac{x(x+2)(x-2)}{x^2(x-2)}=\dfrac{x+2}{x}$

22. $\dfrac{2y-6}{3y-9}\cdot\dfrac{2y}{2y-4}=\dfrac{2(y-3)\cdot 2y}{3(y-3)\cdot 2(y-2)}=\dfrac{2y}{3(y-2)}$

23. $\dfrac{5y^2}{y^2+3y}\cdot\dfrac{y^2-9}{5y-5}=\dfrac{5y^2(y+3)(y-3)}{y(y+3)\cdot 5(y-1)}=\dfrac{y(y-3)}{y-1}$

24. $\dfrac{x^2-8x+16}{x^2-16}\cdot\dfrac{2x+8}{4}=\dfrac{(x-4)^2\cdot 2(x+4)}{4(x+4)(x-4)}=\dfrac{x-4}{2}$

25. $\dfrac{2x^2-x}{6x^2+12x}\div\dfrac{2x-1}{4x^2+8x}=\dfrac{2x^2-x}{6x^2+12x}\cdot\dfrac{4x^2+8x}{2x-1}$

$$=\dfrac{x(2x-1)\cdot 4x(x+2)}{6x(x+2)(2x-1)}$$

$$=\dfrac{2x}{3}$$

26. $\dfrac{x-7}{3}\div\dfrac{x^2-49}{6x+42}=\dfrac{x-7}{3}\cdot\dfrac{6x+42}{x^2-49}$

$$=\dfrac{(x-7)\cdot 6(x+7)}{3(x+7)(x-7)}$$

$$=2$$

27. $\dfrac{x^2+2x-8}{4x}\div\dfrac{2x^2-5x+2}{5x^2}=\dfrac{x^2+2x-8}{4x}\cdot\dfrac{5x^2}{2x^2-5x+2}$

$$=\dfrac{(x+4)(x-2)\cdot 5x^2}{4x(2x-1)(x-2)}$$

$$=\dfrac{5x(x+4)}{4(2x-1)}$$

28. $\dfrac{y-3}{y^2-3y-4}\div\dfrac{y^2-y-2}{y+1}=\dfrac{y-3}{y^2-3y-4}\cdot\dfrac{y+1}{y^2-y-2}$

$$=\dfrac{(y-3)(y+1)}{(y-4)(y+1)(y-2)(y+1)}$$

$$=\dfrac{y-3}{(y-4)(y-2)(y+1)}$$

29. $\dfrac{2a^2-7a+3}{6a^2+5a+1} \div \dfrac{36a^2-1}{2a^2-5a-3} \div \dfrac{18a^2+3a-10}{2a+1}$

$= \dfrac{2a^2-7a+3}{6a^2+5a+1} \cdot \dfrac{2a^2-5a-3}{36a^2-1} \cdot \dfrac{2a+1}{18a^2+3a-10}$

$= \dfrac{(2a-1)(a-3)(2a+1)(a-3)(2a+1)}{(2a+1)(3a+1)(6a+1)(6a-1)(3a-2)(6x+5)}$

$= \dfrac{(a-3)^2(2a-1)(2a+1)}{(3a+1)(6a+1)(6a-1)(3a-2)(6a+5)}$

30. $\dfrac{16x^2}{x^2-4} \cdot \dfrac{4x^2-4x-3}{16x^2-4} \div \dfrac{6x^2+9x}{1-4x^2}$

$= \dfrac{16x^2}{x^2-4} \cdot \dfrac{4x^2-4x-3}{16x^2-4} \cdot \dfrac{1-4x^2}{6x^2+9x}$

$= \dfrac{16x^2(2x-3)(2x+1)(1+2x)(1-2x)}{(x+2)(x-2)\cdot 4(2x+1)(2x-1)\cdot 3x(2x+3)}$

$= \dfrac{-4x(2x+1)(2x-3)}{3(x+2)(x-2)(2x+3)}$

31. By solving the area formula for the width, w, we find that the width is: $w = \dfrac{A}{l}$. Plug the given expressions for area and length into this equation and simplify the resulting rational expression:

$w = \dfrac{10x^2+21x+9}{5x+3} = \dfrac{(5x+3)(2x+3)}{5x+3} = 2x+3$

The width is $2x+3$ meters.

32. a. The area of the original rectangle is

$x(x+5) = x^2+5x$

b. The area of the new rectangle is

$(x+6)(x+11) = x^2+17x+66$

c. The increase in area can be found by taking the difference in area of the new and original rectangles:

$(x^2+17x+66)-(x^2+5x)$

$= x^2-x^2+17x-5x+66$

$= 12x+66$

33. $\dfrac{5x}{x+2} + \dfrac{10}{x+2} = \dfrac{5x+10}{x+2}$

$= \dfrac{5(x+2)}{x+2}$

$= 5$

34. $\dfrac{3x}{x+4} - \dfrac{12}{x-4} = \dfrac{3x}{x+4} \cdot \dfrac{x-4}{x-4} - \dfrac{12}{x-4} \cdot \dfrac{x+4}{x+4}$

$= \dfrac{3x(x-4)-12(x+4)}{(x+4)(x-4)}$

$= \dfrac{3x^2-12x-12x-48}{(x+4)(x-4)}$

$= \dfrac{3x^2-24x-48}{(x+4)(x-4)}$

$= \dfrac{3(x^2-8x-16)}{(x+4)(x-4)}$

35. $\dfrac{3x-2}{2x-6} + \dfrac{x-10}{2x-6} = \dfrac{3x-2+x-10}{2x-6} = \dfrac{4x-12}{2x-6}$

$= \dfrac{4(x-3)}{2(x-3)} = 2$

36. $\dfrac{2x+7}{5x+5} + \dfrac{-x-6}{5(x+1)} = \dfrac{2x+7}{5(x+1)} + \dfrac{-(x+6)}{5(x+1)}$

$= \dfrac{2x+7-(x+6)}{5(x+1)}$

$= \dfrac{2x-x+7-6}{5(x+1)}$

$= \dfrac{x+1}{5(x+1)} = \dfrac{1}{5}$

37. $\dfrac{y+1}{3y-1} + \dfrac{2y}{y+2} = \dfrac{(y+1)(y+2)}{(3y-1)(y+2)} + \dfrac{2y(3y-1)}{(y+2)(3y-1)}$

$= \dfrac{y^2+3y+2+6y^2-2y}{(y+2)(3y-1)}$

$= \dfrac{7y^2+y+2}{(y+2)(3y-1)}$

38. $\dfrac{a}{a+2} - \dfrac{1}{a-1} = \dfrac{a(a-1)}{(a+2)(a-1)} - \dfrac{1(a+2)}{(a-1)(a+2)}$

$= \dfrac{a^2-a-a-2}{(a-1)(a+2)}$

$= \dfrac{a^2-2a-2}{(a-1)(a+2)}$

39. $\dfrac{6}{2x-3}-\dfrac{5}{3-2x}=\dfrac{6(3-2x)}{(2x-3)(3-2x)}-\dfrac{5(2x-3)}{(2x-3)(3-2x)}$

$$=\dfrac{18-12x-10x+15}{(2x-3)(3-2x)}=\dfrac{33-22x}{(2x-3)(3-2x)}$$

$$=\dfrac{11(3-2x)}{(2x-3)(3-2x)}=\dfrac{11}{2x-3}$$

40. $\dfrac{4}{2-x}-\dfrac{3}{x-2}=\dfrac{4}{2-x}-\dfrac{3}{-(2-x)}$

$$=\dfrac{4}{2-x}+\dfrac{3}{2-x}=\dfrac{7}{2-x}$$

41. $\dfrac{x^2}{x^2+4x+4}-\dfrac{2x+8}{x^2+4x+4}=\dfrac{x^2-2x-8}{(x+2)^2}$

$$=\dfrac{(x+2)(x-4)}{(x+2)^2}$$

$$=\dfrac{x-4}{x+2}$$

42. $\dfrac{x+5}{x-5}-\dfrac{100}{x^2-25}=\dfrac{(x+5)(x+5)}{(x-5)(x+5)}-\dfrac{100}{(x-5)(x+5)}$

$$=\dfrac{x^2+10x+25-100}{(x-5)(x+5)}=\dfrac{x^2+10x-75}{(x-5)(x+5)}$$

$$=\dfrac{(x-5)(x+15)}{(x-5)(x+5)}=\dfrac{x+15}{x+5}$$

43. $\dfrac{3x-12}{x^2+x-20}-\dfrac{x^2+5x}{x^2+9x+20}$

$$=\dfrac{3(x-4)}{(x+5)(x-4)}-\dfrac{x(x+5)}{(x+5)(x+4)}$$

$$=\dfrac{3}{x+5}-\dfrac{x}{x+4}$$

$$=\dfrac{3(x+4)}{(x+5)(x+4)}-\dfrac{x(x+5)}{(x+5)(x+4)}$$

$$=\dfrac{3x+12-x^2-5x}{(x+5)(x+4)}=\dfrac{-x^2-2x+12}{(x+5)(x+4)}$$

44. $\dfrac{5x+20}{x^2+8x+16}+\dfrac{3x-12}{x^2-16}=\dfrac{5(x+4)}{(x+4)^2}+\dfrac{3(x-4)}{(x+4)(x-4)}$

$$=\dfrac{5}{x+4}+\dfrac{3}{x+4}$$

$$=\dfrac{8}{x+4}$$

45. $\dfrac{4x}{x^2-2x-15}-\dfrac{10}{x^2-6x+5}$

$$=\dfrac{4x}{(x-5)(x+3)}-\dfrac{10}{(x-5)(x-1)}$$

$$=\dfrac{4x(x-1)}{(x-5)(x+3)(x-1)}-\dfrac{10(x+3)}{(x-5)(x-1)(x+3)}$$

$$=\dfrac{4x^2-4x-10x-30}{(x-5)(x-1)(x+3)}$$

$$=\dfrac{2(2x+3)(x-5)}{(x-5)(x-1)(x+3)}$$

$$=\dfrac{2(2x+3)}{(x-1)(x+3)}$$

46. $\dfrac{4}{y^2-1}+\dfrac{4}{y+1}+\dfrac{2}{1-y}$

$$=\dfrac{4}{y^2-1}+\dfrac{4(y-1)}{(y+1)(y-1)}+\dfrac{2(y+1)(-1)}{(1-y)(y+1)(-1)}$$

$$=\dfrac{4+4y-4-2y-2}{(y+1)(y-1)}=\dfrac{2y-2}{(y+1)(y-1)}$$

$$=\dfrac{2(y-1)}{(y+1)(y-1)}=\dfrac{2}{y+1}$$

47. $\dfrac{x+3}{x^2+x-6}+\dfrac{x-2}{x^2+4x-12}$

$$=\dfrac{x+3}{(x-2)(x+3)}+\dfrac{x-2}{(x-2)(x+6)}$$

$$=\dfrac{1}{x-2}+\dfrac{1}{x+6}$$

$$=\dfrac{(x+6)}{(x-2)(x+6)}+\dfrac{(x-2)}{(x+6)(x-2)}$$

$$=\dfrac{x+6+x-2}{(x-2)(x+6)}$$

$$=\dfrac{2x+4}{(x-2)(x+6)}$$

48. $-4 + \dfrac{1-2x}{x+6} + \dfrac{x^2+1}{x^2+4x-12}$

$$= \dfrac{-4\left(x^2+4x-12\right)}{x^2+4x-12} + \dfrac{(1-2x)(x-2)}{(x+6)(x-2)} + \dfrac{x^2+1}{x^2+4x-12}$$

$$= \dfrac{\left(-4x^2-16x+48\right)+\left(x-2-2x^2+4x\right)+\left(x^2+1\right)}{x^2+4x-12}$$

$$= \dfrac{-5x^2-11x+47}{x^2+4x-12}$$

49. $\dfrac{\frac{3}{4}}{\frac{5}{8}} = \dfrac{3}{4} \cdot \dfrac{8}{5} = \dfrac{3\cdot 8}{4\cdot 5} = \dfrac{6}{5}$

50. $\dfrac{\frac{7}{16}}{\frac{5}{32}} = \dfrac{7}{16} \cdot \dfrac{32}{5} = \dfrac{7\cdot 32}{16\cdot 5} = \dfrac{14}{5}$

51. $\dfrac{\frac{1}{2}+\frac{2}{3}}{\frac{1}{4}-\frac{1}{6}} = \left(\dfrac{1}{2}+\dfrac{2}{3}\right) \div \left(\dfrac{1}{4}-\dfrac{1}{6}\right) = \dfrac{7}{6} \div \dfrac{1}{12} = \dfrac{7}{6} \cdot 12 = 14$

52. $\dfrac{1-\frac{2}{5}}{1+\frac{2}{5}} = \left(1-\dfrac{2}{5}\right) \div \left(1+\dfrac{2}{5}\right) = \dfrac{3}{5} \div \dfrac{7}{5} = \dfrac{3}{5} \cdot \dfrac{5}{7} = \dfrac{3}{7}$

53. $\dfrac{\frac{9x^2}{5y}}{\frac{3x}{y^2}} = \dfrac{9x^2}{5y} \cdot \dfrac{y^2}{3x} = \dfrac{9x^2y^2}{5y\cdot 3x} = \dfrac{3xy}{5}$

54. $\dfrac{\frac{25y^2}{3x}}{\frac{15y}{4x^2}} = \dfrac{25y^2}{3x} \cdot \dfrac{4x^2}{15y} = \dfrac{25y^2\cdot 4x^2}{3x\cdot 15y} = \dfrac{20xy}{9}$

55. $\dfrac{\frac{y^2-25}{y}}{y-5} = \dfrac{y^2-25}{y} \cdot \dfrac{1}{y-5} = \dfrac{y^2-25}{y(y-5)} = \dfrac{y+5}{y}$

56. $\dfrac{\frac{x^2-5x+4}{x+1}}{x-4} = \dfrac{x^2-5x+4}{x+1} \cdot \dfrac{1}{x-4} = \dfrac{x^2-5x+4}{(x+1)(x-4)}$

$$= \dfrac{x-1}{x+1}$$

57. $\dfrac{\frac{3}{x}}{1-\frac{1}{x+1}} = \dfrac{3}{x} \div \left(1-\dfrac{1}{x+1}\right) = \dfrac{3}{x} \div \dfrac{(x+1)-1}{x+1}$

$$= \dfrac{3}{x} \cdot \dfrac{x+1}{x} = \dfrac{3(x+1)}{x^2}$$

58. $\dfrac{3-\frac{6}{x}}{\frac{x^2-4x+4}{3x}} = \left(3-\dfrac{6}{x}\right) \cdot \dfrac{3x}{x^2-4x+4}$

$$= \dfrac{3x-6}{x} \cdot \dfrac{3x}{x^2-4x+4}$$

$$= \dfrac{3(x-2)\cdot 3x}{x(x-2)^2} = \dfrac{9}{x-2}$$

59. $\dfrac{1}{x+2} - \dfrac{5}{2x} \cdot \dfrac{4x}{x+2} = \dfrac{2x}{2x(x+2)} - \dfrac{20x}{2x(x+2)}$

$$= \dfrac{-18x}{2x(x+2)} = \dfrac{-9}{x+2}$$

60. $\dfrac{7}{x} + \dfrac{5}{x^2-3x} \cdot \dfrac{x-3}{10} = \dfrac{7}{x} + \dfrac{5(x-3)}{x(x-3)\cdot 10}$

$$= \dfrac{70}{10x} + \dfrac{5}{10x} = \dfrac{75}{10x} = \dfrac{15}{2x}$$

61. $\left(\dfrac{4}{y} - \dfrac{2}{3y}\right) \div \dfrac{3y+2}{6y} = \left(\dfrac{12}{3y} - \dfrac{2}{3y}\right) \cdot \dfrac{6y}{3y+2}$

$$= \dfrac{10}{3y} \cdot \dfrac{2\cdot 3y}{3y+2} = \dfrac{20}{3y+2}$$

62. $\left(\dfrac{4}{y} + \dfrac{2}{y+3}\right) \div \dfrac{y^2-9}{2y} = \left(\dfrac{4(y+3)}{y(y+3)} + \dfrac{2y}{y(y+3)}\right) \cdot \dfrac{2y}{y^2-9}$

$$= \dfrac{6y+12}{y(y+3)} \cdot \dfrac{2y}{y^2-9}$$

$$= \dfrac{6(y+2)\cdot 2y}{y(y+3)(y^2-9)}$$

$$= \dfrac{12(y+2)}{(y+3)^2(y-3)}$$

63. Restriction: $\quad 5x \neq 0$

$$x \neq 0$$

$$\frac{8}{5x} = \frac{4}{25}$$

$$5x\left(\frac{8}{5x}\right) = 5x\left(\frac{4}{25}\right)$$

$$8 = \frac{4}{5}x$$

$$10 = x$$

64. Restriction: $\quad 3x \neq 0$

$$x \neq 0$$

$$\frac{2}{15} = \frac{16}{3x}$$

$$3x\left(\frac{2}{15}\right) = 3x\left(\frac{16}{3x}\right)$$

$$\frac{2}{5}x = 16$$

$$x = 40$$

65. Restriction: $\quad x + 3 \neq 0$

$$x \neq -3$$

$$\frac{4}{x+3} = \frac{1}{2}$$

$$(x+3)\left(\frac{4}{x+3}\right) = (x+3)\left(\frac{1}{2}\right)$$

$$4 = \frac{x+3}{2}$$

$$8 = x + 3$$

$$5 = x$$

66. Restriction: $\quad x + 4 \neq 0$

$$x \neq -4$$

$$\frac{20}{x+4} = \frac{4}{5}$$

$$(x+4)\left(\frac{20}{x+4}\right) = (x+4)\left(\frac{4}{5}\right)$$

$$20 = \frac{4(x+4)}{5}$$

$$100 = 4x + 16$$

$$84 = 4x$$

$$21 = x$$

67. Restriction: $\quad x - 2 \neq 0$

$$x \neq 2$$

$$\frac{3x}{x-2} + 4 = \frac{8}{x-2}$$

$$(x-2)\left(\frac{3x}{x-2} + 4\right) = (x-2)\left(\frac{8}{x-2}\right)$$

$$3x + 4(x-2) = 8$$

$$3x + 4x - 8 = 8$$

$$7x = 16$$

$$x = \frac{16}{7}$$

68. Restriction: $\quad y - 5 \neq 0$

$$y \neq 5$$

$$\frac{2y}{y-5} + 3 = \frac{4}{y-5}$$

$$(y-5)\left(\frac{2y}{y-5} + 3\right) = (y-5)\left(\frac{4}{y-5}\right)$$

$$2y + 3(y-5) = 4$$

$$2y + 3y - 15 = 4$$

$$5y = 19$$

$$y = \frac{19}{5}$$

69. Restrictions: $\quad x + 3 \neq 0 \quad$ and $\quad x + 7 \neq 0$

$$x \neq -3 \qquad x \neq -7$$

$$\frac{3}{x+3} = 1 - \frac{5}{x+7}$$

$$(x+3)(x+7)\left(\frac{3}{x+3}\right) = (x+3)(x+7)\left(1 - \frac{5}{x+7}\right)$$

$$3(x+7) = (x+3)(x+7) - 5(x+3)$$

$$3x + 21 = x^2 + 10x + 21 - 5x - 15$$

$$0 = x^2 + 2x - 15$$

$$0 = (x-3)(x+5)$$

Set: $\qquad x - 3 = 0 \qquad x + 5 = 0$

Solve: $\qquad x = 3 \qquad x = -5$

70. Restrictions:

$$y - 1 \neq 0 \quad \text{and} \quad y + 1 \neq 0$$
$$y \neq 1 \qquad\qquad y \neq -1$$

$$\frac{5}{y-1} - \frac{3}{y+1} = \frac{28}{9(y+1)}$$

$$9(y+1)(y-1)\left(\frac{5}{y-1} - \frac{3}{y+1}\right) = 9(y+1)(y-1)\left(\frac{28}{9(y+1)}\right)$$

$$45(y+1) - 27(y-1) = 28(y-1)$$
$$45y + 45 - 27y + 27 = 28y - 28$$
$$18y + 72 = 28y - 28$$
$$-10y = -100$$
$$y = 10$$

71. Restrictions:

$$y - 1 \neq 0 \quad \text{and} \quad y + 4 \neq 0$$
$$y \neq 1 \qquad\qquad y \neq -4$$

$$\frac{5}{y+4} + \frac{3}{y-1} = \frac{-11}{2(y+4)}$$

$$2(y+4)(y-1)\left(\frac{5}{y+4} + \frac{3}{y-1}\right) = 2(y+4)(y-1) \cdot \frac{-11}{2(y+4)}$$

$$10(y-1) + 6(y+4) = -11(y-1)$$
$$10y - 10 + 6y + 24 = -11y + 11$$
$$16y + 14 = -11y + 11$$
$$27y = -3$$
$$y = \frac{-1}{9}$$

72. Restrictions:

$$x + 2 \neq 0 \quad \text{and} \quad x + 1 \neq 0$$
$$x \neq -2 \qquad\qquad x \neq -1$$

$$\frac{x}{x+2} = \frac{1}{x+1} - \frac{x}{x^2+3x+2}$$

$$(x+1)(x+2)\left(\frac{x}{x+2}\right) = (x+1)(x+2)\left(\frac{1}{x+1} - \frac{x}{x^2+3x+2}\right)$$

$$x(x+1) = 1 \cdot (x+2) - x$$
$$x^2 + x = 2$$
$$x^2 + x - 2 = 0$$
$$(x+2)(x-1) = 0$$

Set: $\quad x + 2 = 0 \qquad x - 1 = 0$

Solve: $\quad x = -2 \qquad\quad x = 1$

However, $x \neq -2$, so $x = 1$ is the only solution.

73. Let n be the smaller of the numbers. Then $45 - n$ is the other number.

$$\frac{n}{45-n} = \frac{2}{3}$$
$$3n = 2(45 - n)$$
$$3n = 90 - 2n$$
$$5n = 90$$
$$n = 18$$
$$45 - n = 27$$

The two numbers are 18 and 27.

74. Let n be the number of women. Then $36 - n$ is the number of men.

$$\frac{n}{36-n} = \frac{4}{5}$$
$$5n = 4(36 - n)$$
$$5n = 144 - 4n$$
$$9n = 144$$
$$n = 16$$

There are 16 women in the class.

75. Let d be the denominator. Then $2d + 3$ is the numerator.

$$\frac{(2d+3)+5}{d+5} = \frac{11}{6}$$
$$6(2d+8) = 11(d+5)$$
$$12d + 48 = 11d + 55$$
$$d = 7$$
$$2(7) + 3 = 17$$

The original fraction was $\dfrac{17}{7}$.

76. Let r be the rate of the boat in still water.

	Rate	Time	Distance
Downstream	$r + 5$	$\dfrac{10}{r+5}$	10
Upstream	$r - 5$	$\dfrac{5}{r-5}$	5

$$\frac{10}{r+5} = \frac{5}{r-5}$$
$$10(r-5) = 5(r+5)$$
$$10r - 50 = 5r + 25$$
$$5r = 75$$
$$r = 15$$

So the boat travels 15 mph in still water.

77. Let r be the flying speed of the second plane.

	Rate	Time	Distance
1st Plane	150	$\dfrac{450}{150} = 3$	450
2nd Plane	r	$\dfrac{450}{r}$	450

Since the second plane left 30 minutes, or 0.5 hours, after the first, the time to travel 450 miles was $3 - 0.5 = 2.5$ hours.

$$2.5 = \frac{450}{r}$$
$$2.5r = 450$$
$$r = \frac{450}{2.5} = 180$$

The second plane flew at 180 mph on average.

78. Let r be the rate at which he rode up the mountain.

	Rate	Time	Distance
Up	r	$\dfrac{12}{r}$	12
Down	$4r$	$\dfrac{12}{4r} = \dfrac{3}{r}$	12

$$2.5 = \frac{12}{r} + \frac{3}{r}$$
$$2.5 = \frac{15}{r}$$
$$2.5r = 15$$
$$r = \frac{15}{2.5} = 6$$

So Adam rode up the mountain at 6mph.

79. Let x be the number of hours it will take the faster pipe to fill the pool.

	Hours to finish	Part in One Hour
Pipe #1	x	$\dfrac{1}{x}$
Pipe #2	$2x$	$\dfrac{1}{2x}$
Together	2	$\dfrac{1}{2}$

$$\frac{1}{x} + \frac{1}{2x} = \frac{1}{2}$$
$$2x\left(\frac{1}{x} + \frac{1}{2x}\right) = 2x \cdot \frac{1}{2}$$
$$2 + 1 = x$$
$$3 = x$$

So one pipe can fill the pool in three hours and the other can fill it in six hours.

80. Let x be the number of hours the partner needs to complete the job alone.

	Hours to finish	Part in One Hour
Bricklayer	8	$\dfrac{1}{8}$
Partner	x	$\dfrac{1}{x}$
Together	4.8	$\dfrac{1}{4.8} = \dfrac{10}{48}$

$$\frac{1}{8} + \frac{1}{x} = \frac{10}{48}$$
$$48x\left(\frac{1}{8} + \frac{1}{x}\right) = 48x \cdot \frac{10}{48}$$
$$6x + 48 = 10x$$
$$48 = 4x$$
$$12 = x$$

So the partner can complete the job in 12 hours.

81. Let x be the number of hours they need to paint the room together.

	Hours to finish	Part in One Hour
Painter	2	$\dfrac{1}{2}$
Helper	3	$\dfrac{1}{3}$
Together	x	$\dfrac{1}{x}$

$$\frac{1}{2} + \frac{1}{3} = \frac{1}{x}$$
$$6x\left(\frac{1}{2} + \frac{1}{3}\right) = 6x \cdot \frac{1}{x}$$
$$3x + 2x = 6$$
$$5x = 6$$
$$x = \frac{6}{5}$$

So together they can finish in $\dfrac{6}{5}$ hours.

82. Let r be the average speed of the pilot's plane.

	Rate	Time	Distance
Pilot's Plane	r	$\dfrac{600}{r}$	600
Jet	$3r$	$\dfrac{900}{3r} = \dfrac{300}{r}$	900

$$\frac{300}{r} = \frac{600}{r} - 2$$

$$r \cdot \frac{300}{r} = r\left(\frac{600}{r} - 2\right)$$

$$300 = 600 - 2r$$

$$2r = 300$$

$$r = 150$$

$$3r = 450$$

The average speed of the jet is 450 mph.

83. y varies as x varies means that for some number k, $y = kx$. Then for $y = 4$ and $x = 20$

$$4 = k \cdot 20$$

$$\frac{4}{20} = k$$

$$\frac{1}{5} = k$$

So $y = \dfrac{1}{5}x$ and when $x = 30$, $y = \dfrac{1}{5} \cdot 30 = 6$.

84. y inversely proportional to x means that for some number k, $xy = k$. Then for $y = 8$ and $x = 10$

$$10 \cdot 8 = k$$

$$80 = k$$

So $xy = 80$, and when $x = 16$, $16y = 80$, or $y = 5$.

85. Let s be the length of the spring and w be the weight. The two variables are directly proportional, so there is some number k such that $s = kw$. Since $s = 10$ when $w = 14$, we can find k:

$$10 = 14k$$

$$\frac{10}{14} = \frac{5}{7} = k$$

So $s = \dfrac{5}{7}w$ and when $w = 20$, $s = \dfrac{5}{7} \cdot 20 = \dfrac{100}{7}$ in.

86. Circumference C varies directly with diameter d means that $C = kd$ for some number k. For $C = 12.56$, $d = 4$, so $k = 3.14$. Note that these numbers are rounded to the nearest hundredth. Then, for $d = 6$,

$$C = 3.14(6) = 18.84 \text{ inches.}$$

87. Area, A, varies directly with the radius, r, squared indicates that $A = kr^2$ for some number k. For $A = 50.24$, $r = 4$, so $k = 3.14$. (Note: these numbers are rounded to the nearest hundredth.) Then, for $r = 6$, $A = 3.14(6)^2 = 113.04$ in.^{2}.

88. Distance, D, a falling object travels varies directly with the square of the time, t, it falls means that $D = kt^2$ for some number k. For $D = 144$, $t = 3$, so $k = \dfrac{144}{3^2} = 16$. Then, for $t = 5$, $D = 16(5)^2 = 400$ ft.

89. Time, t, and speed, s, are inversely proportional means that $ts = k$ for some number k. When $t = 3$, $s = 40$, so $k = 3(40) = 120$. For $s = 50$,

$$t = \frac{120}{50} = \frac{12}{5} \text{ hours.}$$

90. Using the formula $V = \dfrac{k}{P}$, and given that $P = 1200$ and $V = 3$, then $k = 3 \cdot 1200 = 3600$. Then, for $V = 2$, $P = \dfrac{3600}{2} = 1800$ lb per ft^2.

91. Resistance, R, is inversely proportional to the square of the diameter d means that $Rd^2 = k$ for some number k. For $R = 0.5$, $d = 0.02$, so $k = 0.5(0.02)^2 = 0.0002$. Then, for $d = 0.04$,

$$R = \frac{0.0002}{(0.04)^2} = 0.125 \text{ ohm.}$$

92. Use $T_1 R_1 = T_2 R_2$, where T is the number of teeth in the gear and R is the number of revolutions per minute of the gear. For $T_1 = 48$, $T_2 = 16$, and $R_1 = 54$, you can solve for R_2:

$$48 \cdot 54 = 16R_2$$

$$\frac{48 \cdot 54}{16} = R_2$$

$$162 = R_2$$

So the small gear is rotating at 162 rpm.

[End of Chapter 8 Review]

Chapter 8 Test
Solutions to All Exercises

1. $\dfrac{2x^2+5x-3}{2x^2-x}=\dfrac{(2x-1)(x+3)}{x(2x-1)}=\dfrac{x+3}{x}$

Restrictions:

$x(2x-1)=0$

Set: $\quad x=0 \qquad 2x-1=0$

Solve: $\qquad\qquad\qquad 2x=1$

$$x=\frac{1}{2}$$

So $x\ne 0$ and $x\ne\dfrac{1}{2}$, as the original expression is undefined for these values.

2. $\dfrac{8x^2-2x-3}{4x^2+x-3}=\dfrac{(2x+1)(4x-3)}{(x+1)(4x-3)}=\dfrac{2x+1}{x+1}$

Restrictions:

$(x+1)(4x-3)=0$

Set: $\quad x+1=0 \qquad 4x-3=0$

Solve: $\quad x=-1 \qquad\ 4x=3$

$$x=\frac{3}{4}$$

So $x\ne -1$ and $x\ne\dfrac{3}{4}$, as the original expression is undefined for these values.

3. $\dfrac{15x^4-18x^3}{36x-30x^2}=\dfrac{3x^3(5x-6)}{-6x(5x-6)}=-\dfrac{x^2}{2}$

Restrictions:

$6x(6-5x)=0$

Set: $\qquad 6x=0 \qquad 6-5x=0$

Solve: $\qquad x=0 \qquad -5x=-6$

$$x=\frac{6}{5}$$

So $x\ne 0$ and $x\ne\dfrac{6}{5}$, as the original expression is undefined for these values.

4. For $x=2$, evaluate $\dfrac{x-6}{6-x}$:

Note that $\dfrac{x-6}{6-x}=\dfrac{x-6}{-(x-6)}=-1$ for all valid values of

x $(x\ne 6)$.

5. For $y=-3$, evaluate $\dfrac{2y^3}{y^2-25}$:

$$\frac{2(-3)^3}{(-3)^2-25}=\frac{-54}{-16}=\frac{27}{8}$$

6. Factor and cancel out all factors that are common to both the numerator and denominator:

$\dfrac{16x^2+24x+9}{3x^2+14x+8}\cdot\dfrac{21x+14}{4x^2+3x}$

$\qquad =\dfrac{(4x+3)^2}{(3x+2)(x+4)}\cdot\dfrac{7(3x+2)}{x(4x+3)}$

$\qquad =\dfrac{7(4x+3)}{x(x+4)}$

Restrictions:

$(3x+2)(x+4)=0 \qquad$ and $\qquad x(4x+3)=0$

Set: $\ 3x+2=0 \quad x+4=0 \quad x=0 \quad 4x+3=0$

Solve: $\quad 3x=-2 \qquad x=-4 \qquad\qquad 4x=-3$

$$x=-\frac{2}{3} \qquad\qquad\qquad\qquad x=-\frac{3}{4}$$

So $x\ne-\dfrac{2}{3}$, $x\ne -4$, $x\ne 0$ and $x\ne-\dfrac{3}{4}$, as the original expression is undefined for these values.

7. Factor and invert the divisor. Cancel out all factors that are common to both the numerator and denominator:

$\dfrac{3x-3}{3x^2-x-2}\div\dfrac{24x^2-6}{6x^2+x-2}$

$\qquad =\dfrac{3(x-1)}{(3x+2)(x-1)}\cdot\dfrac{(3x+2)(2x-1)}{6(2x+1)(2x-1)}$

$\qquad =\dfrac{1}{2(2x+1)}$

Restrictions:

$(3x+2)(x-1)=0 \qquad$ and $\qquad (2x+1)(2x-1)=0$

Set: $\ 3x+2=0 \quad x-1=0 \quad 2x+1=0 \quad 2x-1=0$

Solve: $\quad 3x=-2 \qquad x=-1 \qquad 2x=-1 \qquad 2x=1$

$$x=-\frac{2}{3} \qquad\qquad x=-\frac{1}{2} \qquad x=\frac{1}{2}$$

So $x\ne-\dfrac{2}{3}$, $x\ne 1$, $x\ne\dfrac{1}{2}$ and $x\ne-\dfrac{1}{2}$, as the original expression is undefined for these values.

8. Factor the denominators and determine their LCM. Using the LCM as the common denominator, add:

$$\frac{3}{x^2-1}+\frac{7}{x^2-x-2}=\frac{3}{(x+1)(x-1)}+\frac{7}{(x-2)(x+1)}$$

$$=\frac{3(x-2)+7(x-1)}{(x+1)(x-1)(x-2)}$$

$$=\frac{3x-6+7x-7}{(x+1)(x-1)(x-2)}$$

$$=\frac{10x-13}{(x+1)(x-1)(x-2)}$$

Restrictions:

Set: $\quad x+1=0 \qquad x-1=0 \qquad x-2=0$

Solve: $\quad x=-1 \qquad x=1 \qquad x=2$

So $x \neq -1$, $x \neq 1$ and $x \neq 2$, as the original expression is undefined for these values.

9. Factor the denominators and determine their LCM. Using the LCM as the common denominator, combine:

$$\frac{3x}{x-4}+\frac{7x}{x+4}-\frac{x+3}{x^2-16}$$

$$=\frac{3x}{x-4}+\frac{7x}{x+4}-\frac{x+3}{(x-4)(x+4)}$$

$$=\frac{3x(x+4)+7x(x-4)-(x+3)}{(x-4)(x+4)}$$

$$=\frac{3x^2+12x+7x^2-28x-x-3}{(x-4)(x+4)}$$

$$=\frac{10x^2-16x-3}{(x-4)(x+4)}$$

Restrictions:

$(x+4)(x-4)=0$

Set: $\quad x+4=0 \qquad x-4=0$

Solve: $\quad x=-4 \qquad x=4$

So $x \neq -4$ and $x \neq 4$, as the original expression is undefined for these values.

10. Factor the denominators and determine their LCM. Using the LCM as the common denominator, combine:

$$\frac{1}{x-1}+\frac{2x}{x^2-x-12}-\frac{5}{x^2-5x+4}$$

$$=\frac{1}{x-1}+\frac{2x}{(x+3)(x-4)}-\frac{5}{(x-1)(x-4)}$$

$$=\frac{1(x+3)(x-4)+2x(x-1)-5(x+3)}{(x-1)(x+3)(x-4)}$$

$$=\frac{x^2-x-12+2x^2-2x-5x-15}{(x-1)(x+3)(x-4)}$$

$$=\frac{3x^2-8x-27}{(x-1)(x+3)(x-4)}$$

Restrictions:

Set: $\quad x-1=0 \qquad x+3=0 \qquad x-4=0$

Solve: $\quad x=1 \qquad x=-3 \qquad x=4$

So $x \neq 1$, $x \neq -3$ and $x \neq 4$, as the original expression is undefined for these values.

11. Factor all numerators and denominators, and invert the divisor expression. Divide out factors that are common to both numerator and denominator:

$$\frac{x^2+8x+15}{x^2+6x+5}\cdot\frac{x-3}{2x}\div\frac{x+3}{x^2}$$

$$=\frac{(x+5)(x+3)}{(x+5)(x+1)}\cdot\frac{x-3}{2x}\cdot\frac{x^2}{x+3}$$

$$=\frac{x(x-3)}{2(x+1)}$$

Set the factors of the denominator of the original expression equal to 0, and solve, to determine any restrictions on the variable.

So $x \neq -5$, $x \neq -1$, $x \neq 0$ and $x \neq -3$, as the original expression is undefined for these values.

12. Evaluate the expressions in the numerator and denominator first. Then, invert the denominator and multiply:

$$\frac{\dfrac{1}{x}-\dfrac{1}{x^2}}{1-\dfrac{1}{x^2}}=\frac{\dfrac{x-1}{x^2}}{\dfrac{x^2-1}{x^2}}=\frac{x-1}{x^2}\cdot\frac{x^2}{x^2-1}=\frac{x-1}{(x+1)(x-1)}$$

$$=\frac{1}{x+1}$$

13. Invert the fraction within the denominator and set up as a multiplication. Factor the trinomial and divide out common factors:

$$\frac{\dfrac{6}{x-5}}{\dfrac{x+18}{x^2-2x-15}} = \frac{6}{x-5} \cdot \frac{(x-5)(x+3)}{x+18} = \frac{6(x+3)}{x+18}$$

14.
$$\frac{9}{3-x} = \frac{8}{2x+1}$$
$$9(2x+1) = 8(3-x)$$
$$18x+9 = 24-8x$$
$$26x = 15$$
$$x = \frac{15}{26}$$

15.
$$\frac{3}{4} = \frac{2x+5}{5-x}$$
$$3(5-x) = 4(2x+5)$$
$$15-3x = 8x+20$$
$$-5 = 11x$$
$$\frac{-5}{11} = x$$

16. Multiply both sides of the equation by the LCM of all of the denominators:
$$LCM = (x+1)(x-1)$$
$$\frac{2x}{x-1} - \frac{3}{x^2-1} = \frac{2x+5}{x+1}$$
$$\left(x^2-1\right)\left(\frac{2x}{x-1} - \frac{3}{x^2-1}\right) = \left(x^2-1\right)\left(\frac{2x+5}{x+1}\right)$$
$$(x+1)(2x)-3 = (x-1)(2x+5)$$
$$2x^2+2x-3 = 2x^2+3x-5$$
$$2 = x$$
$$x = 2$$

17. **a.** $LCM = 2x(x+1)$

$$\frac{3}{x} - \frac{2}{x+1} + \frac{5}{2x} = \frac{3(2)(x+1)-2(2x)+5(x+1)}{2x(x+1)}$$
$$= \frac{6x+6-4x+5x+5}{2x(x+1)}$$
$$= \frac{7x+11}{2x(x+1)}$$

b. Solve the equation:
$$\frac{3}{x} - \frac{2}{x+1} = \frac{5}{2x}$$

$$2x(x+1)\left(\frac{3}{x} - \frac{2}{x+1}\right) = 2x(x+1)\left(\frac{5}{2x}\right)$$
$$6(x+1)-4x = 5(x+1)$$
$$6x+6-4x = 5x+5$$
$$-3x = -1$$
$$x = \frac{1}{3}$$

18. Let x be the smaller of the two numbers. Then $156 - x$ is the larger number. The numbers are in a ratio of 5 to 7. Solve the equation:
$$\frac{5}{7} = \frac{x}{156-x}$$
$$5(156-x) = 7x$$
$$780-5x = 7x$$
$$780 = 12x$$
$$x = 65$$
$$156-x = 91$$
The two numbers are 65 and 91.

19. Let r be Kim's speed (or rate).

	Rate	Time	Distance
Lisa	$r+15$	$\dfrac{228}{r+15}$	228
Kim	r	$\dfrac{168}{r}$	168

Given that their times are the same, solve the equation:
$$\frac{228}{r+15} = \frac{168}{r}$$
$$228r = 168(r+15)$$
$$228r = 168r+2520$$
$$60r = 2520$$
$$r = 42$$
$$r+15 = 57$$
Kim's speed is 42 mph. Lisa's is 57 mph.

20. Let x be the number of hours for the apprentice to finish the work alone.

	Number of hrs to finish	Part finished in one hour
Plumber	2	$\dfrac{1}{2}$
Apprentice	x	$\dfrac{1}{x}$
Together	$1\dfrac{1}{3}$	$\dfrac{1}{1\frac{1}{3}} = \dfrac{3}{4}$

Given that the sum of the part each can do alone in one hour is equal to the part they can do together in one hour, solve the equation:

$$\frac{1}{2} + \frac{1}{x} = \frac{3}{4}$$

$$4x\left(\frac{1}{2} + \frac{1}{x}\right) = 4x\left(\frac{3}{4}\right)$$

$$2x + 4 = 3x$$

$$x = 4$$

The apprentice can do the job in 4 hours working alone.

21. Pressure, P, varies as depth, D, so there is some number k, such that $P = kD$, or $k = \dfrac{P}{D}$. At a depth of $D = 10$ ft, it is known that $P = 625$ lb.

So $k = \dfrac{625}{10} = 62.5$. Then, for $D = 25$ ft,

$$P = 62.5(25) = 1562.5 \ .$$

The pressure is 1562.5 lb at a depth of 25 ft.

22. Using $P = kAv^2$, and the given values,

$$16,000 = k(180)(120)^2$$

$$k = \frac{16,000}{180(120)^2}$$

$$\approx 0.006172840$$

Now, use this value of k to find P when $A = 180$ and $v = 150$:

$$P \approx 0.006172840(180)(150)^2$$

$$P \approx 25,000$$

The lift pressure $P \approx 25,000$ lb for the given values of A and v.

23. From $R = \dfrac{kL}{d^2}$, you can solve for k.

$$k = \frac{Rd^2}{L}$$

So with $R = 10$, $d = 0.01$, $L = 250$ ft $= 3000$ in.,

$$k = \frac{10(0.01)^2}{3000} = 3.\overline{3} \times 10^{-7}$$

Find R when $L = 300$ feet and $d = 0.02$ inches. First, change 300 ft to 3600 in., so the linear units will be the same.

$$R = \frac{3.\overline{3} \times 10^{-7} \cdot 3600}{(0.02)^2} = 3$$

The resistance is $R = 3$ ohms.

24. Let x be the time in hours for the 2nd pipe to fill the pool working alone.

	Number of hours to fill the pool	Part of pool filled in one hour
1st Pipe	3	$\dfrac{1}{3}$
2nd Pipe	x	$\dfrac{1}{x}$
Together	2	$\dfrac{1}{2}$

Given that the sum of the part filled in one hour for each pipe working alone is equal to the part of the pool that both can fill together, solve the equation:

$$\frac{1}{3} + \frac{1}{x} = \frac{1}{2}$$

$$6x\left(\frac{1}{3} + \frac{1}{x}\right) = 6x\left(\frac{1}{2}\right)$$

$$2x + 6 = 3x$$

$$x = 6$$

The second pipe can fill the pool in 6 hours.

[End of Chapter 8 Test]

Cumulative Review: Chapters 1 – 8
Solutions to All Exercises

1. **a.** $\dfrac{5}{6} + \dfrac{3}{4} = \dfrac{10+9}{12} = \dfrac{19}{12}$

   b. $\dfrac{1}{2} - \dfrac{5}{7} = \dfrac{7-10}{14} = \dfrac{-3}{14}$

   c. $\dfrac{1}{2} \cdot \dfrac{1}{2} = \dfrac{1}{4}$

   d. $\dfrac{7}{32} \div \dfrac{21}{8} = \dfrac{7}{32} \cdot \dfrac{8}{21} = \dfrac{1}{12}$

2. 12.5% of $40 = 0.125(40) = 5$

3. $(3x+7)(x-4) = 3x^2 - 5x - 28$

4. $(2x-5)^2 = 4x^2 - 20x + 25$

5. $-4(x+2) + 5(2x-3) = -4x - 8 + 10x - 15 = 6x - 23$

6. $x(x+7) - (x-1)(x-4) = x^2 + 7x - \left(x^2 - 3x - 4\right)$
   $= 10x + 4$

7. $A = P + Prt$. Solve for t:
   $A - P = Prt$
   $t = \dfrac{A - P}{Pr}$

8. Using $I = Prt$, solve for t. Substitute the values given:
   $P = \$600$, $r = 0.05$, and $I = 615 - 600 = 15$.
   $I = Prt$
   $\dfrac{I}{Pr} = t$
   $\dfrac{15}{600 \cdot 0.05} = t$
   $\dfrac{1}{2} = t$,   or $t = 6$ months

   It will take 6 months, or $\dfrac{1}{2}$ of a year, for the investment to be worth $\$615$.

9. $4x^2 + 16x + 15 = (2x+5)(2x+3)$

10. $2x^2 + 6x - 20 = 2\left(x^2 + 3x - 10\right) = 2(x+5)(x-2)$

11. **a.** $D = \{-1, 0, 5, 6\}$

   b. $R = \{0, 2, 5\}$

   c. Not a function because the domain value of -1 is assigned to two different values in the range.

12. $f(x) = x^3 - 2x^2 - 1$

   a. $f(3) = 3^3 - 2 \cdot 3^2 - 1$
      $= 27 - 18 - 1 = 8$

   b. $f(0) = 0^3 - 2 \cdot 0^2 - 1$
      $= -1$

   c. $f\left(-\dfrac{2}{3}\right) = \left(-\dfrac{2}{3}\right)^3 - 2\left(-\dfrac{2}{3}\right)^2 - 1$
      $= -\dfrac{8}{27} - \dfrac{8}{9} - 1$
      $= -\dfrac{8}{27} - \dfrac{24}{27} - \dfrac{27}{27}$
      $= -\dfrac{59}{27}$

13. Using the point-slope form of a line:
   $y - 4 = \dfrac{3}{4}\left(x - (-2)\right)$
   $y - 4 = \dfrac{3}{4}(x + 2)$
   $y = \dfrac{3}{4}x + \dfrac{11}{2}$,   or $3x - 4y = 22$

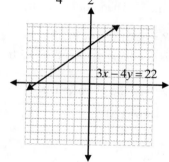

14. Use the formula for slope and the given points:

$$m = \frac{-1-2}{-5-4} = \frac{-3}{-9} = \frac{1}{3}$$

Use the slope with any one of the given points to write the equation in point slope form:

$$y - 2 = \frac{1}{3}(x - 4)$$

$$y - 2 = \frac{1}{3}x - \frac{4}{3}$$

$$y = \frac{1}{3}x + \frac{2}{3}$$

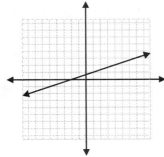

15. A line with undefined slope is a vertical line. Since the given point has an x-coordinate of -3, the equation is $x = -3$.

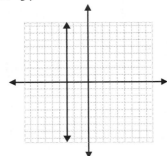

16. The slope of the given line is $\frac{1}{2}$, so the slope of the requested line must be -2. Using the point-slope form of a line and the given point, write an equation for the requested line:

$$y - 0 = -2(x - (-4))$$

$$y = -2x - 8, \quad \text{or} \quad 2x + y = -8$$

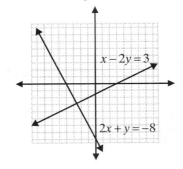

17. The graph represents a function.

18. The graph represents a function.

19. Use a calculator to graph the solution:

$$3x + y > 4$$

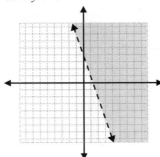

20. Use a calculator to graph the solution:

$$\begin{cases} x + y < 6 \\ x - 2y \le 4 \end{cases}$$

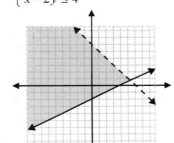

21. Find the solution of the system of equations by using a graphing calculator:

$$\begin{cases} x + y = 5.2 \\ 2x - y = -1.5 \end{cases}$$

The solution is $\left(\frac{37}{30}, \frac{119}{30} \right)$.

22. Find the solution of the system of equations using a graphing calculator:

$$\begin{cases} -x + 2y = 6 \\ 3x + 2y = -4 \end{cases}$$

The solution is $\left(\frac{-5}{2}, \frac{7}{4} \right)$.

23. Solve the system:
$$\begin{cases} 2x - y = 10 \\ x + 2y = 5 \end{cases}$$

Multiply the first equation by 2 and use addition:
$$4x - 2y = 20$$
$$\underline{x + 2y = 5}$$
$$5x = 25$$
$$x = 5$$
$$2(5) - y = 10$$
$$y = 0$$

The solution is $(5,0)$. The system is consistent.

24. Solve the system:
$$\begin{cases} y = 3x + 2 \\ -6x + 2y = 4 \end{cases}$$

Solve by substitution:
$$-6x + 2(3x + 2) = 4$$
$$-6x + 6x + 4 = 4$$
$$4 = 4$$

This result is always true. The system is dependent.
The solution is $(x, 3x + 2)$.

25. Solve the system:
$$\begin{cases} y = \dfrac{1}{2}x + 3 \\ 6x - 2y = 3 \end{cases}$$

Solve by substitution:
$$6x - 2\left(\frac{1}{2}x + 3\right) = 3$$
$$6x - x - 6 = 3$$
$$5x = 9$$
$$x = \frac{9}{5}$$
$$y = \frac{1}{2}\left(\frac{9}{5}\right) + 3$$
$$y = \frac{9}{10} + 3 = \frac{39}{10}$$

The solution is $\left(\dfrac{9}{5}, \dfrac{39}{10}\right)$. The system is consistent.

26. Solve the system:
$$\begin{cases} x + 2y = 5 \\ 2x + 4y = 9 \end{cases}$$

Multiply the first equation by -2 and use the addition method to eliminate y.
$$-2x - 4y = -10$$
$$\underline{2x + 4y = 9}$$
$$0 = -1$$

This statement is never true, so the system is inconsistent. The lines are parallel.

27. Use a graphing calculator to graph the two given functions:
$$f(x) = \frac{2}{5}x + 1$$
$$g(x) = -2x^2 + 6$$

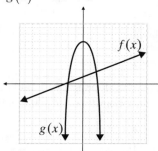

The points of intersection of the two graphs are approximately $(-1.684, 0.326)$ and $(1.484, 1.594)$.

28. $\dfrac{32x^3 y^2}{4xy^2} = \dfrac{4 \cdot 8 \cdot x^2 xy^2}{4xy^2} = 8x^2$

29. $\dfrac{15xy^{-1}}{3x^{-2}y^{-3}} = \dfrac{15}{3} \cdot x^{1-(-2)} y^{-1-(-3)} = 5x^3 y^2$

30. $\dfrac{3x^3 y + 10x^2 y^2 + 5xy^3}{5x^2 y}$

$$= \frac{3x^3 y}{5x^2 y} + \frac{10x^2 y^2}{5x^2 y} + \frac{5xy^3}{5x^2 y}$$
$$= \frac{3x}{5} + 2y + \frac{y^2}{x}$$

31. $\dfrac{7x^2 y^2 - 21x^2 y^3 + 8x^2 y^4}{7x^2 y^2}$

$$= \frac{7x^2 y^2}{7x^2 y^2} - \frac{21x^2 y^3}{7x^2 y^2} + \frac{8x^2 y^4}{7x^2 y^2}$$
$$= 1 - 3y + \frac{8y^2}{7}$$

32. $x - 15 + \dfrac{-1}{x+1}$

$$
\begin{array}{r}
x - 15 \\
x+1\overline{)x^2 - 14x - 16} \\
\end{array}
$$

$-\left(x^2 + x\right)$

$-15x - 16$

$-\left(-15x - 15\right)$

-1

33. $2x^2 - x + 3 + \dfrac{-2}{x+3}$

$$
\begin{array}{r}
2x^2 - x + 3 \\
x+3\overline{)2x^3 + 5x^2 + 0x + 7} \\
\end{array}
$$

$-\left(2x^3 + 6x^2\right)$

$-x^2 + 0x$

$-\left(-x^2 - 3x\right)$

$3x + 7$

$-\left(3x + 9\right)$

-2

34. $\dfrac{8x^3 + 4x^2}{2x^2 - 5x - 3} = \dfrac{4x^2\left(2x+1\right)}{\left(2x+1\right)\left(x-3\right)} = \dfrac{4x^2}{x-3}$

Restrictions: $2x + 1 \neq 0$ and $x - 3 \neq 0$

$x \neq -\dfrac{1}{2} \qquad x \neq 3$

35. $\dfrac{4x^2 + 5x + 9}{x - 2}$ does not simplify.

Restrictions: $x - 2 \neq 0$

$x \neq 2$

36. $\dfrac{x^2 - 9}{x+3} = \dfrac{\left(x-3\right)\left(x+3\right)}{\left(x+3\right)} = x - 3$

Restrictions: $x + 3 \neq 0$

$x \neq -3$

37. $\dfrac{x}{x^2 + x} = \dfrac{x}{x\left(x+1\right)} = \dfrac{1}{x+1}$

Restrictions: $\quad x \neq 0 \quad$ and $\quad x + 1 \neq 0$

$x \neq -1$

38. $\dfrac{4 - x}{3x - 12} = \dfrac{-\left(x-4\right)}{3\left(x-4\right)} = -\dfrac{1}{3}$

Restrictions: $x - 4 \neq 0$

$x \neq 4$

39. $\dfrac{2x + 6}{3x + 9} = \dfrac{2\left(x+3\right)}{3\left(x+3\right)} = \dfrac{2}{3}$

Restrictions: $x + 3 \neq 0$

$x \neq -3$

40. $\dfrac{x^2 + 3x}{x^2 + 7x + 12} = \dfrac{x\left(x+3\right)}{\left(x+3\right)\left(x+4\right)} = \dfrac{x}{x+4}$

Restrictions: $x + 3 \neq 0 \quad$ and $\quad x + 4 \neq 0$

$x \neq -3 \qquad\qquad x \neq -4$

41. $\dfrac{x^2 + 2x - 15}{2x^2 - 12x + 18} = \dfrac{\left(x-3\right)\left(x+5\right)}{2\left(x^2 - 6x + 9\right)}$

$\phantom{\dfrac{x^2+2x-15}{2x^2-12x+18}} = \dfrac{\left(x-3\right)\left(x+5\right)}{2\left(x-3\right)\left(x-3\right)}$

$\phantom{\dfrac{x^2+2x-15}{2x^2-12x+18}} = \dfrac{x+5}{2\left(x-3\right)}$

Restrictions: $x - 3 \neq 0$

$x \neq 3$

42. $\dfrac{x^2}{x+y} - \dfrac{y^2}{x+y} = \dfrac{x^2 - y^2}{x+y}$

$\phantom{\dfrac{x^2}{x+y}-\dfrac{y^2}{x+y}} = \dfrac{\left(x+y\right)\left(x-y\right)}{x+y}$

$\phantom{\dfrac{x^2}{x+y}-\dfrac{y^2}{x+y}} = x - y$

43. $\dfrac{7}{x} + \dfrac{9}{x-4} = \dfrac{7\left(x-4\right) + 9x}{x\left(x-4\right)}$

$\phantom{\dfrac{7}{x}+\dfrac{9}{x-4}} = \dfrac{7x - 28 + 9x}{x\left(x-4\right)}$

$\phantom{\dfrac{7}{x}+\dfrac{9}{x-4}} = \dfrac{16x - 28}{x\left(x-4\right)}$

$\phantom{\dfrac{7}{x}+\dfrac{9}{x-4}} = \dfrac{4\left(4x - 7\right)}{x\left(x-4\right)}$

44. $\dfrac{4}{y+2} - \dfrac{4}{y+3} = \dfrac{4\left(y+3\right) - 4\left(y+2\right)}{\left(y+2\right)\left(y+3\right)}$

$\phantom{\dfrac{4}{y+2}-\dfrac{4}{y+3}} = \dfrac{4y + 12 - 4y - 8}{\left(y+2\right)\left(y+3\right)}$

$\phantom{\dfrac{4}{y+2}-\dfrac{4}{y+3}} = \dfrac{4}{\left(y+2\right)\left(y+3\right)}$

45. $\dfrac{4x}{3x+3} - \dfrac{x}{x+1} = \dfrac{4x-3x}{3(x+1)}$

$$= \dfrac{x}{3(x+1)} \quad \text{or} \quad \dfrac{x}{3x+3}$$

46. $\dfrac{4x}{x-4} \div \dfrac{12x^2}{x^2-16} = \dfrac{4x}{x-4} \cdot \dfrac{x^2-16}{12x^2}$

$$= \dfrac{4x(x+4)(x-4)}{12x^2(x-4)}$$

$$= \dfrac{x+4}{3x}$$

47. $\dfrac{x^2+3x+2}{x+3} \div \dfrac{x+1}{3x^2+6x} = \dfrac{x^2+3x+2}{x+3} \cdot \dfrac{3x^2+6x}{x+1}$

$$= \dfrac{(x+2)(x+1) \cdot 3x(x+2)}{(x+3)(x+1)}$$

$$= \dfrac{3x(x+2)^2}{x+3}$$

48. $\dfrac{2x+1}{x^2+5x-6} \cdot \dfrac{x^2+6x}{x} = \dfrac{(2x+1) \cdot x(x+6)}{(x+6)(x-1)x}$

$$= \dfrac{2x+1}{x-1}$$

49. $\dfrac{8}{x^2+x-6} + \dfrac{2x}{x^2-3x+2}$

$$= \dfrac{8}{(x+3)(x-2)} + \dfrac{2x}{(x-2)(x-1)}$$

$$= \dfrac{8(x-1)+2x(x+3)}{(x+3)(x-2)(x-1)}$$

$$= \dfrac{2x^2+14x-8}{(x+3)(x-2)(x-1)}$$

50. $\dfrac{x}{x^2+3x-4} + \dfrac{x+1}{x^2-1} = \dfrac{x}{(x+4)(x-1)} + \dfrac{x+1}{(x+1)(x-1)}$

$$= \dfrac{x}{(x+4)(x-1)} + \dfrac{1}{(x-1)}$$

$$= \dfrac{x+(x+4)}{(x+4)(x-1)}$$

$$= \dfrac{2x+4}{(x+4)(x-1)}$$

$$\text{or} \quad \dfrac{2(x+2)}{(x+4)(x-1)}$$

51. $\dfrac{x+1}{x^2+4x+4} \div \dfrac{x^2-x-2}{x^2-2x-8}$

$$= \dfrac{x+1}{(x+2)^2} \cdot \dfrac{(x-4)(x+2)}{(x-2)(x+1)}$$

$$= \dfrac{x-4}{(x+2)(x-2)} \quad \text{or} \quad \dfrac{x-4}{x^2-4}$$

52. $\dfrac{\dfrac{3}{x} + \dfrac{1}{6x}}{\dfrac{7}{3x}} = \dfrac{\dfrac{18+1}{6x}}{\dfrac{7}{3x}} \cdot \dfrac{6x}{6x} = \dfrac{19}{14}$

53. $\dfrac{\dfrac{1}{x} - \dfrac{1}{x^2}}{\dfrac{1}{x} + \dfrac{1}{x^2}} = \left(\dfrac{\dfrac{1}{x} - \dfrac{1}{x^2}}{\dfrac{1}{x} + \dfrac{1}{x^2}} \right) \cdot \dfrac{x^2}{x^2} = \dfrac{x-1}{x+1}$

54. $\dfrac{\dfrac{4}{3x} + \dfrac{1}{6x}}{\dfrac{1}{x^2} - \dfrac{1}{2x}} = \dfrac{\dfrac{8+1}{6x}}{\dfrac{2-x}{2x^2}} \cdot \dfrac{6x^2}{6x^2} = \dfrac{9x}{3(2-x)} = \dfrac{3x}{2-x}$

55. $4(x+2)-7 = -2(3x+1)-3$

$$4x+8-7 = -6x-2-3$$

$$4x+1 = -6x-5$$

$$10x = -6$$

$$x = -\dfrac{3}{5}$$

56. $(x+4)(x-5) = 10$

$$x^2-x-20 = 10$$

$$x^2-x-30 = 0$$

$$(x-6)(x+5) = 0$$

Set: $\quad x-6=0 \qquad x+5=0$

Solve: $\quad x=6 \qquad x=-5$

57. $0 = x^2-7x+10$

$$0 = (x-5)(x-2)$$

Set: $\quad x-5=0 \qquad x-2=0$

Solve: $\quad x=5 \qquad x=2$

58. $x^2+8x-4 = 16$

$$x^2+8x-20 = 0$$

$$(x+10)(x-2) = 0$$

Set: $\quad x+10=0 \qquad x-2=0$

Solve: $\quad x=-10 \qquad x=2$

59.
$$\frac{7}{2x-1}=\frac{3}{x+6}$$
$$7(x+6)=3(2x-1)$$
$$7x+42=6x-3$$
$$x=-45$$

60. a.
$$\frac{3}{x}-\frac{5}{x+3}=\frac{3(x+3)-5x}{x(x+3)}$$
$$=\frac{3x+9-5x}{x^2+3x}$$
$$=\frac{-2x+9}{x^2+3x}\quad\text{or}\quad\frac{9-2x}{x(x+3)}$$

b.
$$\frac{3}{x}=\frac{5}{x+3}$$
$$x(x+3)\left(\frac{3}{x}\right)=x(x+3)\left(\frac{5}{x+3}\right)$$
$$3x+9=5x$$
$$-2x=-9$$
$$x=\frac{9}{2}$$

61. a.
$$\frac{3x}{x+2}+\frac{2}{x-4}+3$$
$$=\frac{3x(x-4)+2(x+2)+3(x+2)(x-4)}{(x+2)(x-4)}$$
$$=\frac{3x^2-12x+2x+4+3x^2-6x-24}{(x+2)(x-4)}$$
$$=\frac{6x^2-16x-20}{(x+2)(x-4)}\quad\text{or}\quad\frac{6x^2-16x-20}{x^2-2x-8}$$

b.
$$\frac{3x}{x+2}+\frac{2}{x-4}=3$$
$$3x(x-4)+2(x+2)=3(x+2)(x-4)$$
$$3x^2-12x+2x+4=3x^2-6x-24$$
$$-10x+4=-6x-24$$
$$-4x=-28$$
$$x=7$$

62. Let r be the speed of the train

	Rate	Time	Distance
Airplane	$3r+50$	$\dfrac{1035}{3r+50}$	1035
Train	r	$\dfrac{270}{r}$	270

Using the fact that the times are equal, solve the following equation for r:
$$\frac{1035}{3r+50}=\frac{270}{r}$$
$$1035r=270(3r+50)$$
$$1035r=810r+13,500$$
$$225r=13,500$$
The train's speed is 60 mph.

The airplane's speed is $3(60)+50=230$ mph.

63. Let x be the amount of time, in hours, it takes the father to wax the car. Then, $3x$ is the amount of time it takes the daughter alone.

	Time	Part of job done in 1 hr
Father	x	$\dfrac{1}{x}$
Daughter	$3x$	$\dfrac{1}{3x}$
Together	2	$\dfrac{1}{2}$

The sum of the parts completed by the individuals working alone is equal to the amount completed by the two working together. Solve for x:
$$\frac{1}{x}+\frac{1}{3x}=\frac{1}{2}$$
$$6+2=3x$$
$$3x=8$$
$$x=\frac{8}{3}$$

The father can wax the car in $\dfrac{8}{3}=2\dfrac{2}{3}$ hrs, and the

daughter can wax the car in $3\cdot\dfrac{8}{3}=8$ hrs.

64. Illumination, L, varies inversely as the square of the distance, D, from the source implies that for some number, k, $L = k\left(\dfrac{1}{D^2}\right)$. Find the constant of variation k from the given information: $L = 10$ foot candles when $D = 20$ feet.

$$k = LD^2$$
$$k = 10(20)^2 = 4000$$

Now, find L for $D = 40$ ft:

$$L = 4000\frac{1}{40^2}$$
$$L = 4000\frac{1}{1600} = 2.5 \text{ foot candles}$$

65. Let x be the smaller of the two numbers. Then, $84 - x$ is the larger number.

$$\frac{2}{5} = \frac{x}{84 - x}$$
$$2(84 - x) = 5x$$
$$168 - 2x = 5x$$
$$7x = 168$$
$$x = 24 \quad \text{and} \quad 84 - x = 60$$

The two numbers are 24 and 60.

66. Let t be the time it takes to go downstream. Then, $8 - t$ is the time to go upstream.

	Rate	Time	Distance
Upstream	$\dfrac{18}{8-t}$	$8 - t$	18
Downstream	$\dfrac{18}{t}$	t	18

Let r be the rate of the current. Then, $2r$ is the rate of the boat in still water. Base the system on the rate going upstream and downstream:

Upstream: $\qquad 2r - r = \dfrac{18}{8 - t}$

Downstream: $\qquad 2r + r = \dfrac{18}{t}$

Simplifying yields the system:

$$\begin{cases} r = \dfrac{18}{8 - t} \\ 3r = \dfrac{18}{t} \end{cases}$$

Solve by substitution:

$$3\left(\frac{18}{8 - t}\right) = \frac{18}{t}$$

$$\frac{3}{8 - t} = \frac{1}{t}$$
$$3t = 8 - t$$
$$4t = 8$$
$$t = 2$$
$$r = \frac{18}{8 - (2)} = \frac{18}{6} = 3$$

Going downstream took 2 hours and the rate of the current is 3 mph.

67. The stretch length, L, varies directly as the weight W implies that for some number k, $L = kW$. Find the value of k by solving when $L = 5$ and $W = 13$.

$$k = \frac{L}{W} = \frac{5}{13}$$

Using this value of k, find L when $W = 20$.

$$L = kW = \frac{5}{13}(20) = \frac{100}{13} = 7\frac{9}{13} \text{ cm.}$$

[End of Cumulative Review: Chapters 1 – 8]

Section 9.1
Solutions to Odd Exercises

1. $\sqrt{9} = 3$

3. $\sqrt[3]{125} = 5$

5. $\sqrt{49} = 7$

7. $\sqrt{169} = 13$

9. $\sqrt{81} = 9$

11. $\sqrt[3]{216} = 6$

13. $\sqrt[3]{343} = 7$

15. a. $\sqrt{32} = 5.65685424...$
    b. $-\sqrt{18} = -4.24264068...$
    c. $-\sqrt{\pi} = -1.77245385...$
    d. $\sqrt[3]{20} = 2.71441761...$
    e. $\sqrt[4]{60} = 2.78315768...$

17. $\sqrt{4} = 2$ is rational.

19. $\sqrt{169} = 13$ is rational.

21. $\sqrt[3]{8} = 2$ is rational.

23. $\sqrt{-36}$ is non-real.

25. $\sqrt{1.69} = 1.3$ is rational

27. $-\sqrt[3]{125} = -5$ is rational.

29. $0.1010010001...$ is irrational.

31. $1.\overline{93471}$ is rational.

33. $3.1415926535...$ (non-repeating) is irrational.

35. $3 + 2\sqrt{6} \approx 3 + 2(2.449490) \approx 7.8990$

37. $4 - 3\sqrt{5} \approx 4 - 3(2.236067) \approx -2.7082$

39. $\dfrac{1 + \sqrt{10}}{2} \approx \dfrac{1 + 3.162278}{2} \approx 2.0811$

41. $\dfrac{-10 + \sqrt[3]{16}}{3} \approx \dfrac{-10 + 2.519842}{3} \approx -2.4934$

43. $\dfrac{5 + \sqrt{90}}{6} \approx \dfrac{5 + 9.486833}{6} \approx 2.4145$

45. $\left(\sqrt{25}\right)^2 = 25$

47. $\left(-\sqrt{6}\right)^2 = 6$

49. $\sqrt[3]{27} = 3$

51. $\sqrt[3]{8} = 2$

53. $\sqrt[4]{16} = 2$

55. $\sqrt[3]{1331} = 11$

57. $\sqrt[3]{\dfrac{27}{8}} = \sqrt[3]{\left(\dfrac{3}{2}\right)^3} = \dfrac{3}{2}$

59. $\sqrt{65} \approx 8.0623$

61. $\sqrt{37.9} \approx 6.1563$

63. $\sqrt[3]{39.4} \approx 3.4028$

65. $\sqrt[4]{75} \approx 2.9428$

67. $\sqrt[5]{10,000} \approx 6.3096$

69. $\sqrt[3]{216} = 6$

71. $\sqrt[4]{256} = 4$

73. $\sqrt[3]{512} = 8$

75. $\sqrt{175} \approx 13.2288$

77. $\sqrt[3]{130} \approx 5.0658$

79. The square root of a negative number is not a real number, because the square of any real number is positive. The same is true for any even root.

[End of Section 9.1]

Section 9.2
Solutions to Odd Exercises

For Exercises 1 – 62, assume that all variables are greater than or equal to 0.

1. $\sqrt{36x^2} = \sqrt{6^2 x^2} = 6x$

3. $\sqrt{\dfrac{1}{4}} = \dfrac{1}{2}$

5. $\sqrt{8x^3} = \sqrt{2^2 x^2 \cdot 2 \cdot x} = 2x\sqrt{2x}$

7. $-\sqrt{56} = -\sqrt{2^2 \cdot 14} = -2\sqrt{14}$

9. $\sqrt{\dfrac{5x^4}{9}} = \sqrt{\dfrac{5x^4}{3^2}} = \dfrac{\sqrt{5}x^2}{3}$

11. $\sqrt{12} = \sqrt{2^2 \cdot 3} = 2\sqrt{3}$

13. $\sqrt{288} = \sqrt{144 \cdot 2} = \sqrt{12^2 \cdot 2} = 12\sqrt{2}$

15. $-\sqrt{72} = -\sqrt{6^2 \cdot 2} = -6\sqrt{2}$

17. $-\sqrt{125} = -\sqrt{5^2 \cdot 5} = -5\sqrt{5}$

19. $-\sqrt{\dfrac{11}{64}} = -\sqrt{\left(\dfrac{1}{8}\right)^2 \dfrac{11}{1}} = -\dfrac{1}{8}\sqrt{11}$ or $-\dfrac{\sqrt{11}}{8}$

21. $\sqrt{\dfrac{28}{25}} = \sqrt{\dfrac{2^2 \cdot 7}{5^2}} = \dfrac{2}{5}\sqrt{7}$ or $\dfrac{2\sqrt{7}}{5}$

23. $\sqrt{\dfrac{32a^5}{81b^{16}}} = \sqrt{\dfrac{4^2 \cdot 2 \cdot \left(a^2\right)^2 a}{9^2 \cdot \left(b^8\right)^2}} = \dfrac{4a^2\sqrt{2a}}{9b^8}$

25. $\sqrt{\dfrac{200x^8}{289}} = \sqrt{\dfrac{2^3 \cdot 5^2 \left(x^4\right)^2}{17^2}} = \dfrac{10x^4}{17}\sqrt{2}$

27. $\sqrt[3]{-1} = \sqrt[3]{(-1)^3} = -1$

29. $\sqrt[3]{1} = \sqrt[3]{1^3} = 1$

31. $\sqrt[3]{56} = \sqrt[3]{2^3 \cdot 7} = 2\sqrt[3]{7}$

33. $\sqrt[3]{-250} = \sqrt[3]{(-5)^3 \cdot 2} = -5\sqrt[3]{2}$

35. $\sqrt[3]{64x^9 y^2} = \sqrt[3]{4^3 \cdot \left(x^3\right)^3 \cdot y^2} = 4x^3 \sqrt[3]{y^2}$

37. $\dfrac{\sqrt[3]{81}}{6} = \dfrac{\sqrt[3]{3^3 \cdot 3}}{3 \cdot 2} = \dfrac{3\sqrt[3]{3}}{3 \cdot 2} = \dfrac{\sqrt[3]{3}}{2}$

39. $\sqrt[3]{\dfrac{375}{8}} = \sqrt[3]{\dfrac{5^3 \cdot 3}{2^3}} = \dfrac{5\sqrt[3]{3}}{2}$

41. $\sqrt[3]{125x^4} = 5x\sqrt[3]{x}$

43. $\sqrt[3]{\dfrac{125y^{12}}{27x^6}} = \sqrt[3]{\dfrac{5^3 \left(y^4\right)^3}{3^3 \left(x^2\right)^3}} = \dfrac{5y^4}{3x^2}$

45. $\dfrac{4 + 2\sqrt{6}}{2} = \dfrac{2\left(2 + \sqrt{6}\right)}{2} = 2 + \sqrt{6}$

47. $\dfrac{6 - 2\sqrt{3}}{8} = \dfrac{2\left(3 - \sqrt{3}\right)}{2 \cdot 4} = \dfrac{3 - \sqrt{3}}{4}$

49. $\dfrac{7 - \sqrt{98}}{14} = \dfrac{7 - \sqrt{49 \cdot 2}}{14} = \dfrac{7 - 7\sqrt{2}}{14}$

$= \dfrac{7\left(1 - \sqrt{2}\right)}{7 \cdot 2} = \dfrac{1 - \sqrt{2}}{2}$

51. $\dfrac{4 + \sqrt{288}}{20} = \dfrac{4 + \sqrt{144 \cdot 2}}{20} = \dfrac{4 + 12\sqrt{2}}{20}$

$= \dfrac{4\left(1 + 3\sqrt{2}\right)}{4 \cdot 5} = \dfrac{1 + 3\sqrt{2}}{5}$

53. $\dfrac{12 - \sqrt{18}}{21} = \dfrac{12 - \sqrt{9 \cdot 2}}{21} = \dfrac{12 - 3\sqrt{2}}{21}$

$= \dfrac{3\left(4 - \sqrt{2}\right)}{3 \cdot 7} = \dfrac{4 - \sqrt{2}}{7}$

55. $\dfrac{14 + \sqrt{147}}{35} = \dfrac{14 + \sqrt{49 \cdot 3}}{35} = \dfrac{14 + 7\sqrt{3}}{35}$

$= \dfrac{7\left(2 + \sqrt{3}\right)}{7 \cdot 5} = \dfrac{2 + \sqrt{3}}{5}$

57. $\dfrac{15-\sqrt{75a^2}}{5} = \dfrac{15-\sqrt{25a^2 \cdot 3}}{5} = \dfrac{15-5a\sqrt{3}}{5}$

$\qquad\qquad = \dfrac{5\left(3-a\sqrt{3}\right)}{5} = 3 - a\sqrt{3}$

59. $\dfrac{10+\sqrt{24a^3}}{8} = \dfrac{10+\sqrt{4a^2 \cdot 6a}}{8} = \dfrac{10+2a\sqrt{6a}}{8}$

$\qquad\qquad = \dfrac{2\left(5+a\sqrt{6a}\right)}{2 \cdot 4} = \dfrac{5+a\sqrt{6a}}{4}$

61. $\dfrac{16-\sqrt{32x^4}}{20} = \dfrac{16-\sqrt{16x^4 \cdot 2}}{20} = \dfrac{16-4x^2\sqrt{2}}{20}$

$\qquad\qquad = \dfrac{4\left(4-x^2\sqrt{2}\right)}{4 \cdot 5} = \dfrac{4-x^2\sqrt{2}}{5}$

[End of Section 9.2]

Section 9.3
Solutions to Odd Exercises

1. $3\sqrt{2} + 5\sqrt{2} = (3+5)\sqrt{2} = 8\sqrt{2}$

3. $6\sqrt{5} + \sqrt{5} = (6+1)\sqrt{5} = 7\sqrt{5}$

5. $8\sqrt{10} - 11\sqrt{10} = (8-11)\sqrt{10}$
$= -3\sqrt{10}$

7. $4\sqrt[3]{3} + 9\sqrt[3]{3} = (4+9)\sqrt[3]{3} = 13\sqrt[3]{3}$

9. $6\sqrt{11} - 5\sqrt{11} - 2\sqrt{11} = (6-5-2)\sqrt{11} = -\sqrt{11}$

11. $\sqrt{a} + 4\sqrt{a} - 2\sqrt{a} = (1+4-2)\sqrt{a} = 3\sqrt{a}$

13. $5\sqrt{x} + 3\sqrt{x} - \sqrt{x} = (5+3-1)\sqrt{x} = 7\sqrt{x}$

15. $3\sqrt{2} + 5\sqrt{3} - 2\sqrt{3} + \sqrt{2} = 3\sqrt{2} + \sqrt{2} + 5\sqrt{3} - 2\sqrt{3}$
$= (3+1)\sqrt{2} + (5-2)\sqrt{3}$
$= 4\sqrt{2} + 3\sqrt{3}$

17. $2\sqrt{a} + 7\sqrt{b} - 6\sqrt{a} + \sqrt{b}$
$= 2\sqrt{a} - 6\sqrt{a} + 7\sqrt{b} + \sqrt{b}$
$= (2-6)\sqrt{a} + (7+1)\sqrt{b}$
$= -4\sqrt{a} + 8\sqrt{b}$

19. $6\sqrt[3]{x} - 4\sqrt[3]{y} + 7\sqrt[3]{x} + 2\sqrt[3]{y}$
$= 6\sqrt[3]{x} + 7\sqrt[3]{x} - 4\sqrt[3]{y} + 2\sqrt[3]{y}$
$= (6+7)\sqrt[3]{x} - (4-2)\sqrt[3]{y}$
$= 13\sqrt[3]{x} - 2\sqrt[3]{y}$

21. $\sqrt{12} + \sqrt{27} = \sqrt{4\cdot3} + \sqrt{9\cdot3}$
$= 2\sqrt{3} + 3\sqrt{3}$
$= (2+3)\sqrt{3}$
$= 5\sqrt{3}$

23. $3\sqrt{5} - \sqrt{45} = 3\sqrt{5} - \sqrt{9\cdot5}$
$= 3\sqrt{5} - 3\sqrt{5}$
$= (3-3)\sqrt{5}$
$= 0$

25. $3\sqrt[3]{54} + 8\sqrt[3]{2} = 3\sqrt[3]{27\cdot2} + 8\sqrt[3]{2}$
$= 9\sqrt[3]{2} + 8\sqrt[3]{2}$
$= 17\sqrt[3]{2}$

27. $\sqrt{50} - \sqrt{18} - 3\sqrt{12} = \sqrt{25\cdot2} - \sqrt{9\cdot2} - 3\sqrt{4\cdot3}$
$= 5\sqrt{2} - 3\sqrt{2} - 6\sqrt{3}$
$= 2\sqrt{2} - 6\sqrt{3}$

29. $2\sqrt{20} - \sqrt{45} + \sqrt{36} = 2\sqrt{4\cdot5} - \sqrt{9\cdot5} + 6$
$= 4\sqrt{5} - 3\sqrt{5} + 6$
$= (4-3)\sqrt{5} + 6$
$= \sqrt{5} + 6$

31. $\sqrt{8} - 2\sqrt{3} + \sqrt{27} - \sqrt{72}$
$= \sqrt{4\cdot2} - 2\sqrt{3} + \sqrt{9\cdot3} - \sqrt{36\cdot2}$
$= 2\sqrt{2} - 2\sqrt{3} + 3\sqrt{3} - 6\sqrt{2}$
$= -4\sqrt{2} + \sqrt{3}$

33. $5\sqrt[3]{16} - 4\sqrt[3]{24} + \sqrt[3]{-250}$
$= 5\sqrt[3]{8\cdot2} - 4\sqrt[3]{8\cdot3} + \sqrt[3]{-125\cdot2}$
$= 10\sqrt[3]{2} - 8\sqrt[3]{3} - 5\sqrt[3]{2}$
$= 5\sqrt[3]{2} - 8\sqrt[3]{3}$

35. $6\sqrt{2x} - \sqrt{8x} = 6\sqrt{2x} - 2\sqrt{2x} = 4\sqrt{2x}$

37. $5y\sqrt{2y} - y\sqrt{18y} = 5y\sqrt{2y} - y\sqrt{9\cdot2y}$
$= 5y\sqrt{2y} - 3y\sqrt{2y}$
$= 2y\sqrt{2y}$

39. $4x\sqrt{3xy} - x\sqrt{12xy} - 2x\sqrt{27xy}$
$= 4x\sqrt{3xy} - 2x\sqrt{3xy} - 6x\sqrt{3xy}$
$= -4x\sqrt{3xy}$

41. $\sqrt{36x^3} + \sqrt{81x^3} = 6x\sqrt{x} + 9x\sqrt{x} = 15x\sqrt{x}$

43. $\sqrt{16x^3y^4} - \sqrt{25x^3y^4} = 4xy^2\sqrt{x} - 5xy^2\sqrt{x}$
$= -xy^2\sqrt{x}$

45. $\sqrt{12x^{10}y^{20}} + \sqrt{27x^{10}y^{20}} - \sqrt{3x^{10}y^{20}}$
$= 2x^5y^{10}\sqrt{3} + 3x^5y^{10}\sqrt{3} - x^5y^{10}\sqrt{3}$
$= 4x^5y^{10}\sqrt{3}$

47. $\sqrt[3]{-27x^{24}y^6} + \sqrt[3]{-125x^{24}y^6} = -3x^8y^2 - 5x^8y^2$
$$= -8x^8y^2$$

49. $\sqrt[3]{-16x^9y^{12}} - \sqrt[3]{16x^{12}y^9} + \sqrt[3]{54x^3y^6}$
$$= -2x^3y^4\sqrt[3]{2} - 2x^4y^3\sqrt[3]{2} + 3xy^2\sqrt[3]{2}$$
$$= xy^2\sqrt[3]{2}\left(-2x^2y^2 - 2x^3y + 3\right)$$

51. $\sqrt{2}\left(3 - 4\sqrt{2}\right) = 3\sqrt{2} - 4 \cdot 2 = 3\sqrt{2} - 8$

53. $3\sqrt{18} \cdot \sqrt{2} = 3 \cdot 3\sqrt{2} \cdot \sqrt{2} = 9 \cdot 2 = 18$

55. $-2\sqrt{6} \cdot \sqrt{8} = -2\sqrt{6 \cdot 8} = -2\sqrt{16 \cdot 3} = -8\sqrt{3}$

57. $\sqrt{3}\left(\sqrt{2} + 2\sqrt{12}\right) = \sqrt{6} + 2\sqrt{36} = \sqrt{6} + 12$

59. $\sqrt{y}\left(\sqrt{x} + 2\sqrt{y}\right) = \sqrt{xy} + 2y$

61. $\left(5 + \sqrt{2}\right)\left(3 - \sqrt{2}\right) = 15 - 5\sqrt{2} + 3\sqrt{2} - 2$
$$= 13 - 2\sqrt{2}$$

63. $\left(4\sqrt{3} + \sqrt{2}\right)\left(\sqrt{3} - 2\sqrt{2}\right) = 12 - 8\sqrt{6} + \sqrt{6} - 4$
$$= 8 - 7\sqrt{6}$$

65. $\left(\sqrt{x} + 3\right)\left(\sqrt{x} - 3\right) = x - 9$

67. $\left(\sqrt{2} + \sqrt{7}\right)\left(\sqrt{2} - \sqrt{7}\right) = 2 - 7 = -5$

69. $\left(\sqrt{x} + 5\right)\left(\sqrt{x} - 3\right) = x - 3\sqrt{x} + 5\sqrt{x} - 15$
$$= x + 2\sqrt{x} - 15$$

[End of Section 9.3]

Section 9.4
Solutions to Odd Exercises

1. $\dfrac{5}{\sqrt{2}} = \dfrac{5}{\sqrt{2}} \cdot \dfrac{\sqrt{2}}{\sqrt{2}} = \dfrac{5\sqrt{2}}{2}$

3. $\dfrac{-3}{\sqrt{7}} = \dfrac{-3}{\sqrt{7}} \cdot \dfrac{\sqrt{7}}{\sqrt{7}} = \dfrac{-3\sqrt{7}}{7}$

5. $\dfrac{6}{\sqrt{3}} = \dfrac{6}{\sqrt{3}} \cdot \dfrac{\sqrt{3}}{\sqrt{3}} = \dfrac{6\sqrt{3}}{3} = 2\sqrt{3}$

7. $\dfrac{\sqrt{18}}{\sqrt{2}} = \sqrt{\dfrac{18}{2}} = \sqrt{9} = 3$

9. $\dfrac{\sqrt{27x}}{\sqrt{3x}} = \sqrt{\dfrac{27x}{3x}} = \sqrt{9} = 3$

11. $\dfrac{\sqrt{ab}}{\sqrt{9ab}} = \sqrt{\dfrac{ab}{9ab}} = \sqrt{\dfrac{1}{9}} = \dfrac{1}{3}$

13. $\dfrac{\sqrt{4}}{\sqrt{3}} = \sqrt{\dfrac{4}{3} \cdot \dfrac{3}{3}} = \dfrac{2\sqrt{3}}{3}$

15. $\sqrt{\dfrac{9}{2}} = \sqrt{\dfrac{9}{2} \cdot \dfrac{2}{2}} = \dfrac{3\sqrt{2}}{2}$

17. $\sqrt{\dfrac{1}{x}} = \sqrt{\dfrac{1}{x} \cdot \dfrac{x}{x}} = \dfrac{\sqrt{x}}{x}$

19. $\sqrt{\dfrac{2x}{y}} = \sqrt{\dfrac{2x}{y} \cdot \dfrac{y}{y}} = \dfrac{\sqrt{2xy}}{y}$

21. $\dfrac{2}{\sqrt{2y}} = \dfrac{2}{\sqrt{2y}} \cdot \dfrac{\sqrt{2y}}{\sqrt{2y}} = \dfrac{2\sqrt{2y}}{2y} = \dfrac{\sqrt{2y}}{y}$

23. $\dfrac{21}{5\sqrt{7}} = \dfrac{21}{5\sqrt{7}} \cdot \dfrac{\sqrt{7}}{\sqrt{7}} = \dfrac{21\sqrt{7}}{5 \cdot 7} = \dfrac{3\sqrt{7}}{5}$

25. $\dfrac{-2y}{5\sqrt{2y}} = \dfrac{-2y}{5\sqrt{2y}} \cdot \dfrac{\sqrt{2y}}{\sqrt{2y}} = \dfrac{-2y\sqrt{2y}}{5 \cdot 2y} = \dfrac{-\sqrt{2y}}{5}$

27. $\dfrac{2}{\sqrt{6}-2} = \dfrac{2}{\sqrt{6}-2} \cdot \dfrac{\sqrt{6}+2}{\sqrt{6}+2} = \dfrac{2\left(\sqrt{6}+2\right)}{6-4}$

$\qquad = \dfrac{2\left(\sqrt{6}+2\right)}{2} = \sqrt{6}+2$

29. $\dfrac{1}{\sqrt{5}-3} = \dfrac{1}{\sqrt{5}-3} \cdot \dfrac{\sqrt{5}+3}{\sqrt{5}+3} = \dfrac{\sqrt{5}+3}{5-9}$

$\qquad = \dfrac{\sqrt{5}+3}{-4} = -\dfrac{\sqrt{5}+3}{4} \text{ or } -\dfrac{3+\sqrt{5}}{4}$

31. $\dfrac{-6}{5-3\sqrt{2}} = \dfrac{-6}{5-3\sqrt{2}} \cdot \dfrac{5+3\sqrt{2}}{5+3\sqrt{2}}$

$\qquad = \dfrac{-6\left(5+3\sqrt{2}\right)}{25-18} = \dfrac{-6\left(5+3\sqrt{2}\right)}{7}$

33. $\dfrac{-\sqrt{3}}{\sqrt{2}+5} = \dfrac{-\sqrt{3}}{\sqrt{2}+5} \cdot \dfrac{\sqrt{2}-5}{\sqrt{2}-5} = \dfrac{-\sqrt{6}+5\sqrt{3}}{2-25}$

$\qquad = \dfrac{-\sqrt{6}+5\sqrt{3}}{-23}$

$\qquad = \dfrac{\sqrt{6}-5\sqrt{3}}{23} \text{ or } \dfrac{\sqrt{3}\left(\sqrt{2}-5\right)}{23}$

35. $\dfrac{7}{1-3\sqrt{5}} = \dfrac{7}{1-3\sqrt{5}} \cdot \dfrac{1+3\sqrt{5}}{1+3\sqrt{5}}$

$\qquad = \dfrac{7\left(1+3\sqrt{5}\right)}{1-45}$

$\qquad = -\dfrac{7+21\sqrt{5}}{44} \text{ or } -\dfrac{7\left(1+3\sqrt{5}\right)}{44}$

37. $\dfrac{1}{\sqrt{3}-\sqrt{5}} = \dfrac{1}{\sqrt{3}-\sqrt{5}} \cdot \dfrac{\sqrt{3}+\sqrt{5}}{\sqrt{3}+\sqrt{5}}$

$\qquad = \dfrac{\sqrt{3}+\sqrt{5}}{3-5} = -\dfrac{\sqrt{3}+\sqrt{5}}{2}$

39. $\dfrac{-5}{\sqrt{2}+\sqrt{3}} = \dfrac{-5}{\sqrt{2}+\sqrt{3}} \cdot \dfrac{\sqrt{2}-\sqrt{3}}{\sqrt{2}-\sqrt{3}}$

$\qquad = \dfrac{-5\left(\sqrt{2}-\sqrt{3}\right)}{2-3} = \dfrac{-5\left(\sqrt{2}-\sqrt{3}\right)}{-1}$

$\qquad = 5\left(\sqrt{2}-\sqrt{3}\right)$

41. $\dfrac{4}{\sqrt{x}+1} = \dfrac{4}{\sqrt{x}+1} \cdot \dfrac{\sqrt{x}-1}{\sqrt{x}-1}$

$\qquad = \dfrac{4\left(\sqrt{x}-1\right)}{x-1}$

43. $\dfrac{5}{6+\sqrt{y}} = \dfrac{5}{6+\sqrt{y}} \cdot \dfrac{6-\sqrt{y}}{6-\sqrt{y}}$

$\qquad = \dfrac{5\left(6-\sqrt{y}\right)}{36-y}$

45. $\dfrac{8}{2\sqrt{x}+3} = \dfrac{8}{2\sqrt{x}+3} \cdot \dfrac{2\sqrt{x}-3}{2\sqrt{x}-3}$

$\qquad = \dfrac{8\left(2\sqrt{x}-3\right)}{4x-9}$

47. $\dfrac{\sqrt{4y}}{\sqrt{5y}-\sqrt{3}} = \dfrac{2\sqrt{y}}{\sqrt{5y}-\sqrt{3}} \cdot \dfrac{\sqrt{5y}+\sqrt{3}}{\sqrt{5y}+\sqrt{3}}$

$\qquad = \dfrac{2\sqrt{y}\left(\sqrt{5y}+\sqrt{3}\right)}{5y-3}$

$\qquad = \dfrac{2y\sqrt{5}+2\sqrt{3y}}{5y-3}$

49. $\dfrac{3}{\sqrt{x}-\sqrt{y}} = \dfrac{3}{\sqrt{x}-\sqrt{y}} \cdot \dfrac{\sqrt{x}+\sqrt{y}}{\sqrt{x}+\sqrt{y}}$

$\qquad = \dfrac{3\left(\sqrt{x}+\sqrt{y}\right)}{x-y}$

51. $\dfrac{x}{\sqrt{x}+2\sqrt{y}} = \dfrac{x}{\sqrt{x}+2\sqrt{y}} \cdot \dfrac{\sqrt{x}-2\sqrt{y}}{\sqrt{x}-2\sqrt{y}}$

$\qquad = \dfrac{x\left(\sqrt{x}-2\sqrt{y}\right)}{x-4y}$

53. $\dfrac{\sqrt{3}+1}{\sqrt{3}-2} = \dfrac{\sqrt{3}+1}{\sqrt{3}-2} \cdot \dfrac{\sqrt{3}+2}{\sqrt{3}+2}$

$\qquad = \dfrac{3+3\sqrt{3}+2}{3-4} = \dfrac{5+3\sqrt{3}}{-1}$

$\qquad = -5-3\sqrt{3}$

55. $\dfrac{\sqrt{5}-2}{\sqrt{5}+3} = \dfrac{\sqrt{5}-2}{\sqrt{5}+3} \cdot \dfrac{\sqrt{5}-3}{\sqrt{5}-3}$

$\qquad = \dfrac{5-2\sqrt{5}-3\sqrt{5}+6}{5-9}$

$\qquad = \dfrac{11-5\sqrt{5}}{-4} = \dfrac{-11+5\sqrt{5}}{4}$

57. $\dfrac{\sqrt{x}+1}{\sqrt{x}-1} = \dfrac{\sqrt{x}+1}{\sqrt{x}-1} \cdot \dfrac{\sqrt{x}+1}{\sqrt{x}+1}$

$\qquad = \dfrac{\left(\sqrt{x}+1\right)^2}{x-1}$ or $\dfrac{x+2\sqrt{x}+1}{x-1}$

59. $\dfrac{\sqrt{x}+2}{\sqrt{3x}+y} = \dfrac{\sqrt{x}+2}{\sqrt{3x}+y} \cdot \dfrac{\sqrt{3x}-y}{\sqrt{3x}-y}$

$\qquad = \dfrac{\left(\sqrt{x}+2\right)\left(\sqrt{3x}-y\right)}{3x-y^2}$

or $= \dfrac{\sqrt{3x^2}-y\sqrt{x}+2\sqrt{3x}-2y}{3x-y^2}$

[End of Section 9.4]

Section 9.5
Solutions to Odd Exercises

Solve the equations. Remember that extraneous
solutions might be introduced when solving, so check all
results in the original form of the equation.

1. $\sqrt{x+3} = 6$
Square both sides and solve.
$$x+3 = 36$$
$$x = 33$$
Check:
$$\sqrt{33+3} = \sqrt{36} = 6$$
So, $x = 33$ is a solution.

3. $\sqrt{2x-1} = 3$
Square both sides and solve.
$$2x-1 = 9$$
$$2x = 10$$
$$x = 5$$
Check:
$$\sqrt{2(5)-1} = \sqrt{9} = 3$$
So, $x = 5$ is a solution.

5. $\sqrt{3x+4} = -5$
Square both sides and solve.
$$3x+4 = 25$$
$$3x = 21$$
$$x = 7$$
Check:
$$\sqrt{3(7)+4} = \sqrt{25} = 5 \neq -5$$
So, there is no real number solution.

7. $\sqrt{5x+4} = 7$
Square both sides and solve.
$$5x+4 = 49$$
$$5x = 45$$
$$x = 9$$
Check:
$$\sqrt{5(9)+4} = \sqrt{49} = 7$$
So, $x = 9$ is a solution.

9. $\sqrt{6-x} = 3$
Square both sides and solve.
$$6-x = 9$$
$$-x = 3$$
$$x = -3$$

Check:
$$\sqrt{6-(-3)} = \sqrt{9} = 3$$
So, $x = -3$ is a solution.

11. $\sqrt{x-4} + 6 = 2$
First, add -6 to both sides.
Then, square both sides and solve.
$$\sqrt{x-4} = -4$$
$$x-4 = 16$$
$$x = 20$$
Check:
$$\sqrt{20-4} + 6 = \sqrt{16} + 6 = 10 \neq 2$$
So, there is no real number solution.

13. $\sqrt{4x+1} - 2 = 3$
Add 2 to both sides.
Then, square both sides and solve.
$$\sqrt{4x+1} = 5$$
$$4x+1 = 25$$
$$4x = 24$$
$$x = 6$$
Check:
$$\sqrt{4(6)+1} - 2 = \sqrt{25} - 2 = 5 - 2 = 3$$
So, $x = 6$ is a solution.

15. $\sqrt{2x+11} = 3$
Square both sides and solve.
$$2x+11 = 9$$
$$2x = -2$$
$$x = -1$$
Check:
$$\sqrt{2(-1)+11} = \sqrt{9} = 3$$
So, $x = -1$ is a solution.

17. $\sqrt{5-2x} = 7$
Square both sides and solve.
$$5-2x = 49$$
$$-2x = 44$$
$$x = -22$$
Check:
$$\sqrt{5-2(-22)} = \sqrt{49} = 7$$
So, $x = -22$ is a solution.

19. $\sqrt{5x-4} = 14$
Square both sides and solve.
$$5x - 4 = 196$$
$$5x = 200$$
$$x = 40$$
Check:
$$\sqrt{5(40)-4} = \sqrt{196} = 14$$
So, $x = 40$ is a solution.

21. $\sqrt{2x-1} = \sqrt{x+1}$
Square both sides and solve.
$$2x - 1 = x + 1$$
$$x = 2$$
Check:
$$\sqrt{2(2)-1} = \sqrt{3} \text{ and } \sqrt{2+1} = \sqrt{3}$$
So, $x = 2$ is a solution.

23. $\sqrt{x+2} = \sqrt{2x-5}$
Square both sides and solve.
$$x + 2 = 2x - 5$$
$$-x = -7$$
$$x = 7$$
Check:
$$\sqrt{7+2} = \sqrt{9} = 3 \text{ and } \sqrt{2(7)-5} = \sqrt{9} = 3$$
So, $x = 7$ is a solution.

25. $\sqrt{4x-3} = \sqrt{2x+5}$
Square both sides and solve.
$$4x - 3 = 2x + 5$$
$$2x = 8$$
$$x = 4$$
Check:
$$\sqrt{4(4)-3} = \sqrt{13} \text{ and } \sqrt{2(4)+5} = \sqrt{13}$$
So, $x = 4$ is a solution.

27. $\sqrt{x+6} = x + 4$
Square both sides and solve.
$$x + 6 = x^2 + 8x + 16$$
$$x^2 + 7x + 10 = 0$$
$$(x+5)(x+2) = 0$$
Set each factor equal to 0 and solve.
$$x = -5 \text{ and } x = -2$$
Check: For $x = -2$,
$$\sqrt{-2+6} = \sqrt{4} = 2 \text{ and } -2+4 = 2$$
So, $x = -2$ is a solution.
However, $x = -5$ is not a solution, since
$$\sqrt{-5+6} = \sqrt{1} = 1 \text{ and } -5+4 = -1.$$

29. $\sqrt{x-2} = x - 2$
Square both sides and solve.
$$x - 2 = x^2 - 4x + 4$$
$$x^2 - 5x + 6 = 0$$
$$(x-2)(x-3) = 0$$
Set each factor equal to 0 and solve.
$$x = 2 \text{ and } x = 3$$
Check: For $x = 2$,
$$\sqrt{2-2} = 0 \text{ and } 2-2 = 0$$
So, $x = 2$ is a solution.
For $x = 3$,
$$\sqrt{3-2} = \sqrt{1} = 1 \text{ and } 3-2 = 1$$
So, $x = 3$ is a solution.

31. $x + 1 = \sqrt{x+7}$
Square both sides and solve.
$$x^2 + 2x + 1 = x + 7$$
$$x^2 + x - 6 = 0$$
$$(x+3)(x-2) = 0$$
Set each factor equal to 0 and solve.
$$x = -3 \text{ and } x = 2$$
Check: For $x = 2$,
$$2+1 = 3 \text{ and } \sqrt{2+7} = \sqrt{9} = 3$$
So, $x = 2$ is a solution.
However, $x = -3$ is not a solution, since
$$-3+1 = -2 \text{ and } \sqrt{-3+7} = \sqrt{4} = 2.$$

33. $\sqrt{4x+1} = x - 5$
Square both sides and solve.
$$4x + 1 = x^2 - 10x + 25$$
$$x^2 - 14x + 24 = 0$$
$$(x-12)(x-2) = 0$$
Set each factor equal to 0 and solve.
$$x = 12 \text{ and } x = 2.$$
Check: For $x = 12$,
$$\sqrt{4(12)+1} = \sqrt{49} = 7 \text{ and } 12-5 = 7$$
So, $x = 12$ is a solution.
However, $x = 2$ is not a solution, since
$$\sqrt{4(2)+1} = \sqrt{9} = 3 \text{ and } 2-5 = -3.$$

35. $\sqrt{2-x} = x + 4$
Square both sides and solve.
$$2 - x = x^2 + 8x + 16$$
$$x^2 + 9x + 14 = 0$$
$$(x+2)(x+7) = 0$$

Set each factor equal to 0 and solve.
$x = -2$ and $x = -7$.
Check: For $x = -2$,
$$\sqrt{2-(-2)} = \sqrt{4} = 2 \text{ and } -2+4 = 2$$
So, $x = -2$ is a solution.
However, $x = -7$ is not a solution, since
$$\sqrt{2-(-7)} = \sqrt{9} = 3 \text{ and } -7+4 = -3.$$

37. $x - 3 = \sqrt{27 - 3x}$
Square both sides and solve.
$$x^2 - 6x + 9 = 27 - 3x$$
$$x^2 - 3x - 18 = 0$$
$$(x-6)(x+3) = 0$$
Set each factor equal to 0 and solve.
$x = 6$ and $x = -3$
Check: For $x = 6$,
$$6 - 3 = 3 \text{ and } \sqrt{27 - 3(6)} = \sqrt{9} = 3$$
So, $x = 6$ is a solution.
However, $x = -3$ is not a solution, since
$$-3 - 3 = -6 \text{ and } \sqrt{27 - 3(-3)} = \sqrt{36} = 6.$$

39. $2x + 3 = \sqrt{3x + 7}$
Square both sides and solve.
$$4x^2 + 12x + 9 = 3x + 7$$
$$4x^2 + 9x + 2 = 0$$
$$(4x+1)(x+2) = 0$$
Set each factor equal to 0 and solve.
$$x = -\frac{1}{4} \text{ and } x = -2$$
Check: For $x = -\frac{1}{4}$,
$$2\left(-\frac{1}{4}\right) + 3 = \frac{5}{2} \text{ and } \sqrt{3\left(-\frac{1}{4}\right) + 7} = \sqrt{\frac{25}{4}} = \frac{5}{2}$$
So, $x = -\frac{1}{4}$ is a solution.
However, $x = -2$ is not a solution, since
$$2(-2) + 3 = -1 \text{ and } \sqrt{3(-2) + 7} = \sqrt{1} = 1.$$

41. $x - 2 = \sqrt{3x - 6}$
Square both sides and solve.
$$x^2 - 4x + 4 = 3x - 6$$
$$x^2 - 7x + 10 = 0$$
$$(x-5)(x-2) = 0$$
Set each factor equal to 0 and solve.
$x = 5$ and $x = 2$

Check: For $x = 5$,
$$5 - 2 = 3 \text{ and } \sqrt{3(5) - 6} = \sqrt{9} = 3$$
So, $x = 5$ is a solution.
For $x = 2$,
$$2 - 2 = 0 \text{ and } \sqrt{3(2) - 6} = \sqrt{0} = 0$$
So, $x = 2$ is a solution.

43. $\sqrt{x + 3} = x + 3$
Square both sides and solve.
$$x + 3 = x^2 + 6x + 9$$
$$x^2 + 5x + 6 = 0$$
$$(x+3)(x+2) = 0$$
Set each factor equal to 0 and solve.
$x = -3$ and $x = -2$
Check: For $x = -3$,
$$\sqrt{-3 + 3} = 0 \text{ and } -3 + 3 = 0$$
So, $x = -3$ is a solution.
For $x = -2$,
$$\sqrt{-2 + 3} = \sqrt{1} = 1 \text{ and } -2 + 3 = 1$$
So, $x = -2$ is a solution.

45. $x + 6 = \sqrt{2x + 12}$
Square both sides and solve.
$$x^2 + 12x + 36 = 2x + 12$$
$$x^2 + 10x + 24 = 0$$
$$(x+6)(x+4) = 0$$
Set each factor equal to 0 and solve.
$x = -6$ and $x = -4$
Check: For $x = -6$,
$$-6 + 6 = 0 \text{ and } \sqrt{2(-6) + 12} = \sqrt{0} = 0$$
So, $x = -6$ is a solution.
For $x = -4$,
$$-4 + 6 = 2 \text{ and } \sqrt{2(-4) + 12} = \sqrt{4} = 2$$
So, $x = -4$ is a solution.

[End of Section 9.5]

Section 9.6
Solutions to Odd Exercises

Write each of the expressions in Exercises 1 – 10 as radical expressions, then simplify.

1. $16^{\frac{1}{2}} = \sqrt{16} = \sqrt{4^2} = 4$

3. $-8^{\frac{1}{3}} = -\sqrt[3]{8} = -\sqrt[3]{2^3} = -2$

5. $81^{\frac{1}{4}} = \sqrt[4]{81} = \sqrt[4]{3^4} = 3$

7. $(-32)^{\frac{1}{5}} = \sqrt[5]{-32} = \sqrt[5]{(-2)^5} = -2$

9. $0.0004^{\frac{1}{2}} = \sqrt{0.0004} = \sqrt{(0.02)^2} = 0.02$

Simplify the expressions in Exercises 11 – 30.

11. $\left(\frac{4}{9}\right)^{\frac{1}{2}} = \sqrt{\frac{4}{9}} = \sqrt{\left(\frac{2}{3}\right)^2} = \frac{2}{3}$

13. $\left(\frac{8}{125}\right)^{\frac{1}{3}} = \sqrt[3]{\left(\frac{2}{5}\right)^3} = \frac{2}{5}$

15. $36^{\frac{-1}{2}} = \frac{1}{36^{\frac{1}{2}}} = \frac{1}{\sqrt{6^2}} = \frac{1}{6}$

17. $8^{\frac{-1}{3}} = \frac{1}{8^{\frac{1}{3}}} = \frac{1}{\sqrt[3]{2^3}} = \frac{1}{2}$

19. $81^{\frac{-1}{4}} = \frac{1}{81^{\frac{1}{4}}} = \frac{1}{\sqrt[4]{3^4}} = \frac{1}{3}$

21. $4^{\frac{3}{2}} = \sqrt{4^3} = \sqrt{64} = 8$

23. $(-27)^{\frac{2}{3}} = \left(\sqrt[3]{-27}\right)^2 = (-3)^2 = 9$

25. $9^{\frac{5}{2}} = \left(\sqrt{9}\right)^5 = 3^5 = 243$

27. $16^{\frac{-3}{4}} = \frac{1}{16^{\frac{3}{4}}} = \frac{1}{\left(\sqrt[4]{16}\right)^3} = \frac{1}{2^3} = \frac{1}{8}$

29. $(-8)^{\frac{-2}{3}} = \frac{1}{(-8)^{\frac{2}{3}}} = \frac{1}{\left(\sqrt[3]{-8}\right)^2} = \frac{1}{(-2)^2} = \frac{1}{4}$

Using the properties of exponents, simplify each expression in Exercises 31 – 60. Assume all variables are positive.

31. $x^{\frac{1}{4}} \cdot x^{\frac{1}{3}} = x^{\frac{1}{4}+\frac{1}{3}} = x^{\frac{3}{12}+\frac{4}{12}} = x^{\frac{7}{12}}$

33. $5^{\frac{1}{2}} \cdot 5^{\frac{-3}{2}} = 5^{\frac{1}{2}+\left(\frac{-3}{2}\right)} = 5^{\frac{-2}{2}} = 5^{-1} = \frac{1}{5}$

35. $x^{\frac{1}{5}} \cdot x^{\frac{2}{5}} = x^{\frac{1}{5}+\frac{2}{5}} = x^{\frac{3}{5}}$

37. $\dfrac{x^{\frac{3}{4}}}{x^{\frac{1}{4}}} = x^{\frac{3}{4}-\frac{1}{4}} = x^{\frac{2}{4}} = x^{\frac{1}{2}}$

39. $\dfrac{x^{\frac{4}{5}}}{x^{\frac{2}{5}}} = x^{\frac{4}{5}-\frac{2}{5}} = x^{\frac{2}{5}}$

41. $\dfrac{2^{\frac{2}{3}}}{2^{\frac{-1}{3}}} = 2^{\frac{2}{3}-\left(\frac{-1}{3}\right)} = 2^{\frac{3}{3}} = 2$

43. $a^{\frac{2}{3}} \cdot a^{\frac{1}{2}} = a^{\frac{2}{3}+\frac{1}{2}} = a^{\frac{4}{6}+\frac{3}{6}} = a^{\frac{7}{6}}$

45. $x^{\frac{3}{4}} \cdot x^{\frac{-1}{8}} = x^{\frac{3}{4}+\left(\frac{-1}{8}\right)} = x^{\frac{6}{8}-\frac{1}{8}} = x^{\frac{5}{8}}$

47. $\dfrac{x^{\frac{2}{5}}}{x^{\frac{1}{2}}} = x^{\frac{2}{5}-\frac{1}{2}} = x^{\frac{4}{10}-\frac{5}{10}} = x^{\frac{-1}{10}}$

49. $\dfrac{6^2}{6^{\frac{1}{2}}} = 6^{2-\frac{1}{2}} = 6^{\frac{3}{2}}$

51. $\dfrac{x^{\frac{3}{4}}}{x^{\frac{-1}{2}}} = x^{\frac{3}{4}-\left(\frac{-1}{2}\right)} = x^{\frac{3}{4}+\frac{1}{2}} = x^{\frac{3}{4}+\frac{2}{4}} = x^{\frac{5}{4}}$

53. $\dfrac{x^{\frac{5}{3}}}{x^{\frac{-1}{3}}} = x^{\frac{5}{3}-\left(\frac{-1}{3}\right)} = x^{\frac{5}{3}+\frac{1}{3}} = x^{\frac{6}{3}} = x^2$

55. $\left(7x^{\frac{1}{3}}\right)^2 \cdot \left(4x^{\frac{1}{2}}\right) = 49x^{\frac{2}{3}} \cdot 4x^{\frac{1}{2}}$

$$= 196x^{\frac{2}{3}+\frac{1}{2}}$$

$$= 196x^{\frac{4}{6}+\frac{3}{6}}$$

$$= 196x^{\frac{7}{6}}$$

57. $\left(9x^2\right)^{\frac{1}{2}} \cdot \left(8x^3\right)^{\frac{1}{3}} = \sqrt{9x^2} \cdot \sqrt[3]{8x^3}$

$$= 3x \cdot 2x$$

$$= 6x^2$$

59. $\left(16x^8\right)^{\frac{1}{4}}\left(9x^2\right)^{\frac{-1}{2}} = \sqrt[4]{16x^8} \cdot \frac{1}{\sqrt{9x^2}}$

$$= \frac{2x^2}{3x}$$

$$= \frac{2x}{3}$$

[End of Section 9.6]

Section 9.7
Solutions to Odd Exercises

For Exercises 1 – 24, find the distance between the two given points.

1. Given points: $(-3,6),(-3,2)$

$\text{Distance} = \sqrt{(-3-(-3))^2 + (6-2)^2}$

$\qquad = \sqrt{0^2 + 4^2} = \sqrt{4^2}$

$\qquad = 4$

3. Given points: $(4,-3),(7,-3)$

$\text{Distance} = \sqrt{(4-7)^2 + (-3-(-3))^2}$

$\qquad = \sqrt{(-3)^2 + 0^2}$

$\qquad = \sqrt{9}$

$\qquad = 3$

5. Given points: $(3,1),\left(\dfrac{-1}{2},1\right)$

$\text{Distance} = \sqrt{\left(3-\left(\dfrac{-1}{2}\right)\right)^2 + (1-1)^2}$

$\qquad = \sqrt{\left(\dfrac{7}{2}\right)^2 + 0^2} = \dfrac{7}{2}$

7. Given points: $(3,1),(2,0)$

$\text{Distance} = \sqrt{(3-2)^2 + (1-0)^2}$

$\qquad = \sqrt{1^2 + 1^2}$

$\qquad = \sqrt{2}$

9. Given points: $(1,5),(-1,2)$

$\text{Distance} = \sqrt{(1-(-1))^2 + (5-2)^2}$

$\qquad = \sqrt{2^2 + 3^2}$

$\qquad = \sqrt{13}$

11. Given points: $(2,-7),(-3,5)$

$\text{Distance} = \sqrt{(2-(-3))^2 + (-7-5)^2}$

$\qquad = \sqrt{5^2 + (-12)^2}$

$\qquad = \sqrt{25+144}$

$\qquad = \sqrt{169}$

$\qquad = 13$

13. Given points: $\left(\dfrac{3}{7},\dfrac{4}{7}\right),(0,0)$

$\text{Distance} = \sqrt{\left(\dfrac{3}{7}-0\right)^2 + \left(\dfrac{4}{7}-0\right)^2}$

$\qquad = \sqrt{\left(\dfrac{3}{7}\right)^2 + \left(\dfrac{4}{7}\right)^2}$

$\qquad = \sqrt{\dfrac{9}{49} + \dfrac{16}{49}}$

$\qquad = \sqrt{\dfrac{25}{49}}$

$\qquad = \dfrac{5}{7}$

15. Given points: $(4,1),(7,5)$

$\text{Distance} = \sqrt{(4-7)^2 + (1-5)^2}$

$\qquad = \sqrt{(-3)^2 + (-4)^2}$

$\qquad = \sqrt{9+16}$

$\qquad = \sqrt{25}$

$\qquad = 5$

17. Given points: $(-10,3),(2,-2)$

$\text{Distance} = \sqrt{(-10-2)^2 + (3-(-2))^2}$

$\qquad = \sqrt{(-12)^2 + 5^2}$

$\qquad = \sqrt{144+25}$

$\qquad = \sqrt{169}$

$\qquad = 13$

19. Given points: $(-3,2),(3,-6)$

$\text{Distance} = \sqrt{(-3-3)^2 + (2-(-6))^2}$

$\qquad = \sqrt{(-6)^2 + 8^2}$

$\qquad = \sqrt{36+64}$

$\qquad = \sqrt{100}$

$\qquad = 10$

21. Given points: $(4,0),(0,-3)$

Distance $= \sqrt{(4-0)^2 + (0-(-3))^2}$

$= \sqrt{4^2 + 3^2}$

$= \sqrt{16+9}$

$= \sqrt{25}$

$= 5$

23. Given points: $\left(\dfrac{4}{5},\dfrac{2}{7}\right),\left(\dfrac{-6}{5},\dfrac{2}{7}\right)$

Distance $= \sqrt{\left(\dfrac{4}{5}-\dfrac{-6}{5}\right)^2 + \left(\dfrac{2}{7}-\dfrac{2}{7}\right)^2}$

$= \sqrt{\left(\dfrac{10}{5}\right)^2 + 0^2}$

$= \sqrt{2^2 + 0}$

$= \sqrt{4}$

$= 2$

25. Find the distances between each pair of the three given points. Then, determine whether the sum of the squares of the lengths of the two shortest segments equals the square of the length of the largest segment.

Given points: $A(1,-2)$, $B(7,1)$, and $C(5,5)$

$|AB| = \sqrt{(1-7)^2 + (-2-1)^2} = \sqrt{36+9} = \sqrt{45}$

$|AC| = \sqrt{(1-5)^2 + (-2-5)^2} = \sqrt{16+49} = \sqrt{65}$

$|BC| = \sqrt{(7-5)^2 + (1-5)^2} = \sqrt{4+16} = \sqrt{20}$

$|AB|^2 + |BC|^2 = \left(\sqrt{45}\right)^2 + \left(\sqrt{20}\right)^2$

$= 45 + 20 = 65$

$|AC|^2 = \left(\sqrt{65}\right)^2 = 65$

This confirms that $|AB|^2 + |BC|^2 = |AC|^2$.

So, the triangle ABC is a right triangle.

27. To show that the triangle is isosceles, demonstrate that two of its legs are of equal length.

Given points: $A(1,1)$, $B(5,9)$, and $C(9,5)$

$|AB| = \sqrt{(1-5)^2 + (1-9)^2}$

$= \sqrt{(-4)^2 + (-8)^2}$

$= \sqrt{16+64}$

$= \sqrt{80}$

$= 4\sqrt{5}$

$|AC| = \sqrt{(1-9)^2 + (1-5)^2}$

$= \sqrt{(-8)^2 + (-4)^2}$

$= \sqrt{64+16}$

$= \sqrt{80}$

$= 4\sqrt{5}$

So, $|AB| = |AC| = 4\sqrt{5}$.

Note: $|BC| = \sqrt{(5-9)^2 + (9-5)^2}$

$= \sqrt{16+16} = 4\sqrt{2}$

29. To show the triangle is equilateral, use the distance formula with the given points and determine that the lengths of the sides are the same.

Given points: $A(1,0)$, $B\left(3,\sqrt{12}\right)$, and $C(5,0)$

$|AB| = \sqrt{(1-3)^2 + \left(0-\sqrt{12}\right)^2}$

$= \sqrt{4+12} = \sqrt{16} = 4$

$|AC| = \sqrt{(1-5)^2 + (0-0)^2}$

$= \sqrt{16+0} = \sqrt{16} = 4$

$|BC| = \sqrt{(3-5)^2 + \left(\sqrt{12}-0\right)^2}$

$= \sqrt{4+12} = \sqrt{16} = 4$

The lengths of the sides are each equal to 4, so the triangle is equilateral.

31. Show that $|BD| = |AC|$. Given points:

$A(2,-2)$, $B(2,3)$, $C(8,3)$ and $D(8,-2)$

$|BD| = \sqrt{(2-8)^2 + (3-(-2))^2}$

$= \sqrt{36+25} = \sqrt{61}$

$|AC| = \sqrt{(2-8)^2 + (-2-3)^2}$

$= \sqrt{36+25} = \sqrt{61}$

The diagonals are each $\sqrt{61}$ units long.

For Exercises 33 – 40, verify that the sum of the squares of the two shortest legs equals the square of the longest.

33. $13^2 = 169$

$5^2 + 12^2 = 25 + 144 = 169$
So, this is a right triangle.

35. $10^2 = 100$

$8^2 + 6^2 = 64 + 36 = 100$
So, this is a right triangle.

37. $\left(\sqrt{5}\right)^2 = 5$

$1^2 + 2^2 = 1 + 4 = 5$
So, this is a right triangle.

39. $4^2 = 16$

$\left(\sqrt{10}\right)^2 + 2^2 = 10 + 4 = 14 \neq 16$

So, this is not a right triangle.

For Exercises 41 – 46, use a calculator to find answers accurate to two decimals places.

41. The diagonal of the square that is inscribed in the circle is the same as a diameter of the circle.
   a. To find the circumference of the a circle with a diameter of 30 feet, use
   $$\pi = 3.14 \text{ and } C = \pi d :$$
   $$C = 3.14(30) = 94.20 \text{ feet.}$$

   To find the area of a circle with a diameter of 30 feet, or a radius of 15 feet, use $A = \pi r^2$:
   $$A = 3.14(15)^2 = 706.50 \text{ square feet.}$$

   b. To find the perimeter of the square in the circle, first find the side lengths of the square. Each side is the leg of a right triangle whose hypotenuse is 30 feet. Using the Pythagorean Theorem, you have that
   $$s^2 + s^2 = 30^2$$
   $$2s^2 = 900$$
   $$s^2 = 450$$
   $$s = 15\sqrt{2}$$

   So, the perimeter is $4s = 4\left(15\sqrt{2}\right) \approx 84.85$ ft.

   To find the area of the square, use $A = s^2$:
   $$s^2 = 450 \text{ sq. ft.}$$

43. A baseball "diamond" is a square that is 90 feet on each side. So, the distance between home plate and second base is the length of the hypotenuse of a right, isosceles triangle whose legs are each 90 feet.

Using the Pythagorean Theorem, you can solve for the length of the hypotenuse:
$$h = \sqrt{90^2 + 90^2}$$
$$h = \sqrt{16200}$$
$$h \approx 127.28$$

   a. No, the pitcher's mound is not halfway between home plate and second base; halfway is actually 63.64 feet.
   b. The pitcher's mound is closer to home plate
   c. No, the diagonals intersect at their midway point, which is 63.64 feet.

45. Use the Pythagorean Theorem and let x be the length of each of the legs of one of the triangles:
   $$12^2 = x^2 + x^2$$
   $$144 = 2x^2$$
   Now, solve for x .
   $$x^2 = 72$$
   $$x = \sqrt{72}$$
   $$x = 6\sqrt{2} \approx 8.5$$
   The lengths of the sides of the triangles are approximately 8.5 cm each.

In Exercises 47 – 49, use the distance formula to calculate the perimeter of each triangle with the given vertices.

47. Given the vertices $A(-5,0), B(3,4),$ and $C(0,0)$:
   $$|AB| = \sqrt{(-5-3)^2 + (0-4)^2} = \sqrt{64+16} = \sqrt{80} = 4\sqrt{5}$$
   $$|AC| = \sqrt{(-5-0)^2 + (0-0)^2} = \sqrt{25+0} = \sqrt{25} = 5$$
   $$|BC| = \sqrt{(-3-0)^2 + (4-0)^2} = \sqrt{9+16} = \sqrt{25} = 5$$
   Perimeter $= |AB| + |AC| + |BC| = 10 + 4\sqrt{5}$

49. Given the vertices $A(-2,5), B(3,1),$ and $C(2,-2)$:
   $$|AB| = \sqrt{(-2-3)^2 + (5-1)^2} = \sqrt{25+16} = \sqrt{41}$$
   $$|AC| = \sqrt{(-2-2)^2 + (5-(-2))^2} = \sqrt{16+49} = \sqrt{65}$$
   $$|BC| = \sqrt{(3-2)^2 + (1-(-2))^2} = \sqrt{1+9} = \sqrt{10}$$
   Perimeter $= |AB| + |AC| + |BC| = \sqrt{41} + \sqrt{65} + \sqrt{10}$

[End of Section 9.7]

Chapter 9 Review
Solutions to All Exercises

1. $\sqrt{196} = 14$

2. $\sqrt{361} = 19$

3. $\sqrt[3]{8} = 2$

4. $\sqrt[3]{729} = 9$

5. $16.\overline{03}$ is rational, since it can be written as the quotient of two integers: $16.\overline{03} = \dfrac{529}{33}$.

6. $\sqrt{-20}$ is non-real. There is no real number that is the square root of a negative number.

7. $\sqrt{3}$ is irrational; it cannot be written as the quotient of two integers or a repeating decimal.

8. $\sqrt{\dfrac{1}{9}}$ is rational, since it can be written as the quotient of two integers: $\sqrt{\dfrac{1}{9}} = \dfrac{1}{3}$.

9. $-\sqrt{196}$ is a rational number; in fact, it is an integer: $-\sqrt{196} = -14$.

10. $0.2020020002\ldots$ is irrational; it cannot be written as the quotient of two integers and is not a repeating decimal.

11. $1 + 3\sqrt{2} \approx 5.2426$

12. $-2 - \sqrt{10} \approx -5.1623$

13. $\dfrac{3 + \sqrt{12}}{4} \approx 1.6160$

14. $\dfrac{-16 + \sqrt{30}}{2} \approx -5.2614$

15. $\sqrt{4} + \sqrt{9} = 2 + 3 = 5$

16. $\sqrt{121} + \sqrt[3]{125} = 11 + 5 = 16$

17. $\sqrt[3]{80.5} \approx 4.3178$

18. $\sqrt[4]{100} \approx 3.1623$

19. $\sqrt[5]{-32} = -2$

20. $\sqrt[4]{\dfrac{625}{81}} = \dfrac{5}{3} \approx 1.6667$

21. $\sqrt{\dfrac{1}{16}} = \dfrac{\sqrt{1}}{\sqrt{16}} = \dfrac{1}{4}$

22. $-\sqrt{225} = -15$

23. $\sqrt{9x^3} = 3x\sqrt{x}$

24. $\sqrt{8a^4} = 2a^2\sqrt{2}$

25. $\sqrt{\dfrac{27x^2}{100}} = \dfrac{\sqrt{27x^2}}{\sqrt{100}} = \dfrac{3x\sqrt{3}}{10}$

26. $\sqrt{\dfrac{75a^3}{9}} = \dfrac{\sqrt{75a^3}}{\sqrt{9}} = \dfrac{5a\sqrt{3a}}{3}$

27. $\sqrt[3]{\dfrac{3000x^6}{343}} = \dfrac{\sqrt[3]{3000x^6}}{\sqrt[3]{343}} = \dfrac{10x^2\sqrt[3]{3}}{7}$

28. $\sqrt[3]{\dfrac{8y^{12}}{27x^{15}}} = \dfrac{\sqrt[3]{8y^{12}}}{\sqrt[3]{27x^{15}}} = \dfrac{2y^4}{3x^5}$

29. $\dfrac{2 + 2\sqrt{3}}{2} = 1 + \sqrt{3}$

30. $\dfrac{5 - \sqrt{125}}{5} = \dfrac{5 - 5\sqrt{5}}{5} = 1 - \sqrt{5}$

31. $\dfrac{10 + \sqrt{200a^2}}{5} = \dfrac{10 + 10a\sqrt{2}}{5} = 2 + 2a\sqrt{2}$

32. $\dfrac{4 - \sqrt{16x^3}}{8} = \dfrac{4 - 4x\sqrt{x}}{8} = \dfrac{1 - x\sqrt{x}}{2}$

33. $10\sqrt{3} - 4\sqrt{3} = 6\sqrt{3}$

34. $8\sqrt{5} + \sqrt{5} = 9\sqrt{5}$

35. $\sqrt{a} + 4\sqrt{a} + 2\sqrt{a} = 7\sqrt{a}$

36. $3\sqrt{x} - 4\sqrt{x} - 2\sqrt{x} = -3\sqrt{x}$

37. $\sqrt{6} + \sqrt{4} - 3\sqrt{6} + 3 = -2\sqrt{6} + 2 + 3 = 5 - 2\sqrt{6}$

38. $\sqrt{a} + 4\sqrt{b} + 3\sqrt{a} - \sqrt{b} = 4\sqrt{a} + 3\sqrt{b}$

39. $\sqrt[3]{24} + \sqrt[3]{54} = 2\sqrt[3]{3} + 3\sqrt[3]{2}$

40. $3\sqrt{7} + 4\sqrt{28} = 3\sqrt{7} + 8\sqrt{7} = 11\sqrt{7}$

41. $\sqrt{25x^3} + \sqrt{49x^3} = 5x\sqrt{x} + 7x\sqrt{x} = 12x\sqrt{x}$

42. $\sqrt{16a^2b} + \sqrt{25a^2b} = 4a\sqrt{b} + 5a\sqrt{b} = 9a\sqrt{b}$

43. $\sqrt{75x^{10}y^{13}} + \sqrt{20x^{10}y^{12}} = 5x^5y^6\sqrt{3y} + 2x^5y^6\sqrt{5}$
$= x^5y^6\left(5\sqrt{3y} + 2\sqrt{5}\right)$

44. $\sqrt[3]{64x^{12}y^{16}} - \sqrt[3]{125x^{12}y^{16}} = 4x^4y^5\sqrt[3]{y} - 5x^4y^5\sqrt[3]{y}$
$= -x^4y^5\sqrt[3]{y}$

45. $\sqrt{3}\left(\sqrt{3} + 4\sqrt{2}\right) = \sqrt{3}\cdot\sqrt{3} + 4\sqrt{3}\cdot\sqrt{2} = 3 + 4\sqrt{6}$

46. $2\sqrt{7}\left(\sqrt{7} - 2\sqrt{3}\right) = 2\sqrt{7}\cdot\sqrt{7} - 4\sqrt{7}\cdot\sqrt{3}$
$= 2\cdot 7 - 4\sqrt{21}$
$= 14 - 4\sqrt{21}$

47. $\sqrt{2}\left(\sqrt{2} + 3\sqrt{6}\right) = \sqrt{2}\cdot\sqrt{2} + 3\sqrt{2}\cdot\sqrt{6}$
$= 2 + 3\sqrt{12}$
$= 2 + 6\sqrt{3}$

48. $\sqrt{x}\left(2\sqrt{y} - 3\sqrt{x}\right) = 2\sqrt{x}\cdot\sqrt{y} - 3\sqrt{x}\cdot\sqrt{x} = 2\sqrt{xy} - 3x$

49. $\left(3 + \sqrt{2}\right)\left(5 - \sqrt{2}\right) = 15 - 3\sqrt{2} + 5\sqrt{2} - 2$
$= 15 + 2\sqrt{2} - 2$
$= 13 + 2\sqrt{2}$

50. $\left(2\sqrt{5} + 3\right)\left(\sqrt{5} + 1\right) = 2\cdot 5 + 2\sqrt{5} + 3\sqrt{5} + 3$
$= 10 + 5\sqrt{5} + 3$
$= 13 + 5\sqrt{5}$

51. $\left(\sqrt{2} + \sqrt{5}\right)\left(\sqrt{2} - \sqrt{5}\right) = 2 - \sqrt{2}\cdot\sqrt{5} + \sqrt{2}\cdot\sqrt{5} - 5$
$= 2 - 5$
$= -3$

52. $\left(\sqrt{3} + \sqrt{8}\right)\left(\sqrt{3} + 5\sqrt{2}\right)$
$= 3 + 5\sqrt{3}\cdot\sqrt{2} + \sqrt{8}\cdot\sqrt{3} + 5\sqrt{8}\cdot\sqrt{2}$
$= 3 + 5\sqrt{6} + \sqrt{24} + 5\sqrt{16}$
$= 3 + 5\sqrt{6} + 2\sqrt{6} + 5\cdot 4$
$= 23 + 7\sqrt{6}$

53. $\dfrac{7}{\sqrt{2}} = \dfrac{7}{\sqrt{2}}\cdot\dfrac{\sqrt{2}}{\sqrt{2}} = \dfrac{7\sqrt{2}}{2}$

54. $\dfrac{-8}{\sqrt{5}} = \dfrac{-8}{\sqrt{5}}\cdot\dfrac{\sqrt{5}}{\sqrt{5}} = \dfrac{-8\sqrt{5}}{5}$

55. $\sqrt{\dfrac{7}{12}} = \dfrac{\sqrt{7}}{\sqrt{12}} = \dfrac{\sqrt{7}}{2\sqrt{3}} = \dfrac{\sqrt{7}}{2\sqrt{3}}\cdot\dfrac{\sqrt{3}}{\sqrt{3}} = \dfrac{\sqrt{21}}{2\cdot 3} = \dfrac{\sqrt{21}}{6}$

56. $\sqrt{\dfrac{1}{y}} = \dfrac{\sqrt{1}}{\sqrt{y}} = \dfrac{1}{\sqrt{y}}\cdot\dfrac{\sqrt{y}}{\sqrt{y}} = \dfrac{\sqrt{y}}{y}$

57. $\dfrac{28}{5\sqrt{7}} = \dfrac{28}{5\sqrt{7}}\cdot\dfrac{\sqrt{7}}{\sqrt{7}} = \dfrac{28\sqrt{7}}{5\cdot 7} = \dfrac{4\sqrt{7}}{5}$

58. $\dfrac{a}{3\sqrt{a}} = \dfrac{a}{3\sqrt{a}}\cdot\dfrac{\sqrt{a}}{\sqrt{a}} = \dfrac{a\sqrt{a}}{3a} = \dfrac{\sqrt{a}}{3}$

59. $\dfrac{-10}{\sqrt{6} + 2} = \dfrac{-10}{\sqrt{6} + 2}\cdot\dfrac{\sqrt{6} - 2}{\sqrt{6} - 2} = \dfrac{-10\sqrt{6} + 20}{6 - 4}$
$= \dfrac{-10\sqrt{6} + 20}{2} = -5\sqrt{6} + 10$

60. $\dfrac{1}{\sqrt{3} - 5} = \dfrac{1}{\sqrt{3} - 5}\cdot\dfrac{\sqrt{3} + 5}{\sqrt{3} + 5} = \dfrac{\sqrt{3} + 5}{3 - 25} = -\dfrac{\sqrt{3} + 5}{22}$

61. $\dfrac{5}{\sqrt{y} - 1} = \dfrac{5}{\sqrt{y} - 1}\cdot\dfrac{\sqrt{y} + 1}{\sqrt{y} + 1} = \dfrac{5\sqrt{y} + 5}{y - 1}$

62. $\dfrac{x}{\sqrt{x} + 3} = \dfrac{x}{\sqrt{x} + 3}\cdot\dfrac{\sqrt{x} - 3}{\sqrt{x} - 3} = \dfrac{x\sqrt{x} - 3x}{x - 9}$

63. $\dfrac{\sqrt{x}-1}{\sqrt{x}+1} = \dfrac{\sqrt{x}-1}{\sqrt{x}+1} \cdot \dfrac{\sqrt{x}-1}{\sqrt{x}-1} = \dfrac{x-2\sqrt{x}+1}{x-1}$

64. $\dfrac{2+\sqrt{3}}{2-\sqrt{3}} = \dfrac{2+\sqrt{3}}{2-\sqrt{3}} \cdot \dfrac{2+\sqrt{3}}{2+\sqrt{3}} = \dfrac{4+4\sqrt{3}+3}{4-3} = 7+4\sqrt{3}$

65. $\sqrt{2x-5} = 9$

$\quad 2x-5 = 9^2 = 81$

$\quad\quad 2x = 86$

$\quad\quad\quad x = 43$

Check: $\quad \sqrt{2\cdot43-5} = \sqrt{86-5} = \sqrt{81} = 9$

66. $\sqrt{3x+4} = 7$

$\quad 3x+4 = 7^2 = 49$

$\quad\quad 3x = 45$

$\quad\quad\quad x = 15$

Check: $\quad \sqrt{3\cdot15+4} = \sqrt{45+4} = \sqrt{49} = 7$

67. $\sqrt{x+7} = -5$ has no real solution.

68. $-11 = \sqrt{5x+1}$ has no real solution.

69. $x+3 = \sqrt{x+9}$

$\quad (x+3)^2 = x+9$

$\quad x^2+6x+9 = x+9$

$\quad\quad x^2+5x = 0$

$\quad\quad x(x+5) = 0$

Set: $\quad\quad x=0 \quad\quad x+5=0$

Solve: $\quad\quad\quad\quad\quad\quad x=-5$

Check: $\quad\quad 0+3=3 \quad$ and $\quad \sqrt{0+9} = \sqrt{9} = 3$

$\quad\quad\quad\quad -5+3=-2 \quad$ and $\quad \sqrt{-5+9} = \sqrt{4} = 2$

So $x=0$ is the only solution.

70. $\sqrt{x-4}-6 = 2$

$\quad \sqrt{x-4} = 8$

$\quad\quad x-4 = 8^2 = 64$

$\quad\quad\quad x = 68$

Check: $\quad \sqrt{68-4}-6 = \sqrt{64}-6 = 8-6 = 2$

71. $x-3 = \sqrt{2x-6}$

$\quad (x-3)^2 = 2x-6$

$\quad x^2-6x+9 = 2x-6$

$\quad\quad x^2-8x+15 = 0$

$\quad\quad (x-5)(x-3) = 0$

Set: $\quad x-5=0 \quad x-3=0$

Solve: $\quad\quad x=5 \quad\quad x=3$

Check:

$\quad\quad 5-3=2 \quad$ and $\quad \sqrt{2\cdot5-6} = \sqrt{10-6} = \sqrt{4} = 2$

$\quad\quad 3-3=0 \quad$ and $\quad \sqrt{2\cdot3-6} = \sqrt{6-6} = \sqrt{0} = 0$

72. $\sqrt{x+5} = x+5$

$\quad x+5 = (x+5)^2$

$\quad x+5 = x^2+10x+25$

$\quad\quad 0 = x^2+9x+20$

$\quad\quad 0 = (x+5)(x+4)$

Set: $\quad x+5=0 \quad x+4=0$

Solve: $\quad x=-5 \quad\quad x=-4$

Check: $\quad\quad -5+5=0 \quad$ and $\quad \sqrt{-5+5} = \sqrt{0} = 0$

$\quad\quad\quad\quad -4+5=1 \quad$ and $\quad \sqrt{-4+5} = \sqrt{1} = 1$

73. $\sqrt{2x-7} = 3-x$

$\quad 2x-7 = (3-x)^2$

$\quad 2x-7 = 9-6x+x^2$

$\quad\quad 0 = x^2-8x+16$

$\quad\quad 0 = (x-4)^2$

$\quad\quad 0 = x-4$

$\quad\quad 4 = x$

Check:

$\quad\quad 3-4=-1 \quad$ and $\quad \sqrt{2\cdot4-7} = \sqrt{8-7} = \sqrt{1} = 1$

There is no real solution.

74.
$$\sqrt{5x-9} = 1-x$$
$$5x-9 = (1-x)^2$$
$$5x-9 = 1-2x-x^2$$
$$x^2+7x-10 = 0$$
$$(x-5)(x-2) = 0$$

Set: $\quad x-5=0 \quad x-2=0$

Solve: $\quad x=5 \qquad x=2$

Check: $\quad 1-5=-4$ and $\sqrt{5\cdot5-9} = \sqrt{16} = 4$

$\qquad\qquad 1-2=-1$ and $\sqrt{5\cdot2-9} = \sqrt{1} = 1$

So there is no real solution.

75.
$$2x+1 = \sqrt{2x+3}$$
$$(2x+1)^2 = 2x+3$$
$$4x^2+4x+1 = 2x+3$$
$$4x^2+2x-2 = 0$$
$$2(2x-1)(x+1) = 0$$

Set: $\quad 2x-1=0 \qquad x+1=0$

Solve: $\qquad x=\dfrac{1}{2} \qquad x=-1$

Check:

$\quad 2\cdot\dfrac{1}{2}+1=2$ and $\sqrt{2\cdot\dfrac{1}{2}+3} = \sqrt{4} = 2$

$\quad 2(-1)+1=-1$ and $\sqrt{2(-1)+3} = \sqrt{1} = 1$

So $x=\dfrac{1}{2}$ is the only solution.

76. $\sqrt{-x-7} = x+7$
$$-x-7 = (x+7)^2$$
$$-x-7 = x^2+14x+49$$
$$0 = x^2+15x+56$$
$$0 = (x+8)(x+7)$$

Set: $\quad x+8=0 \quad x+7=0$

Solve: $\quad x=-8 \qquad x=-7$

Check:

$\quad -8+7=-1$ and $\sqrt{-(-8)-7} = \sqrt{1} = 1$

$\quad -7+7=0$ and $\sqrt{-(-7)-7} = \sqrt{0} = 0$

So $x=-7$ is the only solution.

77. $16^{\frac{3}{4}} = \left(\sqrt[4]{16}\right)^3 = 2^3 = 8$

78. $9^{\frac{3}{2}} = \left(\sqrt{9}\right)^3 = 3^3 = 27$

79. $8^{\frac{-2}{3}} = \left(\sqrt[3]{8}\right)^{-2} = 2^{-2} = \dfrac{1}{2^2} = \dfrac{1}{4}$

80. $25^{\frac{-1}{2}} = \left(\sqrt{25}\right)^{-1} = 5^{-1} = \dfrac{1}{5}$

81. $x^{\frac{1}{2}} \cdot x^{\frac{1}{3}} = x^{\frac{1}{2}+\frac{1}{3}} = x^{\frac{5}{6}}$

82. $y^{\frac{3}{4}} \cdot y^{\frac{3}{16}} = y^{\frac{3}{4}+\frac{3}{16}} = y^{\frac{15}{16}}$

83. $\dfrac{y^{\frac{2}{3}}}{y^{\frac{-1}{2}}} = y^{\frac{2}{3}-\left(-\frac{1}{2}\right)} = y^{\frac{2}{3}+\frac{1}{2}} = y^{\frac{7}{6}}$

84. $\dfrac{a^{\frac{2}{5}}}{a^{\frac{2}{3}}} = a^{\frac{2}{5}-\frac{2}{3}} = a^{\frac{-4}{15}} = \dfrac{1}{a^{\frac{4}{15}}}$

85. $\left(2x^{\frac{1}{4}}\right)^2 \cdot \left(4x^{\frac{1}{3}}\right)^2 = 4x^{\frac{1}{2}} \cdot 16x^{\frac{2}{3}} = 64x^{\frac{1}{2}+\frac{2}{3}} = 64x^{\frac{7}{6}}$

86. $\left(8x^6\right)^{\frac{1}{3}} \cdot \left(16x^2\right)^{\frac{-1}{2}} = 2x^2 \cdot 4^{-1}x^{-1} = \dfrac{2}{4}x^{2+(-1)} = \dfrac{x}{2}$

87. $\left(5a^{\frac{1}{6}}\right)^2 \cdot \left(9a^4\right)^{\frac{1}{2}} = 25a^{\frac{1}{3}} \cdot 3a^2 = 75a^{\frac{1}{3}+2} = 75a^{\frac{7}{3}}$

88. $\left(36x^{\frac{2}{3}}\right)^{\frac{3}{2}} \cdot \left(6x^{\frac{3}{2}}\right)^{-2} = 6^3x \cdot 6^{-2}x^{-3} = 6^{3+(-2)}x^{1+(-3)} = \dfrac{6}{x^2}$

89. Given points: $(-4,6),(-4,7)$

$\text{Distance} = \sqrt{\left(-4-(-4)\right)^2 + \left(6-7\right)^2}$

$\qquad\qquad = \sqrt{0+1} = \sqrt{1} = 1$

90. Given points: $(-5,2),(5,2)$

$\text{Distance} = \sqrt{\left(-5-5\right)^2 + \left(2-2\right)^2}$

$\qquad\qquad = \sqrt{100+0} = \sqrt{100} = 10$

91. Given points: $(-4,2),(-1,6)$

$\text{Distance} = \sqrt{\left(-4-(-1)\right)^2 + \left(2-6\right)^2}$

$\qquad\qquad = \sqrt{9+16} = \sqrt{25} = 5$

92. Given points: $(0,-3),(7,2)$

$$\text{Distance} = \sqrt{(0-7)^2 + (-3-2)^2}$$
$$= \sqrt{49+25} = \sqrt{74}$$

93. Verify that the sum of the squares of the two shortest legs is equal to the square of the longest side length:

$$9^2 = 81$$
$$6^2 + 8^2 = 36 + 64 = 100 \neq 81$$

So the triangle is not a right triangle.

94. Verify that the sum of the squares of the two shortest legs is equal to the square of the longest side length:

$$25^2 = 625$$
$$7^2 + 24^2 = 49 + 576 = 625$$

So the triangle is a right triangle.

95. To show that the triangle is isosceles, demonstrate that two of its legs are of equal length:

Given points: $A(1,-2)$, $B(4,4)$ and $C(7,1)$

$$|AB| = \sqrt{(1-4)^2 + (-2-4)^2} = \sqrt{9+36} = \sqrt{45} = 3\sqrt{5}$$
$$|AC| = \sqrt{(1-7)^2 + (-2-1)^2} = \sqrt{36+9} = \sqrt{45} = 3\sqrt{5}$$

So the triangle is isosceles.

96. Use the Pythagorean theorem, where the height of the man is the length of one leg of the triangle, the length of his shadow is the other leg, and the distance from the top of his head to the tip of his shadow is the hypotenuse.

$$x^2 = 6^2 + 8^2$$
$$x^2 = 36 + 64$$
$$x^2 = 100$$
$$x = \sqrt{100} = 10 \text{ feet}$$

So the distance from the top of the man to the end of his shadow is 10 feet.

97. Let s be the length of one side of the square. Then the diagonal of the square forms a right triangle with two of the sides, and the Pythagorean theorem can be used to find s:

$$s^2 + s^2 = 40^2$$
$$2s^2 = 1600$$
$$s^2 = 800$$
$$s = \sqrt{800} \approx 28.3 \text{ feet}$$

So the square has sides of length 28.3 ft, to the nearest tenth.

98. To show that the triangle is equilateral, demonstrate that all of its legs are of equal length:

Given points: $X(2,0)$, $Y(6,0)$ and $Z\left(4,\sqrt{12}\right)$

$$|XY| = \sqrt{(2-6)^2 + (0-0)^2} = \sqrt{16+0} = \sqrt{16} = 4$$
$$|XZ| = \sqrt{(2-4)^2 + \left(0-\sqrt{12}\right)^2} = \sqrt{4+12} = \sqrt{16} = 4$$
$$|YZ| = \sqrt{(6-4)^2 + \left(0-\sqrt{12}\right)^2} = \sqrt{4+12} = \sqrt{16} = 4$$

So the triangle is equilateral.

99. Given points: $A(-2,2)$, $B(-2,5)$, $C(6,2)$ and $D(6,5)$

$$|AD| = \sqrt{(-2-6)^2 + (2-5)^2} = \sqrt{64+9} = \sqrt{73}$$
$$|BC| = \sqrt{(-2-6)^2 + (5-2)^2} = \sqrt{64+9} = \sqrt{73}$$

So the diagonals have the same length.

100. To find the perimeter, first find the distance between each vertex, which is the length of each side, then take the sum.

Given points: $D(-6,0)$, $E(-6,3)$ and $F(0,0)$

$$|DE| = \sqrt{\left(-6-(-6)\right)^2 + (0-3)^2} = \sqrt{0+9} = \sqrt{9} = 3$$
$$|DF| = \sqrt{(-6-0)^2 + (0-0)^2} = \sqrt{36+0} = \sqrt{36} = 6$$
$$|EF| = \sqrt{(-6-0)^2 + (3-0)^2} = \sqrt{36+9} = \sqrt{45} = 3\sqrt{5}$$
$$P = 3 + 6 + 3\sqrt{5} = 9 + 3\sqrt{5}$$

Use the Pythagorean theorem to see if the triangle is a right triangle:

$$3^2 + 6^2 = 9 + 36 = 45$$
$$\left(3\sqrt{5}\right)^2 = 9\cdot 5 = 45$$

So it is a right triangle.

[End of Chapter 9 Review]

Chapter 9 Test
Solutions to All Exercises

1. $\dfrac{3}{7} = 0.\overline{428571}$

2. Accurate to 7 decimal places: $\sqrt{13} \approx 3.6055513$

3. $\sqrt{144} = 12$

4. $\sqrt{25x^2} = 5x$

5. $-\sqrt{81x^4 y^3} = -9x^2 y\sqrt{y}$

6. $\sqrt[3]{27a^3} = 3a$

7. $\sqrt[3]{24x^6 y^{12}} = \sqrt[3]{2^3 \cdot 3 \cdot \left(x^2\right)^3 \left(y^4\right)^3} = 2x^2 y^4 \sqrt[3]{3}$

8. $-\sqrt{\dfrac{16x^2}{y^8}} = -\dfrac{4x}{y^4}$

9. $\sqrt{\dfrac{4a^5}{9b^{10}}} = \dfrac{2a^2 \sqrt{a}}{3b^5}$

10. $5\sqrt{8} + 3\sqrt{2} = 5 \cdot 2\sqrt{2} + 3\sqrt{2} = 13\sqrt{2}$

11. $2\sqrt{27x^2} + 5\sqrt{48x^2} - 4\sqrt{3x^2}$
$$= 2\sqrt{9 \cdot 3x^2} + 5\sqrt{16 \cdot 3x^2} - 4\sqrt{3x^2}$$
$$= 6x\sqrt{3} + 20x\sqrt{3} - 4x\sqrt{3}$$
$$= 22x\sqrt{3}$$

12. $\dfrac{\sqrt{96} + 12}{24} = \dfrac{\sqrt{16 \cdot 6} + 12}{24} = \dfrac{\sqrt{6} + 3}{6}$

13. $\dfrac{21 - \sqrt{98}}{7} = \dfrac{21 - \sqrt{49 \cdot 2}}{7} = 3 - \sqrt{2}$

14. $\sqrt[3]{-40} + \sqrt[3]{108} = \sqrt[3]{-8 \cdot 5} + \sqrt[3]{27 \cdot 4} = -2\sqrt[3]{5} + 3\sqrt[3]{4}$

15. $3\sqrt{5} \cdot 5\sqrt{5} = 15 \cdot 5 = 75$

16. $\sqrt{15} \cdot \left(\sqrt{3} + \sqrt{5}\right) = \sqrt{45} + \sqrt{75} = 3\sqrt{5} + 5\sqrt{3}$

17. $\left(5 + \sqrt{11}\right)\left(5 - \sqrt{11}\right) = 5^2 - \left(\sqrt{11}\right)^2 = 25 - 11 = 14$

18. $\left(6 + \sqrt{2}\right)\left(5 - \sqrt{2}\right) = 30 - 6\sqrt{2} + 5\sqrt{2} - 2 = 28 - \sqrt{2}$

19. $\dfrac{3}{\sqrt{18}} = \dfrac{3}{\sqrt{18}} \cdot \dfrac{\sqrt{18}}{\sqrt{18}} = \dfrac{3\sqrt{18}}{18} = \dfrac{3\sqrt{9 \cdot 2}}{18} = \dfrac{\sqrt{2}}{2}$

20. $\sqrt{\dfrac{4}{75}} = \sqrt{\dfrac{4}{75} \cdot \dfrac{75}{75}} = \sqrt{\dfrac{4 \cdot 25 \cdot 3}{75 \cdot 75}} = \dfrac{10}{75}\sqrt{3} = \dfrac{2\sqrt{3}}{15}$

21. $\dfrac{2}{\sqrt{3} - 1} = \dfrac{2}{\sqrt{3} - 1} \cdot \dfrac{\sqrt{3} + 1}{\sqrt{3} + 1} = \dfrac{2\left(\sqrt{3} + 1\right)}{3 - 1} = \sqrt{3} + 1$

22. $\dfrac{\sqrt{2}}{\sqrt{3} + \sqrt{2}} = \dfrac{\sqrt{2}}{\sqrt{3} + \sqrt{2}} \cdot \dfrac{\sqrt{3} - \sqrt{2}}{\sqrt{3} - \sqrt{2}}$
$$= \dfrac{\sqrt{2}\left(\sqrt{3} - \sqrt{2}\right)}{3 - 2} = \sqrt{6} - 2$$

23. $\sqrt{5x - 1} = 13$
$$5x - 1 = 169$$
$$5x = 170$$
$$x = 34$$
Check by substituting into the original equation:
$$\sqrt{5(34) - 1} = \sqrt{170 - 1} = \sqrt{169} = 13$$
The solution is $x = 34$.

24. $\sqrt{4x + 2} = \sqrt{12 - x}$
$$4x + 2 = 12 - x$$
$$5x = 10$$
$$x = 2$$
Check by substituting into the right and left sides of the original equation independently.
$$\sqrt{4(2) + 2} = \sqrt{8 + 2} = \sqrt{10}$$
$$\sqrt{12 - 2} = \sqrt{10}$$
The solution is $x = 2$.

25. $2x - 5 = \sqrt{13 - 6x}$
$$4x^2 - 20x + 25 = 13 - 6x$$
$$4x^2 - 14x + 12 = 0$$
$$2(2x - 3)(x - 2) = 0$$
Set: $2x - 3 = 0 \qquad x - 2 = 0$
Solve: $2x = 3 \qquad x = 2$
$$x = \dfrac{3}{2}$$
Check these results in the original equation:

Substitute $x = \dfrac{3}{2}$:

$$2\left(\dfrac{3}{2}\right) - 5 = -2$$

$$\sqrt{13 - 6\left(\dfrac{3}{2}\right)} = \sqrt{4} = 2$$

So $x = \dfrac{3}{2}$ is not a solution.

Substitute $x = 2$:

$$2(2) - 5 = -1$$

$$\sqrt{13 - 6(2)} = \sqrt{1} = 1$$

So $x = 2$ is not a solution.
There is no solution to this equation.

26. $16^{\frac{3}{4}} = \left(\sqrt[4]{16}\right)^3 = 2^3 = 8$

27. $\left(\dfrac{4}{9}\right)^{\frac{-3}{2}} = \left(\dfrac{9}{4}\right)^{\frac{3}{2}} = \left(\sqrt{\dfrac{9}{4}}\right)^3 = \left(\dfrac{3}{2}\right)^3 = \dfrac{27}{8}$

28. $a^{\frac{3}{4}} \cdot a^{\frac{-1}{2}} = a^{\frac{3}{4} + \left(\frac{-1}{2}\right)} = a^{\frac{1}{4}}$

29. $\dfrac{y^{\frac{3}{4}}}{y^{\frac{-1}{2}}} = y^{\frac{3}{4} - \left(\frac{-1}{2}\right)} = y^{\frac{5}{4}}$

30. $\left(16x^2\right)^{\frac{1}{2}} \left(27x^3\right)^{\frac{2}{3}} = \sqrt{16x^2} \cdot \left(\sqrt[3]{27x^3}\right)^2$

$$= 4x(3x)^2$$
$$= 4x \cdot 9x^2$$
$$= 36x^3$$

31. Given the points $(-5, 7), (3, -1)$:

Distance $= \sqrt{(-5 - 3)^2 + (7 - (-1))^2}$

$$= \sqrt{64 + 64}$$
$$= \sqrt{64 \cdot 2}$$
$$= 8\sqrt{2}$$

32. Given the points $(2, 5), (-4, -6)$:

Distance $= \sqrt{(2 - (-4))^2 + (5 - (-6))^2}$

$$= \sqrt{36 + 121}$$
$$= \sqrt{157}$$

33. Yes, this is a right triangle:
By the Pythagorean Theorem:

$$20^2 + 21^2 = 400 + 441 = 841 = 29^2$$

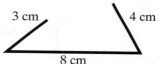

34. Use the Pythagorean Theorem to calculate the unknown leg h of the right triangle:

$$h^2 = 25^2 - 6^2$$
$$h^2 = 625 - 36$$
$$h^2 = 589$$
$$h = \sqrt{589} \approx 24.3$$

The height from the base of the building to the point where the ladder leans against the building is 24.3 ft, to the nearest tenth.

35. A triangle with sides 3 cm, 4 cm, and 8 cm cannot exist because $3 + 4 < 8$. No matter how you try to draw such a triangle, two of the sides will never meet.

3 cm   4 cm

8 cm

[End of Chapter 9 Test]

Cumulative Review: Chapters 1 – 9
Solutions to All Exercises

1. Complete the square:

$$x^2 + 8x + \underline{\left(\frac{8}{2}\right)^2} = \left(x+4\right)^2$$

$$x^2 + 8x + 16 = \left(x+4\right)^2$$

2. Complete the square:

$$49 = 7^2 = \left(\frac{14}{2}\right)^2$$

$$x^2 - \underline{14}x + 49 = \left(\underline{x-7}\right)^2$$

3.
$$\frac{2x+10}{x^2-x-6} \cdot \frac{x^2+6x+8}{x+5} = \frac{2(x+5)\cdot(x+2)(x+4)}{(x-3)(x+2)(x+5)}$$
$$= \frac{2(x+4)}{x-3}$$

4.
$$\frac{x}{x+6} - \frac{2x-5}{x^2+4x-12} = \frac{x}{x+6} - \frac{2x-5}{(x+6)(x-2)}$$
$$= \frac{x(x-2)-(2x-5)}{(x+6)(x-2)}$$
$$= \frac{x^2-2x-2x+5}{(x+6)(x-2)}$$
$$= \frac{x^2-4x+5}{(x+6)(x-2)}$$
$$\text{or} \quad = \frac{x^2-4x+5}{x^2+4x-12}$$

5. Graph $y = \frac{3}{4}x - 2$:

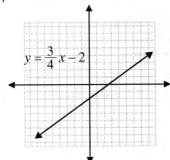

6. Graph $2x + 3y = 12$:

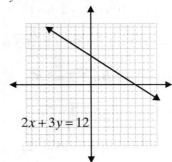

7. $-\sqrt{63x^4} = -3x^2\sqrt{7}$

8. $\sqrt{\dfrac{36a^3}{25b^6}} = \dfrac{6a\sqrt{a}}{5b^3}$

9. $\sqrt{48x^{10}y^{20}} = 4x^5y^{10}\sqrt{3}$

10. $\sqrt{54a^8b^{14}} = 3a^4b^7\sqrt{6}$

11. $\sqrt[3]{40} = \sqrt[3]{8\cdot 5} = 2\sqrt[3]{5}$

12. $\sqrt[3]{-250} = -5\sqrt[3]{2}$

13. $\dfrac{8+\sqrt{12}}{2} = \dfrac{8+2\sqrt{3}}{2} = 4+\sqrt{3}$

14. $\dfrac{20-\sqrt{48}}{8} = \dfrac{20-4\sqrt{3}}{8} = \dfrac{5-\sqrt{3}}{2}$

15. $2\sqrt{50} - \sqrt{98} + 3\sqrt{8} = 2\sqrt{25\cdot 2} - \sqrt{49\cdot 2} + 3\sqrt{4\cdot 2}$
$$= 10\sqrt{2} - 7\sqrt{2} + 6\sqrt{2}$$
$$= 9\sqrt{2}$$

16. $11\sqrt{2} + \sqrt{50} = 11\sqrt{2} + 5\sqrt{2} = 16\sqrt{2}$

17. $4\sqrt{20} - \sqrt{45} = 8\sqrt{5} - 3\sqrt{5} = 5\sqrt{5}$

18. $\sqrt[3]{-54} + 5\sqrt[3]{2} = \sqrt[3]{-27\cdot 2} + 5\sqrt[3]{2}$
$$= -3\sqrt[3]{2} + 5\sqrt[3]{2}$$
$$= 2\sqrt[3]{2}$$

19. $\sqrt{50x^4} - 2\sqrt{8x^4} = 5x^2\sqrt{2} - 4x^2\sqrt{2} = x^2\sqrt{2}$

20. $\sqrt{12} - 5\sqrt{3} + \sqrt{48} = 2\sqrt{3} - 5\sqrt{3} + 4\sqrt{3} = \sqrt{3}$

21. $\sqrt{24} \cdot \sqrt{6} = 2\sqrt{6} \cdot \sqrt{6} = 2 \cdot 6 = 12$

22. $2\sqrt{30} \cdot \sqrt{20} = 2\sqrt{30 \cdot 20} = 2\sqrt{100 \cdot 6} = 20\sqrt{6}$

23. $2\sqrt{48} \cdot \sqrt{2} = 2\sqrt{16 \cdot 3 \cdot 2} = 8\sqrt{6}$

24. $\sqrt{14} \cdot \sqrt{21} = \sqrt{7 \cdot 2 \cdot 7 \cdot 3} = 7\sqrt{6}$

25. $\left(\sqrt{3} - 5\right)\left(\sqrt{3} + 5\right) = 3 - 25 = -22$

26. $\left(\sqrt{2} + 3\right)\left(\sqrt{2} + 5\right) = 2 + 5\sqrt{2} + 3\sqrt{2} + 15 = 17 + 8\sqrt{2}$

27. $\left(\sqrt{x} + \sqrt{2}\right)\left(\sqrt{x} - \sqrt{2}\right) = x - 2$

28. $\left(2\sqrt{x} + 3\right)\left(\sqrt{x} - 8\right) = 2x - 16\sqrt{x} + 3\sqrt{x} - 24$
$$= 2x - 13\sqrt{x} - 24$$

29. $\dfrac{9}{\sqrt{3}} = \dfrac{9}{\sqrt{3}} \cdot \dfrac{\sqrt{3}}{\sqrt{3}} = \dfrac{9\sqrt{3}}{3} = 3\sqrt{3}$

30. $\sqrt{\dfrac{3}{20}} = \sqrt{\dfrac{3}{20} \cdot \dfrac{20}{20}} = \dfrac{2\sqrt{15}}{20} = \dfrac{\sqrt{15}}{10}$

31. $\sqrt{\dfrac{5}{27}} = \sqrt{\dfrac{5}{27} \cdot \dfrac{3}{3}} = \sqrt{\dfrac{15}{81}} = \dfrac{\sqrt{15}}{9}$

32. $-\sqrt{\dfrac{9}{2y}} = -\sqrt{\dfrac{9}{2y} \cdot \dfrac{2y}{2y}} = -\dfrac{3\sqrt{2y}}{2y}$

33. $\dfrac{2}{\sqrt{3} - 5} = \dfrac{2}{\sqrt{3} - 5} \cdot \dfrac{\sqrt{3} + 5}{\sqrt{3} + 5}$
$$= \dfrac{2\left(\sqrt{3} + 5\right)}{3 - 25}$$
$$= \dfrac{2\left(\sqrt{3} + 5\right)}{-22}$$
$$= \dfrac{-\sqrt{3} - 5}{11}$$

34. $\dfrac{1}{\sqrt{2} + 3} = \dfrac{1}{\sqrt{2} + 3} \cdot \dfrac{\sqrt{2} - 3}{\sqrt{2} - 3} = \dfrac{\sqrt{2} - 3}{2 - 9}$
$$= \dfrac{\sqrt{2} - 3}{-7} = \dfrac{-\sqrt{2} + 3}{7}$$

35. $x(x + 4) - (x + 4)(x - 3) = 0$
$$x^2 + 4x - \left(x^2 + x - 12\right) = 0$$
$$x^2 + 4x - x^2 - x + 12 = 0$$
$$3x = -12$$
$$x = -4$$

36. $(2x + 1)(x - 4) = 5$
$$2x^2 - 7x - 4 - 5 = 0$$
$$2x^2 - 7x - 9 = 0$$
$$(2x - 9)(x + 1) = 0$$
Set: $\quad 2x - 9 = 0 \qquad x + 1 = 0$
Solve: $\quad 2x = 9 \qquad\quad x = -1$
$$x = \dfrac{9}{2}$$

37. $\dfrac{x - 1}{2x + 1} + \dfrac{1}{x + 2} = \dfrac{1}{2}$
$$2(x + 2)(x - 1) + 2(2x + 1) = (2x + 1)(x + 2)$$
$$2\left(x^2 + x - 2\right) + 4x + 2 = 2x^2 + 5x + 2$$
$$2x^2 + 2x - 4 + 4x + 2 = 2x^2 + 5x + 2$$
$$6x - 2 = 5x + 2$$
$$x = 4$$

38. $x + 3 = \sqrt{x + 15}$
$$(x + 3)^2 = x + 15$$
$$x^2 + 6x + 9 = x + 15$$
$$x^2 + 5x - 6 = 0$$
$$(x - 1)(x + 6) = 0$$
Set: $\quad x - 1 = 0 \qquad x + 6 = 0$
Solve: $\quad x = 1 \qquad\quad x = -6$
Check each solution:
For $x = -6$: $\quad -6 + 3 = -3$
$$\sqrt{-6 + 15} = \sqrt{9} = 3$$
So, $x = -6$ is not a solution to the original equation.

For $x = 1$: $\quad 1 + 3 = 4$
$$\sqrt{1 + 15} = 4$$
So, $x = 1$ is the solution to the original equation.

39. $\sqrt{7x+2}=3$

$7x+2=9$

$7x=7$

$x=1$

Check: $\sqrt{7(1)+2}=\sqrt{9}=3$

40. $\sqrt{4x-3}=5$

$4x-3=25$

$4x=28$

$x=7$

Check: $\sqrt{4(7)-3}=\sqrt{25}=5$

41. $\sqrt{2x-9}=4-x$

$2x-9=(4-x)^2$

$2x-9=16-8x+x^2$

$0=x^2-10x+25$

$0=(x-5)^2$

Set: $x-5=0$

Solve: $x=5$

Check: $\sqrt{2(5)-9}=\sqrt{10-9}=1$

$4-(5)=-1$

So, $x=5$ is not a solution. There is no solution.

42. $\sqrt{5x-3}=\sqrt{2x+3}$

$5x-3=2x+3$

$3x=6$

$x=2$

Check: $\sqrt{5(2)-3}=\sqrt{10-3}=\sqrt{7}$

$\sqrt{2(2)+3}=\sqrt{4+3}=\sqrt{7}$

So, $x=2$ is the solution.

43. The slope can be found using the given points:

$$m=\frac{3-1}{-5-5}=\frac{2}{-10}=-\frac{1}{5}.$$

Use point slope form to find the equation.

$$y-1=-\frac{1}{5}(x-5)$$

$$y-1=-\frac{1}{5}x+1$$

$$y=-\frac{1}{5}x+2$$

44. To be perpendicular to the given line, the slope must be $-\dfrac{1}{3}$. Use point slope form to find the equation.

$$-\frac{1}{3}(x-6)=y-2$$

$$-\frac{1}{3}x+2=y-2$$

$$y=-\frac{1}{3}x+4$$

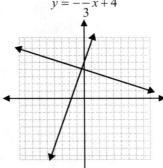

45. $\begin{cases} 2x-y=7 \\ 3x+2y=0 \end{cases}$

Multiply the first equation by 2 and use addition:

$4x-2y=14$

$\underline{3x+2y=\ \ 0}$

$7x\ \ \ \ \ =14$

$x\ \ \ \ \ \ =\ 2$

Substitute into the first equation:

$2(2)-y=7$

$y=4-7$

$y=-3$

The solution is $(2,-3)$.

46. $\begin{cases} 4x+3y=15 \\ 2x-5y=1 \end{cases}$

Multiply the second equation by -2 and use addition:

$4x+\ \ 3y=15$

$\underline{-4x+10y=-2}$

$13y=13$

$y=\ \ 1$

Substitute into the first equation.

$4x+3(1)=15$

$4x=12$

$x=3$

The solution is $(3,1)$.

47. $\begin{cases} y \le 4x+1 \\ y \le -\dfrac{1}{4}x \end{cases}$

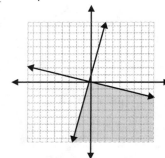

48. $\begin{cases} y > 3 \\ y > \dfrac{1}{2}x \end{cases}$

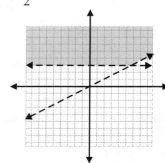

49. Let r be the rate of the slower automobile. Then, $r + 12$ is the rate of the faster one.

$$3.5r + 3.5(r+12) = 350$$
$$35r + 35(r+12) = 10(350)$$
$$35r + 35r + 420 = 3500$$
$$70r = 3080$$
$$r = 44$$
$$r + 12 = 56$$

The slower automobile is going 44 mph and the faster one is going 56 mph.

50. Let x be the number of pounds of the \$1.50 per lb candy. Then $20 - x$ is the number of pounds of the \$2.00 per lb candy.

$$1.50x + 2.00(20-x) = 1.80(20)$$
$$15x + 20(20-x) = 18(20)$$
$$15x + 400 - 20x = 360$$
$$-5x = -40$$
$$x = 8$$
$$20 - x = 12$$

The mixture should contain 8 lb of the \$1.50 per lb candy and 12 lb of the \$2.00 per lb candy.

51. Let x be the number of hours the father would take to finish the job alone.

	Hours alone to finish the job	Amount of job in one hour
Father	x	$\dfrac{1}{x}$
Daughter	$x + 8$	$\dfrac{1}{x+8}$
Together	3	$\dfrac{1}{3}$

The sum of the parts each can complete working alone is equal to the part they can complete together. Set up the equation and solve.

$$\frac{1}{x} + \frac{1}{x+8} = \frac{1}{3}$$
$$3(x+8) + 3x = x(x+8)$$
$$3x + 24 + 3x = x^2 + 8x$$
$$x^2 + 2x - 24 = 0$$
$$(x+6)(x-4) = 0$$

Set: $\qquad x + 6 = 0 \qquad x - 4 = 0$

Solve: $\qquad x = -6 \qquad x = 4$

$$\qquad\qquad\qquad\qquad x + 8 = 12$$

Disregard the negative solution, as it has no meaning here. The father can finish the job in 4 hours, and his daughter can finish in 12 hours.

52. Let x be the price of a taco and y be the price of a burrito.

$$\begin{cases} 2x + y = 5.30 \\ 3x + 2y = 9.25 \end{cases}$$

Multiply the first equation by –2 and use addition.

$$-4x - 2y = -10.60$$
$$\underline{3x + 2y = \quad 9.25}$$
$$-x \qquad = -1.35$$
$$x \qquad = \quad 1.35$$

Substitute into the first equation.

$$2(1.35) + y = 5.30$$
$$y = 5.30 - 2.70$$
$$y = 2.60$$

A taco is \$1.35 and a burrito is \$2.60.

53. Use the Pythagorean Theorem for the right triangle that models the problem situation:

$$x^2 + 30^2 = 40^2$$
$$x^2 = 1600 - 900$$
$$x^2 = 700$$
$$x = 10\sqrt{7}$$

The height of the dock above the deck of the boat is $10\sqrt{7} \approx 26.5$ m.

54. Use $h = -16t^2 + v_0 t + h_0$:

a. The initial velocity $v_0 = 0$, since the ball is dropped. The initial height is $h_0 = 240$.

$$-16t^2 + 240 = 144$$
$$-16t^2 = -96$$
$$t^2 = 6$$
$$t = \sqrt{6} \approx 2.45$$

The ball will be 144 feet above the ground at approximately 2.45 sec.

b. The ball will have dropped $240 - 144 = 96$ feet.

c. The ball will hit the ground when $h = 0$:

$$-16t^2 + 240 = 0$$
$$-16t^2 = -240$$
$$t^2 = 15$$
$$t = \sqrt{15} \approx 3.87$$

The ball will hit the ground in approximately 3.87 seconds.

[End Cumulative Review: Chapters 1 – 9]

Section 10.1
Solutions to Odd Exercises

1.
$$x^2 = 11x$$
$$x^2 - 11x = 0$$
$$x(x-11) = 0$$
Set: $\qquad x = 0 \qquad x - 11 = 0$
Solve: $\qquad\qquad\qquad x = 11$

3.
$$x^2 = -15x - 36$$
$$x^2 + 15x + 36 = 0$$
$$(x+12)(x+3) = 0$$
Set: $\quad x + 12 = 0 \qquad x + 3 = 0$
Solve: $\quad x = -12 \qquad x = -3$

5.
$$9x^2 + 6x - 15 = 0$$
$$3(3x^2 + 2x - 5) = 0$$
$$3(3x+5)(x-1) = 0$$
Set: $\qquad 3x + 5 = 0 \qquad x - 1 = 0$
Solve: $\qquad 3x = -5 \qquad x = 1$
$$x = -\frac{5}{3}$$

7.
$$(x+3)(x-1) = 4x$$
$$x^2 + 2x - 3 = 4x$$
$$x^2 - 2x - 3 = 0$$
$$(x-3)(x+1) = 0$$
Set: $\qquad x - 3 = 0 \qquad x + 1 = 0$
Solve: $\qquad x = 3 \qquad x = -1$

9.
$$(2x-3)(2x+1) = 3x - 6$$
$$4x^2 - 4x - 3 = 3x - 6$$
$$4x^2 - 7x + 3 = 0$$
$$(4x-3)(x-1) = 0$$
Set: $\qquad 4x - 3 = 0 \qquad x - 1 = 0$
Solve: $\qquad 4x = 3 \qquad x = 1$
$$x = \frac{3}{4}$$

11.
$$x^2 = 121$$
$$x = \pm\sqrt{121}$$
$$x = \pm 11$$

13.
$$3x^2 = 108$$
$$x^2 = 36$$
$$x = \pm\sqrt{36}$$
$$x = \pm 6$$

15.
$$x^2 = 35$$
$$x = \pm\sqrt{35}$$

17.
$$x^2 - 62 = 0$$
$$x^2 = 62$$
$$x = \pm\sqrt{62}$$

19.
$$x^2 - 45 = 0$$
$$x^2 = 45$$
$$x = \pm\sqrt{9 \cdot 5}$$
$$x = \pm 3\sqrt{5}$$

21.
$$3x^2 = 54$$
$$x^2 = 18$$
$$x = \pm\sqrt{9 \cdot 2}$$
$$x = \pm 3\sqrt{2}$$

23.
$$9x^2 = 4$$
$$x^2 = \frac{4}{9}$$
$$x = \pm\sqrt{\frac{4}{9}}$$
$$x = \pm\frac{2}{3}$$

25.
$$(x-1)^2 = 4$$
$$x - 1 = \pm 2$$
$$x = \pm 2 + 1$$
$$x = 3 \quad \text{or} \quad x = -1$$

27.
$$(x+2)^2 = -25$$
There is no real number solution. The square of any real number is never negative.

29. $(x+1)^2 = \dfrac{1}{4}$

$$x+1 = \pm\sqrt{\dfrac{1}{4}}$$

$$x+1 = \pm\dfrac{1}{2}$$

$$x = \pm\dfrac{1}{2} - 1$$

$$x = -\dfrac{1}{2} \quad \text{or} \quad x = -\dfrac{3}{2}$$

31. $(x-3)^2 = \dfrac{4}{9}$

$$x-3 = \pm\sqrt{\dfrac{4}{9}}$$

$$x-3 = \pm\dfrac{2}{3}$$

$$x = \pm\dfrac{2}{3} + 3$$

$$x = \dfrac{11}{3} \quad \text{or} \quad x = \dfrac{7}{3}$$

33. $(x-6)^2 = 18$

$$x-6 = \pm\sqrt{18}$$

$$x-6 = \pm 3\sqrt{2}$$

$$x = 6 \pm 3\sqrt{2}$$

35. $2(x-7)^2 = 24$

$$(x-7)^2 = 12$$

$$x-7 = \pm\sqrt{12}$$

$$x = 7 \pm 2\sqrt{3}$$

37. $(3x+4)^2 = 27$

$$3x+4 = \pm\sqrt{27}$$

$$3x = -4 \pm 3\sqrt{3}$$

$$x = \dfrac{-4 \pm 3\sqrt{3}}{3}$$

39. $(5x-2)^2 = 63$

$$5x-2 = \pm\sqrt{63}$$

$$5x = 2 \pm 3\sqrt{7}$$

$$x = \dfrac{2 \pm 3\sqrt{7}}{5}$$

For Exercises 41 – 57, note that only the positive square root is taken, since the negative root does not apply to the context of the problem.

41. Does the sum of the squares of the two shortest sides equal the square of the longest side (the hypotenuse)?

$$8^2 + 15^2 = 64 + 225$$
$$= 289$$
$$= 17^2$$

Yes.

43. Same basic question as #41:

$$\left(\sqrt{2}\right)^2 + \left(\sqrt{2}\right)^2 = 2 + 2$$
$$= 4$$
$$= 2^2$$

Yes.

45. Find c, the hypotenuse.

$$c^2 = 9^2 + 12^2$$
$$c^2 = 81 + 144$$
$$c^2 = 225$$
$$c = 15$$

47. Find b, one of the sides of the right triangle.

$$6^2 + b^2 = 12^2$$
$$36 + b^2 = 144$$
$$b^2 = 144 - 36$$
$$b = \sqrt{108}$$
$$b = 6\sqrt{3}$$

49. Find c, one of the sides of the right triangle.

$$1^2 + c^2 = \left(\sqrt{2}\right)^2$$
$$1 + c^2 = 2$$
$$c^2 = 2 - 1$$
$$c^2 = 1$$
$$c = 1$$

51. Let x be the length of the unknown leg. Then $2x$ is the length of the hypotenuse.

$$\left(2x\right)^2 = x^2 + \left(4\sqrt{3}\right)^2$$
$$4x^2 = x^2 + 16 \cdot 3$$
$$3x^2 = 48$$
$$x^2 = 16$$
$$x = 4$$
$$2x = 8$$

The leg is 4 ft and the hypotenuse is 8 ft.

53. Let x be length of each leg. (Two are of equal length).

$$x^2 + x^2 = 36$$
$$2x^2 = 36$$
$$x^2 = 18$$
$$x = \sqrt{9 \cdot 2}$$
$$x = 3\sqrt{2}$$

The length of each leg is $3\sqrt{2}$ cm.

55. Let x be length of the hypotenuse.

$$x^2 = 40^2 + 10^2$$
$$x^2 = 1700$$
$$x = \sqrt{1700}$$
$$x = 10\sqrt{17} \approx 41.23$$

Adding on 6 ft, the wire should be 47.2 ft, to the nearest tenth.

57. Set the height equal to 0, and solve for t.

$$144 - 16t^2 = 0$$
$$16t^2 = 144$$
$$t^2 = 9$$
$$t = 3 \text{ seconds.}$$

Calculator Problems

59. $x^2 = 647$

$$x = \pm\sqrt{647}$$
$$x \approx \pm 25.44$$

61. $19x^2 = 523$

$$x^2 = \frac{523}{19}$$
$$x = \sqrt{\frac{523}{19}}$$
$$x \approx \pm 5.25$$

63. $6x^2 = 17.32$

$$x^2 = \frac{17.32}{6}$$
$$x = \pm\sqrt{\frac{17.32}{6}} \approx \pm 1.70$$

65. $2.1x^2 = 35.82$

$$x^2 = \frac{35.82}{2.1}$$
$$x = \pm\sqrt{\frac{35.82}{2.1}} \approx \pm 4.13$$

67. $4.2361, -0.2361$

69. $2.3229, -0.3229$

[End of Section 10.1]

Section 10.2
Solutions to Odd Exercises

1. Complete the square:

$$\left(\frac{12}{2}\right)^2 = 6^2 = 36$$

$$x^2 + 12x + \underline{36} = \left(\underline{x+6}\right)^2$$

3. Complete the square:

$$\left(\frac{-8}{2}\right)^2 = \left(-4\right)^2 = 16$$

$$2x^2 - 16x + \underline{32} = 2\left(\underline{x-4}\right)^2$$

5. Complete the square:

$$\left(\frac{-3}{2}\right)^2 = \frac{9}{4}$$

$$x^2 - 3x + \underline{\frac{9}{4}} = \left(\underline{x - \frac{3}{2}}\right)^2$$

7. Complete the square:

$$\left(\frac{1}{2}\right)^2 = \frac{1}{4}$$

$$x^2 + x + \underline{\frac{1}{4}} = \left(\underline{x + \frac{1}{2}}\right)^2$$

9. Complete the square:

$$\left(\frac{2}{2}\right)^2 = 1^2 = 1$$

$$2x^2 + 4x + \underline{2} = 2\left(\underline{x+1}\right)^2$$

11.
$$x^2 + 6x - 7 = 0$$
$$x^2 + 6x = 7$$
$$x^2 + 6x + 3^2 = 7 + 3^2$$
$$\left(x+3\right)^2 = 16$$
$$x + 3 = \pm\sqrt{16}$$
$$x = -3 \pm 4$$
$$x = 1 \quad \text{or} \quad x = -7$$

13.
$$x^2 - 4x - 45 = 0$$
$$x^2 - 4x = 45$$
$$x^2 - 4x + 2^2 = 45 + 2^2$$
$$\left(x-2\right)^2 = 49$$
$$x - 2 = \pm\sqrt{49}$$
$$x = 2 \pm 7$$
$$x = 9 \quad \text{or} \quad x = -5$$

15.
$$x^2 - 3x - 40 = 0$$
$$x^2 - 3x = 40$$
$$x^2 - 3x + \left(\frac{3}{2}\right)^2 = 45 + \left(\frac{3}{2}\right)^2$$
$$\left(x - \frac{3}{2}\right)^2 = 40 + \frac{9}{4}$$
$$x - \frac{3}{2} = \pm\sqrt{\frac{169}{4}}$$
$$x = \frac{3}{2} \pm \frac{13}{2}$$
$$x = \frac{16}{2} = 8 \quad \text{or} \quad x = \frac{-10}{2} = -5$$

17.
$$3x^2 + x - 4 = 0$$
$$3x^2 + x = 4$$
$$x^2 + \frac{1}{3}x = \frac{4}{3}$$
$$x^2 + \frac{1}{3}x + \left(\frac{1}{6}\right)^2 = \frac{4}{3} + \left(\frac{1}{6}\right)^2$$
$$\left(x + \frac{1}{6}\right)^2 = \frac{49}{36}$$
$$x + \frac{1}{6} = \pm\sqrt{\frac{49}{36}}$$
$$x = -\frac{1}{6} \pm \frac{7}{6}$$
$$x = \frac{6}{6} = 1 \quad \text{or} \quad x = \frac{-8}{6} = -\frac{4}{3}$$

19. $4x^2 - 4x - 3 = 0$

$4x^2 - 4x = 3$

$x^2 - x = \dfrac{3}{4}$

$x^2 - x + \left(\dfrac{1}{2}\right)^2 = \dfrac{3}{4} + \left(\dfrac{1}{2}\right)^2$

$\left(x - \dfrac{1}{2}\right)^2 = 1$

$x - \dfrac{1}{2} = \pm\sqrt{1}$

$x = \dfrac{1}{2} \pm 1$

$x = \dfrac{3}{2}$ or $x = -\dfrac{1}{2}$

21. $x^2 + 6x + 3 = 0$

$x^2 + 6x = -3$

$x^2 + 6x + 3^2 = -3 + 3^2$

$(x + 3)^2 = 6$

$x + 3 = \pm\sqrt{6}$

$x = -3 \pm \sqrt{6}$

23. $x^2 + 2x - 5 = 0$

$x^2 + 2x = 5$

$x^2 + 2x + 1^2 = 5 + 1^2$

$(x + 1)^2 = 6$

$x + 1 = \pm\sqrt{6}$

$x = -1 \pm \sqrt{6}$

25. $3x^2 - 2x - 1 = 0$

$3x^2 - 2x = 1$

$x^2 - \dfrac{2}{3}x = \dfrac{1}{3}$

$x^2 - \dfrac{2}{3}x + \left(\dfrac{1}{3}\right)^2 = \dfrac{1}{3} + \left(\dfrac{1}{3}\right)^2$

$\left(x - \dfrac{1}{3}\right)^2 = \dfrac{1}{3} + \dfrac{1}{9}$

$x - \dfrac{1}{3} = \pm\sqrt{\dfrac{4}{9}}$

$x = \dfrac{1}{3} \pm \dfrac{2}{3}$

$x = \dfrac{3}{3} = 2$ or $x = -\dfrac{1}{3}$

27. $x^2 + x - 3 = 0$

$x^2 + x = 3$

$x^2 + x + \left(\dfrac{1}{2}\right)^2 = 3 + \left(\dfrac{1}{2}\right)^2$

$\left(x + \dfrac{1}{2}\right)^2 = 3 + \dfrac{1}{4}$

$x + \dfrac{1}{2} = \pm\sqrt{\dfrac{13}{4}}$

$x = -\dfrac{1}{2} \pm \dfrac{\sqrt{13}}{2}$

$x = \dfrac{-1 \pm \sqrt{13}}{2}$

29. $2x^2 + 3x - 1 = 0$

$2x^2 + 3x = 1$

$x^2 + \dfrac{3}{2}x = \dfrac{1}{2}$

$x^2 + \dfrac{3}{2} + \left(\dfrac{3}{4}\right)^2 = \dfrac{1}{2} + \left(\dfrac{3}{4}\right)^2$

$\left(x + \dfrac{3}{4}\right)^2 = \dfrac{1}{2} + \dfrac{9}{16}$

$x + \dfrac{3}{4} = \pm\sqrt{\dfrac{17}{16}}$

$x = -\dfrac{3}{4} \pm \dfrac{\sqrt{17}}{4} = \dfrac{-3 \pm \sqrt{17}}{4}$

31. $x^2 - 9x + 2 = 0$

$x^2 - 9x + \left(\dfrac{9}{2}\right)^2 = -2 + \left(\dfrac{9}{2}\right)^2$

$\left(x - \dfrac{9}{2}\right)^2 = -2 + \dfrac{81}{4}$

$x - \dfrac{9}{2} = \pm\sqrt{\dfrac{73}{4}}$

$x = \dfrac{9}{2} \pm \dfrac{\sqrt{73}}{2} = \dfrac{9 \pm \sqrt{73}}{2}$

33. $x^2 + 7x - 14 = 0$

$$x^2 + 7x + \left(\frac{7}{2}\right)^2 = 14 + \left(\frac{7}{2}\right)^2$$

$$\left(x + \frac{7}{2}\right)^2 = 14 + \frac{49}{4}$$

$$x + \frac{7}{2} = \pm\sqrt{\frac{105}{4}}$$

$$x = -\frac{7}{2} \pm \frac{\sqrt{105}}{2} = \frac{-7 \pm \sqrt{105}}{2}$$

35. $x^2 - 11x - 26 = 0$

$(x - 13)(x + 2) = 0$

Set: $x - 13 = 0$ $\qquad$ $x + 2 = 0$

Solve: $\qquad x = 13$ $\qquad\qquad$ $x = -2$

37. $2x^2 + x - 2 = 0$

Divide by 2 and solve by completing the square:

$$x^2 + \frac{x}{2} - 1 = 0$$

$$x^2 + \frac{x}{2} + \left(\frac{1}{4}\right)^2 = 1 + \left(\frac{1}{4}\right)^2$$

$$\left(x + \frac{1}{4}\right)^2 = \frac{17}{16}$$

$$x + \frac{1}{4} = \pm\sqrt{\frac{17}{16}}$$

$$x = -\frac{1}{4} \pm \frac{\sqrt{17}}{4}$$

$$x = \frac{-1 \pm \sqrt{17}}{4}$$

39. $3x^2 - 6x + 3 = 0$

Divide both sides by 3 and solve:

$$x^2 - 2x + 1 = 0$$

$$(x - 1)^2 = 0$$

$$x - 1 = 0$$

$$x = 1$$

41. $4x^2 + 20x - 8 = 0$

Divide by 4 and solve by completing the square:

$$x^2 + 5x - 2 = 0$$

$$x^2 + 5x + \left(\frac{5}{2}\right)^2 = 2 + \left(\frac{5}{2}\right)^2$$

$$\left(x + \frac{5}{2}\right)^2 = \frac{33}{4}$$

$$x + \frac{5}{2} = \pm\sqrt{\frac{33}{4}}$$

$$x = -\frac{5}{2} \pm \frac{\sqrt{33}}{2}$$

$$x = \frac{-5 \pm \sqrt{33}}{2}$$

43. $6x^2 - 8x + 1 = 0$

Divide by 6 and solve by completing the square:

$$x^2 - \frac{8}{6}x + \frac{1}{6} = 0$$

$$x^2 - \frac{4}{3}x + \left(\frac{2}{3}\right)^2 = -\frac{1}{6} + \left(\frac{2}{3}\right)^2$$

$$\left(x - \frac{2}{3}\right)^2 = \frac{5}{18}$$

$$x - \frac{2}{3} = \pm\sqrt{\frac{5}{18}}$$

$$x = \frac{2}{3} \pm \sqrt{\frac{5 \cdot 2}{18 \cdot 2}}$$

$$x = \frac{2}{3} \pm \frac{\sqrt{10}}{6}$$

$$x = \frac{4 \pm \sqrt{10}}{6}$$

45. $2x^2 + 7x + 4 = 0$

Divide by 2 and solve by completing the square:

$$\frac{2x^2}{2} + \frac{7}{2}x + \frac{4}{2} = 0$$

$$x^2 + \frac{7}{2}x + \left(\frac{7}{4}\right)^2 = \left(\frac{7}{4}\right)^2 - 2$$

$$\left(x + \frac{7}{4}\right)^2 = \frac{17}{16}$$

$$x + \frac{7}{4} = \pm\sqrt{\frac{17}{16}}$$

$$x = -\frac{7}{4} \pm \frac{\sqrt{17}}{4}$$

$$x = \frac{-7 \pm \sqrt{17}}{4}$$

47. $4x^2 - 2x - 3 = 0$

Divide by 4 and solve by completing the square:

$$\frac{4x^2}{4} - \frac{2}{4}x - \frac{3}{4} = 0$$

$$x^2 - \frac{2}{4}x + \left(\frac{1}{4}\right)^2 = \left(\frac{1}{4}\right)^2 + \frac{3}{4}$$

$$\left(x - \frac{1}{4}\right)^2 = \frac{13}{16}$$

$$x - \frac{1}{4} = \pm\sqrt{\frac{13}{16}}$$

$$x = \frac{1}{4} \pm \frac{\sqrt{13}}{4}$$

$$x = \frac{1 \pm \sqrt{13}}{4}$$

49. $3x^2 + 8x + 5 = 0$

$(3x + 5)(x + 1) = 0$

Set: $\qquad 3x + 5 = 0 \qquad x + 1 = 0$

Solve: $\qquad 3x = -5 \qquad x = -1$

$$x = -\frac{5}{3}$$

51. $2x^2 + 5x + 3 = 0$

$(2x + 3)(x + 1) = 0$

Set: $\qquad 2x + 3 = 0 \qquad x + 1 = 0$

Solve: $\qquad 2x = -3 \qquad x = -1$

$$x = -\frac{3}{2}$$

53. $3x^2 + 24x + 21 = 0$

$3(x + 7)(x + 1) = 0$

Set: $\qquad x + 7 = 0 \qquad x + 1 = 0$

Solve: $\qquad x = -7 \qquad x = -1$

55. $6x^2 - x - 1 = 0$

$(2x - 1)(3x + 1) = 0$

Set: $\qquad 2x - 1 = 0 \qquad 3x + 1 = 0$

Solve: $\qquad 2x = 1 \qquad 3x = -1$

$$x = \frac{1}{2} \qquad x = -\frac{1}{3}$$

[End of Section 10.2]

Section 10.3
Solutions to Odd Exercises

1. $x^2 - 3x - 2 = 0$
$a = 1$, $b = -3$, $c = -2$

3. $2x^2 - x + 6 = 0$
$a = 2$, $b = -1$, $c = 6$

5. $7x^2 - 4x - 3 = 0$
$a = 7$, $b = -4$, $c = -3$

7. $3x^2 - 9x - 4 = 0$
$a = 3$, $b = -9$, $c = -4$

9. $2x^2 + 5x - 3 = 0$
$a = 2$, $b = 5$, $c = -3$

11. $x^2 - 4x - 1 = 0$
$a = 1$, $b = -4$, $c = -1$
$$x = \frac{-(-4) \pm \sqrt{(-4)^2 - 4(1)(-1)}}{2 \cdot 1}$$
$$x = \frac{4 \pm \sqrt{16 + 4}}{2}$$
$$x = \frac{4 \pm \sqrt{20}}{2}$$
$$x = \frac{4 \pm 2\sqrt{5}}{2}$$
$$x = 2 \pm \sqrt{5}$$

13. $x^2 - 3x = 4$
$x^2 - 3x - 4 = 0$
$a = 1$, $b = -3$, $c = -4$
$$x = \frac{-(-3) \pm \sqrt{(-3)^2 - 4(1)(-4)}}{2 \cdot 1}$$
$$x = \frac{3 \pm \sqrt{9 + 16}}{2}$$
$$x = \frac{3 \pm \sqrt{25}}{2}$$
$$x = \frac{3 \pm 5}{2}$$
$$x = 4, \; x = -1$$

15. $-2x^2 + x = -1$
$-2x^2 + x + 1 = 0$
$a = -2$, $b = 1$, $c = 1$
$$x = \frac{-1 \pm \sqrt{1^2 - 4(-2)(1)}}{2(-2)}$$
$$x = \frac{-1 \pm \sqrt{1 + 8}}{-4}$$
$$x = \frac{-1 \pm 3}{-4}$$
$$x = -\frac{1}{2}$$
$$x = 1$$

17. $5x^2 + 3x - 2 = 0$
$a = 5$, $b = 3$, $c = -2$
$$x = \frac{-3 \pm \sqrt{3^2 - 4(5)(-2)}}{2 \cdot 5}$$
$$x = \frac{-3 \pm \sqrt{9 + 40}}{10}$$
$$x = \frac{-3 \pm \sqrt{49}}{10}$$
$$x = \frac{-3 \pm 7}{10}$$
$$x = \frac{2}{5}, \; x = -1$$

19. $9x^2 = 3x$
$9x^2 - 3x = 0$
$a = 9$, $b = -3$, $c = 0$
$$x = \frac{-(-3) \pm \sqrt{(-3)^2 - 4(9)(0)}}{2 \cdot 9}$$
$$x = \frac{3 \pm \sqrt{9}}{18}$$
$$x = \frac{3 \pm 3}{18}$$
$$x = \frac{1}{3}, \; x = 0$$

21. $x^2 - 7 = 0$
$a = 1$, $b = 0$, $c = -7$
$$x = \frac{0 \pm \sqrt{0^2 - 4(1)(-7)}}{2 \cdot 1}$$
$$x = \frac{\pm\sqrt{28}}{2}$$
$$x = \frac{\pm 2\sqrt{7}}{2}$$
$$x = \pm\sqrt{7}$$

23. $x^2 + 4x = x - 2x^2$

$3x^2 + 3x = 0$

$a = 3$, $b = 3$, $c = 0$

$$x = \frac{-3 \pm \sqrt{3^2 - 4(3)(0)}}{2 \cdot 3}$$

$$x = \frac{-3 \pm \sqrt{9}}{6}$$

$$x = \frac{-3 \pm 3}{6}$$

$x = 0$, $x = -1$

25. $3x^2 + 4x = 0$

$a = 3$, $b = 4$, $c = 0$

$$x = \frac{-4 \pm \sqrt{4^2 - 4(3)(0)}}{2 \cdot 3}$$

$$x = \frac{-4 \pm \sqrt{16}}{6}$$

$$x = \frac{-4 \pm 4}{6}$$

$x = 0$, $x = -\dfrac{4}{3}$

27. $\dfrac{2}{5}x^2 + x - 1 = 0$

$$5\left(\frac{2}{5}x^2 + x - 1\right) = 5(0)$$

$2x^2 + 5x - 5 = 0$

$a = 2$, $b = 5$, $c = -5$

$$x = \frac{-5 \pm \sqrt{5^2 - 4(2)(-5)}}{2 \cdot 2}$$

$$x = \frac{-5 \pm \sqrt{25 + 40}}{4}$$

$$x = \frac{-5 \pm \sqrt{65}}{4}$$

29. $\dfrac{-x^2}{2} - 2x + \dfrac{1}{3} = 0$

$$-6\left(\frac{-x^2}{2} - 2x + \frac{1}{3}\right) = -6(0)$$

$3x^2 + 12x - 2 = 0$

$a = 3$, $b = 12$, $c = -2$

$$x = \frac{-12 \pm \sqrt{12^2 - 4(3)(-2)}}{2 \cdot 3}$$

$$x = \frac{-12 \pm \sqrt{144 + 24}}{6}$$

$$x = \frac{-12 \pm \sqrt{168}}{6}$$

$$x = \frac{-12 \pm 2\sqrt{42}}{6}$$

$$x = \frac{-6 \pm \sqrt{42}}{3}$$

31. $(2x+1)(x+3) = 2x + 6$

$x^2 + 7x + 3 = 2x + 6$

$x^2 + 5x - 3 = 0$

$a = 2$, $b = 5$, $c = -3$

$$x = \frac{-5 \pm \sqrt{5^2 - 4(2)(-3)}}{2 \cdot 2}$$

$$x = \frac{-5 \pm \sqrt{25 + 24}}{4}$$

$$x = \frac{-5 \pm \sqrt{49}}{4}$$

$$x = \frac{-5 \pm 7}{4}$$

$x = \dfrac{1}{2}$, $x = -3$

33. $\dfrac{5x + 2}{3x} = x - 1$

$5x + 2 = 3x(x - 1)$

$5x + 2 = 3x^2 - 3x$

$3x^2 - 8x - 2 = 0$

$a = 3$, $b = -8$, $c = -2$

$$x = \frac{-(-8) \pm \sqrt{(-8)^2 - 4(3)(-2)}}{2 \cdot 3}$$

$$x = \frac{8 \pm \sqrt{64 + 24}}{6}$$

$$x = \frac{8 \pm \sqrt{88}}{6}$$

$$x = \frac{8 \pm 2\sqrt{22}}{6}$$

$$x = \frac{4 \pm \sqrt{22}}{3}$$

35. $4x^2 = 7x - 3$

$4x^2 - 7x + 3 = 0$

$a = 4$, $b = -7$, $c = 3$

$$x = \frac{-(-7) \pm \sqrt{(-7)^2 - 4(4)(3)}}{2 \cdot 4}$$

$$x = \frac{7 \pm \sqrt{49-48}}{8} = \frac{7 \pm 1}{8}$$

$$x = 1, \quad x = \frac{3}{4}$$

37. $2x^2 + 7x + 3 = 0$

$(2x+1)(x+3) = 0$

Set: $\quad 2x+1 = 0 \qquad x+3 = 0$

Solve: $\quad 2x = -1 \qquad x = -3$

$$x = -\frac{1}{2}$$

39. $3x^2 - 7x + 1 = 0$

$a = 3$, $b = -7$, $c = 1$

$$x = \frac{-(-7) \pm \sqrt{(-7)^2 - 4(3)(1)}}{2(3)}$$

$$x = \frac{7 \pm \sqrt{37}}{6}$$

41. $\qquad 10x^2 = x + 24$

$10x^2 - x - 24 = 0$

$(5x-8)(2x+3) = 0$

Set: $\quad 5x-8 = 0 \qquad 2x+3 = 0$

Solve: $\quad 5x = 8 \qquad 2x = -3$

$$x = \frac{8}{5} \qquad\qquad x = -\frac{3}{2}$$

43. $\qquad 3x^2 - 11x = 4$

$3x^2 - 11x - 4 = 0$

$(3x+1)(x-4) = 0$

Set: $\quad 3x+1 = 0 \qquad x-4 = 0$

Solve: $\quad 3x = -1 \qquad x = 4$

$$x = -\frac{1}{3}$$

45. $-4x^2 + 11x - 5 = 0$

$a = -4$, $b = 11$, $c = -5$

$$x = \frac{-11 \pm \sqrt{11^2 - 4(-4)(-5)}}{2(-4)}$$

$$x = \frac{-11 \pm \sqrt{121-80}}{-8}$$

$$x = \frac{11 \pm \sqrt{41}}{8}$$

47. $10x^2 + 35x + 30 = 0$

$5(x+2)(2x+3) = 0$

Set: $\qquad x+2 = 0 \qquad 2x+3 = 0$

Solve: $\qquad x = -2 \qquad 2x = -3$

$$x = -\frac{3}{2}$$

49. $\qquad 25x^2 + 4 = 20$

$25x^2 - 16 = 0$

$(5x-4)(5x+4) = 0$

Set: $\qquad 5x-4 = 0 \qquad 5x+4 = 0$

Solve: $\qquad 5x = 4 \qquad 5x = -4$

$$x = \frac{4}{5} \qquad\qquad x = -\frac{4}{5}$$

51. $\frac{3}{4}x^2 - 2x + \frac{1}{8} = 0$

$6x^2 - 16x + 1 = 0$

$a = 6$, $b = -16$, $c = 1$

$$x = \frac{-(-16) \pm \sqrt{(-16)^2 - 4(6)(1)}}{2(6)}$$

$$x = \frac{16 \pm \sqrt{256-24}}{12}$$

$$x = \frac{16 \pm \sqrt{232}}{12}$$

$$x = \frac{8 \pm \sqrt{58}}{6}$$

53. $\qquad \frac{3}{7}x^2 = \frac{1}{2}x + 1$

$\frac{3}{7}x^2 - \frac{1}{2}x - 1 = 0$

$6x^2 - 7x - 14 = 0$

$a = 6$, $b = -7$, $c = -14$

$$x = \frac{-(-7) \pm \sqrt{(-7)^2 - 4(6)(-14)}}{2(6)}$$

$$x = \frac{7 \pm \sqrt{49+336}}{12}$$

$$x = \frac{7 \pm \sqrt{385}}{12}$$

55.
$$x^2 - x - 1 = 0$$
$$x^2 - x + \left(\frac{1}{2}\right)^2 = 1 + \left(\frac{1}{2}\right)^2$$
$$\left(x - \frac{1}{2}\right)^2 = \frac{5}{4}$$
$$x - \frac{1}{2} = \pm\sqrt{\frac{5}{4}}$$
$$x = \frac{1}{2} \pm \frac{1}{2}\sqrt{5}$$
$$x = \frac{1 \pm \sqrt{5}}{2}$$
$$x \approx \frac{1 \pm 2.2361}{2}$$
$$x \approx 1.6180 \text{ or } x \approx -0.6180$$

57.
$$(2x+1)(2x-1) = 3x$$
$$4x^2 - 3x - 1 = 0$$
$$(4x+1)(x-1) = 0$$

Set: $\quad 4x+1 = 0 \qquad x-1 = 0$

Solve: $\quad 4x = -1 \qquad x = 1$

$$x = -\frac{1}{4} = -0.25$$

59.
$$5x^2 - 100 = 0$$
$$5x^2 = 100$$
$$x^2 = 20$$
$$x = \pm 2\sqrt{5} \approx \pm 4.4721$$

[End of Section 10.3]

Section 10.4
Solutions to Odd Exercises

1. Let n be the number.
$$n + n^2 = 132$$
$$n^2 + n - 132 = 0$$
$$(n + 12)(n - 11) = 0$$
Set: $\qquad n + 12 = 0 \qquad n - 11 = 0$
Solve: $\qquad n = -12 \qquad n = 11$
Disregard -12, since $n > 0$. The number is 11.

3. Let w be the width of the rectangle. Then $2w - 5$ is the length. Use $A = lw = 63$:
$$(2w - 5)w = 63$$
$$2w^2 - 5w - 63 = 0$$
$$(2w + 9)(w - 7) = 0$$
Set: $\qquad 2w + 9 = 0 \qquad w - 7 = 0$
Solve: $\qquad 2w = -9 \qquad w = 7$
$$w = \frac{-9}{2} \qquad 2w - 5 = 9$$
Disregard $w = -\dfrac{9}{2}$, because $w > 0$. The width is 7 meters and the length is 9 meters.

5. The perimeter is 58 meters, so for width w and length l, $2w + 2l = 58$, or $w + l = 29$. This means $l = 29 - w$. Using $A = lw = 198$:
$$(29 - w)w = 198$$
$$-w^2 + 29w - 198 = 0$$
$$w^2 - 29w + 198 = 0$$
$$(w - 11)(w - 18) = 0$$
Set: $\qquad w - 11 = 0 \qquad w - 18 = 0$
Solve: $\qquad w = 11 \qquad w = 18$
$$29 - w = 18 \qquad 29 - w = 11$$
Since width and length are interchangeable here, the field is 11m by 18m.

7. Let w be the width in feet. Then $w + 10$ is the length in feet. The area of the deck and pool is 1344 ft^2. The figure shows that the deck and pool space is a rectangle with dimensions $w + 12$ by $w + 22$. Using $A = lw$:
$$(w + 12)(w + 22) = 1344$$
Simplify and factor:
$$w^2 + 34w + 264 = 1344$$
$$w^2 + 34w - 1080 = 0$$
$$(w - 20)(w + 54) = 0$$

Set: $\qquad w - 20 = 0 \qquad w + 54 = 0$
Solve: $\qquad w = 20 \qquad w = -54$
Disregard $w = -54$, since width must be positive. The pool is 20 ft by 30 ft.

9. Let x be the smaller of two positive integers. Then $x + 1$ is the next integer.
$$x^2 + (x + 1)^2 = 221$$
$$2x^2 + 2x + 1 = 221$$
$$2x^2 + 2x - 220 = 0$$
$$x^2 + x - 110 = 0$$
$$(x + 11)(x - 10) = 0$$
Set: $\qquad x + 11 = 0 \qquad x - 10 = 0$
Solve: $\qquad x = -11 \qquad x = 10$
Disregard $x = -11$, since x is positive. The integers are 10 and 11.

11. Let n be the number.
$$n - 3 = 4\left(\frac{1}{n}\right)$$
$$(n - 3)n = 4$$
$$n^2 - 3n - 4 = 0$$
$$(n - 4)(n + 1) = 0$$
Set: $\qquad n - 4 = 0 \qquad n + 1 = 0$
Solve: $\qquad n = 4 \qquad n = -1$
Disregard -1, since n is positive. The number is 4.

13. Let s be the side length of the original square.
$$(s + 10)^2 = 9s^2$$
$$s^2 + 20s + 100 = 9s^2$$
$$8s^2 - 20s - 100 = 0$$
$$2s^2 - 5s - 25 = 0$$
$$(s - 5)(2s + 5) = 0$$
Set: $\qquad s - 5 = 0 \qquad 2s + 5 = 0$
Solve: $\qquad s = 5 \qquad s = -\dfrac{5}{2}$

Disregard $-\dfrac{5}{2}$, since the side length must be positive. The original square has sides of length 5.

15. Use the Pythagorean Theorem. Let w be the width of the rectangle. Then, $2w + 2$ is the length.

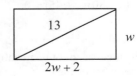

$2w + 2$

$$w^2 + (2w+2)^2 = 13^2$$
$$w^2 + 4w^2 + 8w + 4 = 169$$
$$5w^2 + 8w - 165 = 0$$
$$(5w + 33)(w - 5) = 0$$

Set: $\qquad 5w + 33 = 0 \qquad w - 5 = 0$

Solve: $\qquad w = -\dfrac{33}{5} \qquad w = 5$

$$l = 2w + 2 = 12$$

Disregard $-\dfrac{33}{5}$, since the width must be positive.

The rectangle is 5 m by 12 m.

17. Let x be the length of one of the congruent legs of the right triangle. Then $x^2 + x^2 = 12^2$ by the Pythagorean Theorem,

$$2x^2 = 144$$
$$x^2 = 72$$
$$x = \pm\sqrt{36 \cdot 2}$$
$$x = \pm 6\sqrt{2}$$

Only the positive solution applies to this problem, so $x = 6\sqrt{2}$. So, the length of each of the two legs of the triangle is $6\sqrt{2}$ cm.

19. Let x be the number of members of the club.

Sharing equally, the cost per person is $\dfrac{900}{x}$ dollars.

We assume that $900 is a fixed price no matter how many members go. If the club had 15 more members, the price would be $10 less per person, or $\dfrac{900}{x} - 10$. If 15 more go, the price per person is also equal to $\dfrac{900}{x+15}$. So,

$$\frac{900}{x} - 10 = \frac{900}{x+15}$$
$$\frac{900 - 10x}{x} = \frac{900}{x+15}$$
$$(x+15)(900 - 10x) = 900x$$
$$900x - 10x^2 + 13500 - 150x = 900x$$
$$-10x^2 - 150x + 13500 = 0$$
$$x^2 + 15x - 1350 = 0$$
$$(x - 30)(x + 45) = 0$$

Set: $\qquad x - 30 = 0 \qquad x + 45 = 0$
$\qquad\qquad x = 30 \qquad\quad x = -45$

Disregard -45, since x cannot be negative.
There are 30 members of the club.

21. Let x be the rate going. Then $x - 10$ is the rate returning. Use the formula time $= \dfrac{\text{distance}}{\text{rate}}$:

	Rate	Time	Distance
Going	x	$\dfrac{200}{x}$	200
Returning	$x - 10$	$\dfrac{200}{x-10}$	200

Since the total time was 9 hours,

$$\frac{200}{x} + \frac{200}{x-10} = 9$$
$$200(x-10) + 200x = 9x(x-10)$$
$$200x - 2000 + 200x = 9x^2 - 90x$$
$$9x^2 - 490x + 2000 = 0$$
$$(x - 50)(9x - 40) = 0$$

Set: $\qquad x - 50 = 0 \qquad 9x - 40 = 0$

Solve: $\qquad x = 50 \qquad\quad x = \dfrac{40}{9}$

Disregard $\dfrac{40}{9}$, since this would yield a negative returning rate. He traveled to the city at 50 mph.

23. Let c be the rate of the current.

	Rate	Time	Distance
Down	$c + 12$	$\dfrac{45}{c+12}$	45
Up	$12 - c$	$\dfrac{45}{12-c}$	45

Since it takes 2 hours longer to travel upstream than downstream,

$$\frac{45}{12-c} - \frac{45}{c+12} = 2$$
$$45(c+12) - 45(12-c) = 2(12-c)(c+12)$$
$$45c + 540 - 540 + 45c = 2(144 - c^2)$$
$$90c = 288 - 2c^2$$
$$2c^2 + 90c - 288 = 0$$
$$2(c - 3)(c + 48) = 0$$

Set: $\qquad c - 3 = 0 \qquad c + 48 = 0$
Solve: $\qquad c = 3 \qquad\quad c = -48$

Disregard -48, since the current must be positive.
The rate of the current is 3 mph.

25. Let x be number of club members initially.

Cost per member (initially) is $\dfrac{120}{x}$. When 5 members dropped out, the new cost per member became \$2 more than the original cost per member, or $\dfrac{120}{x}+2$. The cost per member with a loss of 5 members is also $\dfrac{120}{x-5}$. So,

$$\frac{120}{x}+2=\frac{120}{x-5}$$

$$\frac{120+2x}{x}=\frac{120}{x-5}$$

$$(x-5)(120+2x)=120x$$

$$120x-600-10x+2x^2=120x$$

$$2x^2-10x-600=0$$

$$x^2-5x-300=0$$

$$(x+15)(x-20)=0$$

Set: $\quad x+15=0 \qquad x-20=0$

Solve: $\quad x=-15 \qquad x=20$

Disregard -15, since x must be positive. There were 20 members in the club initially.

27. Let L be the length of the side of the box to be made. Use the volume formula $V=LWH$ with $W=L$ (since the base is square) and $H=2$.

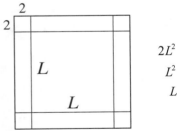

$$2L^2=162$$
$$L^2=81$$
$$L=9$$

The square cardboard was $9+2+2=13$ inches on each side.

29. Let t be the number of minutes for the small pipe to fill the tank.

	Minutes needed to fill the tank	Amount of job completed in 1 minute
Small Pipe	t	$\dfrac{1}{t}$
Large Pipe	$t-30$	$\dfrac{1}{t-30}$
Together	8	$\dfrac{1}{8}$

The sum of the amounts completed seperately by each pipe in one minute is equal to the amount completed by the pipes working together, so:

$$\frac{1}{t}+\frac{1}{t-30}=\frac{1}{8}$$

$$t-30+t=\frac{t(t-30)}{8}$$

$$(2t-30)8=t^2-30t$$

$$16t-240=t^2-30t$$

$$t^2-46t+240=0$$

$$(t-6)(t-40)=0$$

Set: $\quad t-6=0 \qquad t-40=0$

Solve: $\quad t=6 \qquad t=40$

Disregard $t=6$ because this value would yield a negative value for the larger pipe. Then, the small pipe would take 40 minutes to fill the tank.

31. a. Set $h=66$:

$$66=-16t^2+32t+50$$

$$16t^2-32t+16=0$$

$$t^2-2t+1=0$$

$$(t-1)^2=0$$

$$t-1=0$$

$$t=1$$

At $t=1$ second, the ball is at 66 ft.

b. Set $h=30$:

$$30=-16t^2+32t+50$$

$$16t^2-32t-20=0$$

$$4t^2-8t-5=0$$

$$(2t+1)(2t-5)=0$$

Set: $\quad 2t+1=0 \qquad 2t-5=0$

Solve: $\quad t=-\dfrac{1}{2} \qquad t=\dfrac{5}{2}$

Disregard $-\dfrac{1}{2}$ since t must be positive.

Then, at $t=\dfrac{5}{2}$, or 2.5 seconds, the ball is 30 feet above the beach.

c. Set $h=0$ to find when the ball hits the beach:

$$0=-16t^2+32t+50$$

$$16t^2-32t-50=0$$

$$8t^2-16t-25=0$$

Using the quadratic formula to solve for t:

$$t = \frac{-(-16) \pm \sqrt{(-16)^2 - 4(8)(-25)}}{2 \cdot 8}$$

$$t = \frac{16 \pm \sqrt{256 + 800}}{16}$$

$$t = \frac{16 \pm \sqrt{1056}}{16}$$

$$t \approx \frac{16 \pm 32.50}{16}$$

$$t \approx -1.031 \approx 1 \ \text{ or } \ t \approx 3.031 \approx 3$$

Disregard -1, since time cannot be negative.
The ball hits the ground after about 3 seconds.

33. Use the Pythagorean Theorem to find the equation: $10^2 + h^2 = 30^2$, where h is the point at which the ladder reaches the building. Solve this equation:

$$100 + h^2 = 900$$

$$h^2 = 800$$

$$h = \sqrt{800} \approx \pm 28.3$$

Disregard -28.3, since distance cannot be negative. The ladder reaches the building at approximately 28.3 feet.

Writing and Thinking About Mathematics

35. There are two major parts of the quadratic equation that might lead to having no solution.
   a. The denominator cannot be zero, so the coefficient of the x^2 term denoted by $'a'$ cannot be equal to 0.
   b. The numerical result under the radical sign, $b^2 - 4ac$, called the discriminant, must be greater than or equal to zero.
   In addition, when a quadratic equation arises in an application, each "solution" must be checked for its validity to the problem at hand.

[End of Section 10.4]

Section 10.5
Solutions to Odd Exercises

In Exercises 1 – 20, use the general quadratic

equation: $y = ax^2 + bx + c$, the vertex x-value of $\dfrac{-b}{2a}$, and

the axis of symmetry formula $x = \dfrac{-b}{2a}$. To find the x-

intercepts, set $ax^2 + bx + c = 0$ and solve for x.

1. Given $y = x^2 + 4$, $a = 1$, $b = 0$ $c = 4$.

   a. Vertex: $x = -\dfrac{0}{2(1)} = 0$

   Substitute $x = 0$ in $y = x^2 + 4$:

   $$y = 0^2 + 4$$

   The vertex is $(0, 4)$.

   b. Line of Symmetry: From part **a**:

   $$x = -\dfrac{0}{2(1)} = 0$$

   c. x-intercepts:

   Set $y = x^2 + 4$ equal to 0 and solve:

   $$x^2 + 4 = 0$$
   $$x^2 = -4$$

   There are no real number solutions to this equation. This means that the parabola does not cross the x-axis.

   d. Graph:

   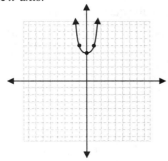

3. Given $y = 8 - x^2$; $a = -1$, $b = 0$, $c = 8$.

   a. Vertex: $x = -\dfrac{0}{2(-1)} = 0$

   Substitute $x = 0$ in $y = 8 - x^2$:

   $$y = 8 - 0^2 = 8$$

   The vertex is $(0, 8)$.

   b. Line of Symmetry: From part **a**:

   $$x = -\dfrac{0}{2(-1)} = 0.$$

   c. x-intercepts:

Set $y = 8 - x^2$ equal to 0, and solve:

$$0 = 8 - x^2$$
$$x^2 = 8$$
$$x = \pm\sqrt{8} = \pm 2\sqrt{2}$$

For graphing purposes, the intercepts are approximately ± 2.8.

d. Graph:

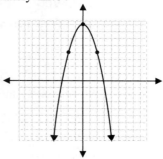

5. Given $y = x^2 - 2x - 3$; $a = 1$, $b = -2$, $c = -3$.

   a. Vertex: $x = -\dfrac{-2}{2(1)} = 1$

   Substitute $x = 1$ in $y = x^2 - 2x - 3$:

   $$y = 1^2 - 2(1) - 3 = -4$$

   The vertex is $(1, -4)$.

   b. Line of Symmetry: From part **a**:

   $$x = -\dfrac{-2}{2(1)} = 1.$$

   c. x-intercepts:

   Set $y = x^2 - 2x - 3$ equal to 0, and solve:

   $$x^2 - 2x - 3 = 0$$
   $$(x - 3)(x + 1) = 0$$
   $$x - 3 = 0 \quad \text{or} \quad x + 1 = 0$$
   $$x = 3 \qquad\qquad x = -1$$

   d. Graph:

   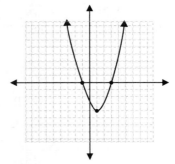

7. Given $y = x^2 + 6x$; $a = 1$, $b = 6$, $c = 0$.

a. <u>Vertex:</u> $x = -\dfrac{6}{2(1)} = -3$

Substitute $x = -3$ in $y = x^2 + 6x$:
$$y = (-3)^2 + 6(-3) = 9 - 18 = -9$$
The vertex is $(-3, -9)$.

b. <u>Line of Symmetry:</u> From part **a**:
$$x = -\dfrac{6}{2(1)} = -3$$

c. <u>x-intercepts:</u>
Set $y = x^2 + 6x$ equal to 0, and solve:
$$x^2 + 6x = 0$$
$$x(x + 6) = 0$$
$$x = 0 \text{ or } x + 6 = 0$$
$$x = -6$$

d. <u>Graph:</u>

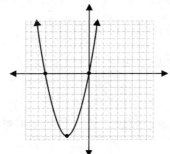

9. Given $y = -x^2 - 4x + 2$; $a = -1$, $b = -4$, $c = 2$.

a. <u>Vertex:</u> $x = -\dfrac{-4}{2(-1)} = -2$

Substitute $x = -2$ in $y = -x^2 - 4x + 2$:
$$y = -(-2)^2 - 4(-2) + 2 = 6$$
The vertex is $(-2, 6)$.

b. <u>Line of Symmetry:</u> From part **a**:
$$x = -\dfrac{-4}{2(-1)} = -2$$

c. <u>x-intercepts:</u>
Set $y = -x^2 - 4x + 2$ equal to 0, and solve. Use the quadratic equation:
$$x = \dfrac{-(-4) \pm \sqrt{(-4)^2 - 4(-1)(2)}}{2(-1)}$$
$$x = \dfrac{4 \pm \sqrt{24}}{-2}$$
$$x = \dfrac{4 \pm 2\sqrt{6}}{-2}$$
$$x = -2 \pm \sqrt{6}$$

For graphing purposes, the intercepts are approximately $x = 0.45$ and at $x = -4.44$.

d. <u>Graph:</u>

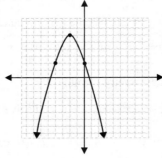

11. Given $y = 2x^2 - 10x + 3$; $a = 2$, $b = -10$, $c = 3$.

a. <u>Vertex:</u> $x = -\dfrac{-10}{2(2)} = \dfrac{5}{2}$

Substitute $x = \dfrac{5}{2}$ in $y = 2x^2 - 10x + 3$:
$$y = 2\left(\dfrac{5}{2}\right)^2 - 10\left(\dfrac{5}{2}\right) + 3 = -\dfrac{19}{2}$$
The vertex is $\left(\dfrac{5}{2}, -\dfrac{19}{2}\right)$.

b. <u>Line of Symmetry:</u> From part **a**:
$$x = \dfrac{5}{2}$$

c. <u>x-intercepts:</u>
Set $y = 2x^2 - 10x + 3$ equal to 0, and solve.
Use the quadratic equation:
$$x = \dfrac{-(-10) \pm \sqrt{(-10)^2 - 4(2)(3)}}{2(2)}$$
$$x = \dfrac{10 \pm \sqrt{100 - 24}}{4}$$
$$x = \dfrac{10 \pm \sqrt{76}}{4}$$
$$x = \dfrac{10 \pm \sqrt{4 \cdot 19}}{4}$$
$$x = \dfrac{5 \pm \sqrt{19}}{2}$$

For graphing purposes, the intercepts are approximately $x = 4.7$ and $x = 0.3$.

d. Graph:

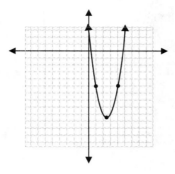

13. Given $y = x^2 + 7x - 4$; $a = 1$, $b = 7$, $c = -4$.

a. Vertex: $x = -\dfrac{7}{2(1)} = -\dfrac{7}{2}$

Substitute $x = -\dfrac{7}{2}$ in $y = x^2 + 7x - 4$:

$$y = \left(-\frac{7}{2}\right)^2 - 7\left(-\frac{7}{2}\right) - 4$$

$$y = \frac{49}{4} - \frac{49}{2} - 4$$

$$y = \frac{49}{4} - \frac{98}{4} - \frac{16}{4} = -\frac{65}{4}$$

The vertex is $\left(-\dfrac{7}{2}, -\dfrac{65}{4}\right)$.

b. Line of Symmetry: From part **a**:

$$x = -\frac{7}{2}$$

c. x-intercepts:

Set $y = x^2 + 7x - 4$ equal to 0, and solve. Use the quadratic equation:

$$x = \frac{-7 \pm \sqrt{7^2 - 4(1)(-4)}}{2(1)}$$

$$x = \frac{-7 \pm \sqrt{49 + 16}}{2}$$

$$x = \frac{-7 \pm \sqrt{65}}{2}$$

For graphing purposes, the x-axis intersects are approximately 0.53 and −7.53.

d. Graph:

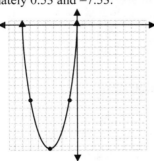

15. Given $y = -x^2 + x - 3$; $a = -1$, $b = 1$, $c = -3$.

a. Vertex: $x = -\dfrac{1}{2(-1)} = \dfrac{1}{2}$

Substitute $x = \dfrac{1}{2}$ in $y = -x^2 + x - 3$:

$$y = -\left(\frac{1}{2}\right)^2 + \left(\frac{1}{2}\right) - 3$$

$$y = -\frac{1}{4} + \frac{1}{2} - 3$$

The vertex is $\left(\dfrac{1}{2}, -\dfrac{11}{4}\right)$.

b. Line of Symmetry: From part **a**:

$$x = \frac{1}{2}$$

c. x-intercepts:

Set $y = -x^2 + x - 3$ equal to 0, and solve. Use the quadratic equation:

$$x = \frac{-1 \pm \sqrt{1^2 - 4(-1)(-3)}}{2(-1)}$$

$$x = \frac{-1 \pm \sqrt{-11}}{-2}$$

There are no real number solutions, so there are no x-intercepts.

d. Graph:

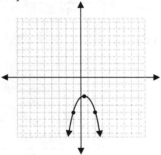

17. Given $y = 2x^2 + 7x - 4$; $a = 2$, $b = 7$, $c = -4$.

a. Vertex: $x = -\dfrac{7}{2(2)} = -\dfrac{7}{4}$

Substitute $x = -\dfrac{7}{4}$ in $y = 2x^2 + 7x - 4$:

$$y = 2\left(-\frac{7}{4}\right)^2 + 7\left(-\frac{7}{4}\right) - 4$$

$$y = 2\left(\frac{49}{16}\right) - \frac{49}{4} - 4$$

$$y = \frac{49}{8} - \frac{98}{8} - \frac{32}{8} = -\frac{81}{8}$$

The vertex is $\left(-\dfrac{7}{4},-\dfrac{81}{8}\right)$.

b. Line of Symmetry: From part **a**:

$$x=-\frac{7}{4}$$

c. x-intercepts:

Set $y=2x^2+7x-4$ equal to 0 and solve.

$$2x^2+7x-4=0$$
$$(2x-1)(x+4)=0$$
$$2x-1=0 \quad \text{or} \quad x+4=0$$
$$2x=1 \qquad\qquad x=-4$$
$$x=\frac{1}{2}$$

d. Graph:

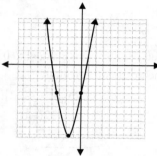

19. Given $y=3x^2+5x+2$; $a=3$, $b=5$, $c=2$.

a. Vertex: $x=-\dfrac{5}{2(3)}=-\dfrac{5}{6}$

Substitute $x=-\dfrac{5}{6}$ in $y=3x^2+5x+2$:

$$y=3\left(-\frac{5}{6}\right)^2+5\left(-\frac{5}{6}\right)+2$$
$$y=3\left(\frac{25}{36}\right)-\frac{25}{6}+2$$
$$y=\frac{25}{12}-\frac{50}{12}+\frac{24}{12}=-\frac{1}{12}$$

The vertex is $\left(-\dfrac{5}{6},-\dfrac{1}{12}\right)$.

b. Line of Symmetry: From part **a**:

$$x=-\frac{5}{6}$$

c. x-intercepts:

Set $y=3x^2+5x+2$ equal to 0 and solve.

$$3x^2+5x+2=0$$
$$(3x+2)(x+1)=0$$
$$3x+2=0 \quad \text{or} \quad x+1=0$$
$$3x=-2 \qquad\qquad x=-1$$
$$x=\frac{-2}{3}$$

d. Graph:

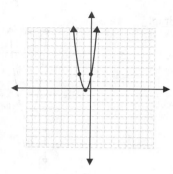

21. A rectangle with perimeter 60 means that, for width w and length l, $2l+2w=60$, or $l+w=30$. So $l=30-w$. Then, with area $A=lw$,

$$A=(30-w)w$$
$$A=30w-w^2$$

The area A is represented by a parabola, which has a maximum (or minimum) at its vertex. This parabola curves downward since the coefficient of w^2 is negative. Therefore, this parabola has a maximum value, and the y-value of the vertex is the maximum value of the area.

$$\text{Vertex:} \quad w=\frac{-30}{2(-1)}=15.$$

Therefore, the maximum area occurs when $w=15$ and $l=15$. This implies that the rectangle is a square.

23. Use $h=-16t^2+v_0t+h_0$, where t is in seconds, v_0 is the initial velocity and h_0 is the initial height above the ground. You are given that $v_0=112$ ft/sec and $h_0=0$ ft. So this application of the formula simplifies to $h=-16t^2+112t$.

a. The ball will reach its maximum height at the vertex of the parabola represented by h:

$$t=-\frac{112}{2(-16)}=\frac{7}{2}=3.5$$

The ball will reach its maximum height after 3.5 seconds.

b. Plug the result from part **a** into the given equation to find the maximum height:

$$h=-16(3.5)^2+112(3.5)$$
$$h=-196+392$$
$$h=196$$

The maximum height is 196 feet.

25. Use $h = -16t^2 + v_0 t + h_0$, where t is in seconds, v_0 is initial velocity and h_0 is the initial height above the ground. You are given that $h_0 = 32$ ft and $v_0 = 128$ ft/sec.

a. The stone will reach its maximum height at the vertex of the parabola represented by h:

$$t = -\frac{128}{2(-16)} = 4$$

The stone reaches its maximum height after 4 sec.

b. Plug the result from part **a.** into the given equation to find the maximum height:

$$h = -16(4)^2 + 128(4) + 32$$
$$h = -256 + 512 + 32$$
$$h = 288$$

The maximum height is 288 feet.

27. **a.** The revenue R from $40 - x$ lamps at x dollars each would be

$$R = (40 - x)x$$
$$R = 40x - x^2$$

b. The maximum occurs at the vertex of the parabola $R = 40x - x^2$.

$a = -1$, $b = 40$ and $c = 0$

$$x = -\frac{40}{2(-1)} = 20$$

The maximum value of R occurs when the price of each lamp is \$20.

29. **a.** The revenue R from selling $80 - 2x$ items each week at x dollars each is

$$R = (80 - 2x)x$$
$$R = 80x - 2x^2$$

b. Find the vertex of the parabola R to determine the price x that yields the maximum revenue:

$a = -2$, $b = 80$ and $c = 0$

$$x = \frac{-80}{2(-2)} = 20$$

The maximum revenue occurs at $x = \$20$

c. Plug the result of part **b** into the revenue function to determine the maximum revenue:

$$R = 80(20) - 2(20)^2 = 1600 - 800$$
$$R = 800$$

The maximum revenue is \$800.

Writing and Thinking About Mathematics

31. Enter each equation into the calculator. Activate its graphing utility.

a. $y = x^2 + 1$

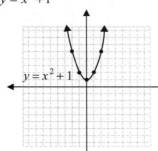

$y = x^2 + 3$

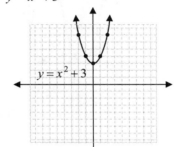

$y = x^2 - 4$

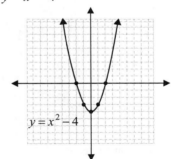

b. All three equations together:

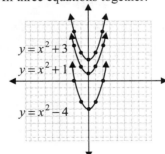

33. Enter each equation into the calculator. Activate its graphing utility.

   a. $y = (x - 3)^2$

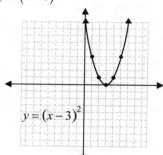

$$y = (x - 5)^2$$

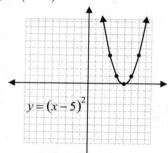

$$y = (x + 2)^2$$

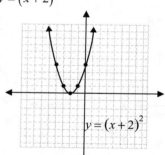

   b. All three equations together:

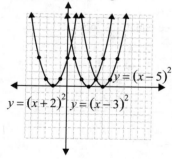

35. The graph of $(x - h)^2 + k$ is the same as the graph of $y = x^2$ shifted by h units horizontally from the line $x = 0$ and k units vertically from the line $y = 0$.

[End of Section 10.5]

Chapter 10 Review
Solutions to All Exercises

1. $x^2 = 400$

    $x = \pm\sqrt{400}$

    $x = \pm 20$

2. $2x^2 = 144$

    $x^2 = 72$

    $x = \pm\sqrt{72}$

    $x = \pm 6\sqrt{2}$

3. $9x^2 = -289$

    $x^2 = -\dfrac{289}{9}$

    There is no real number solution.
    The square of a real number is always positive.

4. $(x-6)^2 = 49$

    $x-6 = \pm\sqrt{49}$

    $x-6 = \pm 7$

    $x = 6 \pm 7$

    $x = -1$ or $x = 13$

5. $(x+3)^2 = 20$

    $x+3 = \pm\sqrt{20}$

    $x+3 = \pm 2\sqrt{5}$

    $x = -3 \pm 2\sqrt{5}$

6. $2(x+1)^2 = 80$

    $(x+1)^2 = 40$

    $x+1 = \pm\sqrt{40}$

    $x+1 = \pm 2\sqrt{10}$

    $x = -1 \pm 2\sqrt{10}$

7. $(3x-1)^2 = 16$

    $3x-1 = \pm\sqrt{16}$

    $3x-1 = \pm 4$

    $3x = 1 \pm 4$

    $x = \dfrac{1 \pm 4}{3}$

    $x = \dfrac{5}{3}$ or $x = \dfrac{-3}{3} = -1$

8. $(4x+9)^2 = 81$

    $4x+9 = \pm\sqrt{81}$

    $4x+9 = \pm 9$

    $4x = -9 \pm 9$

    $x = \dfrac{-9 \pm 9}{4}$

    $x = \dfrac{0}{4} = 0$ or $x = \dfrac{-18}{4} = \dfrac{-9}{2}$

9. $(2x+5)^2 = 75$

    $2x+5 = \pm\sqrt{75}$

    $2x+5 = \pm 5\sqrt{3}$

    $2x = -5 \pm 5\sqrt{3}$

    $x = \dfrac{-5 \pm 5\sqrt{3}}{2}$

10. $3(x-1)^2 = 240$

    $(x-1)^2 = 80$

    $x-1 = \pm\sqrt{80}$

    $x-1 = \pm 4\sqrt{5}$

    $x = 1 \pm 4\sqrt{5}$

11. Does the sum of the squares of the two shortest sides equal the square of the hypotenuse?

    $6^2 + 8^2 = 36 + 64$

    $\qquad = 100$

    $\qquad = 10^2$

    Yes, it is a right triangle.

12. Does the sum of the squares of the two shortest sides equal the square of the hypotenuse?

    $6^2 + 6^2 = 36 + 36$

    $\qquad = 72$

    $\qquad = \left(6\sqrt{2}\right)^2$

    Yes, it is a right triangle.

13. Find c, the hypotenuse. Note that a negative root does not apply in this context, so we may consider only the positive root.

    $c^2 = 8^2 + 8^2$

    $c^2 = 64 + 64$

    $c^2 = 128$

    $c = \sqrt{128}$

    $c = 8\sqrt{2}$

14. Find a, one of the legs of the right triangle. Again, we only need to consider positive roots in this problem.

$$a^2 + 20^2 = 29^2$$
$$a^2 + 400 = 841$$
$$a^2 = 441$$
$$a = \sqrt{441}$$
$$a = 21$$

15. Let c be the length of the hypotenuse. Then the length of the unknown leg is $c - 2$.

$$c^2 = (c-2)^2 + 10^2$$
$$c^2 = c^2 - 4c + 4 + 100$$
$$4c = 104$$
$$c = 26$$
$$c - 2 = 24$$

So the hypotenuse is 26 inches and the legs are 24 and 10 inches. Then the perimeter is:
$$26 + 24 + 10 = 60 \text{ inches.}$$

16. Set the height h equal to zero and solve the given equation:

$$0 = -4.9t^2 + 100$$
$$4.9t^2 = 100$$
$$t^2 = \frac{100}{4.9}$$
$$t = \sqrt{\frac{100}{4.9}} \approx 4.5$$

So the ball will hit the ground in 4.5 seconds, to the nearest tenth.

17. $x^2 = 250$

$$x = \pm\sqrt{250} \approx \pm 15.81$$

18. $20x^2 = 52.5$

$$x^2 = 2.625$$
$$x = \pm\sqrt{2.625} \approx \pm 1.62$$

19. $3.2x^2 = 100$

$$x^2 = 31.25$$
$$x = \pm\sqrt{31.25} \approx \pm 5.59$$

20. $5.6x^2 = -200$

$$x^2 = -35.71$$

There is no real number solution.

21.
$$x^2 - 8x - 9 = 0$$
$$x^2 - 8x = 9$$
$$x^2 - 8x + 4^2 = 9 + 4^2$$
$$(x-4)^2 = 25$$
$$x - 4 = \pm\sqrt{25}$$
$$x - 4 = \pm 5$$
$$x = 4 \pm 5$$
$$x = 9 \text{ or } x = -1$$

22.
$$x^2 + 3x + 2 = 0$$
$$x^2 + 3x = -2$$
$$x^2 + 3x + \left(\frac{3}{2}\right)^2 = -2 + \left(\frac{3}{2}\right)^2$$
$$\left(x + \frac{3}{2}\right)^2 = -2 + \frac{9}{4}$$
$$\left(x + \frac{3}{2}\right)^2 = \frac{1}{4}$$
$$x + \frac{3}{2} = \pm\sqrt{\frac{1}{4}}$$
$$x = -\frac{3}{2} \pm \frac{1}{2}$$
$$x = \frac{-3 \pm 1}{2}$$
$$x = -1 \text{ or } x = -2$$

23.
$$x^2 + x - 56 = 0$$
$$x^2 + x = 56$$
$$x^2 + x + \left(\frac{1}{2}\right)^2 = 56 + \left(\frac{1}{2}\right)^2$$
$$\left(x + \frac{1}{2}\right)^2 = 56 + \frac{1}{4}$$
$$\left(x + \frac{1}{2}\right)^2 = \frac{225}{4}$$
$$x + \frac{1}{2} = \pm\sqrt{\frac{225}{4}}$$
$$x = -\frac{1}{2} \pm \frac{15}{2}$$
$$x = \frac{-1 \pm 15}{2}$$
$$x = 7 \text{ or } x = -8$$

24. $4x^2 - 1 = 0$

$$4x^2 = 1$$

$$x^2 = \frac{1}{4}$$

$$x = \pm\sqrt{\frac{1}{4}}$$

$$x = \pm\frac{1}{2}$$

25. $9x^2 - 4 = 0$

$$9x^2 = 4$$

$$x^2 = \frac{4}{9}$$

$$x = \pm\sqrt{\frac{4}{9}}$$

$$x = \pm\frac{2}{3}$$

26. $x^2 - 4x + 1 = 0$

$$x^2 - 4x = -1$$

$$x^2 - 4x + 2^2 = -1 + 2^2$$

$$(x - 2)^2 = 3$$

$$x - 2 = \pm\sqrt{3}$$

$$x = 2 \pm \sqrt{3}$$

27. $x^2 + 8x + 4 = 0$

$$x^2 + 8x = -4$$

$$x^2 + 8x + 4^2 = -4 + 4^2$$

$$(x + 4)^2 = 12$$

$$x + 4 = \pm\sqrt{12}$$

$$x + 4 = \pm 2\sqrt{3}$$

$$x = -4 \pm 2\sqrt{3}$$

28. $x^2 + x - 4 = 0$

$$x^2 + x = 4$$

$$x^2 + x + \left(\frac{1}{2}\right)^2 = 4 + \left(\frac{1}{2}\right)^2$$

$$\left(x + \frac{1}{2}\right)^2 = 4 + \frac{1}{4}$$

$$\left(x + \frac{1}{2}\right)^2 = \frac{17}{4}$$

$$x + \frac{1}{2} = \pm\sqrt{\frac{17}{4}}$$

$$x = -\frac{1}{2} \pm \frac{\sqrt{17}}{2}$$

$$x = \frac{-1 \pm \sqrt{17}}{2}$$

29. $3x^2 - 4x - 1 = 0$

$$x^2 - \frac{4}{3}x - \frac{1}{3} = 0$$

$$x^2 - \frac{4}{3}x = \frac{1}{3}$$

$$x^2 - \frac{4}{3}x + \left(\frac{4}{6}\right)^2 = \frac{1}{3} + \left(\frac{4}{6}\right)^2$$

$$\left(x - \frac{4}{6}\right)^2 = \frac{1}{3} + \frac{16}{36}$$

$$\left(x - \frac{4}{6}\right)^2 = \frac{28}{36}$$

$$x - \frac{4}{6} = \pm\sqrt{\frac{28}{36}}$$

$$x = \frac{4}{6} \pm \frac{2\sqrt{7}}{6}$$

$$x = \frac{4 \pm 2\sqrt{7}}{6}$$

$$x = \frac{2 \pm \sqrt{7}}{3}$$

30.
$$2x^2 + 3x - 2 = 0$$
$$x^2 + \frac{3}{2}x - 1 = 0$$
$$x^2 + \frac{3}{2}x = 1$$
$$x^2 + \frac{3}{2}x + \left(\frac{3}{4}\right)^2 = 1 + \left(\frac{3}{4}\right)^2$$
$$\left(x + \frac{3}{4}\right)^2 = 1 + \frac{9}{16}$$
$$\left(x + \frac{3}{4}\right)^2 = \frac{25}{16}$$
$$x + \frac{3}{4} = \pm\sqrt{\frac{25}{16}}$$
$$x = -\frac{3}{4} \pm \frac{5}{4}$$
$$x = \frac{-3 \pm 5}{4}$$
$$x = \frac{2}{4} = \frac{1}{2} \text{ or } x = \frac{-8}{4} = -2$$

31.
$$16x^2 - 25 = 0$$
$$16x^2 = 25$$
$$x^2 = \frac{25}{16}$$
$$x = \pm\sqrt{\frac{25}{16}}$$
$$x = \pm\frac{5}{4}$$

32.
$$25x^2 - 16 = 0$$
$$25x^2 = 16$$
$$x^2 = \frac{16}{25}$$
$$x = \pm\sqrt{\frac{16}{25}}$$
$$x = \pm\frac{4}{5}$$

33.
$$x^2 + 11x - 26 = 0$$
$$(x + 13)(x - 2) = 0$$
Set: $\quad x + 13 = 0 \quad x - 2 = 0$
Solve: $\quad\quad x = -13 \quad x = 2$

34.
$$x^2 + 8x - 20 = 0$$
$$(x + 10)(x - 2) = 0$$
Set: $\quad x + 10 = 0 \quad\quad x - 2 = 0$
Solve: $\quad\quad x = -10 \quad\quad x = 2$

35.
$$6x^2 - 12x + 5 = 0$$
$$x^2 - 2x + \frac{5}{6} = 0$$
$$x^2 - 2x = -\frac{5}{6}$$
$$x^2 - 2x + 1^2 = -\frac{5}{6} + 1^2$$
$$(x - 1)^2 = \frac{1}{6}$$
$$x - 1 = \pm\sqrt{\frac{1}{6}}$$
$$x = 1 \pm \frac{\sqrt{6}}{6}$$
$$x = \frac{6 \pm \sqrt{6}}{6}$$

36.
$$3x^2 - 4x - 3 = 0$$
$$x^2 - \frac{4}{3}x - 1 = 0$$
$$x^2 - \frac{4}{3}x = 1$$
$$x^2 - \frac{4}{3}x + \left(\frac{4}{6}\right)^2 = 1 + \left(\frac{4}{6}\right)^2$$
$$\left(x - \frac{4}{6}\right)^2 = 1 + \frac{16}{36}$$
$$\left(x - \frac{4}{6}\right)^2 = \frac{52}{36}$$
$$x - \frac{4}{6} = \pm\sqrt{\frac{52}{36}}$$
$$x = \frac{4}{6} \pm \frac{2\sqrt{13}}{6}$$
$$x = \frac{4 \pm 2\sqrt{13}}{6}$$
$$x = \frac{2 \pm \sqrt{13}}{3}$$

37. $4x^2 + 2x - 3 = 0$

$$x^2 + \frac{1}{2}x - \frac{3}{4} = 0$$

$$x^2 + \frac{1}{2}x = \frac{3}{4}$$

$$x^2 + \frac{1}{2}x + \left(\frac{1}{4}\right)^2 = \frac{3}{4} + \left(\frac{1}{4}\right)^2$$

$$\left(x + \frac{1}{4}\right)^2 = \frac{3}{4} + \frac{1}{16}$$

$$\left(x + \frac{1}{4}\right)^2 = \frac{13}{16}$$

$$x + \frac{1}{4} = \pm\sqrt{\frac{13}{16}}$$

$$x = -\frac{1}{4} \pm \frac{\sqrt{13}}{4}$$

$$x = \frac{-1 \pm \sqrt{13}}{4}$$

38. $x^2 + 7x - 10 = 0$

$$x^2 + 7x = 10$$

$$x^2 + 7x + \left(\frac{7}{2}\right)^2 = 10 + \left(\frac{7}{2}\right)^2$$

$$\left(x + \frac{7}{2}\right)^2 = 10 + \frac{49}{4}$$

$$\left(x + \frac{7}{2}\right)^2 = \frac{89}{4}$$

$$x + \frac{7}{2} = \pm\sqrt{\frac{89}{4}}$$

$$x = -\frac{7}{2} \pm \frac{\sqrt{89}}{2}$$

$$x = \frac{-7 \pm \sqrt{89}}{2}$$

39. $5x^2 - 10x + 1 = 0$

$$x^2 - 2x + \frac{1}{5} = 0$$

$$x^2 - 2x = -\frac{1}{5}$$

$$x^2 - 2x + 1^2 = -\frac{1}{5} + 1^2$$

$$(x - 1)^2 = \frac{4}{5}$$

$$x - 1 = \pm\sqrt{\frac{4}{5}}$$

$$x = 1 \pm \frac{2\sqrt{5}}{5}$$

$$x = \frac{5 \pm 2\sqrt{5}}{5}$$

40. $3x^2 + 2x - 5 = 0$

$(x - 1)(3x + 5) = 0$

Set: $\quad x - 1 = 0 \quad\quad 3x + 5 = 0$

Solve: $\quad\quad x = 1 \quad\quad\quad 3x = -5$

$$x = \frac{-5}{3}$$

41. $2x^2 - x - 2 = 0$

$a = 2, b = -1, c = -2$

$$x = \frac{-(-1) \pm \sqrt{(-1)^2 - 4(2)(-2)}}{2(2)}$$

$$x = \frac{1 \pm \sqrt{1 + 16}}{4}$$

$$x = \frac{1 \pm \sqrt{17}}{4}$$

42. $3x^2 - 6x + 2 = 0$

$a = 3, b = -6, c = 2$

$$x = \frac{-(-6) \pm \sqrt{(-6)^2 - 4(3)(2)}}{2(3)}$$

$$x = \frac{6 \pm \sqrt{36 - 24}}{6}$$

$$x = \frac{6 \pm \sqrt{12}}{6} = \frac{6 \pm 2\sqrt{3}}{6}$$

$$x = \frac{3 \pm \sqrt{3}}{3}$$

43. $2x^2 - 49 = 0$

$a = 2, b = 0, c = -49$

$$x = \frac{-0 \pm \sqrt{0^2 - 4(2)(-49)}}{2(2)}$$

$$x = \frac{\pm 14\sqrt{2}}{4}$$

$$x = \frac{\pm 7\sqrt{2}}{2}$$

44. $\dfrac{1}{3}x^2 - x + \dfrac{1}{2} = 0$ Multiply by 6 to remove fractions.

$2x^2 - 6x + 3 = 0$

$a = 2, b = -6, c = 3$

$$x = \frac{-(-6) \pm \sqrt{(-6)^2 - 4(2)(3)}}{2(2)}$$

$$x = \frac{6 \pm \sqrt{36 - 24}}{4}$$

$$x = \frac{6 \pm \sqrt{12}}{4}$$

$$x = \frac{6 \pm 2\sqrt{3}}{4}$$

$$x = \frac{3 \pm \sqrt{3}}{2}$$

45. $\dfrac{1}{4}x^2 + 2x - \dfrac{1}{3} = 0$ Multiply by 12 to remove fractions.

$3x^2 + 24x - 4 = 0$

$a = 3, b = 24, c = -4$

$$x = \frac{-24 \pm \sqrt{24^2 - 4(3)(-4)}}{2(3)}$$

$$x = \frac{-24 \pm \sqrt{576 + 18}}{6}$$

$$x = \frac{-24 \pm 4\sqrt{39}}{6}$$

$$x = \frac{-12 \pm 2\sqrt{39}}{3}$$

46. $(x - 2)(x - 3) = 6$

$x^2 - 5x + 6 = 6$

$x^2 - 5x = 0$

$a = 1, b = -5, c = 0$

$$x = \frac{-(-5) \pm \sqrt{(-5)^2 - 4(1)(0)}}{2(1)}$$

$$x = \frac{5 \pm \sqrt{25}}{2}$$

$$x = \frac{5 \pm 5}{2}$$

$$x = \frac{10}{2} = 5 \quad \text{or} \quad x = \frac{0}{2} = 0$$

47. $(2x + 1)(x - 1) = 8$

$2x^2 - x - 1 = 8$

$2x^2 - x - 9 = 0$

$a = 2, b = -1, c = -9$

$$x = \frac{-(-1) \pm \sqrt{(-1)^2 - 4(2)(-9)}}{2(2)}$$

$$x = \frac{1 \pm \sqrt{1 + 72}}{4}$$

$$x = \frac{1 \pm \sqrt{73}}{4}$$

48. $(3x - 1)(x + 2) = 6x$

$3x^2 + 5x - 2 = 6x$

$3x^2 - x - 2 = 0$

$a = 3, b = -1, c = -2$

$$x = \frac{-(-1) \pm \sqrt{(-1)^2 - 4(3)(-2)}}{2(3)}$$

$$x = \frac{1 \pm \sqrt{1 + 24}}{6}$$

$$x = \frac{1 \pm \sqrt{25}}{6} = \frac{1 \pm 5}{6}$$

$$x = \frac{6}{6} = 1 \quad \text{or} \quad x = \frac{-4}{6} = \frac{-2}{3}$$

49. $4x^2 - 12x + 9 = 0$

$a = 4, b = -12, c = 9$

$$x = \frac{-(-12) \pm \sqrt{(-12)^2 - 4(4)(9)}}{2(4)}$$

$$x = \frac{12 \pm \sqrt{144 - 144}}{8}$$

$$x = \frac{12 \pm 0}{8}$$

$$x = \frac{3}{2}$$

50. $-3x^2 + 4 = 0$

$a = -3, b = 0, c = 4$

$$x = \frac{-0 \pm \sqrt{0^2 - 4(-3)(4)}}{2(-3)}$$

$$x = \frac{\pm\sqrt{48}}{-6}$$

$$x = \frac{\pm 4\sqrt{3}}{-6} = \pm\frac{4\sqrt{3}}{6}$$

$$x = \pm\frac{2\sqrt{3}}{3}$$

51. $x^2 + x - 2 = 0$

$(x + 2)(x - 1) = 0$

Set: $x + 2 = 0 \quad x - 1 = 0$

Solve: $x = -2 \quad\quad x = 1$

52. $7x^2 - 3x = 0$

$x(7x - 3) = 0$

Set: $x = 0 \quad\quad 7x - 3 = 0$

Solve: $\quad\quad\quad\quad 7x = 3$

$$x = \frac{3}{7}$$

53. $x^2 + 6x = 10$

$x^2 + 6x + 3^2 = 10 + 3^2$

$(x + 3)^2 = 19$

$x + 3 = \pm\sqrt{19}$

$x = -3 \pm \sqrt{19}$

54. $\dfrac{11}{2}x - 1 = 2x^2$

$11x - 2 = 4x^2$

$0 = 4x^2 - 11x + 2$

$a = 4, b = -11, c = 2$

$$x = \frac{-(-11) \pm \sqrt{(-11)^2 - 4(4)(2)}}{2(4)}$$

$$x = \frac{11 \pm \sqrt{121 - 32}}{8}$$

$$x = \frac{11 \pm \sqrt{89}}{8}$$

55. $\dfrac{3}{4}x^2 + 2x = 3$

$3x^2 + 8x = 12$

$3x^2 + 8x - 12 = 0$

$a = 3, b = 8, c = -12$

$$x = \frac{-8 \pm \sqrt{8^2 - 4(3)(-12)}}{2(3)}$$

$$x = \frac{-8 \pm \sqrt{64 + 144}}{6}$$

$$x = \frac{-8 \pm \sqrt{208}}{6}$$

$$x = \frac{-8 \pm 4\sqrt{13}}{6}$$

$$x = \frac{-4 \pm 2\sqrt{13}}{3}$$

56. $-4x^2 + 3x + 1 = 0$

$-(4x^2 - 3x - 1) = 0$

$-(x - 1)(4x + 1) = 0$

Set: $x - 1 = 0 \quad 4x + 1 = 0$

Solve: $x = 1 \quad\quad 4x = -1$

$$x = \frac{-1}{4}$$

57. $x^2 - 4x - 1 = 0$

$a = 1, b = -4, c = -1$

$$x = \frac{-(-4) \pm \sqrt{(-4)^2 - 4(1)(-1)}}{2(1)}$$

$$x = \frac{4 \pm \sqrt{16 + 4}}{2}$$

$$x = \frac{4 \pm \sqrt{20}}{2}$$

$$x = \frac{4 \pm 2\sqrt{5}}{2}$$

$$x = 2 \pm \sqrt{5}$$

$$x \approx 4.2361 \quad \text{or} \quad x \approx -0.2361$$

58. $4x^2 - 8x - 11 = 0$

$a = 4, b = -8, c = -11$

$$x = \frac{-(-8) \pm \sqrt{(-8)^2 - 4(4)(-11)}}{2(4)}$$

$$x = \frac{8 \pm \sqrt{64 + 176}}{8}$$

$$x = \frac{8 \pm \sqrt{240}}{8}$$

$$x = \frac{8 \pm 4\sqrt{15}}{8}$$

$$x = \frac{2 \pm \sqrt{15}}{2}$$

$$x \approx 2.9365 \quad \text{or} \quad x \approx -0.9365$$

59. $(3x + 1)(3x - 1) = 6x$

$$9x^2 - 1 = 6x$$

$$9x^2 - 6x - 1 = 0$$

$a = 9, b = -6, c = -1$

$$x = \frac{-(-6) \pm \sqrt{(-6)^2 - 4(9)(-1)}}{2(9)}$$

$$x = \frac{6 \pm \sqrt{36 + 36}}{18}$$

$$x = \frac{6 \pm 6\sqrt{2}}{18}$$

$$x = \frac{1 \pm \sqrt{2}}{3}$$

$$x \approx 0.8047 \quad \text{or} \quad x \approx -0.1381$$

60. $6x^2 = 300$

$$6x^2 - 300 = 0$$

$a = 6, b = 0, c = -300$

$$x = \frac{-0 \pm \sqrt{0^2 - 4(6)(-300)}}{2(6)}$$

$$x = \frac{\pm\sqrt{7200}}{12}$$

$$x = \frac{\pm 60\sqrt{2}}{12}$$

$$x = \pm 5\sqrt{2}$$

$$x \approx \pm 7.0711$$

61. Let n be the first integer. Then $n+1$ is the second integer.

$$n^2 + (n+1)^2 = 85$$

$$n^2 + n^2 + 2n + 1 = 85$$

$$2n^2 + 2n - 84 = 0$$

$$n^2 + n - 42 = 0$$

$a = 1, b = 1, c = -42$

$$x = \frac{-1 \pm \sqrt{1^2 - 4(1)(-42)}}{2(1)}$$

$$x = \frac{-1 \pm \sqrt{1 + 168}}{2}$$

$$x = \frac{-1 \pm \sqrt{169}}{2}$$

$$x = \frac{-1 \pm 13}{2}$$

$$x = \frac{12}{2} = 6 \quad \text{or} \quad x = \frac{-14}{2} = -7$$

$$x + 1 = 7$$

Disregard -7, as the integers are positive.
So the integers are 6 and 7.

62. Let w be the width. Then $w + 6$ is the length.

$$w(w + 6) = 720$$

$$w^2 + 6w = 720$$

$$w^2 + 6w + 3^2 = 720 + 3^2$$

$$(w + 3)^2 = 729$$

$$w + 3 = \pm\sqrt{729}$$

$$w = -3 \pm 27$$

$$w = 24 \quad \text{or} \quad w = -30$$

$$l = w + 6 = 24 + 6 = 30$$

Disregard -30, as width must be positive. The width is 24 and the length is 30.

63. Let w be the width. Then $2w-5$ is the length.

$$84 = (w-3)\big((2w-5)-3\big)$$
$$84 = (w-3)(2w-8)$$
$$84 = 2w^2 - 14w + 24$$
$$0 = 2w^2 - 14w - 60$$
$$0 = w^2 - 7w - 30$$
$$0 = (w-10)(w+3)$$

Set: $\quad w-10=0 \qquad w+3=0$

Solve: $\quad\quad w=10 \qquad\quad w=-3$

$$2(10)-5 = 15$$

Disregard -3, as width must be positive. The rectangle is 10 meters by 15 meters.

64. Let n be the first integer. Then $n+2$ is the second.

$$n(n+2) = 5(n+n+2)-10$$
$$n^2 + 2n = 5(2n+2)-10$$
$$n^2 + 2n = 10n + 10 - 10$$
$$n^2 - 8n = 0$$
$$n(n-8) = 0$$

Set: $\quad n-8=0 \qquad\qquad n=0$

Solve: $\quad\quad n=8$

$$8+2 = 10 \qquad 0+2=2$$

The integers are either 0 and 2, or 8 and 10.

65. Let s be the length of the two legs of the triangle.

$$s^2 + s^2 = 20^2$$
$$2s^2 = 400$$
$$s^2 = 200$$
$$s = \pm\sqrt{200}$$
$$s = \pm 10\sqrt{2}$$

The side length must be positive, so disregard the negative root. So the sides are $10\sqrt{2}$ meters.

66. Let x be her speed going to the conference. Then $x-8$ is her speed returning home. Use the formula:

$$\text{time} = \frac{\text{distance}}{\text{rate}}.$$

	Rate	Time	Distance
Going	x	$\dfrac{288}{x}$	288
Returning	$x-8$	$\dfrac{288}{x-8}$	288

$$8.5 = \frac{288}{x} + \frac{288}{x-8}$$
$$x(x-8)(8.5) = x(x-8)\left(\frac{288}{x}+\frac{288}{x-8}\right)$$
$$(x^2 - 8x)8.5 = 288(x-8)+288x$$
$$8.5x^2 - 68x = 288x - 2304 + 288x$$
$$0 = 8.5x^2 - 644x + 2304$$
$$0 = 85x^2 - 6440x + 2304$$
$$0 = 5(17x^2 - 1288x + 4608)$$
$$0 = 5(x-72)(17x-64)$$

Set: $\quad x-72=0 \qquad 17x-64=0$

Solve: $\quad\quad x=72 \qquad\quad 17x=64$

$$x = \frac{64}{17} \approx 3.76$$

Disregard the smaller root, as it would yield a negative returning speed. So her average speed going to the conference was 72 mph.

67. Let r be the rate at which they can row in still water.

	Rate	Time	Distance
Down River	$r+2$	$\dfrac{12}{r+2}$	12
Up River	$r-2$	$\dfrac{12}{r-2}$	12

$$8 = \frac{12}{r+2} + \frac{12}{r-2}$$
$$8(r+2)(r-2) = 12(r-2)+12(r+2)$$

Set: $\quad r-4=0 \qquad r+1=0$

Solve: $\quad r=4 \qquad\quad r=-1$

Disregard the negative root, as it has no meaning in this context. They can row at 4 mph in still water.

68. Let t be the amount of time it takes for the small pipe to fill the pool.

	Hours taken to fill the pool	Amount of job completed in 1 hr.
Small Pipe	t	$\dfrac{1}{t}$
Larger Pipe	$t-2$	$\dfrac{1}{t-2}$
Together	2.4	$\dfrac{1}{2.4} = \dfrac{5}{12}$

$$\frac{5}{12} = \frac{1}{t} + \frac{1}{t-2}$$

$$12t(t-2)\left(\frac{5}{12}\right) = 12t(t-2)\left(\frac{1}{t} + \frac{1}{t-2}\right)$$

$$5t(t-2) = 12(t-2) + 12t$$

$$5t^2 - 10t = 12t - 24 + 12t$$

$$5t^2 - 10t = 24t - 24$$

$$5t^2 - 34t + 24 = 0$$

Solve using the quadratic formula:
$a = 5, \ b = -34, \ c = 24$

$$x = \frac{-(-34) \pm \sqrt{(-34)^2 - 4(5)(24)}}{2(5)}$$

$$x = \frac{34 \pm \sqrt{1156 - 480}}{10}$$

$$x = \frac{34 \pm \sqrt{676}}{10}$$

$$x = \frac{34 \pm 26}{10}$$

$$x = \frac{60}{10} = 6 \quad \text{or} \quad x = \frac{-2}{10} = \frac{-1}{5}$$

Disregard the negative root, as it has no meaning in this context. The small pipe will fill the pool in 6 hours and the large pipe will fill it in 4 hours.

69. Let L be the side length of the box that is being made. Use the formula $V = LWH$ where $L = W$, since the base is a square, and $H = 5$.

$$L \cdot L \cdot 5 = 40,500$$

$$L^2 = 8100$$

$$L = \pm\sqrt{8100}$$

$$L = \pm 90$$

Disregard the negative root, as it has no meaning in this context. So the dimensions of the box are $90 \times 90 \times 5$ cm.

70. Let r be the rate of the wind.

	Rate	Time	Distance
Against the wind	$200 - r$	$\dfrac{396}{200 - r}$	396
With the wind	$200 + r$	$\dfrac{396}{200 + r}$	396

Note that 24 minutes must be converted to $\dfrac{2}{5}$ hrs.

$$\frac{396}{200 - r} - \frac{2}{5} = \frac{396}{200 + r}$$

$$0 = \frac{396}{200 + r} - \frac{396}{200 - r} + \frac{2}{5}$$

$$0 = (200 + r)(200 - r)\left(\frac{396}{200 + r} - \frac{396}{200 - r} + \frac{2}{5}\right)$$

$$0 = 396(200 - r) - 396(200 + r) + \frac{2}{5}(40,000 - r^2)$$

$$0 = -\frac{2}{5}r^2 - 792r + 16,000$$

$$0 = r^2 + 1980r - 40,000$$

$$0 = (r - 20)(r + 2000)$$

Set: $\quad r - 20 = 0 \quad\quad r + 2000 = 0$

Solve: $\quad\quad r = 20 \quad\quad\quad\quad r = -2000$

Disregard the negative result, as it has no meaning in this context. The rate of the wind is 20 mph.

71. Given $y = x^2 - 4; \ a = 1, \ b = 0, \ c = -4$.

a. <u>Vertex:</u> $\quad x = -\dfrac{0}{2(1)} = 0$

Substitute $x = 0$ in $y = x^2 - 4$:

$$y = 0^2 - 4 = -4$$

The vertex is $(0, -4)$.

b. <u>Line of Symmetry:</u> From part **a**:

$$x = 0$$

c. <u>x-intercepts:</u>

Set $x^2 - 4$ equal to zero and solve:

$$x^2 - 4 = 0$$

$$x^2 = 4$$

$$x = \pm\sqrt{4}$$

$$x = \pm 2$$

d. <u>Graph:</u>

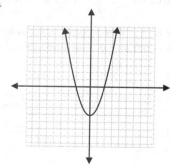

72. Given $y = 16 - x^2$; $a = -1$, $b = 0$, $c = 16$.

   a. <u>Vertex:</u> $x = -\dfrac{0}{2(-1)} = 0$

     Substitute $x = 0$ in $y = 16 - x^2$:
$$y = 16 - 0^2 = 16$$
     The vertex is $(0, 16)$.

   b. <u>Line of Symmetry:</u> From part **a**:
$$x = 0$$

   c. <u>x-intercepts:</u>

     Set $16 - x^2$ equal to zero and solve:
$$0 = 16 - x^2$$
$$x^2 = 16$$
$$x = \pm\sqrt{16}$$
$$x = \pm 4$$

   d. <u>Graph:</u>

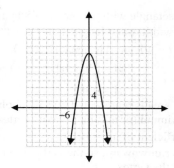

73. Given $y = x^2 - 6x + 5$; $a = 1$, $b = -6$, $c = 5$.

   a. <u>Vertex:</u> $x = -\dfrac{-6}{2(1)} = 3$

     Substitute $x = 3$ in $y = x^2 - 6x + 5$:
$$y = 3^2 - 6 \cdot 3 + 5 = -4$$
     The vertex is $(3, -4)$.

   b. <u>Line of Symmetry:</u> From part **a**:
$$x = 3$$

   c. <u>x-intercepts:</u>

     Set $x^2 - 6x + 5$ equal to zero and solve:
$$x^2 - 6x + 5 = 0$$
$$(x - 5)(x - 1) = 0$$
     Set: $x - 5 = 0$ $x - 1 = 0$

     Solve: $x = 5$ $x = 1$

   d. <u>Graph:</u>

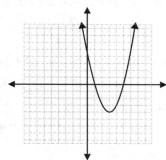

74. Given $y = x^2 - 10x$; $a = 1$, $b = -10$, $c = 0$.

   a. <u>Vertex:</u> $x = -\dfrac{-10}{2(1)} = 5$

     Substitute $x = 5$ in $y = x^2 - 10x$:
$$y = 5^2 - 10 \cdot 5 = -25$$
     The vertex is $(5, -25)$.

   b. <u>Line of Symmetry:</u> From part **a**:
$$x = 5$$

   c. <u>x-intercepts:</u>

     Set $x^2 - 10x$ equal to zero and solve:
$$x^2 - 10x = 0$$
$$x(x - 10) = 0$$
     Set: $x - 10 = 0$ $x = 0$

     Solve: $x = 10$

   d. <u>Graph:</u>

     Note that this graph has a scale of 5.

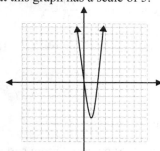

75. Given $y = -x^2 + 2x - 3$; $a = -1$, $b = 2$, $c = -3$.

  a. <u>Vertex:</u> $x = -\dfrac{2}{2(-1)} = 1$

    Substitute $x = 1$ in $y = -x^2 + 2x - 3$:
$$y = -(1)^2 + 2 \cdot 1 - 3 = -2$$
    The vertex is $(1, -2)$.

  b. <u>Line of Symmetry:</u> From part **a**:
$$x = 1$$

  c. <u>x-intercepts:</u>
    Set $-x^2 + 2x - 3$ equal to zero and solve:
$$-x^2 + 2x - 3 = 0$$
$$x^2 - 2x + 3 = 0$$
    This trinomial cannot be factored. There are no x-intercepts.

  d. <u>Graph:</u>

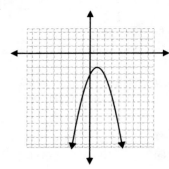

76. Given $y = 3x^2 - 10x + 2$; $a = 3$, $b = -10$, $c = 2$.

  a. <u>Vertex:</u> $x = -\dfrac{-10}{2(3)} = \dfrac{5}{3}$

    Substitute $x = \dfrac{5}{3}$ in $y = 3x^2 - 10x + 2$:
$$y = 3\left(\frac{5}{3}\right)^2 - 10\left(\frac{5}{3}\right) + 2 = -\frac{19}{3}$$
    The vertex is $\left(\dfrac{5}{3}, -\dfrac{19}{3}\right)$.

  b. <u>Line of Symmetry:</u> From part **a**:
$$x = \frac{5}{3}$$

  c. <u>x-intercepts:</u>
    Set $3x^2 - 10x + 2$ equal to zero and solve:
$$x = \frac{-(-10) \pm \sqrt{(-10)^2 - 4(3)(2)}}{2(3)}$$
$$x = \frac{10 \pm \sqrt{100 - 24}}{6}$$

$$x = \frac{10 \pm \sqrt{76}}{6} = \frac{10 \pm 2\sqrt{19}}{6}$$
$$x = \frac{5 \pm \sqrt{19}}{3}$$
$$x \approx 0.2137 \text{ or } x \approx 3.1196$$

  d. <u>Graph:</u>

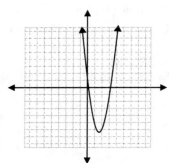

77. A rectangle with perimeter of 80 yards means that for width w and length l, $2w + 2l = 80$, or $w + l = 40$. So $l = 40 - w$. The area is given by:
$$A = (40 - w)w$$
$$A = 40w - w^2$$
The area is represented by a parabola, which has a maximum (or minimum) at its vertex. Since the coefficient of w^2 is negative, this parabola will curve downward, so the vertex is a maximum. To find the vertex, use:
$$w = -\frac{40}{2(-1)} = 20$$
The maximum area occurs when the width is 20 yards. Then the length is $40 - 20 = 20$ yards.

78. **a.** The revenue R from selling $80 - x$ items each week at x dollars is
$$R = (80 - x)x$$
$$R = 80x - x^2$$

  b. Find the vertex of the parabola R to determine the price x that yields the maximum revenue.
$$a = -1, \ b = 80, \ c = 0$$
$$x = -\frac{80}{2(-1)} = 40$$
    So the maximum revenue occurs when the price is \$40.

  c. Plug the result of part **b.** into the revenue function to determine the maximum revenue:
$$R = 80(40) - (40)^2$$
$$R = 3200 - 1600$$
$$R = 1600$$
    So the maximum revenue will be \$1600.

79. Given $h = -16t^2 + 48t + 0 = -16t^2 + 48t$.

    a. The rocket will reach its maximum height at the vertex of the parabola represented by h:

$$t = -\frac{48}{2(-16)} = \frac{3}{2} = 1.5$$

    So the rocket will reach its maximum height after 1.5 seconds.

    b. Plug the result from part **a.** into the given equation to find the maximum height:

$$h = -16(1.5)^2 + 48(1.5)$$
$$h = -36 + 72$$
$$h = 36$$

    So the maximum height is 36 feet.

    c. Set the height equal to zero and solve the equation for t:

$$0 = -16t^2 + 48t$$
$$0 = -16t(t - 3)$$

    Set: $-16t = 0$ $t - 3 = 0$

    Solve: $t = 0$ $t = 3$

    The zero result indicates that the rocket starts at the ground, a fact we already know. The rocket hits the ground again after 3 seconds.

80. Given $h = -16t^2 + 64t + 80$.

    a. The ball will reach its maximum height at the vertex of the parabola represented by h:

$$t = -\frac{64}{2(-16)} = 2$$

    The ball will reach its maximum height after 2 seconds.

    b. Plug the result from part **a.** into the given equation to find the maximum height:

$$h = -16(2)^2 + 64(2) + 80$$
$$h = -64 + 128 + 80$$
$$h = 144$$

    The maximum height is 144 feet.

    c. Set the height equal to zero and solve the equation for t:

$$0 = -16t^2 + 64t + 80$$
$$0 = -16(t^2 - 4t - 5)$$
$$0 = -16(t - 5)(t + 1)$$

    Set: $0 = t - 5$ $0 = t + 1$

    Solve: $5 = t$ $-1 = t$

    The ball will reach the ground after 5 seconds.

[End of Chapter 10 Review]

Chapter 10 Test
Solutions to All Exercises

1. Solve: $(x+5)^2 = 49$

$$x+5 = \pm\sqrt{49}$$
$$x+5 = \pm 7$$
$$x = -5 \pm 7$$
$$x = -12, \; x = 2$$

2. Solve: $8x^2 = 96$

$$x^2 = \frac{96}{8}$$
$$x^2 = 12$$
$$x = \pm\sqrt{12}$$
$$x = \pm 2\sqrt{3}$$

3. Use the Pythagorean Theorem:

$$x^2 + 5^2 = 13^2$$
$$x^2 = 169 - 25$$
$$x^2 = 144$$
$$x = \pm\sqrt{144}$$
$$x = \pm 12$$

Disregard negative values, so the side is 12.

4. Complete the square:

$$x^2 - 24x + \left(\frac{24}{2}\right)^2 = (x-12)^2$$
$$x^2 - 24x + 144 = (x-12)^2$$

5. Complete the square:

$$x^2 + 9x + \left(\frac{9}{2}\right)^2 = \left(x+\frac{9}{2}\right)^2$$
$$x^2 + 9x + \frac{81}{4} = \left(x+\frac{9}{2}\right)^2$$

6. First, divide each side of the equation by 3. Then, complete the square:

$$3x^2 + 9x = 3(\underline{\quad})^2$$
$$x^2 + 3x = (\underline{\quad})^2$$
$$x^2 + 3x + \left(\frac{3}{2}\right)^2 = \left(x+\frac{3}{2}\right)^2$$
$$x^2 + 3x + \frac{9}{4} = \left(x+\frac{3}{2}\right)^2$$
$$3x^2 + 9x + \frac{27}{4} = 3\left(x+\frac{3}{2}\right)^2$$

Remember to multiply through by 3 after completing the square to obtain the original trinomial.

7. Solve $x^2 - 3x + 2 = 0$ by factoring:

$$(x-2)(x-1) = 0$$

Set: $\quad x-2 = 0 \quad\quad x-1 = 0$

Solve: $\quad\quad x = 2 \quad\quad\quad x = 1$

8. Solve $5x^2 + 12x = 0$ by factoring:

$$x(5x+12) = 0$$

Set: $\quad\quad x = 0 \quad\quad 5x+12 = 0$

Solve: $\quad\quad\quad\quad\quad\quad 5x = -12$

$$x = -\frac{12}{5}$$

9. Solve $3x^2 - x - 10 = 0$ by factoring:

$$(x-2)(3x+5) = 0$$

Set: $\quad\quad x-2 = 0 \quad 3x+5 = 0$

Solve: $\quad\quad x = 2 \quad\quad 3x = -5$

$$x = -\frac{5}{3}$$

10. Solve $x^2 + 6x + 8 = 0$ by completing the square:

$$x^2 + 6x \underline{\quad\quad} = -8$$
$$x^2 + 6x + 3^2 = 3^2 - 8$$
$$(x+3)^2 = 1$$
$$x+3 = \pm\sqrt{1}$$
$$x = -3 \pm 1$$
$$x = -2; \; x = -4$$

11. Solve $x^2 - 5x = 6$ by completing the square:

$$x^2 - 5x \underline{\qquad} = 6$$

$$x^2 - 5x + \left(\frac{5}{2}\right)^2 = \left(\frac{5}{2}\right)^2 + 6$$

$$\left(x - \frac{5}{2}\right)^2 = \frac{25}{4} + 6 = \frac{49}{4}$$

$$x - \frac{5}{2} = \pm\sqrt{\frac{49}{4}}$$

$$x = \frac{5}{2} \pm \frac{7}{2}$$

$$x = -1 \; ; \; x = 6$$

12. Solve $2x^2 - 8x - 4 = 0$ by completing the square:

$$2x^2 - 8x \underline{\qquad} = 4$$

$$2\left(x^2 - 4x \underline{\qquad}\right) = 4$$

$$x^2 - 4x \underline{\qquad} = 2$$

$$x^2 - 4x + \left(2\right)^2 = 2 + 2^2$$

$$\left(x - 2\right)^2 = 6$$

$$x - 2 = \pm\sqrt{6}$$

$$x = 2 \pm \sqrt{6}$$

13. Solve $3x^2 + 8x + 2 = 0$ by using the quadratic formula: $a = 3$, $b = 8$ and $c = 2$

$$x = \frac{-8 \pm \sqrt{8^2 - 4(3)(2)}}{2(3)}$$

$$x = \frac{-8 \pm \sqrt{40}}{6}$$

$$x = \frac{-4 \pm \sqrt{10}}{3}$$

14. Solve $2x^2 = 4x + 3$ by using the quadratic formula:

$$2x^2 - 4x - 3 = 0$$

$$a = 2, \ b = -4 \ \text{and} \ c = -3$$

$$x = \frac{-(-4) \pm \sqrt{(-4)^2 - 4(2)(-3)}}{2(2)}$$

$$x = \frac{4 \pm \sqrt{16 + 24}}{4}$$

$$x = \frac{4 \pm \sqrt{40}}{4}$$

$$x = \frac{2 \pm \sqrt{10}}{2}$$

15. Solve $\frac{1}{2}x^2 - x = 2$ by using the quadratic formula.

First, you can simplify the calculations by multiplying both sides of the equation by 2:

$$x^2 - 2x - 4 = 0$$

$$a = 1, \ b = -2 \ \text{and} \ c = -4$$

$$x = \frac{-(-2) \pm \sqrt{(-2)^2 - 4(1)(-4)}}{2(1)}$$

$$x = \frac{2 \pm \sqrt{4 + 16}}{2}$$

$$x = \frac{2 \pm \sqrt{20}}{2} = \frac{2 \pm 2\sqrt{5}}{2}$$

$$x = 1 \pm \sqrt{5}$$

For Exercises 16 – 18, you could use any one of three methods that would solve the problem. Only one method is shown.

16. Solve $x^2 + 6x = 2$ by completing the square:

$$x^2 + 6x + 3^2 = 3^2 + 2$$

$$\left(x + 3\right)^2 = 11$$

$$x + 3 = \pm\sqrt{11}$$

$$x = -3 \pm \sqrt{11}$$

17. Solve $3x^2 - 7x + 2 = 0$ by factoring:

$$\left(x - 2\right)\left(3x - 1\right) = 0$$

Set: $\qquad x - 2 = 0 \qquad 3x - 1 = 0$

Solve: $\qquad x = 2 \qquad 3x = 1$

$$x = \frac{1}{3}$$

18. Solve $2x^2 - 3x = 1$ by using the quadratic formula:

$$2x^2 - 3x - 1 = 0$$

$$a = 2, \ b = -3 \ \text{and} \ c = -1$$

$$x = \frac{-(-3) \pm \sqrt{(-3)^2 - 4(2)(-1)}}{2(2)}$$

$$x = \frac{3 \pm \sqrt{9 + 8}}{4}$$

$$x = \frac{3 \pm \sqrt{17}}{4}$$

19. Given $y = x^2 - 5$: $a = 1$, $b = 0$ and $c = -5$.

   a. <u>Vertex</u>: $x = -\dfrac{0}{2(1)} = 0$

     Substitute $x = 0$ in $y = x^2 - 5$:

$$y = 0^2 - 5 = -5$$

     The vertex is $(0, -5)$.

   b. <u>Line of Symmetry</u>: $x = 0$

   c. <u>x-intercepts</u>: Set $y = x^2 - 5$ equal to 0 and solve:

$$x^2 - 5 = 0$$
$$x^2 = 5$$
$$x = \pm\sqrt{5}$$

     The parabola crosses the x-axis at $x = \pm\sqrt{5}$.

   d. Graph:

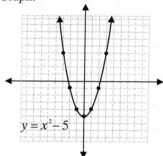

20. Given $y = -x^2 + 4x$: $a = -1$, $b = 4$ and $c = 0$.

   a. <u>Vertex</u>: $x = -\dfrac{4}{2(-1)} = 2$

     Substitute $x = 2$ in $y = -x^2 + 4x$:

$$y = -2^2 + 4(2) = -4 + 8 = 4$$

     The vertex is $(2, 4)$.

   b. <u>Line of Symmetry</u>: $x = 2$

   c. <u>x-intercepts</u>: Set $y = -x^2 + 4x$ equal to 0 and solve:

$$-x^2 + 4x = 0$$
$$-x(x - 4) = 0$$

     Set: $-x = 0$ $x - 4 = 0$

     Solve: $x = 0$ $x = 4$

     The parabola crosses the x-axis at $x = 0$ and $x = 4$.

d. Graph:

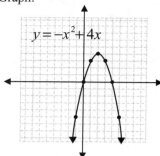

21. Given $y = 2x^2 + 3x + 1$: $a = 2$, $b = 3$ and $c = 1$.

   a. <u>Vertex</u>: $x = -\dfrac{3}{2(2)} = -\dfrac{3}{4}$

     Substitute $x = -\dfrac{3}{4}$ in $y = 2x^2 + 3x + 1$:

$$y = 2\left(-\frac{3}{4}\right)^2 + 3\left(-\frac{3}{4}\right) + 1$$

$$y = 2\left(\frac{9}{16}\right) - \frac{9}{4} + 1$$

$$y = -\frac{1}{8}$$

     The vertex is $\left(\dfrac{-3}{4}, \dfrac{-1}{8}\right)$.

   b. <u>Line of Symmetry</u>: $x = -\dfrac{3}{4}$

   c. <u>x-intercepts</u>: Set $y = 2x^2 + 3x + 1$ equal to 0 and solve by factoring:

$$(x + 1)(2x + 1) = 0$$

     Set: $x + 1 = 0$ $2x + 1 = 0$

     Solve: $x = -1$ $2x = -1$

$$x = \frac{-1}{2}$$

     The parabola crosses the x-axis at $x = \dfrac{-1}{2}$ and $x = -1$.

   d. Graph:

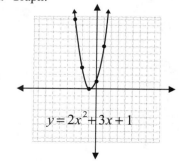

22.

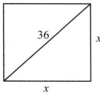

Use the Pythagorean Theorem.

$$x^2 + x^2 = 36^2$$
$$2x^2 = 36^2$$
$$x^2 = 18 \cdot 36 = 18^2 \cdot 2$$
$$x = 18\sqrt{2}$$

Only consider the positive root here, as a negative root has no meaning in this context. The side of the square is $18\sqrt{2}$ inches.

23. Given $h = -16t^2 + 144t$:

a. Set: $h = 96$ and solve.

$$-16t^2 + 144t = 96$$
$$-16t^2 + 144 - 96 = 0$$
$$-16\left(t^2 - 9t + 6\right) = 0$$
$$t^2 - 9t + 6 = 0$$

Solve by using the quadratic formula:

$$t = \frac{-(-9) \pm \sqrt{(-9)^2 - 4(1)(6)}}{2(1)}$$
$$t = \frac{9 \pm \sqrt{81 - 24}}{2}$$
$$t = \frac{9 \pm \sqrt{57}}{2} \approx \frac{9 \pm 7.5498}{2}$$

The rocket is at a height of 96 ft at $t = 0.725$ seconds and again at $t = 8.275$ seconds.

b. Maximum height occurs at the vertex of the parabola. The x-value of the vertex is

$$x = -\frac{b}{2a} = -\frac{144}{2(-16)} = \frac{9}{2}$$

To find the height at the vertex, which is the highest point for the rocket, substitute this x-value in $h = -16t^2 + 144t$:

$$h = -16\left(\frac{9}{2}\right)^2 + 144\left(\frac{9}{2}\right)$$
$$h = -324 + 648$$
$$h = 324$$

The maximum height of the rocket is 324 ft.

c. The rocket will hit the ground when $h = 0$:

$$-16t^2 + 144t = 0$$
$$t\left(-16t + 144\right) = 0$$

Set: $t = 0$ $-16t + 144 = 0$

Solve: $-16t = -144$

                             $t = 9$

The rocket will hit the ground after 9 seconds.

24. Let c be the rate of the current. Recall that $d = rt$.

	Rate	Time	Distance
Down	$8 + c$	$\dfrac{60}{8+c}$	60
Up	$8 - c$	$\dfrac{60}{8-c}$	60

The basis for the equation is the fact that going down stream is 10 hours less than going upstream:

$$\frac{60}{8+c} + 10 = \frac{60}{8-c}$$
$$\frac{60 + 10(8+c)}{8+c} = \frac{60}{8-c}$$
$$\frac{140 + 10c}{8+c} = \frac{60}{8-c}$$
$$(140 + 10c)(8 - c) = 60(8 + c)$$
$$1120 - 60c - 10c^2 = 480 + 60c$$
$$10c^2 + 120c - 640 = 0$$
$$c^2 + 12c - 64 = 0$$

Solve by factoring:

$$(c - 4)(c + 16) = 0$$

Set: $c - 4 = 0$ $c + 16 = 0$

Solve: $c = 4$ $c = -16$

Disregard the negative root, as it has no meaning here. The rate of the current is 4 mph.

25. Let w be the width of the original sheet of metal. Then $w + 3$ is the length of the original sheet. Four corners are cut from the original sheet. Each corner is 4 inches by 4 inches.

Use the formula for volume of a rectangular solid $V = lwh$, where $h = 4$ in. The width of the box, after the corners are cut out, is $w - 8$. The length of the box would then be $w + 3 - 8 = w - 5$.

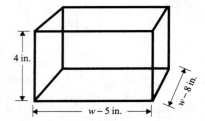

Using the formula when $V = 720$,

$$720 = (w - 5)(w - 8) \cdot 4$$
$$180 = w^2 - 13w + 40$$
$$0 = w^2 - 13w - 140$$
$$0 = (w - 20)(w + 7)$$

Set: $w - 20 = 0$ $w + 7 = 0$

Solve: $w = 20$ $w = -7$

Disregard the negative result, as it has no meaning here. The original sheet of metal is 20 in. by 23 in.

[End of Chapter 10 Test]

Cumulative Review: Chapters 1 – 10
Solutions to All Exercises

1. $\left(-3x^2 y\right)^2 = 9x^4 y^2$

2. $\dfrac{\left(2x^2\right)\left(-6x^3\right)}{3x^{-1}} = \dfrac{-12x^5}{3x^{-1}} = -4x^6$

3. $4x + 3 - 2(x+4) + (2-x) = 4x + 3 - 2x - 8 + 2 - x$
$$= x - 3$$

4. $4x - \left[3x - 2(x+5) + 7 - 5x\right]$
$$= 4x - (3x - 2x - 10 + 7 - 5x)$$
$$= 4x - (-4x - 3)$$
$$= 4x + 4x + 3$$
$$= 8x + 3$$

5. $5(x-8) - 3(x-6) = 0$
$$5x - 40 - 3x + 18 = 0$$
$$2x - 22 = 0$$
$$2x = 22$$
$$x = 11$$

6. $\dfrac{5}{4}x + 4 = \dfrac{3}{4}x + 7$
$$5x + 16 = 3x + 28$$
$$2x = 12$$
$$x = 6$$

7. $x^2 - 8x + 15 = 0$
$(x-3)(x-5) = 0$
Set: $\quad x - 3 = 0 \qquad x - 5 = 0$
Solve: $\qquad x = 3 \qquad\quad x = 5$

8. $x^2 + 5x - 36 = 0$
$(x-4)(x+9) = 0$
Set: $\quad x - 4 = 0 \qquad x + 9 = 0$
Solve: $\quad\; x = 4 \qquad\quad x = -9$

9. $\dfrac{1}{x} + \dfrac{2}{3x} = 1$
$$3 + 2 = 3x$$
$$5 = 3x$$
$$\dfrac{5}{3} = x$$

10. $\dfrac{3}{x-1} + \dfrac{5}{x+1} = 2$
$$3(x+1) + 5(x-1) = 2(x-1)(x+1)$$
$$3x + 3 + 5x - 5 = 2x^2 - 2$$
$$8x - 2 = 2x^2 - 2$$
$$0 = 2x^2 - 8x$$
$$0 = 2x(x-4)$$
Set: $\qquad 2x = 0 \qquad x - 4 = 0$
Solve: $\qquad\; x = 0 \qquad\quad x = 4$

11. $(x-4)^2 = 9$
$$x - 4 = \pm\sqrt{9}$$
$$x = 4 \pm 3$$
$$x = 7 \quad \text{or} \quad x = 1$$

12. $(x+3)^2 = 6$
$$x + 3 = \pm\sqrt{6}$$
$$x = -3 \pm \sqrt{6}$$

13. $x^2 + 4x - 2 = 0$
$$x^2 + 4x = 2$$
$$x^2 + 4x + 4 = 2 + 4$$
$$(x+2)^2 = 6$$
$$x + 2 = \pm\sqrt{6}$$
$$x = -2 \pm \sqrt{6}$$

14. $2x^2 - 3x - 4 = 0$
$$a = 2, \ b = -3 \ \text{and} \ c = -4$$
$$x = \dfrac{-(-3) \pm \sqrt{(-3)^2 - 4(2)(-4)}}{2 \cdot 2}$$
$$x = \dfrac{3 \pm \sqrt{9 + 32}}{4}$$
$$x = \dfrac{3 \pm \sqrt{41}}{4}$$

15. $x - 2 = \sqrt{x + 18}$
$$(x-2)^2 = x + 18$$
$$x^2 - 4x + 4 = x + 18$$
$$x^2 - 5x - 14 = 0$$
$$(x-7)(x+2) = 0$$
Set: $\qquad x - 7 = 0 \qquad x + 2 = 0$
Solve: $\qquad\; x = 7 \qquad\quad x = -2$

Check each solution:

For $x = 7$: $\quad (7) - 2 = 5$

$$\sqrt{(7)+18} = \sqrt{25} = 5$$

For $x = -2$: $\quad (-2) - 2 = -4$

$$\sqrt{(-2)+18} = \sqrt{16} = 4$$

So, $x = 7$ is a solution, but $x = -2$ is not.

16. $\qquad x + 3 = 2\sqrt{2x+6}$

$$(x+3)^2 = 4(2x+6)$$

$$x^2 + 6x + 9 = 8x + 24$$

$$x^2 - 2x - 15 = 0$$

$$(x+3)(x-5) = 0$$

Set: $\qquad x + 3 = 0 \qquad x - 5 = 0$

Solve: $\qquad x = -3 \qquad x = 5$

Check each solution:

For $x = -3$: $\qquad (-3) + 3 = 0$

$$2\sqrt{2(-3)+6} = 2\sqrt{0} = 0$$

For $x = 5$: $\qquad (5) + 3 = 8$

$$2\sqrt{2(5)+6} = 2\sqrt{16} = 8$$

So, $x = -3$ and $x = 5$ are both solutions.

17. $\quad 15 - 3x + 1 \geq x + 4$

$$12 \geq 4x$$

$$x \leq 3$$

18. $\quad 4(6 - x) \geq -2(3x + 1)$

$$24 - 4x \geq -6x - 2$$

$$2x \geq -26$$

$$x \geq -13$$

19. $\quad 2(x - 5) - 4 < x - 3(x - 1)$

$$2x - 10 - 4 < x - 3x + 3$$

$$4x < 17$$

$$x < \frac{17}{4}$$

20. $\quad x - (2x + 5) \leq 9 - (4 - x)$

$$x - 2x - 5 \leq 9 - 4 + x$$

$$-x - 5 \leq 5 + x$$

$$-2x \leq 10$$

$$x \geq -5$$

21. $\begin{cases} x + 5y = 9 \\ 2x - 3y = 1 \end{cases}$

Multiply the first equation by -2.

Solve by the addition method:

$$-2x - 10y = -18$$

$$\underline{2x - 3y = \quad 1}$$

$$-13y = -17$$

$$y = \frac{17}{13}$$

Substitute into the first equation:

$$x + 5\left(\frac{17}{13}\right) = 9$$

$$x = 9 - \frac{85}{13}$$

$$x = \frac{32}{13}$$

The solution is $\left(\dfrac{32}{13}, \dfrac{17}{13}\right)$.

22. $\begin{cases} 6x - y = 11 \\ 2x - 3y = -11 \end{cases}$

Multiply the second equation by -3.

Solve by the addition method:

$$6x - \quad y = 11$$

$$\underline{-6x + 9y = 33}$$

$$8y = 44$$

$$y = \frac{11}{2}$$

Substitute into the first equation:

$$6x - \left(\frac{11}{2}\right) = 11$$

$$6x = \frac{33}{2}$$

$$x = \frac{11}{4}$$

The solution is $\left(\dfrac{11}{4}, \dfrac{11}{2}\right)$.

23. $\begin{cases} 2x-5y=11 \\ 5x-3y=-1 \end{cases}$

Multiply the first equation by -5, and the second by 2. Use the addition method.

$$-10x + 25y = -55$$
$$\underline{10x - 6y = -2}$$
$$19y = -57$$
$$y = -3$$

Substitute into the first equation:

$$2x-5(-3) = 11$$
$$2x+15 = 11$$
$$2x = -4$$
$$x = -2$$

The solution is $(-2,-3)$.

24. $\begin{cases} x+2y=6 \\ \quad y = -\dfrac{1}{2}x+3 \end{cases}$

Solve by the substitution method:

$$x+2\left(-\frac{1}{2}x+3\right) = 6$$
$$x-x+6 = 6$$
$$6 = 6$$

This statement is always true. The system is dependent. The solution is $\left(x, -\dfrac{1}{2}x+3\right)$.

25. $\begin{cases} \quad y=-3x+4 \\ 3x+y=7 \end{cases}$

Solve by the substitution method:

$$3x+(-3x+4) = 7$$
$$3x-3x+4 = 7$$
$$4 = 7$$

This statement is never true, so the system is inconsistent. The lines are parallel.

26. $\begin{cases} y < \dfrac{1}{3}x \\ y > -x+4 \end{cases}$

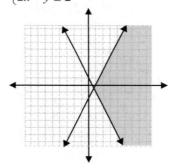

27. $\begin{cases} y \le 4x-1 \\ y \le -2x+3 \end{cases}$

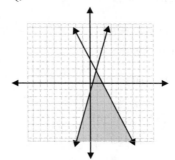

28. $\begin{cases} x > -2 \\ y \le 5 \end{cases}$

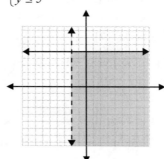

29. $\begin{cases} 2x+y \ge 1 \\ 2x-y \ge 2 \end{cases}$

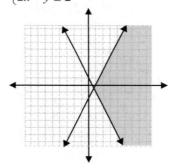

30. $P = a+2b$, solve for b.
$$P-a = 2b$$
$$b = \frac{P-a}{2}$$

31. $A = \dfrac{h}{2}(a+b)$, solve for a.
$$2A = h(a+b)$$
$$\frac{2A}{h} = a+b$$
$$a = \frac{2A}{h}-b$$

32. $x^2 - 12x + 36 = (x-6)^2$

A perfect square trinomial

33. $25x^2 - 49 = (5x+7)(5x-7)$

A difference of two squares

34. $3x^2 + 10x + 3 = (3x+1)(x+3)$

35. $3x^2 + 6x - 72 = 3(x^2 + 2x - 24)$

$= 3(x+6)(x-4)$

36. $x^3 + 4x^2 - 8x - 32 = x^2(x+4) - 8(x+4)$

$= (x^2 - 8)(x+4)$

37. $(5x^2 + 3x - 7) + (2x^2 - 3x - 7)$

$= (5x^2 + 2x^2) + (3x - 3x) + (-7 - 7)$

$= 7x^2 - 14$

38. $(2x^2 - 7x + 3) + (5x^2 - 3x + 4)$

$= (2x^2 + 5x^2) + (-7x - 3x) + (3 + 4)$

$= 7x^2 - 10x + 7$

39. $\dfrac{x^2}{x^2 - 4} \cdot \dfrac{x^2 - 3x + 2}{x^2 - x} = \dfrac{x^2}{(x-2)(x+2)} \cdot \dfrac{(x-2)(x-1)}{x(x-1)}$

$= \dfrac{x}{x+2}$

40. $\dfrac{3x}{x^2 - 6x - 7} \div \dfrac{2x^2}{x^2 - 8x + 7} = \dfrac{3x}{x^2 - 6x - 7} \cdot \dfrac{x^2 - 8x + 7}{2x^2}$

$= \dfrac{3x(x-7)(x-1)}{2x^2(x-7)(x+1)}$

$= \dfrac{3(x-1)}{2x(x+1)}$

41. $\dfrac{2x}{x^2 - 25} - \dfrac{x+1}{x+5} = \dfrac{2x - (x+1)(x-5)}{x^2 - 25}$

$= \dfrac{-x^2 + 6x + 5}{x^2 - 25}$

42. $\dfrac{4x}{x^2 - 3x - 4} + \dfrac{x - 2}{x^2 + 3x + 2}$

$= \dfrac{4x}{(x-4)(x+1)} + \dfrac{x-2}{(x+2)(x+1)}$

$= \dfrac{4x(x+2) + (x-2)(x-4)}{(x-4)(x+1)(x+2)}$

$= \dfrac{4x^2 + 8x + x^2 - 6x + 8}{(x-4)(x+1)(x+2)}$

$= \dfrac{5x^2 + 2x + 8}{(x-4)(x+1)(x+2)}$

43. $\sqrt{63} = \sqrt{9 \cdot 7} = 3\sqrt{7} \approx 7.9373$

44. $\sqrt{243} = \sqrt{81 \cdot 3} = 9\sqrt{3} \approx 15.5885$

45. $\sqrt[3]{250} = \sqrt[3]{125 \cdot 2} = 5\sqrt[3]{2} \approx 6.2996$

46. $\sqrt{\dfrac{25}{12}} = \sqrt{\dfrac{25}{12} \cdot \dfrac{3}{3}} = \dfrac{5\sqrt{3}}{6} \approx 1.4434$

47. $\sqrt{147} + \sqrt{48} = \sqrt{49 \cdot 3} + \sqrt{16 \cdot 3}$

$= 7\sqrt{3} + 4\sqrt{3}$

$= 11\sqrt{3}$

≈ 19.0526

48. $(2\sqrt{3} + 1)(\sqrt{3} - 4) = 2 \cdot 3 - 8\sqrt{3} + \sqrt{3} - 4$

$= 2 - 7\sqrt{3} \approx -10.1244$

49. $4x^{\frac{2}{3}} \cdot 2x^{\frac{1}{3}} = 8x$

50. $(4x^3)^2 (2x^2)^{-1} = \dfrac{16x^6}{2x^2} = 8x^4$

51. $\left(\dfrac{m^{\frac{1}{2}} n^{\frac{3}{4}}}{m^{\frac{-1}{2}} n^{\frac{1}{4}}} \right)^2 = \left(\dfrac{m^1 n^{\frac{3}{2}}}{m^{-1} n^{\frac{1}{2}}} \right) = m^2 n$

52. $\left(\dfrac{25x^2 y^4}{36x^4 y^6} \right)^{\frac{1}{2}} = \dfrac{5xy^2}{6x^2 y^3} = \dfrac{5}{6xy}$

53. $y = \dfrac{2}{3}x + 4$

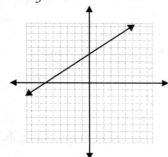

$m = \dfrac{2}{3}$ and the y-intercept is $(0, 4)$.

54. $4x - 3y = 6$

$y = \dfrac{4}{3}x - 2$

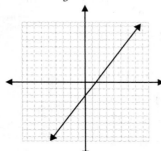

$m = \dfrac{4}{3}$ and the y-intercept is $(0, -2)$.

55. $2x + 4y = -5$

$y = -\dfrac{1}{2}x - \dfrac{5}{4}$

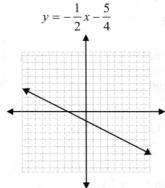

$m = -\dfrac{1}{2}$ and the y-intercept is $\left(0, -\dfrac{5}{4}\right)$.

56. Given $y = x^2 - 8x + 7$: $a = 1$, $b = -8$, $c = 7$.

a. <u>Vertex</u>: $x = -\dfrac{-8}{2(1)} = 4$

Substitute $x = 4$ in $y = x^2 - 8x + 7$:

$$y = (4)^2 - 8(4) + 7 = -9$$

The vertex is $(4, -9)$.

b. <u>Line of symmetry</u>: $x = 4$

c. <u>x-intercepts</u>: Set $y = x^2 - 8x + 7$ equal to 0 and solve:

$$x^2 - 8x + 7 = 0$$
$$(x - 7)(x - 1) = 0$$
$$x = 7 \quad \text{and} \quad x = 1$$

The intercepts are $x = 7$ and $x = 1$.

d. <u>Graph</u>:

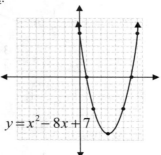

$y = x^2 - 8x + 7$

57. Given $y = 2(x + 3)^2 - 6$:

$$y = 2x^2 + 12x + 12$$
$$a = 2, \ b = 12, \ c = 12$$

a. <u>Vertex</u>: $x = -\dfrac{12}{2(2)} = -3$

Substitute $x = -3$ in $y = 2(x + 3)^2 - 6$:

$$y = 2(-3 + 3)^2 - 6 = -6$$

The vertex is $(-3, -6)$.

b. <u>Line of symmetry</u>: $x = -3$

c. <u>x-intercepts</u>: Set $y = 2x^2 + 12x + 12$ equal to 0 and solve:

$$2x^2 + 12x + 12 = 0$$
$$x^2 + 6x + 6 = 0$$

$$x = \dfrac{-6 \pm \sqrt{6^2 - 4(1)(6)}}{2 \cdot 1}$$

$$x = \dfrac{-6 \pm \sqrt{36 - 24}}{2}$$

$$x = \dfrac{-6 \pm \sqrt{12}}{2}$$

$$x = -3 \pm \sqrt{3}$$

The intercepts are $x = -3 + \sqrt{3}$ and $x = -3 - \sqrt{3}$.

d. <u>Graph</u>:

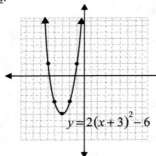

$y = 2(x+3)^2 - 6$

58. Given $y = x^2 + x + 1$: $a = 1$, $b = 1$ and $c = 1$.

a. <u>Vertex</u>: $x = -\dfrac{1}{2(1)} = -\dfrac{1}{2}$

Substitute $x = -\dfrac{1}{2}$ in $y = x^2 + x + 1$:

$$y = \left(-\dfrac{1}{2}\right)^2 - \dfrac{1}{2} + 1 = \dfrac{3}{4}$$

The vertex is $\left(-\dfrac{1}{2}, \dfrac{3}{4}\right)$.

b. <u>Line of symmetry</u>: $x = -\dfrac{1}{2}$

c. <u>x-intercepts</u>: Set $y = x^2 + x + 1$ equal to 0 and solve:

$$x^2 + x + 1 = 0$$

$$x = \dfrac{-1 \pm \sqrt{1^2 - 4(1)(1)}}{2 \cdot 1}$$

$$x = \dfrac{-1 \pm \sqrt{-3}}{2}$$

These roots are not real numbers, so the parabola does not cross the x-axis.

d. <u>Graph</u>:

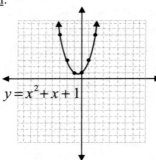

$y = x^2 + x + 1$

59. Given $5x + 2y = 6$, write in slope-intercept form:

$$y = -\dfrac{5}{2}x + 3 \qquad \text{Slope: } m = -\dfrac{5}{2}$$

y-intercept: $(0, 3)$

60. Given the two points $(-5, 2)$ and $(3, 5)$.

a. The slope of the line through the points is

$$m = \dfrac{5 - 2}{3 - (-5)} = \dfrac{3}{8}.$$

b. The distance between the two points is

$$d = \sqrt{(-5 - 3)^2 + (2 - 5)^2}$$

$$d = \sqrt{64 + 9}$$

$$d = \sqrt{73}$$

c. The midpoint of the line segment joining the two points:

$$\text{Midpoint} = \left(\dfrac{-5 + 3}{2}, \dfrac{2 + 5}{2}\right) = \left(-1, \dfrac{7}{2}\right)$$

d. The equation of the line through the two points, using point-slope form:

$$y - 5 = \dfrac{3}{8}(x - 3), \text{ or}$$

$$y - 2 = \dfrac{3}{8}(x + 5)$$

61. Given the function $f(x) = 5x - 3$:

a. $f(-1) = 5(-1) - 3 = -8$

b. $f(6) = 5(6) - 3 = 27$

c. $f(0) = 5(0) - 3 = -3$

62. Given the function $F(x) = 3x^2 - 2x + 7$:

a. $F(0) = 3(0)^2 - 2(0) + 7 = 7$

b. $F(-2) = 3(-2)^2 - 2(-2) + 7 = 23$

c. $F(5) = 3(5)^2 - 2(5) + 7 = 72$

63. Let x be the amount invested at 5%. Then, $x - 800$ is the amount invested at 8%.

$$0.05x + 0.08(x - 800) = 209$$

$$5x + 8(x - 800) = 20900$$

$$13x - 6400 = 20900$$

$$13x = 27300$$

$$x = 2100$$

There is $2100 at 5% and $1300 at 8%.

64. Let x be the amount of time it takes them together.

	Time to finish job	Part finished in 1 hour
Ben	10	$\dfrac{1}{10}$
Beth	8	$\dfrac{1}{8}$
Together	x	$\dfrac{1}{x}$

The sum of the parts each can do alone is equal to the part they can do together. Setup the equation and solve.

$$\frac{1}{10}+\frac{1}{8}=\frac{1}{x}$$
$$4x+5x=40$$
$$9x=40$$
$$x=\frac{40}{9}$$

Together they can do the work in $\dfrac{40}{9}\approx 4.4$ hours.

65. Let c be the rate of the current.

	Rate	Time	Distance
Up	$8-c$	$\dfrac{6}{8-c}$	6
Down	$8+c$	$\dfrac{6}{8+c}$	6

Note: 1 hr 36 min. $=1\dfrac{6}{10}$ hours. The total time is

$1\dfrac{6}{10}$ hours. Setup the equation and solve.

$$\frac{6}{8-c}+\frac{6}{8+c}=1\frac{6}{10}$$
$$60(8+c)+60(8-c)=16(8-c)(8+c)$$
$$480+60c+480-60c=16\left(64-c^2\right)$$
$$960=16\left(64-c^2\right)$$
$$60=64-c^2$$
$$c^2=4$$
$$c=\pm2$$

Disregard the negative solution, as it has no meaning here. The rate of the current is 2 mph.

66. Let x be the number of liters of 40% insecticide. Then, $x+40$ is the total number of liters in the final mixture.

$$0.4x+0.25(40)=0.3(40+x)$$
$$4x+100=120+3x$$
$$x=20$$

20 liters of the 40% insecticide solution are needed to make the 30% solution.

67. Let w be the width of the rectangle. Then, $2w-1$ is the length. The diagonal is the hypotenuse of a right triangle with legs of length w and $2w-1$. Use the Pythagorean Theorem:

$$w^2+(2w-1)^2=17^2$$
$$w^2+4w^2-4w+1=289$$
$$5w^2-4w-288=0$$

Now, use the quadratic formula:

$$w=\frac{-(-4)\pm\sqrt{(-4)^2-4(5)(-288)}}{2\cdot5}$$
$$w=\frac{4\pm\sqrt{16+5760}}{10}$$
$$w=\frac{4\pm76}{10}$$
$$w=8\quad\text{and}\quad w=-\frac{36}{5}$$

Disregard $-\dfrac{36}{5}$, as width cannot be negative. The width is 8 inches and the length is $2(8)-1=15$ in.

68. Let x be the number of \$2.50 socks. Then $7-x$ is the number of \$3.00 socks.

$$2.50x+3.00(7-x)=19.50$$
$$25x+30(7-x)=195$$
$$-5x+210=195$$
$$-5x=-15$$
$$x=3$$
$$7-x=4$$

He bought 3 pairs at \$2.50 and 4 pairs at \$3.00.

69. Let W be the width of the rectangle. Then, $2W+5$ is the length. Use the area formula for a rectangle.

$$W(2W+5)=52$$
$$2W^2+5W=52$$
$$2W^2+5W-52=0$$
$$(2W+13)(W-4)=0$$

Set: $\quad 2W + 13 = 0 \qquad W - 4 = 0$

Solve: $\qquad 2W = -13 \qquad\quad W = 4$

$$W = -\frac{13}{2} \qquad 2W + 5 = 13$$

Disregard the negative solution, as width must be positive. The rectangle is 4 cm by 13 cm.

70. Let W be the number of women. Then, $6400 - W$ is the number of men. Because the ratio of women to men is 9:7, the equation is:

$$\frac{W}{6400 - W} = \frac{9}{7}$$

$$7W = 9(6400 - W)$$

$$7W = 57600 - 9W$$

$$16W = 57600$$

$$W = 3600$$

$$6400 - W = 2800$$

There are 3600 women and 2800 men.

71. Let x be the amount invested at 5.5%. Then, $2700 - x$ is the amount invested at 7.2%.

$$0.055x + 0.072(2700 - x) = 168.90$$

$$55x + 72(2700 - x) = 168900$$

$$55x + 194400 - 72x = 168900$$

$$-17x = -25500$$

$$x = 1500$$

$$2700 - x = 1200$$

$1500 is invested at 5.5% and $1200 is invested at 7.2%.

72. Let r be his upstream speed. Then, $r + 5$ is his downstream speed.

	Rate	Time	Distance
Down	$r+5$	$\dfrac{6}{r+5}$	6
Up	r	$\dfrac{6}{r}$	6

The total time is 1 hour. Setup the equation and solve.

$$\frac{6}{r+5} + \frac{6}{r} = 1$$

$$6r + 6(r+5) = r(r+5)$$

$$6r + 6r + 30 = r^2 + 5r$$

$$r^2 - 7r - 30 = 0$$

$$(r-10)(r+3) = 0$$

Set: $\quad r - 10 = 0 \qquad r + 3 = 0$

Solve: $\qquad r = 10 \qquad\quad r = -3$

$$r + 5 = 15$$

Disregard the negative solution, as it has no meaning here. His speed upstream was 10 mph, and his speed downstream was 15 mph.

73. Given $h = -16t^2 + 96t + 120$:.

a. The ball will be 200 feet above the ground:

$$200 = -16t^2 + 96t + 120$$

$$16t^2 - 96t + 80 = 0$$

$$16(t - 5)(t - 1) = 0$$

Set: $\quad r - 5 = 0 \qquad r - 1 = 0$

Solve: $\qquad r = 5 \qquad\quad r = 1$

The ball is at 200 ft above the ground after 1 second and again after 5 seconds.

b. The ball will be 60 feet above the ground:

$$60 = -16t^2 + 96t + 120$$

$$16t^2 - 96t - 60 = 0$$

$$4t^2 - 24t - 15 = 0$$

Using the quadratic formula:

$$t = \frac{-(-24) \pm \sqrt{(-24)^2 - 4(4)(-15)}}{2 \cdot 4}$$

$$t = \frac{24 \pm \sqrt{576 + 240}}{8}$$

$$t = \frac{24 \pm 4\sqrt{51}}{8}$$

$$t = \frac{6 \pm \sqrt{51}}{2}$$

$$t \approx 6.571 \quad \text{or} \quad t \approx -0.571$$

Disregard the negative solution, since time must be positive. The ball is at 60 ft above the ground after approximately 6.57 seconds.

c. The ball will hit the ground:

$$0 = -16t^2 + 96t + 120$$

$$0 = 2t^2 - 12t - 15$$

$$t = \frac{-(-12) \pm \sqrt{(-12)^2 - 4(2)(-15)}}{2 \cdot 2}$$

$$t = \frac{12 \pm \sqrt{144 + 120}}{4}$$

$$t = \frac{12 \pm \sqrt{264}}{4}$$

$$t = \frac{12 \pm 2\sqrt{66}}{4}$$

$$t = \frac{6 \pm \sqrt{66}}{2}$$

$$t \approx 7.062 \quad \text{or} \quad t \approx -1.062$$

Disregard the negative solution, since time must be positive. The ball will hit the ground after approximately 7.06 seconds.

74. Use the Pythagorean Theorem on this problem:

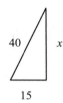

Let x be the vertical length (or height).

$$x^2 + 15^2 = 40^2$$

$$x^2 = 1600 - 225$$

$$x^2 = 1375$$

$$x = \sqrt{1375}$$

$$x \approx 37.1$$

The approximate vertical distance is 37.1 ft.

[End of Cumulative Review: Chapters 1 – 10]

Section A.1
Solutions to Odd Exercises

1. $(x+2)(x^2-2x+4) = x^3+8$
   The sum of two cubes

3. $(x-5)(x^2+5x+25) = x^3-125$
   The difference of two cubes

5. $(x+y)(x^2-xy+y^2) = x^3+y^3$
   The sum of two cubes

7. $(y+1)(y^2-y+1) = y^3+1$
   The sum of two cubes

9. $(xy+2)(x^2y^2-2xy+4) = x^3y^3+8$
   The sum of two cubes

11. $(10x+1)(100x^2-10x+1) = 1000x^3+1$
    The sum of two cubes

13. $(5x+y)(25x^2-5xy+y^2) = 125x^3+y^3$
    The sum of two cubes

15. $(2x-5y)(4x^2+10xy+25y^2) = 8x^3-125y^3$
    The difference of two cubes

17. $(3x+2)(9x^2-6x+4) = 27x^3+8$
    The sum of two cubes

19. $(2-3y)(4+6y+9y^2) = 8-27y^3$
    The difference of two cubes

21. $(x^4+3)(x^8-3x^4+9) = x^{12}+27$
    The sum of two cubes

23. $(x^2+4)(x^4-4x^2+16) = x^6+64$
    The sum of two cubes

25. $(x^3-10)(x^6+10x^3+100) = x^9-1000$
    The difference of two cubes

27. $x^3-64 = \left(x^3-(4)^3\right) = (x-4)(x^2+4x+16)$
    The difference of two cubes

29. $x^3-y^3 = (x-y)(x^2+xy+y^2)$
    The difference of two cubes

31. $x^3+125 = \left(x^3+(5)^3\right) = (x+5)(x^2-5x+25)$
    The sum of two cubes

33. $x^3+27y^3 = \left(x^3+(3y)^3\right) = (x+3y)(x^2-3xy+9y^2)$
    The sum of two cubes

35. $8x^3+1 = \left((2x)^3+(1)^3\right) = (2x+1)(4x^2-2x+1)$
    The sum of two cubes

37. $5x^3+40 = 5\left(x^3+(2)^3\right) = 5(x+2)(x^2-2x+4)$
    The sum of two cubes

39. $54x^3-2y^3 = 2\left((3x)^3-y^3\right)$
    $= 2(3x-y)(9x^2+3xy+y^2)$
    The difference of two cubes

41. $x^3y+y^4 = y(x^3+y^3) = y(x+y)(x^2-xy+y^2)$
    The sum of two cubes

43. $x^2y^2-x^2y^5 = x^2y^2(1-y^3) = x^2y^2(1-y)(1+y+y^2)$
    The difference of two cubes.

45. $24x^4y+81xy^4 = 3xy\left((2x)^3+(3y)^3\right)$
    $= 3xy(2x+3y)(4x^2-6xy+9y^2)$
    The sum of two cubes

47. $x^6-y^9 = \left((x^2)^3-(y^3)^3\right)$
    $= (x^2-y^3)(x^4+x^2y^3+y^6)$
    The difference of two cubes

49. $x^3-1000 = \left(x^3-(10)^3\right) = (x-10)(x^2+10x+100)$
    The difference of two cubes

51. $(x+2)^3+(y-3)^3$
    $= (x+2+y-3)\left((x+2)^2-(x+2)(y-3)+(y-3)^2\right)$
    $= (x+y-1)(x^2-xy+7x+y^2-8y+19)$

53. $27x^3 + (y-1)^3 = (3x)^3 + (y-1)^3$

$= (3x+y-1)\left(9x^2 - 3x(y-1) + (y-1)^2\right)$

$= (3x+y-1)\left(9x^2 - 3xy + 3x + y^2 - 2y + 1\right)$

55. $(x+1)^3 - 1 = (x+1-1)\left((x+1)^2 + (x+1) + 1\right)$

$= x\left(x^2 + 3x + 3\right)$

[End of Section A.1]